Everyday Physics

Waves — From Sounds and Light to Tsunamis and Gravitation

Everyday Physics

Waves — From Sounds and Light to Tsunamis and Gravitation

Michel A. Van Hove

World Scientific

NEW JERSEY • LONDON • SINGAPORE • BEIJING • SHANGHAI • HONG KONG • TAIPEI • CHENNAI • TOKYO

Published by

World Scientific Publishing Co. Pte. Ltd.
5 Toh Tuck Link, Singapore 596224
USA office: 27 Warren Street, Suite 401-402, Hackensack, NJ 07601
UK office: 57 Shelton Street, Covent Garden, London WC2H 9HE

British Library Cataloguing-in-Publication Data
A catalogue record for this book is available from the British Library.

First published 2024 (hardcover)
Reprinted 2026 (in paperback edition)
ISBN 978-981-98-3243-9 (pbk)

EVERYDAY PHYSICS
Waves — From Sounds and Light to Tsunamis and Gravitation

ISBN 978-981-12-7965-2 (hardcover)
ISBN 978-981-12-7964-5 (ebook for institutions)
ISBN 978-981-12-7963-8 (ebook for individuals)

For any available supplementary material, please visit
https://www.worldscientific.com/worldscibooks/10.1142/13501#t=suppl

Desk Editor: Rhaimie Wahap

Typeset by Stallion Press
Email: enquiries@stallionpress.com

Dedicated to

my dear mother Jenny

for her 100th birthday

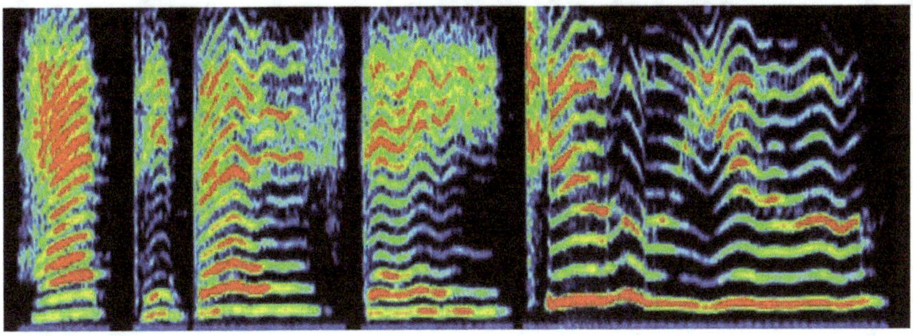

Preface

This book welcomes all those who are curious about nature and have a desire to understand how nature works. The emphasis is on conceptual ideas of physics and their interconnections, while avoiding mathematics entirely. The approach is to explore intriguing topics by asking and discussing questions: thereby the reader can participate in developing answers, which enables a more satisfactory and long-term understanding than is achievable with memorization. This book follows my recent volume on "*Everyday Physics: Colors, Light and Optical Illusions*",[1] which is written with the same goal in mind.

Part A (Chapters 1–7) discusses waves in the context of musical instruments (such as string and wind instruments) and voices (both human and animal): it exhibits the simple properties and great relevance of various kinds of waves.

[1] Michel A. Van Hove, "*Everyday Physics — Colors, Light and Optical Illusions*", World Scientific Publishing Co., 2022, ISBN-13 978-9811238338

Part B (Chapters 8–15) shows how the simple properties described in Part A extend to other waves of great importance, such as: waves in and on water or rocks (including earthquakes, beach waves and tsunamis), as well as light (including rainbows and lasers), quantum waves (responsible for atoms, molecules and color), and gravitational waves (which were only recently observed for the first time).

Since waves are very active and dynamic, five free online videos help the reader with vivid animations of their motion: the videos are indicated in the boxes labelled "ANIMATION" (see further below and in the section References and Resources near the end of this book).

You may jump around this book as you please: there is no need to read from beginning to end, especially in the first reading, and you may skip all details. I have highlighted the important questions and answers for easy guidance: you may use them to locate topics of interest to you. You will find more guidance in the Table of Contents and in the Index. Lists of intriguing questions are also provided in Section 1.1 and Chapter 8: they direct you to relevant sections.

To help you along, I ask ***many questions, highlighted in this manner (bold italic)***: I highly recommend that you first try to answer them on your own before reading my discussion following the questions and before jumping to the **answers, highlighted in this manner (bold)**. By critically thinking on your own, you will understand the physics of waves much faster and better. So, take the time to ponder my questions: they may seem simple at first sight, but they contain important physics that is easy to understand, and which we will discuss after you have had a chance to think about them. The physics will then gradually emerge from our discussion. This approach of investigating on your own will improve your understanding and your recollection of the interesting physical principles involved. **So, let's think along and enjoy the discovery!**

Each chapter is also concluded by a convenient summary section called "What have we learned in this chapter?"

Interestingly, much of science has developed historically through exactly such questioning. Over coffee or a meal, scientists often ask each other such questions out of pure curiosity; also, scientists will ask themselves such questions. For example: How does this work? Does it always work this way? Can I change this? What if I changed that? Why does this happen? How come that does not happen? And so on! Scientists often ask straightforward questions like children do: science is a fun game of exploration, best enjoyed with an open mind. Besides entertainment, the rewards include satisfying your curiosity and understanding daily observations around you. Also, you will better appreciate all the modern machines and devices invented by man to exploit such understanding. Section 15.5 discusses how science is actually done by asking and answering: "Is doing science like doing sports, or playing video games, or reading science fiction?"

You will certainly have more questions as you read. To help you answer those questions, consult the **terms** that are highlighted **in bold letters**: these terms are contained in the Index at the end of this book, which leads to other sections of the book. Furthermore, a list of general References and Resources (books, videos, websites) is provided at the end of this book, so you can extend your reading. In particular, animations made specifically for this book are available online and announced as "see Animation 0*1", for example:

> ANIMATION 0*1 — Note: Waves move continuously: therefore, animations provide vivid insights into the behavior of waves. Such animations are included in my free online YouTube videos WA1, WA2, WA3, WB1 and WB2: see details in the section References and Resources near the end of this book. Viewing these videos is strongly encouraged as an introduction and overview before and during the reading of this book.

Interesting additional information on various topics is announced as "see Box 0-1", for example:

> BOX 0-1 — Note: Boxed text like this one adds some details, or a deeper discussion, or another point of view which may interest you; these boxes also give a more in-depth flavor of physics. If you wish, you may skip these boxes, as such information is usually a bit more technical or advanced.

This book grew out of a popular General Education course that I gave for six years at two universities in Hong Kong, entitled *"Everyday Physics for Future Executives"*. It was inspired by a highly successful course given for many years at the University of California at Berkeley by Professor Richard A. Muller and others, initially entitled *"Physics for Future Presidents"*, and available in book form.[2] As Professor Muller wrote: *"Can physics be taught without math? Of course! Math is a tool for computation, but it is not the essence of physics."*

The approach of this book recognizes that math is a hurdle which frightens many people away from physics. Actually, by avoiding math altogether, everyone, regardless of prior knowledge, can feel and enjoy physics (and other sciences), thereby understanding a multitude of daily experiences, from sound and light to earthquakes and biophysics. That is the principle of this book. While it shows a few sums, subtractions and multiplications to illustrate what math can contribute, these simple calculations are not at all needed to understand the physics. Math is needed for rigor, precision and prediction, but not for understanding the essence of physics.

I chose for this book the topic of "waves" because we face waves every minute of our lives, in the form of sound, light, ocean waves, atoms, biology, color and more.

First and foremost, I acknowledge both the City University of Hong Kong and the Hong Kong Baptist University for the opportunity to develop and fine-tune the abovementioned General Education course, open to all students at those universities. I was most fortunate to benefit from incisive discussions and substantive assistance from Klaus E. Hermann (Berlin), including ideas, graphics and detailed scrutiny. I also benefited from very valuable suggestions and/or voice recordings from: Walter Gander (Switzerland); Carlo Schirru (Italy); John Pendry, Akari Chan (UK); Pradeep Kumar, John Rehr, Miquel Salmeron (USA); Klaus Heinz (Germany); Jean-Paul Bibérian, Christian Minot (France); Che Ting Chan, Christopher J. Keyes, Guancong Ma, Xue-Cheng Tai, Leihan Tang, Sheng-Wei Wang, Sasa Xiao, Martin Ho Yin Yeung, Changsong Zhou, and dog Lucky (Hong Kong). All remaining errors are mine and mine alone.

I have mainly used — or referred to — the following software in preparing this book: Balsac (https://www.fhi.mpg.de/1012355/Balsac), GNU-Octave, Gimp, Excel, PowerPoint, IrfanView, Word, Praat

[2] Richard A. Muller, *"Physics and Technology for Future Presidents: An Introduction to the Essential Physics Every World Leader Needs to Know"*, Princeton University Press, 2010, ISBN-13: 978-0691135045

(https://www.fon.hum.uva.nl/praat/), WaveSurfer (https://sourceforge.net/projects/wavesurfer/), A.L.M.S. (Audio Literacy for Music Students by Prof. Christopher J. Keyes, with his permission to publish images produced with this app, http://palms.polyu.edu.hk/a-l-m-s-hkbu/), OnlineToneGenerator.com, szynalski.com/tone-generator/, and PhET Interactive Simulations (https://phet.colorado.edu/sims/html/wave-on-a-string/latest/wave-on-a-string_en.html). Sounds were downloaded from Mixkit (https://mixkit.co/free-sound-effects/), University of Iowa Musical Instrument Samples (MIS, in public domain, http://theremin.music.uiowa.edu/MIS.html), freeSFX (https://www.freesfx.co.uk/), Inossi (https://soundcloud.com/inossi), or graciously provided by contributors listed above, or taken from my own recordings. Images are credited individually; otherwise they are under my copyright.

Michel A. Van Hove

Emeritus Chair Professor, Hong Kong Baptist University
Retired Chair Professor, City University of Hong Kong
Retired Senior Scientist, Lawrence Berkeley National Laboratory,
University of California

Hong Kong, June 2023

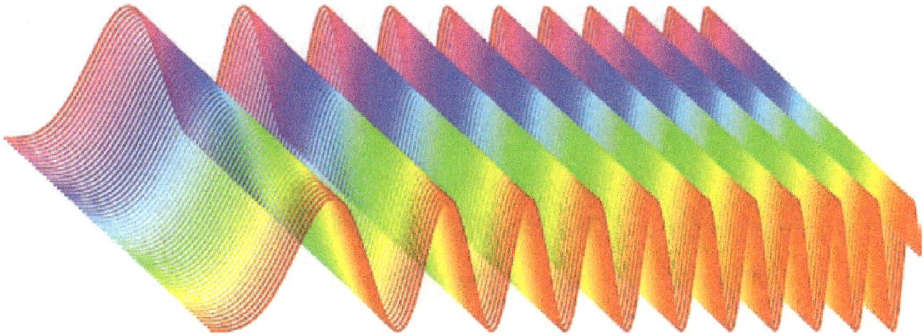

Contents

WAVES – PART A:

From Musical Instruments and Sounds to Human and Animal Voices

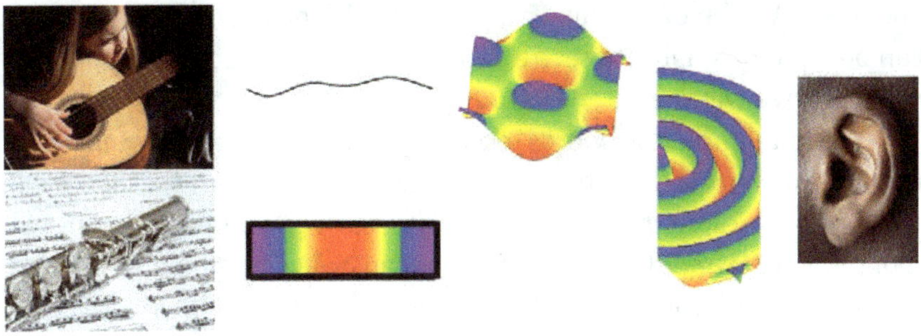

1

Waves: What and Why

We start by looking at various kinds of waves that we see around us in daily life. They include the wave of the hand, waves on water surfaces, waves on strings (especially in musical instruments), and waves on flags. Other kinds of waves are largely invisible, such as sound waves, waves on drumheads, earthquakes, light, quantum waves and more. This overview of waves will give us a useful impression of what a wave is like, and also how widespread and important waves are in our daily experience. In particular, resonances in waves are fundamental to speaking and making music.

1.1 Familiar Waves

What does the word "wave" make you think of? Perhaps **waving** with the hand? What is physically special about waving the hand? Think about this question (and all my questions in this book) before reading on! We can describe waving as a smooth motion which is often repeated back and forth. A wave of the hand can also be quick or slow, and the motion can be large or small. Those are in fact important aspects of all kinds of waves, as we will see!

We often use the words **oscillation**, **vibration** and **swinging** to express the back-and-forth motion that is normal in waves, especially if the motion is rapid. Another word related to waves, less common but more poetic, is **undulation**: it literally means wavy motion and can also be used for any type of wave that we will discuss.

Can you think of other kinds of waves besides waving your hand? How about waves on the surface of water?

Have you thought of **waves on the strings** of violins, guitars, pianos, *etc.*? You may watch a video showing more waves on strings[1]: don't worry about the details and the terminology used in this and other videos; just look for and enjoy the wavy motion. We will explain such waves later.

How about waves on a flyfishing line? Or on **whips**[2] used to control horses or other animals? Whips are similar to flyfishing lines. Or waves on **slinkies**[3,4]? Have you thought of the wavy motion of **snakes**?

Maybe you think of waves on **flags** fluttering in the wind?

Examples of all the above waves are shown in Figure 1-1.

Are you aware of waves on **drumheads?** Usually waves on drumheads move so fast that they look like a blur, somewhat like the vibrating strings in Figure 1-1b, but slow-motion movies can make their shape and motion visible.[5,6]

---

[1] See video "AP Physics 2: Waves 10: Resonance and Standing Waves on a String" by Yau-Jong Twu: https://www.youtube.com/watch?v=7xCmtYXewdk

[2] See video "Steve Harvey Learns How to Use a Whip" by Steve TV Show: https://www.youtube.com/watch?v=peEvYhvWc1U

[3] See video with slinky "Transverse and Longitudinal Wave Demonstration — A level and IGCSE Physics" by Chris Gozzard: https://www.youtube.com/watch?v=iT4KAc0Ag1E

[4] See video "Transverse Waves on a Slinky" by Shyam Srinivasan: https://www.youtube.com/watch?v=y66PSaiGH7Y

[5] See video "Snare Drum in slow motion 3000 FPS" by UltraSlo: https://www.youtube.com/watch?v=STSWLX23xqc

[6] See video "Vibration. See the unseen: Cymbal at 1,000 frames per second" by Fluke Corporation: https://www.youtube.com/watch?v=kpoanOlb3-w

Figure 1-1: Examples of different kinds of waves. a: Waves in Sydney harbor. b: Waves on a vibrating hanging string. The time exposure of this pair of photos smears the string out horizontally. The photos capture several left-right oscillations of the string: the oscillations are less regular in the left photo than in the right photo. Note the smooth, rounded, and snaky shape of the waves, which is typical of many waves. This string was rotating around a vertical axis, so it also had front-back oscillations. c: A wave on a string: flyfishing. Again, we see a smooth rounded shape of the string. d: Wave on a slinky. e: A sidewinder rattlesnake. f: The Olympic flag (the narrow ripples are probably fixed stretch marks due to the colored rings, and not moving waves). (*Sources*: a: the Author. b: Prof. Andrew Davidhazy, http://www.davidhazy.org/andpph/, with permission. c: https://upload.wikimedia.org/wikipedia/commons/0/06/Flyfishing.jpg in the public domain. d: Nic McPhee, under CC BY-SA 2.0, https://commons.wikimedia.org/wiki/File:Standing_wave_on_a_slinky.jpg. e: Extracted from HCA, under CC-BY-SA 4.0, https://en.wikipedia.org/wiki/Snake#/media/File:Neonate_sidewinder_sidewinding_with_tracks_unlabeled.jpg. f: Sam, under CC BY-SA 2.0, https://commons.wikimedia.org/wiki/File:Olympic_flag.jpg.)

All the above are very good examples of waves! And there are many more:

Do you know that **sound** is composed of **waves in air**, including when **speaking**, **singing**, making **music**, making **noise** and **hearing**? And that **earthquakes** are also waves?

Are you aware that **light** is made of **electromagnetic waves**, as are **radio** and **television** signals and **x-rays**?

There are still more kinds of waves: **quantum mechanical waves, gravitational waves, waves in road traffic, human waves** in stadiums, as well as **waves in hair**.

We will discuss interesting behaviors of all these waves in this book. Some common aspects that we already see for all these waves are the following:

- Waves have a generally smooth motion which is often repeated back and forth as a vibration.
- Waves of similar appearance exist in many supporting substances (on a string, in air, in water, on a drumhead, *etc.*).
- Waves have a snaking shape that can travel over long distances and with little change of shape.
- Waves do not carry their supporting substance (string, air, water, *etc.*) with them, so waves move through the substance, and not with the substance.

I thus invite you to open your mind to the fascinating world of waves, which will lead us into various interesting parts of physics and daily life, from sounds and telecommunications to tsunamis and atoms.

Think of those exciting days when you, maybe as a child, entered a new home or new playground and rushed around to uncover its mysteries, secrets and pleasures. Try to do the same here with waves: explore and enjoy nature and modern technologies!

You may start by looking for a topic that strikes your fancy in the following list of frequent questions concerning waves (feel free to jump directly to the indicated chapters and sections). You will find a similar list of interesting topics in Chapter 8, covering Part B: it includes waves in and on water, earthquakes, light and other electromagnetic waves, quantum waves, gravitational waves, and many others.

- Many objects suddenly start to vibrate, often loudly, such as water pipes, machines, cables, and leaves in trees: what is going on? These vibrations are so-called resonances (see Section 1.3 for a preview of such resonances). They are also responsible for a lot of music (think of the vibrations of strings in guitars and pianos; see Chapter 2), including voices (both in speaking and in singing; see Chapter 6), but can be disastrous for bridges and other structures (see Section 1.3).
- How can you best cause a vibration? You will need to push or pull, but with what timing and at what rate? We discuss these questions in Section 2.1.

- What happens when two waves cross each other? This question is of great importance in vibrations (resonances): it produces interference and superposition of waves. See Section 2.5.
- You may have heard of beats and their use in tuning musical instruments: what are they? We discuss them in Section 2.6.
- How are waves reflected, for example by the end of a string? This question is important for understanding vibrations and resonances, in particular the making of music and voices. See Section 2.7.
- How does air vibrate to make sounds, such as music and voices? We address this question in Chapter 3.
- How weak is the weakest sound we can hear? See Section 3.4.
- What is a sonic boom, usually created by a supersonic airplane? See Section 3.6.
- What is the Doppler effect, which causes the pitch of sirens to change as they race by you? See Section 3.6.
- Can sounds turn around corners, unlike balls that you throw? See Section 3.9.
- How do waves behave inside the tubes of wind instruments like a saxophone or trumpet? See Section 3.10.
- What is the role of the large "box" of the violin and other wind instruments? These resonance chambers are discussed in Section 3.14.
- How different are waves on sticks (as in a xylophone) compared to waves on strings (as in a guitar)? See Section 4.1.
- How do cymbals, drums and bells vibrate to make music? See Sections 4.2 and 4.3.
- How do strings in musical instruments correspond to notes in music? See Section 5.2.
- How does bowing a violin string or plucking a guitar string make the string vibrate? See Section 5.5.
- Why do pianos have 230 (or more) strings while violins have only 4? See Section 5.6.
- Why do different musical instruments sound so different? And how different are they? See Section 5.7.
- How are music, voices and other sounds recorded and played back? See Section 5.8.

- What is synthetic sound? See Section 5.9.
- How do we pronounce different vowels and consonants? See Sections 6.4, 6.5 and 6.6.
- How different are tonal languages, such as Chinese, compared to most other languages? See Section 6.7.
- What distinguishes accents, dialects and languages? See Section 6.8.
- How different is singing from speaking? See Section 6.9.
- What are whistling, yodeling, whispering, humming, laughing, *etc.*? See Section 6.10.
- What sounds do babies make? See Section 6.10.
- How good is synthetic speech? See Section 6.11.
- How different are animal sounds from human sounds? See Section 6.12.

1.2 The Importance of Waves

How important are waves in our daily lives? From the kinds of waves mentioned above, you already get a good impression of their relevance to us:

- Sounds and light are essential to humans for communication and observation, through mouth, ears and eyes.
- Light also brings most of our energy from the Sun to the Earth, in the form of heat that keeps us from freezing and as photosynthesis that produces plants and food, as well as coal, oil and gas.
- X-rays are beneficial in medicine and for security checks.
- Electromagnetic waves are also important in telecommunications and computing.
- Water surface waves affect all ship navigation, and provide entertainment at the beach, in the waterpark and in the bathtub.
- Tides in the oceans and seas complicate navigation in harbors and cause tidal bores.
- Earthquakes can be terribly destructive, also through tsunami waves on oceans.
- Less obvious is the fundamental role of quantum waves in materials and hi-tech, such as all sorts of electronic devices,

including telecommunications and computing, as well as modern photography and detection.

Thus, waves provide a fascinating and important topic for this book on everyday physics. There are many interesting aspects to explore. On the one hand, there are common aspects that are universal among all types of waves. For example, most waves do not "collide" with each other but "flow through" each other without changing each other, unlike objects that collide.

On the other hand, each type of wave has its own unique properties because each type is based on very different physics. For example, waves on strings rely on the strings' elastic resistance to stretching; sound waves in air depend on the air's resistance to compression; electromagnetic waves rely on the interaction between electric and magnetic fields, and water surface waves depend on gravity and surface tension.

Before we discuss all these kinds of waves, we can ask: ***How do we know they are actually waves?*** For some waves, our own eyes clearly show the wavy motion and shape; that is the case of water surface waves, as well as some waves on strings and waves on flags. So we can visually confirm that **some of these are really waves from their visible motion and shape**.

However, we cannot see the wavy character of most other kinds of waves mentioned above.

Think of waves on guitar strings or on drumheads, for example: such waves move so fast that we only see a blur (as in Figure 1-1b), unless we use fast photography or other means to "freeze" the motion so as to make it more visible, such as for the flyfishing shown in Figure 1-1c. Of course we can see light, but we can't see its undulating wave shape like we do with water surface waves.

Thus, **the wave character of many waves is invisible to our eyes**. That is especially true of sound waves, light waves, radio waves, x-rays and other electromagnetic waves, as well as quantum waves and gravity waves. Since we cannot see them, it is not obvious that they are actually waves: we will need to justify calling them waves. For example, how do we know that sound is waves? We will have to show that sound indeed is made of waves!

1.3 A Preview of Resonances: Danger and Beauty

A fascinating behavior of waves is called **resonance**: resonances are vibrations of objects; waves also can resonate. You have encountered resonances in many different circumstances. Our speaking and singing are largely based on air that resonates inside our mouth. Most **instrumental music** is due to resonances of strings (see Figure 1-1b), of drumheads, of air in tubes, of air in resonance chambers (which are the bulkiest part of a guitar and violin), of electric currents in electronic circuits, *etc*. A flag that flaps in the wind also resonates (see Figure 1-1f). Water pipes tend to rattle or even "sing" when water flows at certain speeds. You may know that sound, including your voice, can break a wine glass through resonance by selecting the sound pitch that matches and amplifies a "natural" vibration of the wine glass.[7]

If you have ever walked over a light bridge or over a raised wooden board (such as a diving board), you probably noticed it resonating: the bridge or board can swing "in synch" with your footsteps, making walking difficult and dangerous. You may have noticed that the bridge or board only swings strongly if you walk at just the right rhythm. That particular rhythm amplifies the motion at each step. By contrast, if you walk "out of synch" with the rhythm of the bridge or board, it will not swing so much. Such behavior is typical of resonances.

If the swinging of a bridge or board grows, it can literally break up, like the wine glass mentioned above. This happened in a spectacular fashion in 1940 with a very large, brand-new bridge of novel design (a suspension bridge): the Tacoma Narrows Bridge in the US Northwest collapsed in front of cameras, see Figure 1-2[8]; in this case, turbulent wind caused it to twist repeatedly and dramatically sideways until it broke.

Because of such resonances, armies of foot soldiers in the past were told not to march in step as they walked across bridges: too many bridges had collapsed as soldiers crossed them because they

[7] See video "breaking a wine glass using resonance" by iflamenko: https://www.youtube.com/watch?v=17tqXgvCN0E

[8] See story and video of the "Tacoma Narrows Bridge (1940)": https://en.wikipedia.org/wiki/Tacoma_Narrows_Bridge_(1940)#Film_of_collapse

Figure 1-2: *Left*: Snapshot from a movie of the wind-driven twisting swinging of the Tacoma Narrows Bridge (a car stands on the bridge deck which at that moment slopes about 23° sideways). *Right*: Collapse of the bridge. (*Sources*: "Tacoma Narrows Bridge (1940)", Wikipedia, in the public domain, https://en.wikipedia.org/wiki/Tacoma_Narrows_Bridge_(1940); movie by Barney Elliot, in the public domain, https://en.wikipedia.org/wiki/File:Tacoma_Narrows_Bridge_destruction.ogv.)

walked in step and thereby amplified a resonance of the bridge until it broke.

More recently, in the year 2000, the newly designed Millennium Bridge in London, UK (shown in Figure 1-3), suffered a similar problem immediately after its inauguration.[9] It started swaying sideways by about 10 centimeters (4 inches) when people walked across this slim pedestrian bridge. The pedestrians actually worsened the problem by walking in synch with the swaying, thereby amplifying it: this is typical of a resonance. It took the engineers over a year to figure out how to dampen that swaying.

Another example from 2020 is the up-and-down swaying of a large new suspension bridge in China.[10]

An entertaining (and slightly scary) online video shows a resonating footbridge, activated by students repeatedly jumping on it: see a few snapshots in Figure 1-4[11]. It shows how a flexible footbridge bends up

[9] See video "How Did Engineers Fix London's Wobbly Millennium Bridge? | Massive Engineering Mistakes" by Quest TV: https://www.youtube.com/watch?v=t6O43mrc1kA

[10] See video "Just happen! 5.5.20 Scary Bridge Resonance Vibration" by Travel 360: https://www.youtube.com/watch?v=5pgCuLPzG5Q

[11] See video "Resonan Bridge" by Bob Barrett: https://www.youtube.com/watch?v=uWoiMMLIvco (this bridge, named Messiah College Swinging Bridge, apparently still exists at Messiah College, Grantham, Pennsylvania, USA, see https://www.bridgemeister.com/pic.php?pid=40, after being rebuilt in 1983 and 2003)

Figure 1-3: The Millennium Bridge in London. The yellow line exaggerates a possible side-to-side swinging mode. (*Source*: Ibex73, under CC BY-SA 4.0, https://commons.wikimedia.org/wiki/File:Millenium_bridge_2015.jpg.)

and down in a repetitive wave-like motion as the students continue jumping. That motion is sketched in Figure 1-5: by jumping at the right locations and with the right rhythm (repetition rate and timing), the students make the bridge swing up and down in different patterns. I highly recommend that you watch this very informative video to get a better feeling for resonances.

From time 0:15 in the footbridge video, as well as from time 3:30 and again from time 4:42, the students cause the so-called "second

Figure 1-4: Three snapshots of a movie showing the second harmonic resonance of the Messiah College Swinging Bridge (you can see it from time 3:30 to 4:27 in the video). The added horizontal dashed white lines mark the lowest level reached during the swinging. (*Source*: extracted from video "Resonan Bridge", by Bob Barrett, in the public domain, https://www.youtube.com/watch?v=uWoiMMLIvco.)

harmonic" pattern with two "**loops**": it is sketched in dark blue in Figure 1-5. From time 1:00, they cause the "third harmonic" pattern with three "loops", as sketched in purple in Figure 1-5. The students go on to create the "fourth harmonic" and "fifth harmonic" patterns as well. The "first harmonic" pattern (sketched in black in Figure 1-5) is not shown at first in the video, but you can see it from time 3:12 with side-to-side swinging of the bridge (instead of up-and-down swinging): this sideways swinging is similar to the swinging pattern of the Millennium Bridge mentioned above (see Figure 1-3). The students also create a "torsional harmonic" from time 2:56; now the bridge twists around its long axis, similar to you twisting your hips left and right while keeping feet and shoulders immobile; this is like the resonance that destroyed the Tacoma Narrows Bridge (see Figure 1-2).

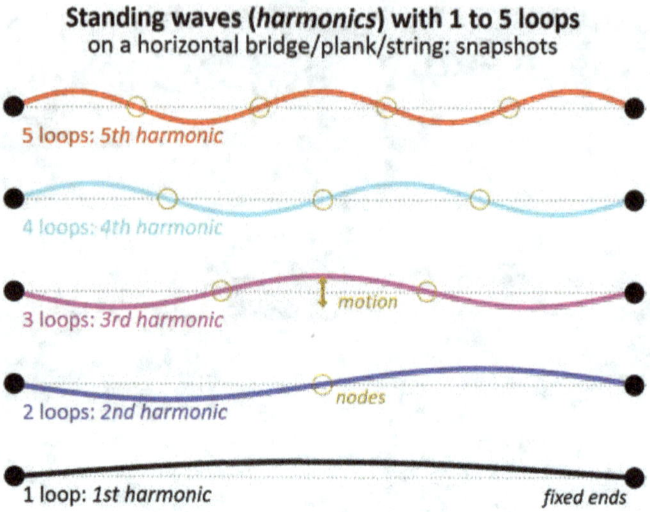

Figure 1-5: The five sketches show snapshots of five different standing waves (resonances) of a horizontal bridge, or of a board such as a wood board, or of a stretched string, with fixed ends (the ends are shown as black points). The strictly up-and-down direction of motion is shown by an arrow. No motion occurs at the open circles: these points are called "nodes"; they separate the "loops" of motion.

The term **harmonic** comes from music and reminds us that resonances are important for music, singing and voices: indeed, harmonic vibrations are common in beautiful music.

All these vibration patterns are **resonances**. **Resonance is a very important concept with waves, and not only for music**. We will therefore encounter harmonics and resonances many times in this book.

Resonances occur not only with waves. The back-and-forth motion of a child sitting on a swing is a resonance, as is the swinging of your legs and arms as you walk comfortably. The pendulum of a grandfather clock is a classic example of a resonating system. If you rhythmically push a small tree or branch, you may get it to sway so strongly that it will topple over or break.

You can easily model a resonating bridge with a string or rope or slinky (the longer and heavier the string, the better, as that slows down the vibrations and thereby makes them more visible): stretch a string between two firm points like door handles or heavy chairs. With your hand, you can now push that string down repeatedly and cause resonances like those seen on the footbridge or in Figure 1-5. You can search for the best location and timing of your pushing to create large vibrations. It will become clear that you must push at the right positions, moments and intervals to create the different harmonics: we will discuss this further in Section 2.1.

If you prefer, you can watch and simulate all this in an excellent and convenient online animation[12]: it animates a string that you can control as if it were a real string. In the following, I will use this online animation to illustrate how resonances behave on strings as well as on bridges: I recommend that you run that animation yourself because it is very instructive to get a feeling for how such strings, bridges, *etc.* behave. (This same online animation will again help us later, in Chapter 2, with general waves on strings.)

In the following, I assume that you have started the online animation "Wave on a String" mentioned in the caption of Figure 1-6. You then see the start-up screen shown in Figure 1-6: a string made of balls is held fixed at right, while you can shake its left end up and down by dragging the wrench with your mouse. As you move the left end of the string, it sends out waves to the right, where they get reflected back to the left (crossing your outgoing wave).

After you stop shaking the string, the waves die out because damping is built into the simulation (damping in reality is due to friction with air and friction inside the string and in the supports; you can cancel damping by setting the Damping slider at the bottom to None: you will then see your wave going back and forth for a long time). You may pause/resume the waves with the large blue Pause button, watch frame by frame with the small gray/blue button, or restart the animation with the gray Restart button at the top.

Now try to simulate the patterns shown in Figure 1-5 by using the "Wave on a String" animation (include a little damping so the animation forgets and erases failed attempts). Try to shake the left end of the

---

[12] See online "Wave on a String" by PhET Interactive Simulations, University of Colorado Boulder: https://phet.colorado.edu/sims/html/wave-on-a-string/latest/wave-on-a-string_en.html (covered by a CC BY 4.0 License, https://phet.colorado.edu/en/licensing/html; this website also offers many other excellent animations in physics, chemistry, mathematics, earth science and biology)

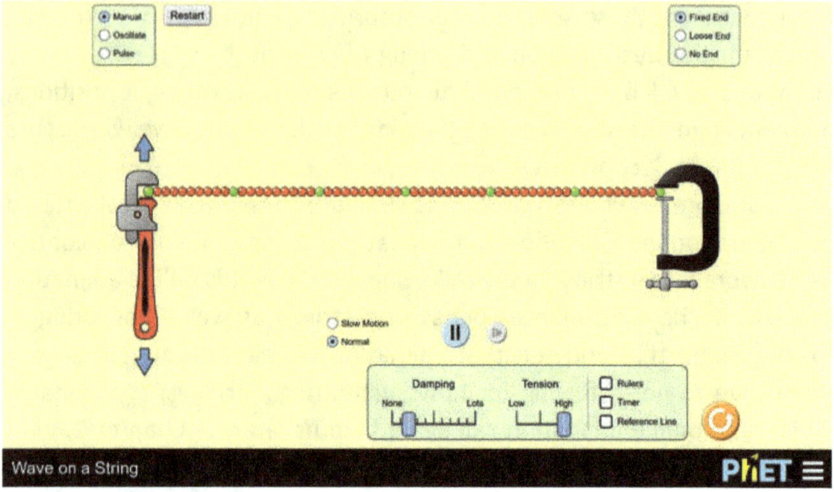

Figure 1-6: Start-up screen of the online animation "Wave on a String", in its Manual mode. (*Source*: PhET Interactive Simulations, University of Colorado Boulder, under CC BY 4.0 License, https://phet.colorado.edu/sims/html/wave-on-a-string/latest/wave-on-a-string_en.html.)

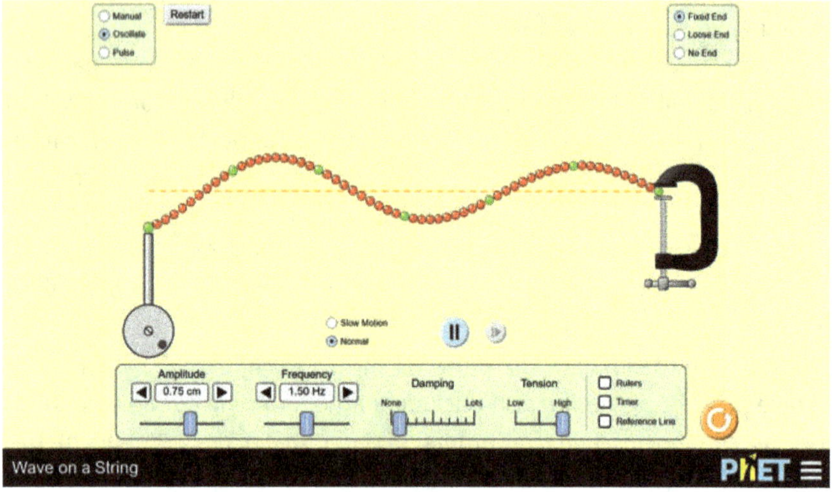

Figure 1-7: Snapshot of the initial "Oscillation" screen of the online animation "Wave on a String". (*Source*: as Figure 1-6.)

string in such a way that you get the first **harmonic** pattern of Figure 1-5 (that is the black shape with one "loop"). You should find that, by moving the wrench up and down smoothly and repeatedly, you can create something resembling the first harmonic, even though it may be rather rough. However, you must find the right rhythm to achieve this: swinging too fast or too slow gives poor results. Fast swinging creates many loops rather than the one loop of the first harmonic; slow swinging just drags the string up and down without forming a nice loop.

Once you have created a nice first harmonic pattern (with one loop), do the following experiment: delay your repetitive pushing so you push each time against the motion of the string. What do you expect to happen? Just like pushing against a moving cart causes it to slow down, pushing against a moving string also slows it down. As a result, the resonance will also decrease and may even totally disappear. This observation teaches us an important lesson: to create a resonance, we need to push at the right time. More precisely: **To create a resonance, we must push only in the direction of the motion of the string.**

Next, try to produce the second harmonic pattern (with two loops in Figure 1-5). You will find that you need a faster rhythm to do so. But it may be difficult to get a clean pattern by hand. You can also produce the third and other harmonics with even faster rhythms, but it gets harder and harder to do this by hand.

Fortunately, the animation can help you produce cleaner patterns: select "Oscillate" at the top left. Now the wrench is replaced by an automatic "driver" that moves the left end of the string up and down smoothly and repeatedly: see a snapshot in Figure 1-7. As a result, the wave on the string becomes much smoother as well and looks a bit more like the patterns of Figure 1-5.

With this "Oscillate" option, you can control the "driver" to change the wave that it produces on the string. Try to generate the first harmonic again (with one loop), by slowing down the driver with a slower rhythm: you do this by reducing the Frequency (which tells you how many times the driver goes up-and-down in each second; Hz is the physical unit for frequency, short for "hertz"). Try frequencies in the range from 0.25 to 0.5 Hz. You can also make the string softer, by reducing the Tension, and you may reduce the Damping. You should be able to create a wave with a single loop. But you will not get exactly the shape sketched in Figure 1-5. The reason is that the left end of the string in the animation is not fixed: it moves up and down with the driver, unlike the fixed left end of the bridge, board or string in Figure 1-5.

There is a simple way to get closer to that "fixed left end" in the animation: reduce the "Amplitude", which is the amount of motion of the driver. Try an Amplitude of 0.1 cm, with a Frequency of 0.41 or 0.42 Hz, no Damping and High Tension; Restart it to stop old vibrations. Wait and see what happens: the string will swing more and more, even off the page, although the driver is only injecting a small amount of motion! **This is a strong resonance! The string swings more and more, despite minimal driving: a real string, board or bridge would break!** (Restart or add Damping to bring the string back into view.) You can also create such a large resonance with Amplitude 0.1 cm, Frequency 0.25 Hz, no Damping and medium Tension.

Here is another interesting "experiment" you can do with the online animation: change the frequency a little bit and see whether you get the same resonance. For example, instead of 0.41 or 0.42 Hz in the above case, use 0.40 Hz or 0.43 Hz (use Restart to stop the old motion). Amazingly, that small change in frequency greatly reduces the resonance! This beautifully illustrates the concept of "in synch" *versus* "out of synch": **an in-synch driver produces a strong resonance, but an out-of-synch driver does not**. This tells us that **resonances can be extremely sensitive to the conditions used to create them**.

One common example of this in-synch resonance behavior (now with electromagnetic waves) is that your television set can "tune in" to a single TV channel without being disturbed by other TV channels: that is due to a resonance which "picks out" only the TV channel that you desire; a TV set can resonate with the "carrier frequency" of your desired channel while avoiding the nearby carrier frequency of another channel; the same is true of radio transmissions.

Next, try to create a 2-loop resonance ("second harmonic"), by changing the frequency in the last online animation: you will have to increase the frequency beyond 0.8 Hz; make a note of the frequency that gives the largest resonance. Repeat this exercise to create a 3-loop resonance: you will have to increase the frequency beyond 1.2 Hz. For 4 loops, you will need more than 1.6 Hz; and for 5 loops, you will need to go beyond 2.0 Hz. Can you create 6, 7, 8, *etc.* loops? Since the animation's frequency is limited to 3.0 Hz, we can't go higher than 7 loops. However, with medium Tension (which makes the string more flexible), we can create up to 12 loops, even with a very small Amplitude of 0.01 cm (use the Slow Motion setting to better see those fast oscillations).

If you have noted the frequencies that make the strongest resonances, you can observe the following significant result: those frequencies are close to whole multiples of the first one. The first frequency (giving the first harmonic with one loop) is called the **fundamental frequency**; in the animation discussed above, the fundamental frequency is about 0.41 Hz. The second harmonic frequency which you should have found is near $2 \times 0.41 = 0.82$ Hz; the third is near $3 \times 0.41 = 1.23$ Hz, *etc.* We will encounter this whole-multiple relationship repeatedly in dealing with resonances in general, and sound in particular. (However, you may have observed that the higher frequencies in the animation are in fact not exactly whole multiples of the fundamental frequency 0.41 Hz: the reason is that the left end of the string is still not fixed. It is only with fixed ends that the frequencies are exactly whole multiples of the fundamental frequency.)

From your attempts with the online animation, you may conclude that it is not very easy to produce a perfect, clean and strong resonance. Indeed, a number of details have to be controlled "just right" to get a good resonance: this is a characteristic property of resonances, which helps to make them very useful and important, for example in television transmission.

Another interesting observation is the following: the swinging patterns only have up-or-down motions of the horizontal bridge, board or string. You will see this if you follow the motion of any piece of the bridge, board or string: no piece will move <u>along</u> the length of the bridge, *etc.*

Moreover, the rounded <u>shape</u> of the wave also does not move along the length of the bridge, *etc.*; for example, the wave's peak remains at the same place on the bridge from one oscillation to the next; the wave shape does not slip along the bridge. We therefore call this a **standing wave**: it stays in place, unlike, for example, a sound wave traveling from

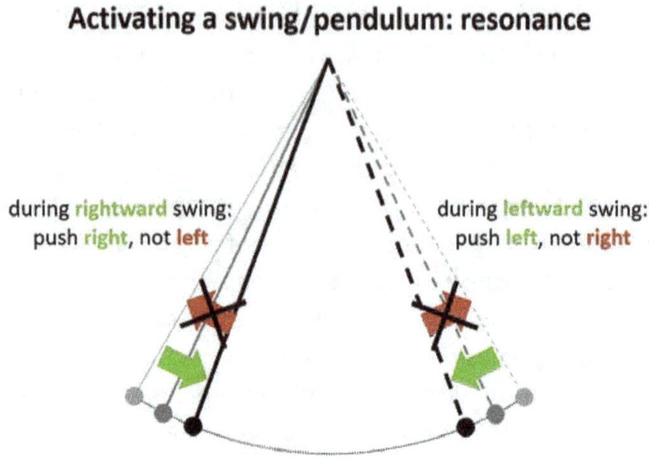

Figure 1-8: The resonating swinging of a swing (or pendulum) under the force of pushing it repeatedly (the gray lines show the direction of motion as fading snapshots). The arrows show external forces acting to make the swing move: green forces increase the swinging; red forces decrease the swinging. (A swing has only one swinging pattern: there are no harmonics.)

a loudspeaker to your ear or a water surface wave that travels away from a ship (see Figure 1-1a). Traveling waves do not resonate; instead, they propagate.

You may be more familiar with the very similar resonating behavior of a **swing** or **pendulum**, illustrated in Figure 1-8. Although this swinging is not a wave (and therefore does not have harmonic vibrations), it has a repetitive motion that also relies on pushing at the right time: this also makes it a resonance.

You can easily make a simple swing or pendulum yourself, like that sketched in Figure 1-8: all you need is a string hanging from a fixed support with a weight at its lower end; you can even use your arm as the string, while holding a heavy bag in your hand. If you push the string (or arm) to one side and let it loose, it will swing back and forth for a while (friction will gradually slow it down). This "free" swinging will have a particular rhythm, meaning that each individual back-and-forth swing will take the same amount of time. This rhythm is a natural property of the swing or pendulum: it is not man-made. Therefore, physicists like to call it **natural frequency** or **resonance frequency**.

To increase the swinging, you can repeatedly push it along at the right moment, as shown by green arrows in Figure 1-8. First, this means pushing in the direction of the swinging. Second, this implies pushing at the same rhythm or frequency as the free swinging of the swing or pendulum.

We will look more closely at how resonances work in Chapter 2. There we will focus more on strings, but resonances are so universal among all sorts of waves that our discussion will help us throughout this book. We will indeed return to the topic of resonances several times in this book.

1.4 Our Presentation of Waves

In this chapter, we have illustrated and examined the behavior of waves on bridges and on strings because they can easily be seen with our eyes, and also because they are physically the simplest waves. In particular, we have discussed resonances: we will see that resonances are important well beyond bridges and strings, such as in music. In Chapter 2 we will consider primarily waves on strings: we will review resonances, which are basically "standing waves", and connect them with the "traveling waves" that are equally important, for example for carrying sound through air or light through space.

Next, in Chapter 3, we will address sound waves in air. In Chapter 4, we consider waves on sticks (as in a xylophone), plates (such as a cymbal or a bell), drums and flags. We will make the connection between all those waves and musical instruments in Chapter 5. Similarly, in Chapter 6, we will connect waves in air with human speech, as well as other human and animal sounds.

Part B will continue the exploration of various types of waves. First, in Chapter 9, we extend sound beyond air to liquids and solids, including the case of earthquakes. Chapter 10 will look more closely at how waves are reflected and/or transmitted at boundaries between different substances. The fascinating case of waves on the surface of water will be discussed in Chapter 11, including topics like beach waves, rogue waves, tsunamis, tides and boat wakes.

Chapter 12 covers the vast subject of electromagnetic (EM) waves, which includes light: we will consider such topics as mirrors, prisms, lenses, mirages, optical fibers, the solar spectrum, rainbows, lasers and holograms. Chapter 13 will address the mysterious quantum world, which is based in large part on quantum mechanical (QM) waves: it includes the structure of atoms, molecules and solids, as well as the color of substances and radioactivity.

Chapter 14 is concerned with gravitational waves, which were only shown to exist in 2015, even though they were predicted to exist a century earlier. Chapter 15 briefly covers several other types of waves, including the string theory of elementary particles, neural and brain waves, waves in human queues and road traffic, and many more.

1.5 What have We Learned in this Chapter?

In this introductory chapter, we have reminded ourselves of the many kinds of waves that exist around us. Some are clearly visible: water surface waves are the most obvious ones, from bathrooms to oceans; waves on bridges and ropes are often also visible; waves on surfaces such as flags are very obvious, while those on drumheads and loudspeakers may be harder to spot. Other waves are normally invisible, so we need to prove that they are really "waves": sound waves; earthquakes; electromagnetic waves such as light, radio waves and x-rays; quantum waves; gravitational waves; brainwaves; and others.

Many of these waves are of great importance in our lives, in particular: sounds and light; x-rays; electromagnetic waves; and quantum waves. A common aspect of all kinds of waves is continuous smooth motion that often is repeated back and forth. Another is resonance, which is the basis for much music, singing, speaking, lasers, atomic structure, *etc*. The following chapters will discuss the different characters and uses of all these waves.

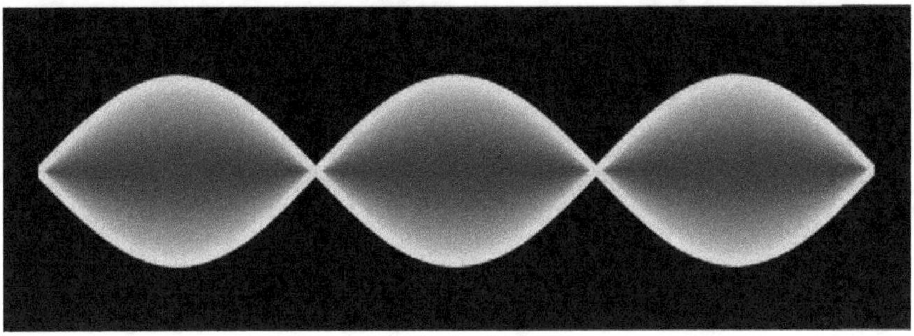

2

Waves on Strings

We start our exploration of waves with a type of wave that we can easily make and observe at home: waves on strings. They are also fundamental to making music with string instruments (guitars, violins, pianos), in particular in the form of resonances and so-called standing waves. We will see how such waves are generated and how they move along strings (or ropes, or toys like slinkies). Very important is the behavior when waves are reflected from the ends of a string.

Also very interesting is the "collision" between waves, called "superposition" and "interference": "colliding" waves can apparently briefly "destroy" each other, but then be resurrected and continue traveling as if nothing had happened!

Such behavior leads us then to a better understanding of standing waves. These also give rise to resonance on strings with fixed ends, as well as quantization, a property that is also crucial for lasers and quantum mechanics.

These various concepts are fundamental to all kinds of waves, as well as to their applications in daily life: we therefore will present and discuss them in this chapter in some detail.

I recommend that you first watch my accompanying video to get a feeling for the subject of this chapter.[1] *It includes many animations that cannot be presented in a printed book: these are especially helpful with a dynamic subject like waves!*

2.1 A Special Wave: The Standing Wave

In Section 1.3, we have illustrated the resonance behavior of bridges, boards and strings. In this section, we intend to learn further interesting aspects of such resonances. These will also be important for other kinds of waves: sound in air, sound in liquids and solids, water surface waves, electromagnetic waves, quantum waves, *etc.*

In Chapter 1, we saw that resonances on a bridge, board or string have the shape of **standing waves**. Good examples are waves on a hanging string (see Figure 1-1b) and on a vibrating bridge (see Section 1.3, including Figure 1-4). You can see the actual motion of standing waves in videos[2–5]: videos give a very lively impression of the behavior of these important waves; see Animation 1*1.

> ANIMATION 1*1 — See my video WA1 at time 3:38 in its section "**Standing waves**" under the title "**What are standing waves?**" (See details in the section References and Resources below.)

---

[1] See my video "Waves on Strings: from resonances to music" (WA1) on Everyday Physics by Michel A. Van Hove: https://www.youtube.com/watch?v=eCKigLmmBZc

[2] See video "Resonan Bridge" by Bob Barrett: https://www.youtube.com/watch?v=uWoiMMLlvco

[3] See my video "Waves on Strings: from resonances to music" (WA1) on Everyday Physics by Michel A. Van Hove: https://www.youtube.com/watch?v=eCKigLmmBZc

[4] See video "AP Physics 2: Waves 10: Resonance and Standing Waves on a String" by Yau-Jong Twu: https://www.youtube.com/watch?v=7xCmtYXewdk

[5] See video "Resonance and the Sounds of Music" by Walter Lewin: https://www.youtube.com/watch?v=f4M-6tWtkoA

Both Figures 1-5 and 2-1 sketch the typical shape of standing waves:

- **a standing wave is smoothly curved.**
- **it is fixed at its two ends.**
- **it may have immobile points, called nodes, between loops (or lobes).**
- **the loops oscillate across the length of the bridge, board or string.**

The standing wave is called "standing" because there is no visible motion along the length of the bridge or string; in particular, neither the nodes nor the loops move along the bridge or string.

In Section 1.3, with the help of online animation, we found that standing waves are quite sensitive to how they are created. For example, we saw that you should time the pushing correctly to create a resonance. Let's consider this and related aspects next, using a string as an example.

How should you time your pushing? To create a resonance, can you push a string any time you like, or should you push only at certain times? More precisely, at what moments in the back-and-forth cycle of the string should you push the string? We have seen that you must push in the direction of the motion of the string: if you push against that motion, the swinging dies out. This is sketched at the right in Figure 2-1 as green *versus* red arrows: the green arrows show forces pushing the string in the direction it is swinging, while the red arrows show forces pushing the string opposite to the direction it is swinging. If you push in the opposite direction from the swinging direction, you slow down the swinging and reduce the resonance.

Where on the string should you push? The preceding paragraph points us to the answer to this question: we should push wherever the string moves in the same direction in which you are pushing. The green arrows in Figure 2-1 show where that is: anywhere away from the immobile points (away from the ends and the nodes of the string), but only where and when the string moves in the direction of your pushing.

What is the "correct rhythm" to create resonance? With online animation, we have seen that, when a string is free to move by itself, it oscillates with a certain rhythm: we can call this the "natural rhythm" of the string. This rhythm is the frequency of oscillation when you

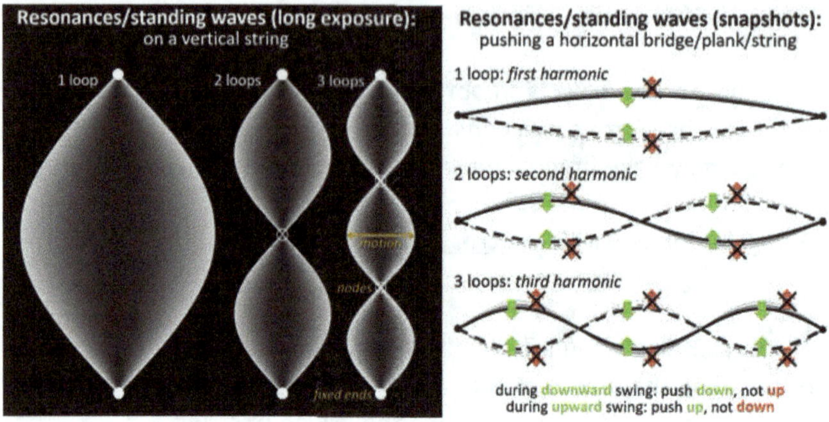

Figure 2-1: Standing waves are shown on a string (or bridge or board) with fixed length and fixed ends (in a simple planar simulation). *Left*: Harmonics with 1, 2 or 3 loops on a vertical string, mimicking a rapid oscillation or a long camera exposure, similar to the photographs in Figure 1-1b. These waves swing to left and right over time, as shown by the orange arrow. *Right*: Several snapshots of a horizontal string, board or bridge with 1, 2, or 3 loops. The green arrows indicate the up-and-down direction of motion (which is emphasized by the gray lines as fading snapshots). The green arrows also show pushing directions that increase the oscillation, while the red arrows show pushing directions that decrease the oscillation.

don't push, namely the rhythm of free swinging. It is called the **natural frequency** or **resonance frequency**. We can thus say: **An object's resonance frequency is the natural rhythm of that object's swinging when no repetitive force is applied to it**. For example, the resonance frequency of a guitar string is the frequency with which it vibrates after you pluck it and while you let it swing freely; the resonance frequency of a bridge is the frequency with which it continues to vibrate after you stop jumping on it; the resonance frequency of a child on a swing is the frequency with which it continues to swing after you stop pushing it; and the resonance of the pendulum in a grandfather clock is its natural rhythm, which is used to tell us the time of day.

We have also seen in Section 1.3 that, in order to increase the swinging, it is necessary to push **in synch** with the swinging (meaning in synchrony). This means pushing with the same frequency as the natural or resonance frequency of the string or bridge. Another way to say this is: **resonance requires that the driving frequency must be equal to the resonance frequency**. Why is this so? If you push "out of synch" with the swinging, namely with a frequency different from the resonance frequency, you will sooner or later find yourself pushing opposite the swinging direction, and you will therefore reduce the swinging. The reason for this is that each swing of the string will get a bit ahead of, or fall slightly behind, your pushing, so you will end up pushing when the string is swinging in the wrong direction.

Pushing in synch allows you to apply many pushes in the same direction as the swinging: the pushes accumulate over time to give large vibrations; even small pushes can then add up to give large vibrations.

How smoothly should the string be pushed? With the online animation in Section 1.3, it became clear that it is not easy to produce a smooth wave by hand. If you push the wrench (visible in Figure 1-6) abruptly, you produce a wave with a nearly square shape, forming a sharp step that travels back and forth. If you move the wrench at a constant speed and then suddenly reverse its direction, you produce a wave with a nearly triangular shape, forming a sharp peak that goes back and forth. To produce a smoothly curved wave like those shown in Figures 1-5 and 2-1, it is best to gently reverse the wrench's motion from up to down and from down to up, by gradually slowing it down and speeding it up each time. This was done in the animation with the "Oscillation" option (as in Figure 1-6): it produces a very smooth up-and-down motion of the "driver" and thereby also a very smoothly curved wave. This wave then has the smooth shape of the standing waves of Figures 1-5 and 2-1.

The standing waves, which we have discussed here, are special cases of more general "traveling waves": these are waves that do move along the length of the string; other examples are sound waves traveling from a mouth or musical instrument to your ears, or waves traveling along the surface of water. As we will see later (in Section 2.8), standing waves can be composed of two traveling waves that cross each other in opposite directions. So let's now turn our attention to traveling waves, starting with wave pulses, then wave packets and finally wave trains: that will bring us back to standing waves!

2.2 Traveling Waves: Wave Pulses

We saw in Figure 1-1c a wave on a fishing line used for flyfishing. You can imagine the fishing line flying around as the fisherman swings his fishing rod back and forth. Such a fishing line, although very thin and speedy, will generally retain a smoothly curved shape. That curved shape travels from the tip of the fishing rod to the end of the line as a wave, usually quite fast: we call this a **traveling wave** or **propagating wave**. A similar wavy motion exists in whips and slinkies (Figure 1-1d). You may review some videos to look more closely at the behavior of slinkies.[6,7] You may also view my Animation 2*1.

ANIMATION 2*1 — See my video WA1 at times 4:07 and 4:55 in its section "**What are waves?**" under the titles "**What are waves? Animation**" and "**Different kinds of waves**". (See details in the section References and Resources below.)

You may also make and watch such waves with the online animation "Wave on a String"; we used it in Section 1.3 to produce resonances and standing waves (see Figure 1-6). If you start up that animation and select Pulse (at the upper left), you can produce waves like that shown in Figure 2-2: click on the green button. With the option No End, that wave will exit through a window and never come back. We call this a **wave pulse** because it is brief (the Pulse Width or duration is only 0.50 seconds) and it goes up and down only once. If you slow down the animation (with option Slow Motion), you can see that the driver rises at constant speed and then abruptly turns around and descends at constant speed, giving this pulse its pointed triangular shape.

You can also launch a few short pulses right after each other, by pressing the green button once for each, as shown in Figure 2-3 for three pulses. We call them a **wave packet** because they travel together as a rigid group. When many such pulses follow each other closely, we call them a **wave train**, illustrated in Figure 2-4. In this example of a wave train, the pulses are rounded like those of standing waves; however,

[6] See video with slinky "Transverse and Longitudinal Wave Demonstration — A level and IGCSE Physics" by Chris Gozzard: https://www.youtube.com/watch?v=iT4KAc0Ag1E

[7] See video "Transverse Waves on a Slinky" by Shyam Srinivasan: https://www.youtube.com/watch?v=y66PSaiGH7Y

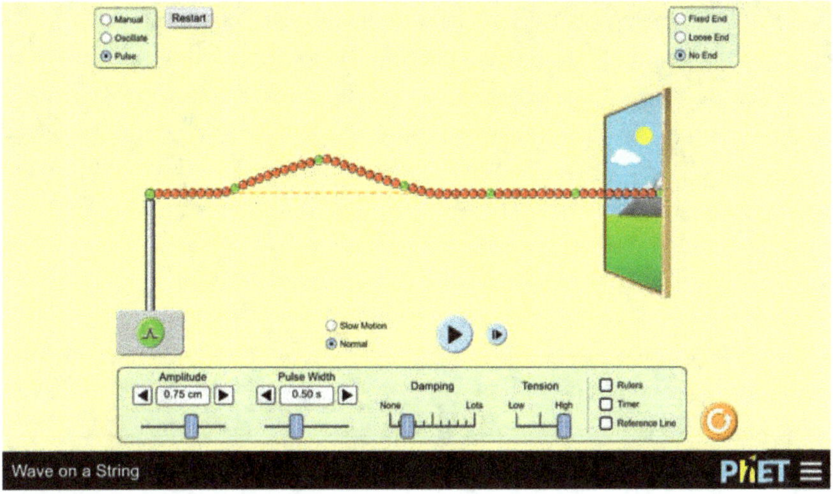

Figure 2-2: Snapshot of the "Pulse" screen of the online animation "Wave on a String", using the option No End (at the top right). This screenshot was taken shortly after clicking the green button, which launched a "triangular wave pulse" along the string; its duration is 0.5 seconds. (*Source*: as Figure 1-6.)

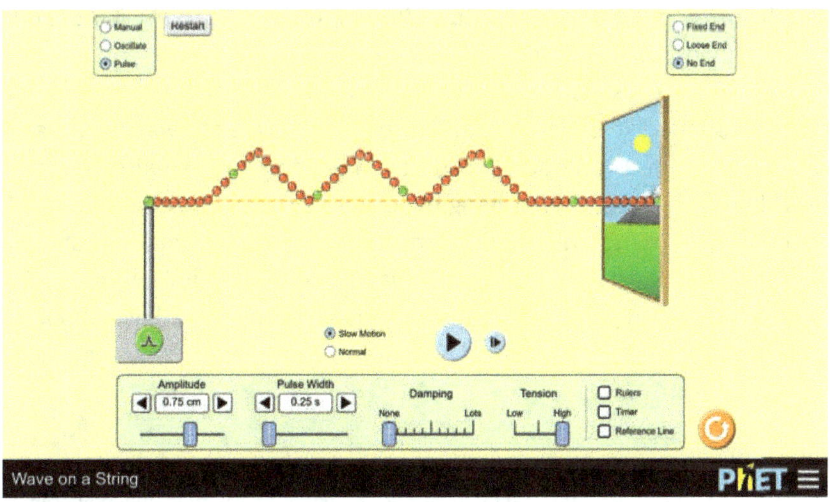

Figure 2-3: As Figure 2-2 but showing a wave packet composed of three triangular wave pulses (they are shortened compared to Figure 2-2 by setting Pulse Width to 0.25 second; damping has been turned off to produce equal pulses, and Slow Motion has been selected to more easily launch the pulses immediately after each other). (*Source*: as Figure 1-6.)

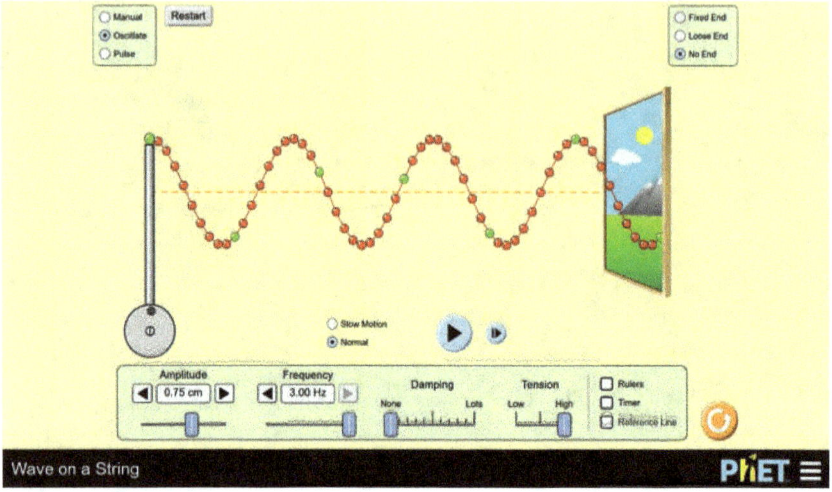

Figure 2-4: Like Figures 2-2 and 2-3 but showing a wave train. The Oscillate mode produces a "sine wave" that is smoother than the triangular wave pulses of Figures 2-2 and 2-3. (*Source*: as Figure 1-6.)

these are not standing: they travel (to the right in this example). We will often use this type of rounded wave train later on since it closely matches the shape of many natural waves, unlike the triangular shape; this rounded shape is called a sine wave.[8]

Figure 2-5 also shows wave pulses, packets and trains, with a rounded shape like a sine wave. You can produce such waves yourself with a string (or a rope or a slinky), as in Section 1.3: see also Box 2-1. Here I suggest using a long hanging string or rope with an unattached lower end, to match the endless string in the online animations shown in Figures 2-2, 2-3 and 2-4. A simple way to do this is to allow the string to hang straight down from your hand (maybe down to a lower floor in a stairwell or outside a window upstairs to gain length). You may view my Animation 2*2.

[8] The up-down driver's motion in the online animation "Wave on a String" is a "sine wave" as a function of time, named after the mathematical sine function. Many oscillating systems in nature, including waves and resonances, move like sine waves. The standing waves in Figures 1-5 and 2-1 are all sine waves. In the case of the online animation in the Oscillate mode, the rotating wheel under the driver makes the driver move up and down like a piston: the constantly rotating wheel generates a sine wave motion of the piston. In the Pulse mode of that animation software, the driver does not move like a sine wave, but has a jagged "triangular" motion.

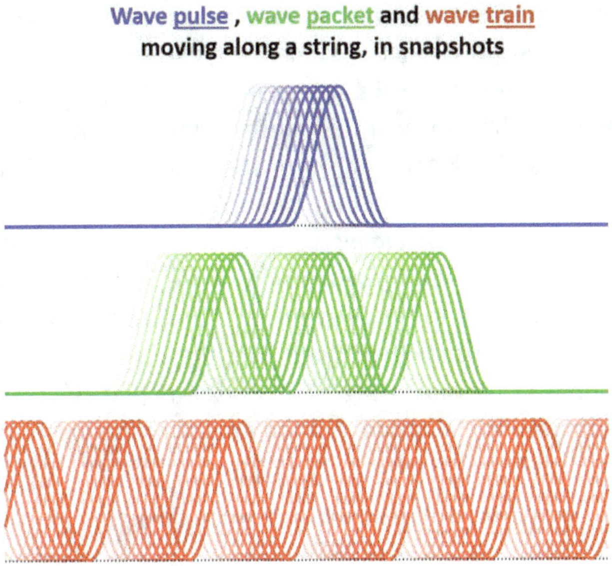

Figure 2-5: A wave pulse (blue), a wave packet (green) and a wave train (red) are shown traveling from left to right along a string, in superposed snapshots. Old snapshots are shown fading out (at any one time you only see one curve of each color, such as the three latest, most intensely colored curves). The straight horizontal lines show the string's position at rest when there is no wave.

ANIMATION 2*2 — See my video WA1 at time 4:55 in its section "**What are waves?**" under the title "**Different kinds of waves**". (See details in the section References and Resources below.)

BOX 2-1 — THERE ARE ALSO OTHER WAYS TO MAKE SUCH WAVES YOURSELF: One way is to lay a string/rope/slinky flat on a slippery floor, firmly attached at the far end (for example to the bottom of a table leg); your hand grasps the near end, keeping it close to the floor, and pulls it taut, so it is under tension; then your hand gives it a sideways push. You may instead attach the far end above the ground to any firm object (door handle, tabletop, *etc.*), while pulling the string/rope/slinky taut by grasping the near end, so it does not touch the ground; you can also now push the near end sideways with your hand. While light strings are readily available, the pulses and waves tend to move quite fast on these and are thus not easily followed with the eyes: thicker and heavier strings as well as ropes are helpful. A slinky is also a good choice because it is usually heavier; choose a long one.

Let's now look more closely at the mechanism that makes waves move. For that purpose, let the string hang as in Figure 2-6. We start with a simple question: ***Are the waves in* Figure 2-6 *falling downward due to gravity?*** To help answer this question, let's also ask these related questions: ***If you shook the <u>bottom</u> of the string, would these waves climb up the string, and would they climb differently than when falling downward? Would these waves behave differently on a horizontal string?*** You can try answering by yourself with a string! The answers to these three questions are: "No, waves don't fall down"; "Yes, waves can climb without falling back down"; and "No, these waves do not behave differently on a horizontal string". The reason is that the waves move due to the tension in the string, not the gravity of the Earth; therefore,

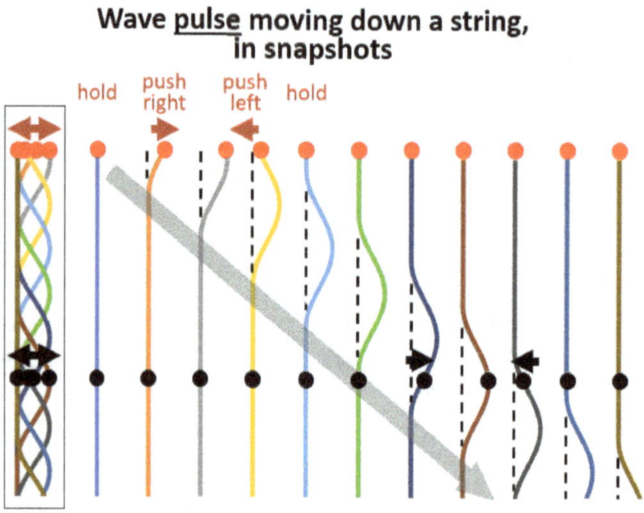

Figure 2-6: A wave pulse on a hanging string is shown traveling from top to bottom, in snapshots. All the snapshots are superposed in the box at the left, and also spread out to the right, using the same colors. The red dots mark a hand holding the top of the string, while the bottom of the string hangs freely, out of sight. Initially, the string is straight and at rest (blue line at the left). The hand then pushes the top of the string first a bit to the right (orange line), and then back (gray and yellow lines) to the initial position (lighter blue line). The pulse continues downward (green line and lines to its right), as suggested by the large gray arrow. The black dots represent one piece of the string: when the pulse passes, it also moves to the right and then returns to its initial position, mimicking the motion of the hand, but delayed in time. The dashed lines show the string's position at rest.

the motion of a wave does not depend on the orientation of the string it travels on. So a wave will climb up a string just as fast as it descends that same string.

Why and how do waves depend on the tension of the string? Remember that we are concerned with strings that are stretched by pulling on both ends and are thus put under tension. Let's take any straight, stretched string in a musical instrument: if we now give it the shape of a wavy curve, it becomes longer; this lengthening increases the tension further. Since all strings have some elasticity under stretching, they pull back against such stretching, which also pulls back the wavy curve toward the original straight line. However, as it straightens out, the string will also overshoot: that overshooting will lengthen the string again (on the other side), so the wave motion will repeat over and over in an oscillating fashion.

Now, if we increase the tension of the string (by pulling more strongly on its two ends), it will be stretched further, which increases the elastic force pulling back: that increased force will speed up the straightening out and the overshooting, resulting in a shorter repeat time between oscillations and thus an increased vibration frequency. This behavior applies to traveling waves as well as standing waves.

We have seen that **strings vibrate due to the tension applied to their two ends, through elastic stretching**. Another mechanism exists for thicker and stiffer strings, such as cables and sticks: elastic bending. If you take a thick cable or a stick, it is obvious that it resists bending: this bending elasticity also restores the original shape and can also propel waves. We will further explore this situation in Chapter 4, for example for the musical triangle and the xylophone.

In the remainder of this section, we focus on the behavior of the **wave pulse**. Section 2.3 will look at wave packets and wave trains.

Figure 2-6 expands the snapshots of the pulse shown in Figure 2-5, now on a hanging string: here we can better see the motion of the string while the hand (the red dot) jerks it right-then-left. The black dots represent one small piece of the string: it will also jerk right-then-left as the pulse passes it, but this motion is delayed by the time it takes for the pulse to reach the black dot. The black dots illustrate the traveling of a simple signal (the single back-and-forth jerking of the top of the string, which may mean "I am here!") down along the string.

What is the shape of a wave pulse? At least initially, the pulse which you created has the shape given by your hand's motion, as we saw in the online animation. For example, you can move your hand smoothly back and forth, giving a gently rounded shape like that seen in Figures 2-5 and 2-6. You could also move your hand farther, creating a stronger wave pulse, having a greater **amplitude**. But you could instead move your hand abruptly, yanking it left and right: that will give the pulse a less rounded and more square or triangular shape. The shape you give the pulse can change the meaning of a signal you send along the string.

Indeed, **your hand's motion can give the wave pulse any shape you wish**. That is true initially, but during its travel, the wave may change its shape gradually due to the physical properties of the string. For example, if you give the pulse a square shape, that shape will gradually become rounder, like that of Figure 2-6, because the string strongly resists being folded with sharp corners. In addition, in such cases, the pulse will tend to broaden out as it moves; its front end goes faster than its back end, unlike the smooth case drawn in Figure 2-6, where the pulse retains its shape. Another way to send a different signal is to send a wave packet or wave train (see Section 2.3).

In a wave pulse, what moves and how? In Figure 2-6, the line of red dots indicates that the top end of the string stays at the top: it only moves a bit to the right and left when you move your hand. Now look at the line of black dots, representing another piece of the string; that piece also stays at that level (it does not go down or up), and also moves a bit to the right and left as the wave pulse passes by. **So, each individual piece of the string stays near its original position.**

We see that the motion of each piece of the string is perpendicular to the length of the string (in this case, the back-and-forth motion is horizontal): for that reason, we call it a **transverse wave**.

Nevertheless, we see a shape, the wave pulse, moving along the string, but no piece of the string follows that pulse! **This is a very important observation: no part of the string moves along with the shape of the wave.** In other words, **the wave pulse does not transport any piece of the string in the direction along the string; only the shape of the string moves along.**

Compare this with waves on the surface of water. Such waves travel, often fast and far, without water flowing along with them: a floating leaf or stick stays in the same place, bobbing up and down. (We will discuss waves on water in Chapter 11.)

What makes a wave pulse move? There is tension in the string due to two opposing forces: in the case of a string hanging from your hand, these two forces are gravity (or a knot) pulling down and your hand pulling up. This causes a tension that straightens the string, as shown at the left in Figure 2-6 (straight blue line). The straight shape is the string's equilibrium shape in the absence of a wave.

Now, as you push the top of the string sideways to the right, you create a curve in the string, which lengthens it and therefore increases its tension. The tension tries to straighten out that curve: that straightening pulls the next piece of the string also to the right, but with a slight time delay due to its inertia (the string resists speeding up because of its mass). This repeats itself from one piece of string to the next: the deformation thus moves down the string.

How does the string return to its original position? It is the same story in reverse: you pull to the left, which causes a curve in the opposite direction; tension then also results in leftward motion in the next piece of string; this again repeats itself down the string, bringing it back to its original straight shape while the pulse moves farther. We may thus describe the mechanism of wave propagation as follows: **in a wave, equilibrium is disrupted and then restored**, and **a disturbance is passed along.**

We can also conclude: **tension is the cause of the motion of a wave pulse in a string** (or rope, slinky, *etc.*). **That is why musical instruments using strings need tension on their strings**: for example, guitars, violins and pianos.

Compare a string under tension with a loose string that you let fall to the ground: what shape does the loose string have after landing? Is it straight? Does it move? No, it is flat, it has a randomly curved shape that depends on how it landed, and it does not move. Why? There are no forces pushing or pulling any of its parts horizontally and thus parallel to the string, so it is not under tension, and no waves move along this loose string.

When looking at pulses (and other waves as well), we seem to have a contradiction or paradox. Since a pulse at any moment extends for some distance along the string, it looks like the pulse's shape is a coordinated effect of the entire string. On the other hand, we saw that the shape is the result of local tension in each little piece of the string.

Think of this analogy: a well-ordered queue of people stands at a bus stop. A rude newcomer slams into the last person in the queue, violently pushing him forward. This last person bumps into the next person, who in turn bumps into the following person, and so on until the head of the queue: it is just the **domino effect**, and therefore easy to understand. This forms a wave pulse (the disturbance between neighbors) that travels all the way along the queue; multiple pulses can build up a wave train (if the people in the queue don't wake up to what is happening!). But here is the crucial part: nobody in the queue feels what is happening beyond their two neighbors. As with a string or dominoes, we have only very local contacts (between neighbors only), but the effect propagates along the entire line: the effect seems coordinated in both time and space to produce a well-choreographed pulse. The underlying reason for this behavior is simply that the queue is the same everywhere along its length, which we can call homogeneous: what happens to the first person to be hit is also exactly what happens to every other person down the line, so the same effect travels all the way unchanged.

One way to summarize this apparent paradox is: **"local" contacts in homogeneous strings cause well-coordinated collective "global" wave shapes**.

An interesting and important consequence of this "local" behavior of pulses and waves is the following: **every piece of the string serves as the source of the wave traveling further down the string** (as in the domino effect). Here is another way to put it: **to each piece of the string, it is the same whether the pulse or wave is due to the neighboring piece or to a source much further back along the string**. We will encounter this surprising fact again as the **Huygens principle**, for example in Section 3.9; this principle is named after Dutch scientist **Christiaan Huygens**, who proposed it. Very important is that this behavior exists for all types of waves.

What affects the speed of a pulse on a string? You can explore this question yourself by changing the length or thickness or tension of a string. If you stretch a string between your hand and a fixed attachment (such as your foot, a door handle or a chair), you can also change the tension in the string by pulling more or less on the string. Here are the results: **a wave pulse on a string moves faster if the string is under more tension and if it is lighter**. The reason is that a larger tension makes the deformation of the string return faster to the straight shape of the string, and so does a lighter mass.

The case of a hanging string (or rope or slinky), like that of Figures 2-5 and 2-6, is especially interesting in this respect! Because the string hangs in gravity, its tension is higher near the top than near the bottom (near the top, the string has to support the weight of the entire string, while near the bottom, it only has to support the weight of the shorter part below it). This changing tension along the string changes the speed of the wave pulse: it is largest near the top, slowing down to practically zero speed near the bottom! (For simplicity, I have not taken this effect into account in drawing Figures 2-5, 2-6 and other figures below; I have assumed the string to be very long, so the tension varies only slightly along its upper segments.)

Does a wave pulse carry energy? To answer this, we must first ask what energy is: that is a fascinating and big topic deserving a separate book, so here we will only say that motion is one among many forms of energy, called **kinetic energy** (other forms of energy include gravitational energy, heat, electrical and magnetic energy, nuclear energy, chemical energy, potential energy, *etc.*).

We know from daily experience that motion carries energy. You notice that when you run into a wall or witness a collision: the energy of motion leads to pain, damage or bouncing. Thus, since the far end of the string will move back and forth when the pulse reaches it, we can certainly say: **a wave pulse carries energy from your moving hand along the string to its far end**.

What is most remarkable here is that energy is transported along the entire length of the string even though the motion of each piece of the string is strictly local: as we saw, the pieces of the string do not travel along the string. At any one moment, the energy is nevertheless present in one part of the string in the form of sideways motion within the pulse: that energy is passed along the string from one piece to the next.

We will see that this property of transporting energy remains true for most other kinds of waves, such as water surface waves, sound waves, electromagnetic waves, *etc.* Thus, **waves are an important method of transporting energy! Indeed, most of our energy is transported from the Sun by light and other electromagnetic waves.**

2.3 Traveling Waves: Wave Packets and Wave Trains

We now look more closely at **a wave packet** and a **wave train** traveling down a hanging string, like those in Figure 2-5. See their motion in Animation 2*3. Remember: a wave packet contains a few wave pulses, while a wave train contains many wave pulses; so a wave packet is formed when you wave your hand a few times, while a wave train is created when you wave your hand many times with a regular repetition. After we spread out the snapshots from left to right, the result looks like Figure 2-7.

> ANIMATION 2*3 — See my video WA1 at time 4:08 in its section "**What are waves?**" under the title "**What are waves? Animation**" (See details in the section References and Resources below.)

Every piece of the string, like the black dots, is shaken back and forth: it repeats the motion of your hand, but its motion is again delayed by the time needed by the wave to reach it. You can see these oscillations very well in the photos of Figure 1-1b: the horizontal stripes are irregularities in the string, and they are clearly moving primarily horizontally, perpendicular to the string itself, just like the black dots in Figure 2-7.

Much of what we wrote in Section 2.2 for wave pulses remains valid for wave packets and wave trains, for example:

- **the shape of the wave train is that given by your hand's motion.**
- **the individual pieces of the string oscillate and stay near their original positions.**
- **the wave does not transport any piece of the string from one place to another.**
- **a wave on a string moves faster if the string is lighter and if it is under higher tension.**

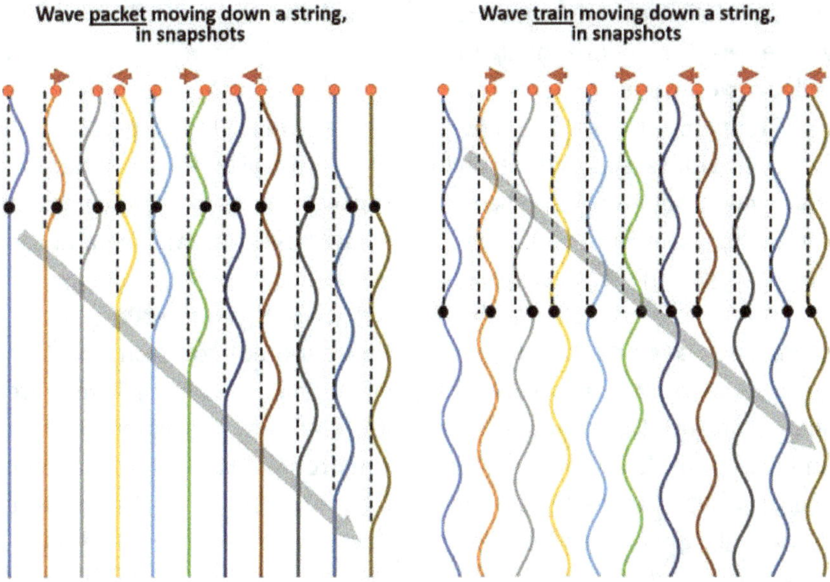

Figure 2-7: On a hanging string, a wave packet (*at the left*) and a wave train (*at the right*) are shown traveling from top to bottom, similar to the wave pulse of Figure 2-6. Here, starting again from the left, a hand (the red dot) continues shaking the string back and forth, creating (*at the left*) a packet of three pulses or (*at the right*) a train of many pulses that follow each other, as indicated by the large gray arrows. Pieces of the string, marked as black dots, shake back and forth as the waves pass by: they repeat the motion of the hand, but are delayed in time.

- **energy is carried by a wave from your hand along the string to its far end.**
- **these statements are true not only for strings but also for ropes, slinkies, *etc.***

Since wave pulses, wave packets and wave trains behave very similarly, we will from here on often simply call them **waves**.

2.4 Wave Shape, Period and Frequency

What are the most common wave shapes? In Figures 2-5 to 2-7, I have used a smooth, snaky shape for a wave that can repeat endlessly. It is a common shape for many types of waves, including waves on strings, ropes and slinkies. This shape is shown again at the top of Figure 2-8, where it is drawn horizontally instead of vertically, as in Figure 2-4: it is called a **sine wave**.

The name sine wave comes from a mathematical function, the sine function, whose shape is very close to many waves found in nature, from waves on strings to electromagnetic waves; the English word sine comes from the Latin word sinus, meaning curve. We will see and use this common sine-wave shape repeatedly in this book!

Figure 2-8 also shows three other wave shapes: **square wave**, **triangular wave** and **sawtooth wave**. These are not common in nature because sharp corners are generally unfavorable in motion. Such wave shapes are used primarily in electronics (electrons are extremely light and can therefore change directions extremely fast); for example, square waves can be used to make lights blink by turning an electric current abruptly on and off; sawtooth waves are much used in music synthesizers, digital photography, computer displays, and television displays, where they can sweep an image electronically; triangular waves allow equal and alternating sweeping from left to right and from right to left. You can listen online[9] to sounds having these four wave shapes: you will notice large differences and realize that sounds with sudden jumps (square and sawtooth shapes) produce much harsher tones due to additional very high frequencies. The sine shape is the smoothest and clearly the most musical.

Also in Figure 2-8 are labeled several quantities that are frequently used when describing waves. Of special importance are the **wavelength** and the **wave amplitude**. The wavelength is the repeat distance in the travel direction (called distance in the figure), while the wave amplitude is the maximum deviation from the rest position, such as the deviation of a string from its position at rest. The "top" of a wave is often called the **wave crest** or **wave peak**, while its "bottom" is often called the

[9] Online Tone Generator, The 432 Hz Frequency — start by clicking on Play, then select Sine, Square, Sawtooth or Triangle; stop by pressing Stop (be careful not to produce too loud sounds, as they may damage your hearing): https://onlinetonegenerator.com/432Hz.html.

You can do the same thing with another Online Tone Generator; change the wave shape with the button to the left of the button saying "COPY LINK": https://www.szynalski.com/tone-generator/

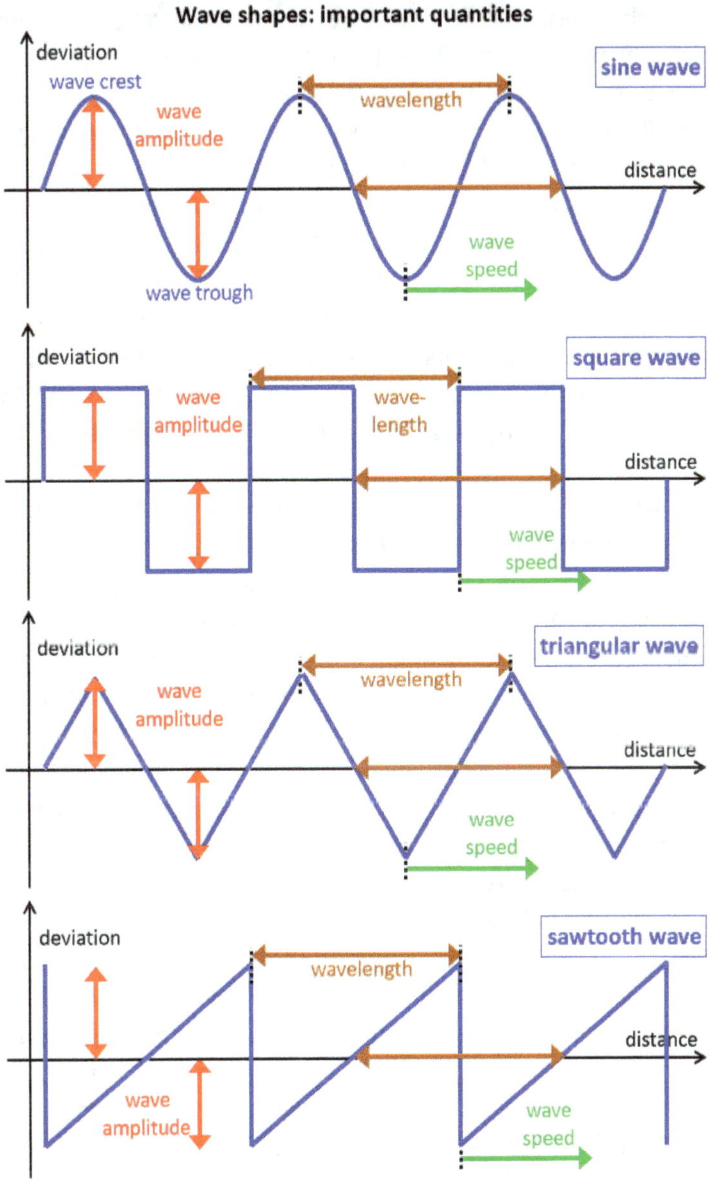

Figure 2-8: Simple wave shapes are profiled in one dimension called "distance", such as along a string (the distance is here drawn from left to right, unlike in some other figures in this book where the distance is drawn from top to bottom). The wave shape does not indicate which way the wave travels: it could travel either to the right or to the left; a wave speed for motion to the right is shown as a green arrow. Frequently used words are marked to quantify important properties of such waves ("deviation" means change relative to the absence of the wave, which is a straight line, for example the sideways displacement of a string as drawn in Figures 2-5 to 2-7).

wave trough or **wave valley**. The **wave speed** indicates the constant speed at which the wave shape travels, as opposed to the back-and-forth speed of pieces of the string, for example.

We will see later that many waves in nature are in fact combinations of smooth waves, like the sine waves in Figures 2-5 to 2-8. We can indeed combine sine waves that have different wavelengths, different amplitudes, different speeds, different directions, *etc.* This happens a lot in music, voices, *etc.* How to combine waves is described in Section 2.5.

What is a wave's period? When an event repeats itself on a regular schedule, we can ask what its period of repetition is. For instance, the Earth turns around its axis once a day relative to the Sun, so the Earth's period of rotation is 1 day or 24 hours. And the week repeats itself every 7 days, so the week's period is 7 days.

Similarly, a wave train has a repetition period: **the wave period is the amount of time between the passage of one wave pulse and the next at any given spot**. In Figure 2-7, we can see a repetition at the black dot, in both the wave packet (at the left) and the wave train (at the right). If you follow a wave crest or trough passing through a black dot in Figure 2-7, you will find that it takes about 3.5 snapshots to see the next wave trough passing by; so, if there is one second between snapshots, the wave period is about 3.5 seconds.

What is a wave's frequency? Another way to think of repetition is to ask: how often does the event repeat itself in a given time interval? For instance, the Earth turns around its axis once per day, which is also about 365.25 revolutions per year (since there are about 365.25 days in a year), or 1/24th of a full revolution per hour; a car engine may be turning at 1,000 revolutions per minute (often written as 1,000 rpm or 1,000 RPM). Such **a rate of repetition is called frequency** in physics.

For a wave, we can say: **the wave frequency is the number of repetitions of the wave passing any given spot per second**. So we may write that as "frequency = number of repetitions/period", using the division symbol "/". For the wave in Figure 2-7, the frequency is therefore about 1/3.5 repetitions per second, which is about 0.29 repetitions per second (assuming 1 second between snapshots).

We choose one second as the time interval, which is common in physics and in electrical as well as electronic technology. Frequency then counts repetitions per second. You have likely seen the word hertz, as in 50 hertz or 60 hertz, which are typical repetition frequencies of electricity supplied by electrical power plants, and you may know the abbreviation of hertz to Hz. **The hertz (or Hz) is used to count frequency in repetitions per second**, so the hertz is a common unit of frequency (just as km/h, or kilometers per hour, is a common unit of speed). It is named after the German physicist **Heinrich Hertz**.

Thus, **1 hertz = 1 Hz is a frequency of 1 repetition per second, also often called 1 cycle per second**.

For example, if you clap your hands once a second, you are clapping with a frequency of 1 hertz; clapping twice per second gives a frequency of 2 hertz; and clapping once every 5 seconds gives a frequency of 1/5 = 0.2 hertz.

You probably also have seen the words **kilohertz**, **megahertz**, **gigahertz** and **terahertz**, often abbreviated as **kHz**, **MHz**, **GHz** and **THz**, respectively (they are analogous to meters and kilometers, which are abbreviated to m and km, respectively, but we rarely use megameters, gigameters or terameters). These units of frequency are often used in electronics and computing. We will often encounter them in this book, especially with electromagnetic waves. Their meaning is: 1 kHz = 1,000 hertz; 1 MHz = 1,000,000 hertz; 1 GHz = 1,000,000,000 hertz; and 1 THz = 1,000,000,000,000 hertz (these large numbers are one thousand, one million, one billion and one trillion, respectively, in US English). For more units and abbreviations, please see the section References and Resources.

2.5 Interference and Superposition of Waves

Here, we explore one of the most unique, characteristic, remarkable and important features of waves: ***how do two waves behave when they collide with each other?*** We will see that waves don't actually "collide" with each other, but "pass through" each other without changing each other, unlike objects such as stones, human bodies, liquids, *etc*. This is called **interference** or **superposition** of waves.

Imagine a string stretched horizontally, on which two identical wave pulses move toward each other. You may try this experiment yourself, but it will be much easier to look at the movie frames shown in Figure 2-9. In the left set of frames, from top to bottom, we see two peaks "colliding" with each other and then separating again. At the right, we see a peak and a valley (dip) "colliding" and separating.

For comparison, try to imagine what would happen if these were two beads sliding along a string instead of two wave pulses: two colliding beads would either bounce back from each other or shatter into pieces; they certainly would not cross each other intact.

How do the colliding wave pulses behave? Notice the following in the white-on-black movie frames of Figure 2-9:

- the pulses look different during the collision (frames 3 to 7), but recover their original shapes afterward, and then continue as if nothing happened; each pulse appears to somehow cross "through" the other pulse and re-appears unchanged on the other side.
- in particular, the pulses are not reflected, and they keep their direction and speed; that is very clear for the peak and dip at the right: they definitely cross each other; you could argue that the two peaks at the left bounce back, but the right peak is slightly taller than the left peak and re-appears on the other side of the smaller peak, confirming a cross-over instead of a bounce back.
- the two peaks overlap to produce a taller peak: we call this **constructive interference**; at the center of the collision (frame 5 at the left), the two peaks somehow form a peak that appears to have double the height of each peak (or, more precisely, a combined height that is the sum of the two slightly different pulse heights); we describe peaks that point in the same direction (here up and up) as being **in phase**.
- the peak and dip cancel each other: we call this **destructive interference**; at the center of the collision (frame 5 at the right), the peak and dip largely disappear, only to reappear shortly thereafter; it's as if the peak momentarily filled in the dip (similar to filling a hole to make the ground flat, or digging a hole to flatten a mountain); however, the peak is slightly taller than the depth of the dip, so the peak overfills the dip a little bit; we call peaks that point in opposite directions (here up and down, like peak and dip) as being **in antiphase**.

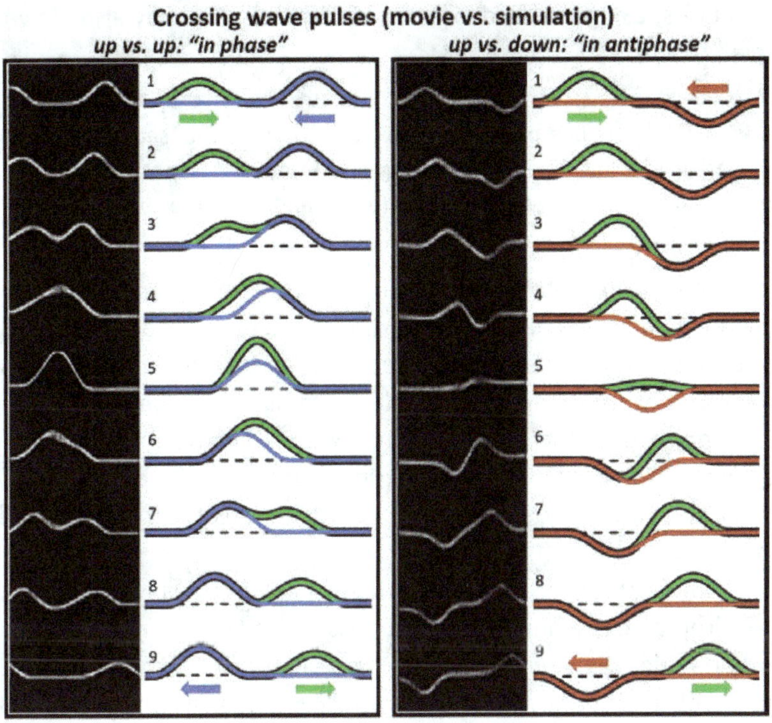

Figure 2-9: *In black/white*: from top to bottom, movie frames of two wave pulses crossing each other on a string. *In color*: simulation using rounded wave pulses of slightly unequal heights/depths; the green pulse is drawn "flowing over" the blue and red pulses; the actual string shape is shown as the black contour, which is the sum of the green+blue and green+red pulses, respectively, and which matches the movie frames. *At the left*: both pulses point up, "in phase". *At the right*: one pulse points up, the other down, "in antiphase". Arrows show the directions of motion of the pulses. (*Source*: movie frames extracted from Lectures on Physics by Benjamin Crowell, under CC BY 2.0 License, http://www.vias.org/physics/bk3_03_02.html.)

The movie frames shown in Figure 2-9 suggest a very simple model for what is happening: ***could it be that the two "colliding" pulses are simply stacking themselves onto each other without changing each other?*** In other words, ***could we simulate the string shape by simply adding one pulse on top of the other?*** To test this idea, the color graphs in Figure 2-9 show matching simulations of the "collision" of two simple wave pulses, using rounded sine-wave shapes.[10] You may also view my Animation 2*4.

ANIMATION 2*4 — See my video WA1 at time 6:36 in its section "**Collisions of waves**" under the title "**An amazing and important aspect of pulses and waves**". (See details in the section References and Resources below.)

At the left in Figure 2-9, a green peak travels toward the right and "flows over" a blue peak traveling in the opposite direction. At the right, a green peak also travels toward the right as it "flows through" a red valley/dip that travels to the left. We could also describe the "flowing" of one pulse past the other in other ways: "climbing over", "sliding over" or "rolling over". The actual string shape is the "sum" of the pairs of colored pulses, shown as black contours: this black wave can be compared directly with the movie frames.

An analogy for this model of crossing wave pulses is the flow of a water drop across a bump or dip: it passes over or inside the obstacle. Another analogy is a bulldozer pushing a pile of dirt across a mound or a hole. Yet another is walking over a hill or across a valley: your head follows the shape of the terrain, going up over hills and down through valleys. (But these analogies are only visual: the physics of these different situations cannot be compared directly.)

In Figure 2-9, the very good match between the movie frames and the simulations supports the idea that "colliding" pulses simply "climb" or "slide" or "roll" on top of each other without changing each other.

To best match the movie frames, the simulations in Figure 2-9 also have slightly unequal heights/depths for the peaks and dips. As a result, in frame 5 at the right, the green peak overfills the red dip, giving the small green hump. With equal heights and dips, the green peak would exactly fill the dip, resulting in a perfectly flat string for a very brief moment. This situation is simulated in Figure 2-10: see the flat green curve at the right (this simulation, which only shows the actual string shape, is otherwise identical to that in Figure 2-9).

---

[10] The simulation uses a sine-squared function for each peak or dip, which is a shifted sine function. The two crossing pulses are mathematically simply summed together.

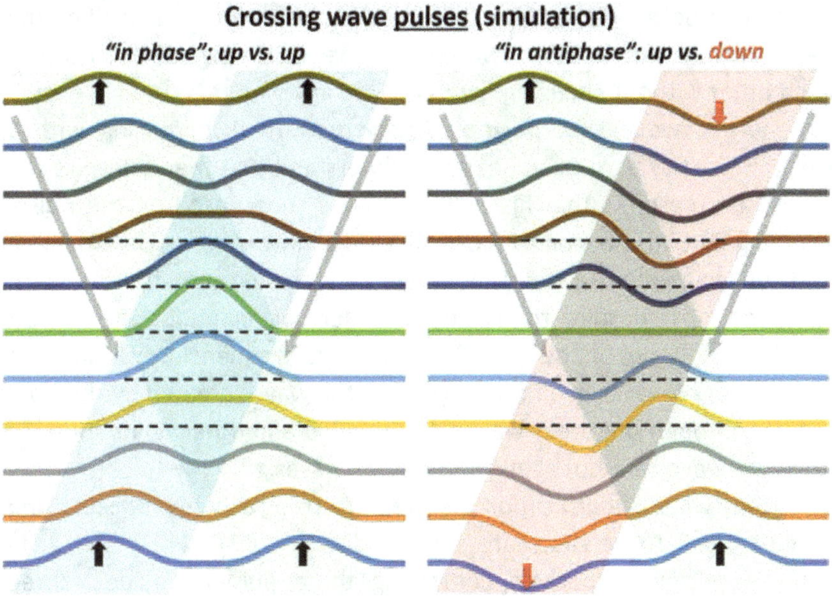

Figure 2-10: Snapshots (from *top to bottom*) of the collision of two identically shaped wave pulses on a string (simulation). *At the left*: one wave pulse is moving toward the right along the string, as in Figure 2-9 (shown by the green shaded band and the gray arrow pointing to the lower right), while another identical wave pulse moves toward the left along the same string (shown by the blue shaded band and the gray arrow pointing to the lower left). *At the right*: the same situation, except that the second wave pulse is "in antiphase": it points down instead of up (as indicated by the red *versus* black arrows).

Note that in Figure 2-9 we could equally well have drawn the blue or red pulse "flowing over" the green or red pulse, instead of the green pulse "flowing over" the blue or red pulse: either way, the result is the same black-contoured wave. In fact, in the overlap region, there is no way to tell which part comes from which pulse: in reality, as in the movie frames, we only see the black-contoured wave, not the individual waves (blue or green or red waves). **Only by understanding the underlying physical mechanism can we represent the apparently complex combined wave as the sum of two simpler waves.**

The most remarkable observation about "colliding" wave pulses is that the pulses keep their shape and direction of motion despite the "collision": the pulses simply go "through" each other, as if they didn't notice the other pulse; so there is no actual "collision". Superposition is clearly a better name than collision! This behavior thus justifies the name **wave superposition**: two crossing waves simply stand on top of each other without changing each other in any way.

Why is wave superposition important? Think of talking with a friend in a noisy party: you hear and understand that friend despite the noise; the reason is that his or her voice is superposed on the noise, meaning that it is not changed or scattered away by that noise. Similarly, in a conversation via a smartphone, you understand your friend's voice loud and clear despite thousands of other conversations racing past your smartphone on other electromagnetic frequencies. Or look at Figure 1-1a of waves in a harbor: you see the waves created by the boat moving through the other waves due to wind and other boats. Here as well, the waves are not changed by other waves.

Does wave superposition also remain true for wave packets and wave trains? These cases are simulated in Figures 2-11 and 2-12. In Figure 2-11, wave packets containing three pulses "collide": three pulses cross three other pulses traveling in the opposite direction. We may hope that each pulse-pulse collision will behave just as in Figures 2-9 and 2-10: that would make it convenient to go from simple pulses to

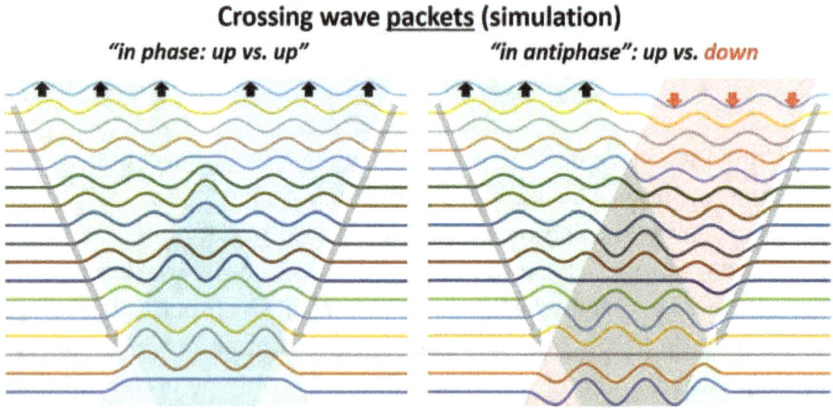

Figure 2-11: As Figure 2-10, but for the collision of two identically-shaped wave packets containing three pulses each, on a string. The gray arrows, as well as the green, blue and red shaded bands, show the motion of the packets. Where these bands overlap, the wave packets produce combined waves that do not move to the left or right: they only oscillate up and down (perpendicular to the string). Note the resemblance to the central portion of Figure 2-10.

more complex waves. Indeed, the packets behave the same way as the individual pulses: they mount each other pulse-on-pulse. You can also see this in my Animation 2*5.

ANIMATION 2*5 — See the behavior of crossing waves in my video WA1 at time 9:59 in its section "**Standing waves on long strings**" under the title "**How do traveling waves form standing waves on long strings?**" (See details in the section References and Resources below.)

Let's look more carefully within the "collision" zone of the two packets, marked as the crossing of the colored bands in Figure 2-11. We notice three special features there. First, there are flat regions where the string is straight (at least momentarily), both in the "in phase" case and in the "antiphase" case: we have already seen such straight sections with the single pulses (see the green straight line at the right in Figure 2-10). Second, we see double-height peaks: these we saw in the "in phase" case of Figure 2-10; now we also see them in the "antiphase" case. Third, the wave shapes in the "collision" zone are identical for the "in phase" case and the "antiphase" case, except for some shifting and ignoring the artificial colors.

Another, more subtle observation is the following. Looking again in the "collision" zone, let's follow a peak from frame to frame (going down the figure): it will diminish, flatten out, become a dip and then rise again <u>without</u> moving to the left or right. This is true of all peaks in the "collision" zone, both "in phase" and "in antiphase". Such behavior reminds us of standing waves and resonances: see Figures 1-1b, 1-5 and 2-1, for example. Indeed, we are beginning to see that we can make standing waves by combining traveling waves. Let's confirm this next with "colliding" wave trains.

Figure 2-12 simulates the "collision" of two wave <u>trains</u> on a string; this figure simply extends Figure 2-11 to very long packets. If you look carefully, you can recognize a part of the wave pattern (including the colors) of Figure 2-12 within the "collision" zone of the left part of Figure 2-11: in particular you should be able to recognize the light-gray, double-height peaks, as well as the light-blue, straight string. More surprisingly, we find the exact same wave pattern also in the "antiphase" case: it makes no difference whether we use up-up (in phase) pairs of

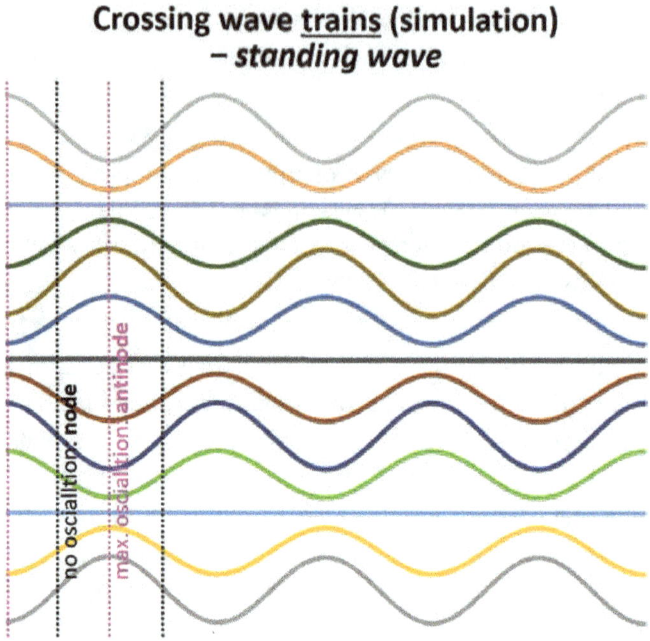

Figure 2-12: As Figures 2-10 and 2-11, but for the collision of two identically-shaped wave trains on a string. These form a standing wave. Now there is no motion to the left or right, even though the pulses still move to the left and right, as in Figures 2-10 and 2-11: however, that motion of the pulses is invisible. Some pieces of the string have no oscillation: they are called nodes, marked in black; other pieces have maximum oscillation: they are called antinodes, marked in purple. Note the resemblance to the central portions of Figures 2-10 and 2-11.

wave trains or up-down (antiphase) pairs of wave trains; hence, Figure 2-12 shows only one pattern, valid for both cases. You can also see the crossing of wave trains in my Animation 2*5 (mentioned above).

We find again with wave trains two special kinds of pieces of the string. Marked with black dotted lines in Figure 2-12 are pieces of the string that do not move at all: that earns them the name **nodes**. And marked with purple dotted lines are pieces that oscillate the most: they are therefore called **antinodes**. We already mentioned nodes in Section 1.3 (see Figure 1-5).

The nodes and antinodes tell us that the wave on the string does not move along the string, whether to the left or to the right. The motion is identical to that of the **standing waves** which we saw earlier: each piece of the string moves only across the string, not along the string. This is exactly like the footbridge that we explored in Section 1.3 (Figure 1-4): its parts only move up and down (or side to side), but not along the length of the bridge.

The most important result is this: **We have formed a standing wave with two identical traveling waves.**

I draw attention to one important aspect: **the distance between neighboring nodes (and also between neighboring antinodes) is exactly one half of the wavelength**. This will become particularly important in connection with music!

Do all wave "collisions" produce standing waves? No, standing waves are a special case that relies on two waves crossing each other with equal wavelengths and nearly equal heights (amplitudes). Another special case, that of beats due to waves with slightly different wavelengths, will be discussed in Section 2.6. Especially with unequal wavelengths, superposition causes many different wave forms: we show a few examples among many possibilities in Figure 2-13, where the black curves simulate the actual string shape obtained by simple addition of the colored waves.[11] We combine wave pulses, packets and trains with different wavelengths.

The bottom example combines one wave train (green) with one wave pulse and two wave packets of varying wavelength (all dark blue); you may imagine these blue waves to be due to ships moving over long green waves in the sea, resulting in the black combined wave. The middle three examples in Figure 2-13 combine the same green wave train with other wave trains (light blue, purple and red). In the top example we combine all four colored waves into one, showing that we are not limited to combining pairs of simple waves: we can combine any number of waves (and they don't actually have to be simple).

---

[11] Here I use a slightly different representation of waves, closer to that of Figure 2-5: now the sine-wave shape oscillates between positive and negative values (above and below the dotted lines drawn in Figure 2-13). For example, the green wave train starts at zero at left, then grows to a positive maximum, before dipping through zero, followed by a negative minimum, all of which is then repeated. Now, adding two waves simply means adding positive and/or negative numbers. The result is identical to the case of "rounded" waves used so far, except that pulses now start and end sharply instead of smoothly, as we see at the bottom of Figure 2-13.

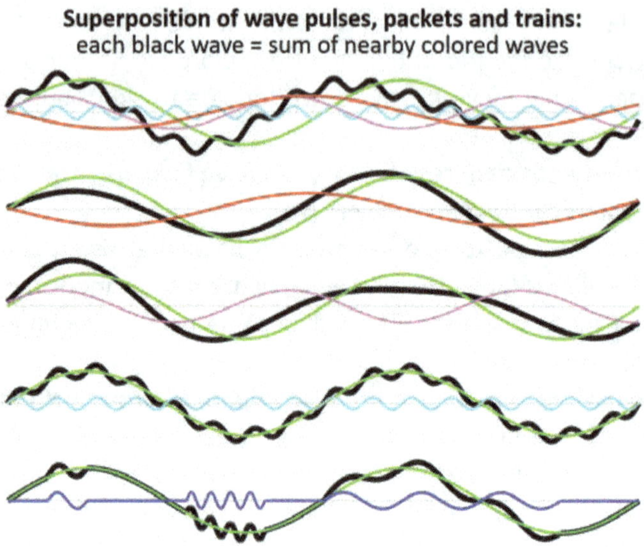

Figure 2-13: Snapshots of five different superpositions of wave pulses, packets and trains. Each of the colored waves can travel independently to the left or the right, rigidly and at constant speed (caution: these snapshots are not successive frames in a movie). In each of the five examples, the thick black wave is the sum of the nearby colored waves; thus, the thick black wave is what we would see. Common to the five examples is the green wave train. In the bottom example, the dark-blue curve is added, which consists of one peak-and-valley wave pulse at the left, a shorter-wavelength wave packet in the middle and a longer-wavelength wave packet to the right. The middle three examples add the light-blue, purple or red wave train to the same green wave train. The top example combines all four colored wave trains.

In Figure 2-13, notice that you can easily recognize the quickly oscillating light-blue wave against the other waves: think in terms of waves on water; it is like a tight ripple on a gently swelling sea; the narrow ripples might be due to a small boat, while the long swells might be due to a distant storm. But it has become very difficult to recognize the other waves (purple and red) that were mixed together! In fact, the top black wave looks rather random and disorganized. This is a further important observation: **mixing multiple simple waves can create very irregular complex waves**! Again, this is true not only for waves on strings, but also for all other types of waves. A good example is the surface of water in oceans, lakes, *etc*. (see Figure 2-14): they generally look very complex, although they actually can be composed of simple waves. You may view this in my Animation 2*6.

Figure 2-14: A boat's bow waves on a calm bay with short wind-driven water ripples.

ANIMATION 2*6 — See the behavior of superposed waves in my video WA1 at times 7:19 and 7:47 in its section "**Superposition**" under the titles "**How do waves superpose and interfere?**" and "**Wave superposition and interference, illustrated with 3 simple sine waves**". (See details in the section References and Resources below.)

The possibility of combining more than two simple waves is very important. For instance, it allows a string of a musical instrument to vibrate with many different waves (resonances) at the same time, giving each different instrument a different sound character: it allows you to tell whether the instrument is a guitar or a piano, as we will discuss in Section 5.7. This example also highlights that the sound which you hear is composed of very many different sounds that are themselves different combinations of wavelengths: in normal life, you may hear, at the same time, a conversation nearby, footsteps, a TV, waterflow, wind, traffic, *etc*. For the same reason, a fiber-optic cable across the ocean can carry large numbers of conversations and data simultaneously. The point is that these "signals" pass through each other without changing each other.[12]

[12] A clarification: I wrote that waves "collide" with each other without consequences, as if they did not notice each other. This is strictly true only for "weak" waves, namely those having small amplitudes. A counterexample is provided by water surface waves when they grow so high that they break up or crash and produce "whitecaps" (drops, bubbles and foam), as is frequently seen in windy conditions or at the beach. Nonetheless, most waves in our daily lives are "weak" and do ignore collisions with other waves. Examples are sounds (you can understand someone speaking despite others speaking nearby at the same time, because sound waves cross each other without changing each other) and radio waves (you can conduct a mobile phone conversation without being disturbed by the millions of other mobile phone conversations conducted over the same network).

We have seen, in Figures 2-10 and 2-11, how waves cross each other: there is no collision. To realize how surprising this "non-collision" of waves is, think again of any two objects that collide, such as beads on a string, stones, drops of water, people's bodies, *etc.*: two such objects will bounce off each other in other directions, or change their shapes, or break into pieces. Waves do none of these things!

Thus, **waves seem to simply ignore each other, passing through each other without any trace of a collision, as if nothing had happened! This behavior of non-collision is true of all kinds of waves, but only if they are weak enough**.

Most waves that we deal with are indeed weak enough to avoid collision, such as sound waves, light and other electromagnetic waves. **A very visible exception is waves on water**: they often "break" into "whitecaps", namely they explode into drops and bubbles, especially when such waves cross each other. This behavior of waves on water is most common at beaches and in wild rivers, as we will discuss in Section 11.1.

2.6 Beats and Tuning

Have you heard of "beats" and their use for tuning musical instruments? If you heard a tuner tune a piano, you must have heard **beats**! You may also have heard a strong "wha-wha-wha-wha-..." sound when two loud machines turn at almost the same frequency, for example in a two-propeller airplane (pilots remove that disturbing sound by making the engines turn at the same frequency, which is an example of tuning). You may listen to beats in my Animation 2*7.

ANIMATION 2*7 — Listen to my video WA1 at times 16:42, 17:13 and 17:43 in its section "**Beats**" under the titles "**What are beats between waves of slightly different frequencies (wavelengths)?**" and "**Beats with other frequencies**" and "**Beats and tuning — listen to sounds**". (See details in the section References and Resources below.)

You can also easily produce beats online as follows. On the webpage in this footnote,[13] enter the frequency 440 hertz (Hz) in both boxes, then

[13] Online Tone Generator, Binaural Beats (be careful not to produce too loud sounds, as they may damage your hearing): https://onlinetonegenerator.com/binauralbeats.html

press Play; you hear a constant, simple sound. Now enter the frequency 441 Hz in the righthand box, then press Play; you will hear the "wha-wha-wha-wha-..." sound typical of beats. Try to time those beats: you should find one beat (one "wha") per second, which is 1 hertz. Now change the second frequency to 442 Hz and press Play again: you should hear beats repeating twice as fast, which is two beats per second or 2 hertz. Try the same with 443 Hz (giving beats at 3 Hz) and 450 Hz (giving rapid beats at 10 Hz, 10 "whas" per second). How far can you go like this? 460 Hz, 470 Hz, 480 Hz, ...? Gradually, you will have more difficulty hearing these very rapid beats. But remember, these beats are still very slow compared to the initial frequency of 440 Hz (still in the left box). Notice the surprisingly simple result: **the beat frequency is just the difference between the two input frequencies**: 1 Hz = 441 – 440 Hz, 10 Hz = 450 – 440 Hz, *etc.*! This remains true if you decrease the second frequency to 439, 438, ... Hz.[14]

What are beats? Beats are a special case of the superposition of two waves: special is that the two frequencies are close together. Looking at Figure 2-13, we would give, for example the green and purple wave trains almost the same frequency and therefore almost the same wavelength (we also give them the same direction of travel and the same amplitude).

The result is illustrated and explained in Figure 2-15. You can also see and hear these beats in my Animation 2*7 (mentioned above). There, two waves repeatedly "beat" against each other. In Figure 2-15, the two waves that beat against each other are the green and light-blue waves at the bottom. They superpose to give the black wave. Imagine that you place your ear near the purple box of Figure 2-15 and listen to the vibration of the string there. The "wha-wha-wha" sound is simply the repeated growing and shrinking of the amplitude of the combined black wave as it passes by; it is just like repeatedly turning the volume knob on your loudspeakers up and down.

The "beating" varies as we vary the wavelength (and therefore the frequency) of one of the two waves: compare the black, red and dark blue waves. In the example of Figure 2-15, if the 40 oscillations of the green wave take 1 second, that green wave has a frequency of 40 Hz,

[14] It doesn't matter which is the lower frequency: a negative frequency is considered to be the same as a positive frequency.

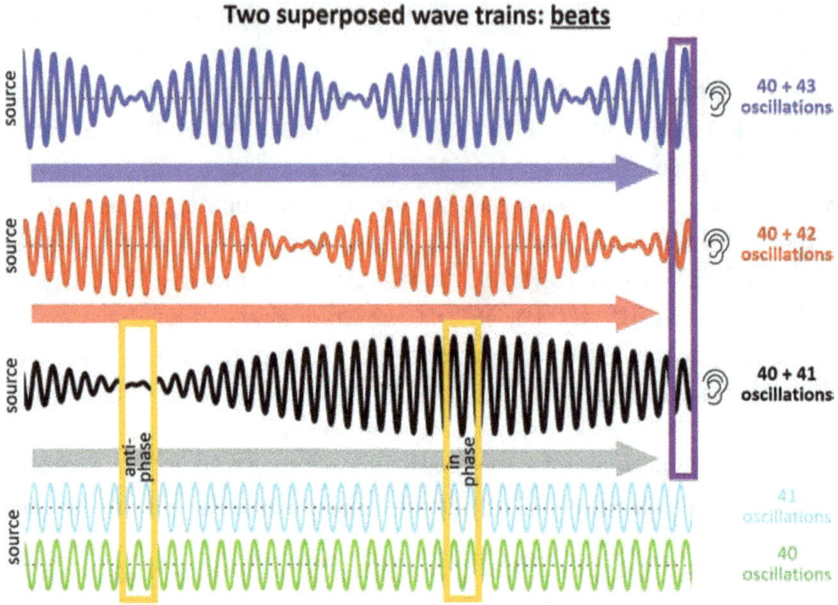

Figure 2-15: Snapshots of two different wave trains beating against each other as they travel together along one string; all the resulting waves travel along the string at the same speed (shown by large arrows). They are heard by the ear when they arrive near the purple box. The bottom input wave (green) has 40 oscillations in this graph. The next input wave (light blue) has 41 oscillations over the same distance, so its wavelength is 1/40 or 2.5% shorter (giving a 2.5% higher frequency). These two waves add up to make the black wave. As highlighted in the righthand yellow box, where they are in phase (meaning that peak adds to peak), the black wave has double amplitude by constructive interference; where these two waves are in antiphase (peak adds to trough in the lefthand yellow box), the black wave has zero amplitude by destructive interference. The red wave is produced by adding the green wave to another wave that has 42 oscillations over the same distance (not shown because it would look very similar to the light blue wave): its wavelength is smaller by 2/40 or 5% (giving a 5% higher frequency). The dark blue wave results from adding the green wave to yet another wave that has 43 oscillations over the same distance (also not shown): its wavelength is smaller by 3/40 or 7.5% (giving a 7.5% higher frequency). The ear hears the black, red or dark blue wave as it passes through the purple box: the frequency of these waves is almost the same at all times (close to 40 Hz), but their amplitude is "beating" up and down at a much slower frequency, oscillating between double amplitude and zero amplitude compared to the green and light blue waves.

and the light blue wave has a frequency of 41 Hz. The beat frequency becomes 41 – 40 = 1 Hz, meaning one beat per second: indeed, in 1 second, one black "wha" will pass by the ear.[15]

If the light blue wave had a frequency of 42 Hz, the beat frequency would become 42 – 40 = 2 Hz: now a red pair "wha-wha" would pass the ear in one second, giving double the beat frequency. And with 43 Hz, the beat frequency would become 43 – 40 = 3 Hz: a dark blue triplet "wha-wha-wha" would pass the ear in one second, giving triple the beat frequency. This large change due to small frequency variations allows fine-tuned strings in musical instruments.

So how does a tuner tune a musical instrument? One common method for **tuning** is to use a **tuning fork** (or an app on a smartphone) that produces a sound with a reliable frequency: this is like the bottom (green) frequency in our example above. Tuners normally use 440 hertz, which is the worldwide standard for the note A_4, as we will see in Section 5.1. The tuning fork (or app) will produce beats against the piano's note A_4. If that piano is out of tune, the beat will be faster for worse tuning. By adjusting the string's tension, the tuner can bring the piano's note A_4 to have the same frequency as the tuning fork, which is confirmed by the absence of beats.

For another note produced by the piano, the tuner may compare the higher harmonics of the tuned A_4 note and that other note because higher harmonics often coincide in frequency (this should become clearer when we discuss music in Chapter 5): then beats between those harmonic frequencies should be eliminated by tuning the tension of the other string. In other words, the already tuned note A_4 then acts as the reference frequency, replacing the tuning fork or app. This is repeated for every string of the piano: there are about 230 strings in a piano!

---

[15] Figure 2-15 uses 40 oscillations per second, namely 40 hertz, instead of the 440 hertz used in the online example above: that is simply because it would be difficult to draw 440 oscillations on one page! You can, on the other hand, try the online webpage with frequencies 40 hertz and 41 hertz, *etc.*, but not everyone will hear these low frequencies well enough.

2.7 Reflection of Waves: Standing Waves

One important use of waves on strings is to make **music** with string instruments. To produce the desired musical **notes**, the strings must be put under tension. That requires fixing the ends of the strings. We therefore need to understand the behavior of waves when they reach the fixed ends of strings. We include the case of a free (loose) end because that case is important with wind instruments (like flutes), to be discussed in Chapter 3: wind instruments are air tubes, whose ends can be open or closed, corresponding to the "free" or "fixed" ends of a string.

What happens when a wave pulse or wave train reaches the end of a string? Does it simply die out? Think of this analogy: if you make a sound (which we will later see is also a wave), and when that sound reaches a wall, what does it do? You know that usually the sound bounces back: it is reflected (as an echo). You could do the experiment with a string yourself, although it is easier to watch someone else do it in an online video[16]: a wave on a string is also reflected back from the end of the string. You may view this in my Animation 2*8.

ANIMATION 2*8 — See my video WA1 at times 8:36 and 9:08 in its section "**Ends of strings**" under the titles "**How are waves reflected by free and fixed ends of strings?**" and "**This is how waves are reflected by fixed and free ends of strings**". (See details in the section References and Resources below.)

You may also use the online animation that we used earlier. Views from that animation are shown in Figure 2-16: a triangular pulse travels to the right end of the string and is reflected as a pulse traveling in the opposite direction. An important aspect is that the reflected pulse is equally triangular: **reflection** has not changed its basic shape.

However, there are two different situations: the right end of the string is either fixed or free. We observe that the pulse flips upside down at the fixed end, but stays "upside up" at the free end. ***How does this happen?*** Unfortunately, things happen very quickly during

[16] See video "Wave Reflection — xmdemo 138" by xmdemo: https://www.youtube.com/watch?v=1PsGZq5sLrw

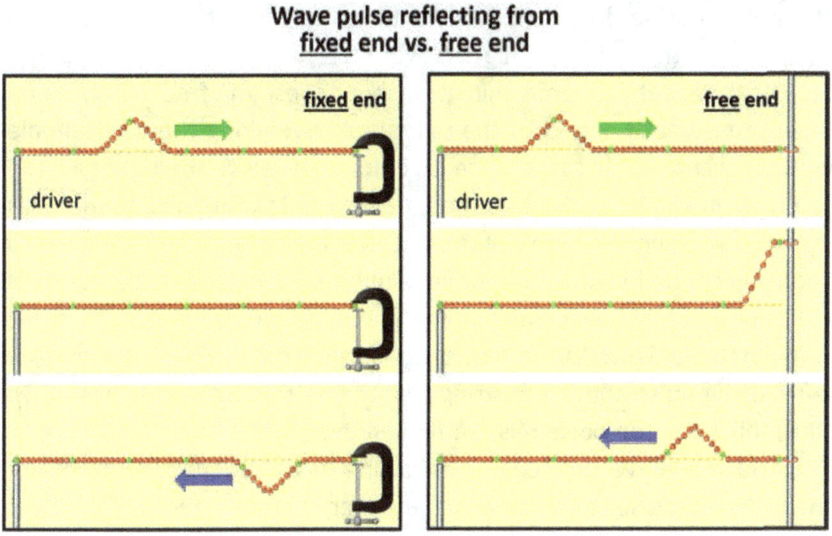

Figure 2-16: Snapshots of the "Pulse" screen of the online animation "Wave on a String", showing a short pulse (green arrows, 0.25 second) traveling to the right with No Damping, and a reflected pulse (blue arrows) traveling to the left. *At the left*: the right end is fixed. *At the right*: the right end is free. (*Source*: as Figure 1-6.)

reflection. However, the online animation allows us to slow down that rapid motion, giving an excellent view of how reflection happens on a string.

Start the animation (using the link in the caption of Figure 1-6), with the settings: Slow Motion, Pulse, Fixed End, Pulse Width 0.25 s, and Damping None (keep Amplitude 0.75 cm and Tension High). The pulse will look like the top left graph of Figure 2-16, moving to the right. Pause the animation before the triangular pulse reaches the fixed end (click on the large blue button).

Now step forward frame by frame (with the small blue button). We notice that the front of the pulse would like to rise (so as to move the triangular shape forward), but the fixed end prevents it from doing so. The string is now pulled down by the fixed end and overshoots below the string. Pay close attention: at one specific moment (see the center left in Figure 2-16), the string actually becomes completely flat and straight, as if there were no pulse! Actually, at that moment, the string is swinging fast downward, so there is still motion (and therefore energy) in that pulse. Next, we see the original triangular shape building up below the string, but upside down and moving to the left (see the bottom left in Figure 2-16): the pulse has flipped over.

Next, repeat the same animation, but choose the free or Loose End (now the ring at the end of the string can slide along the vertical pole without friction). The same pulse, as before, travels to the right (see the top right in Figure 2-16). This time, however, its triangular shape pulls up the ring along the pole at the right. Not only that: the string rises to about twice the height of the original pulse's peak (see the center right in Figure 2-16), before coming down again; also, at no time is the string completely straight. As the end of the string comes down, it creates a left-moving pulse above the string (see the bottom right in Figure 2-16): thus, this time, the pulse has not flipped over.

You may have noticed the following in the animation: if you let the reflected pulse reach back to the driver, it is reflected by the driver toward the right, but flips over again. The reason is clear: the driver is now a fixed end of the string, so it flips the pulse upside down (or downside up). As the pulse continues being reflected back and forth along the string, it flips each time it reflects from a fixed end, but not from a free end.

How is a wave packet reflected? Figure 2-17 simulates the reflection of a 3-pulse triangular wave packet from the fixed end of a string, showing multiple screenshots (numbered 1–13) from the abovementioned online animation. First, focus your attention on the string itself (ignoring the green and blue zigzag lines): the red and green balls. The first snapshot (1) has three complete triangular pulses: the front pulse is just reaching the fixed end of the string, in the red circle. In the next snapshots, the shape of the string is increasingly distorted as the front end of the wave packet turns back on itself. Since the end is fixed, each of the pulses is flipped upside down, as shown at the left in Figure 2-16, but there is now a mixing of the initial packet and the reflected packet. By snapshot 7, the front of the packet has reached its own tail (we will explain later why the result looks flat). By snapshot 8, the front of the reflected wave emerges from the tail of the initial packet: we start to see an upside-down copy of the initial packet; in snapshot 13, the whole packet is seen traveling upside down to the left.

To help understand how the reflection takes place, I have overlaid in Figure 2-17 the initial and final shapes of the wave packet as green and

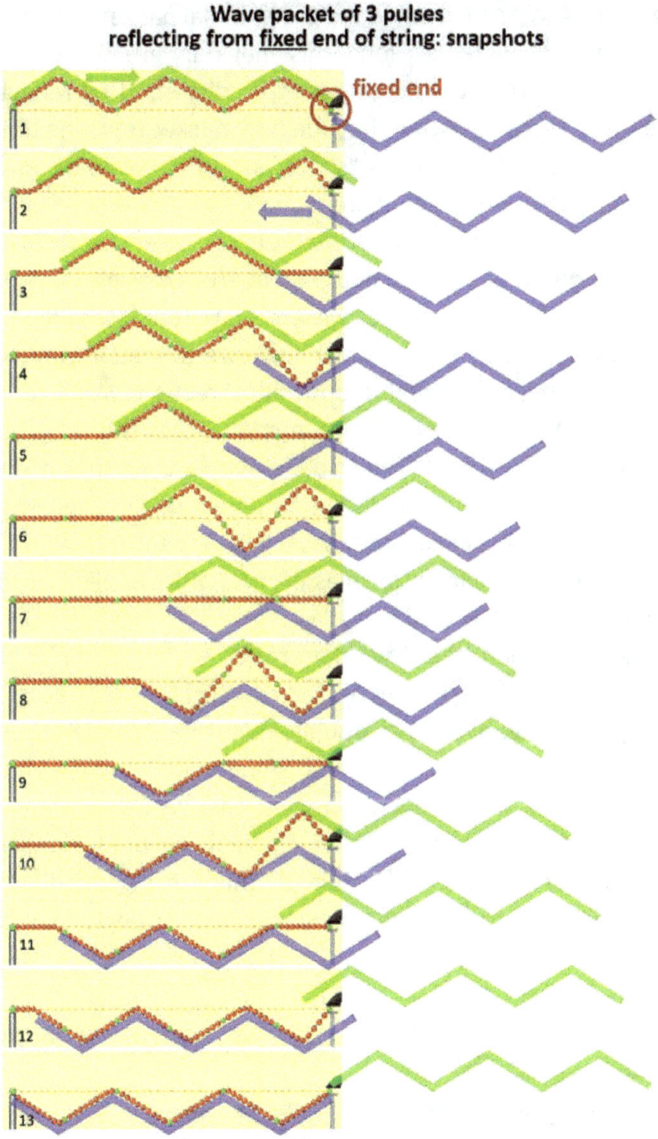

Figure 2-17: From top to bottom, we show snapshots of the "Pulse" screen of the online animation "Wave on a String". A wave packet of three pulses travels to the right (green arrow) with No Damping, and is reflected by a fixed end into a packet traveling to the left (blue arrow). The string exists in the yellow region only (green and red balls). The green zigzag line imagines the original packet traveling undisturbed beyond the fixed end of the string. The blue zigzag line imagines an identical wave packet that travels upside down in the opposite direction. These "add up" by superposition to the actual wave of green-and-red balls. This animation uses a Pulse Width of 0.4 seconds, No Damping and High Tension. (*Source*: as Figure 1-6.)

blue zigzag lines. We can imagine the green initial packet traveling to the right as if the string were very long: this imaginary green packet "slides" out beyond the end of the actual string, undisturbed by reflection. At the same time, we can imagine the blue final packet traveling to the left, again as if on a very long string: it "slides" in from beyond the end of the actual string, also undisturbed by reflection.

This model of two undisturbed packets may seem artificial and odd at first sight, but it should remind you of the crossing of two waves that we discussed in Section 2.5: we are recreating that crossing with its superposition of two waves, which now are the initial wave and the reflected wave that also travel in opposite directions. A key point that allows us to apply that superposition model is that the reflected wave shape is identical to that of the initial wave (except that it travels in the opposite direction and will be upside down for the fixed end of the string, which we have called "antiphase"). This identity of wave shapes comes from the underlying principles of mechanics applied to the ends of strings.

Now let's compare the wave patterns in Figure 2-17 with those at the right in Figure 2-11. They look very much alike: we recognize flat regions of the string, as well as double-height peaks (even though the waves are triangular in Figure 2-17 but rounded in Figure 2-11). We also see that the peaks in the overlap region change from having double height to being flat to being dips, then flat again before becoming peaks once more. And, most importantly, we see that these peaks do not move along the string: they form **standing waves**! Interestingly, the case of a free end, which is illustrated in Figure 2-18, behaves just like the case of the "in phase" crossing of Figure 2-11. In other words: reflection creates an identical reflected wave that can then be superposed on the initial wave as described in Section 2.5.

We may conclude: **Reflected waves on a string create, by superposition, standing waves identical to those caused by the crossing of pairs of waves. It's as if the end of the string created a mirror image of the incoming wave, moving in the opposite direction and superposing on top of the incoming wave.**

What happens with a wave train that is being reflected? Figure 2-19 shows this case, using wave trains made of six pulses identical

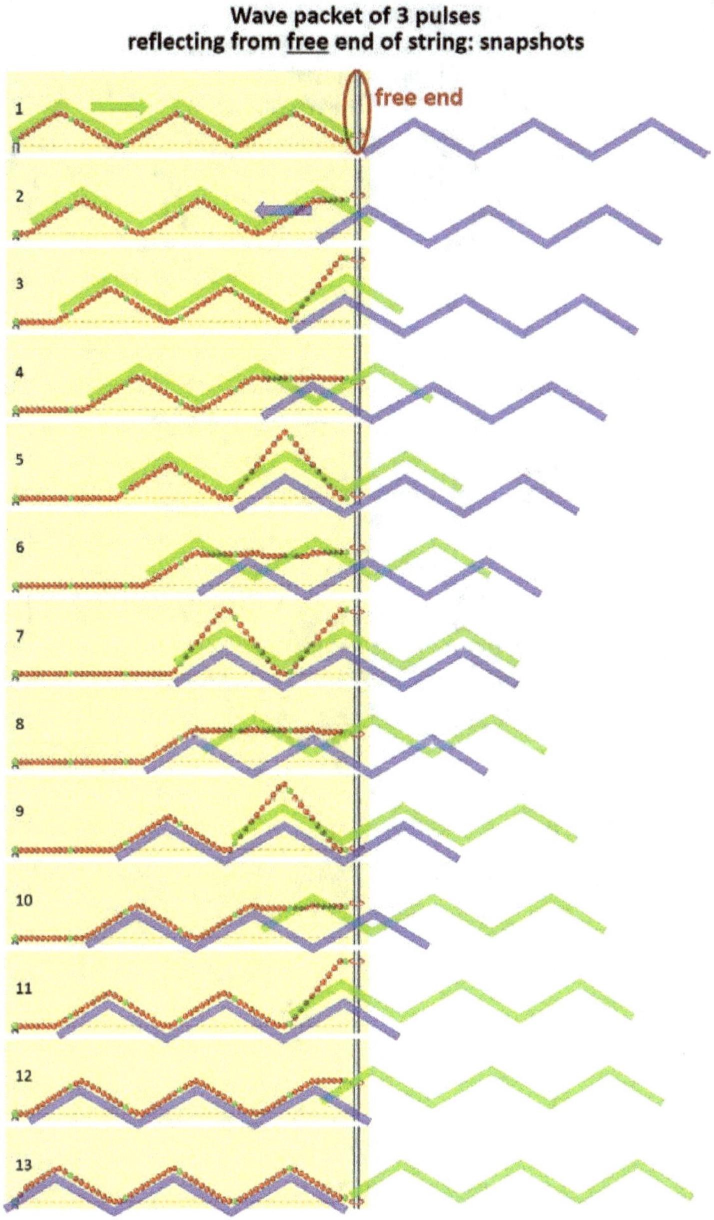

Figure 2-18: As Figure 2-17, but with a free end of the string. (*Source*: as Figure 1-6.)

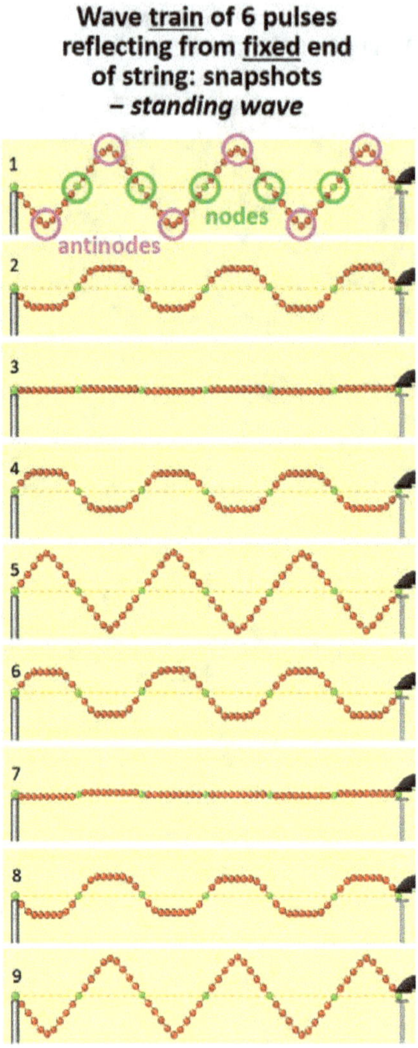

Figure 2-19: Wave train reflecting from two fixed ends of the string, presented as in Figures 2-17 and 2-18. Snapshot 9 repeats snapshot 1 and starts a new identical cycle. The green balls never move: they are nodes; the purple-circled balls move most, but only up and down: they are antinodes. The result is a standing wave. (*Source*: as Figure 1-6.)

to those in Figures 2-17 and 2-18: the six pulses travel back and forth between the immobile driver and the fixed right end of the string. Importantly, I chose the pulse length to be exactly 1/3 of the string length, so that the sixth pulse links up seamlessly to the tail of the first pulse, in effect making an infinitely long wave train (to achieve this, I used a Pulse Width of 0.39 seconds and good manual timing to start each new pulse); as a result, the six pulses wrap around the string, so that at any given time three pulses are moving to the right, while the three other pulses are moving to the left.

In Figure 2-19 we recognize several features seen in Figures 2-17 and 2-18: high steep peaks (in snapshots 1, 5 and 9) and flat straight segments (in snapshots 2 to 4, and 6 to 8). We also see snapshots (3 and 7) in which the string is flat and straight along its full length.

Notice that the green balls of the string in Figure 2-19 do not move at all: they remain at rest and are therefore **nodes**. The string segments midway between nodes move most (but only up and down): they are therefore **antinodes**. Looking back at Figures 2-17 and 2-18, we already had nodes and antinodes in the region where the initial and reflected waves overlap: so this behavior also happens with the reflection of wave packets.

Furthermore, the wave in Figure 2-19 should remind you of Figure 2-12, showing standing waves. We find again the **resonances** and **standing waves** that we saw in Sections 1.3 and 2.1, in particular in Figures 1-1b, 1-5 and 2-1. Indeed, the wave in Figure 2-19 is a resonance and a standing wave, even though it has a more triangular shape than the earlier rounded waves. You may also view my Animation 2*9.

> ANIMATION 2*9 — See my video WA1 at time 10:35 in its section "**Standing waves on long strings**" under the title "**How do traveling waves form standing waves on long strings with fixed ends?**" (See details in the section References and Resources below.)

Figure 2-20 shows in more detail how a reflected wave train "rides over" the original wave train to form a standing wave. Importantly, **a fixed end of the string imposes a node on the standing wave** (shown as black dots in the left sketch), **while a free end imposes an antinode** (shown as purple dots in the right sketch). We also notice that **the standing wave pattern is identical for the fixed and free ends**, apart from the shift by a quarter wavelength.

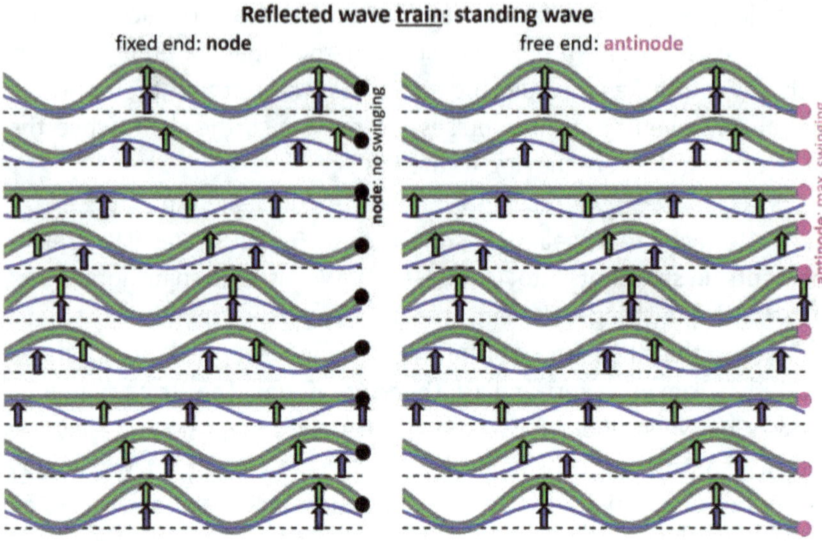

Figure 2-20: The left graphs simulate the crossing of a green wave train (traveling to the right) and its blue reflection, as in Figure 2-9. The reflection is due to a fixed end of the string, marked by black dots, which must be a node. The right graphs are similar but for a free end of the string. Now the end is free to swing (purple dots) and must be an antinode. Each arrow follows a wave peak from snapshot to snapshot, from top to bottom. The actual string shape is shown as the gray contour.

Here we have constructed a standing wave by making a reflected traveling wave superpose itself onto the original traveling wave. We may say it slightly differently and more generally: **Two identical traveling waves crossing each other in opposite directions "add up" to a standing wave.** This remarkable conclusion is fundamental to phenomena such as music and voices.

We may also say that the ends of the strings act like mirrors. Just like mirrors that reflect light, the ends of these strings create, behind the "mirror", a virtual (imagined) mirrored copy of the incoming waves. This makes it very easy to understand what happens: **During the reflection of a wave, the incoming wave is simply added to a mirrored copy that travels in the opposite direction.** See more discussion of this point in Box 2-2.

BOX 2-2 — YOU MAY WONDER, AND RIGHTLY SO: ***How can we say that "a mirrored copy of the incoming wave travels in the opposite direction", when it is obvious that there is no such copy behind the end of the string?*** This is similar to the artificial mirrored source of light that you see in a mirror (see also Section 3.8): there is no such source behind the mirror! Such a mirrored copy and such an artificial source of light are beautiful examples of a scientific **model**: nature behaves just like the model, but this does not mean that nature is identical to that model!

Many scientific explanations are like that: they appear to exactly reproduce natural reality, but they remain abstract models that only happen to behave in the same way as does nature. Models may have to be abandoned or updated when new information is discovered. For example, Newton's theory of gravitation needed to be replaced by Einstein's theory of relativity to properly reproduce the behavior of very fast objects and light.

Similarly, you may wonder: ***How is it possible that the wave near the end of the string (whether fixed or free) behaves exactly like two crossing waves, despite the presence of the end of the string? Shouldn't the end of the string deform the wave in some special way?*** The answer lies in the "local" character of wave motion. As we discussed in Section 2.2, each little piece of a string affects only its two immediate neighboring pieces of the string (one on each side), just like people in a queue or dominoes in a line of standing dominoes.

Let's again use the queue of people as an analogy. Suppose the queue ends at a closed door, so the front person is blocked from moving forward. If a disturbance (pulse) comes from the back of the queue, that front person will bounce from the door and then bump into the person behind him or her: now the pulse travels backward along the queue. To everyone else in the queue, the backward pulse looks the same as the original forward pulse: for them, a pulse arrives (from back or front), and they transmit it to their next neighbor as if the queue were infinitely long. Only the front person is different.

Similarly, only the very last piece of the end of the string is different from the rest. As a result, the wave behaves as if the string were endless, except for that very last little piece, where the reflection takes place. Thus, the model of two opposite waves on endless strings is valid everywhere away from the very end of the string.

You probably will think that this result is not intuitive, and you are right! The reasoning given above shows the power of logic applied to physics. We can reach remarkable, even counterintuitive, conclusions knowing that each piece of a string only interacts with its immediate neighbors (since there are no significant forces between more distant pieces of the string).

(*Continued*)

(*Continued*)

In practice, physicists must do more than reason: they must also check that physical reality in fact behaves as their theories predict, by making suitable measurements (experiments); after all, physicists can and do make mistakes! In the end, nature is the ultimate "judge" of whether an explanation is right or wrong. The most remarkable aspect of this is that nature appears to behave in a logical way; moreover, nature's logic closely matches human logic.

It is interesting to note here that for every new physical observation, physicists produce multiple theories that appear to explain that observation (similar to multiple scenarios proposed to explain a murder). Then, each of those theories is tested against physical reality: this is done by asking each theory to make predictions, which are then tested experimentally. Such testing usually shows that most of the proposed theories are wrong! Normally only one theory survives such experimental verification, no matter whether it is intuitive to humans or not: the only criterion is that the theory must agree with nature, the ultimate and only arbiter; this removes human bias.

2.8 Waves on a String with Two Fixed Ends

In musical string instruments, the strings are fixed at both ends. Therefore, any wave on that string must remain immobile at both ends of the string: they must be nodes. This will have important consequences for the musical tones that the string can produce: see Chapter 5.

Examples of strings fixed at both ends are shown in Figures 1-5, 2-1 and 2-19. We assume that the wavelength of a wave on the string is set and that it will not change: it is set by the material of the string, the tension on the string and the frequency of the wave.

How can we now choose where to fix the other end of the string such that the string also remains immobile there? The answer is simple: **the other end of the string must be at another node of the wave**. Then both ends of the string are immobile, while the string still carries a standing wave. You may view this in my Animation 2*10.

ANIMATION 2*10 — See my video WA1 at time 10:35 in its section "**Standing waves on long strings**" under the title "**How do traveling waves form standing waves on long strings with fixed ends?**" (See details in the section References and Resources below.)

At which node should we fix the other end of the string? **We can choose any node**: the nearest node will have one antinode between the fixed ends of the string, as shown at the left in Figure 2-21: this creates a standing wave with one **loop**. If we choose the next node, we get two nodes at the ends, one node in the center, and two antinodes in between, as shown in the middle of Figure 2-21: we get a standing wave with two loops. We can also choose the next node, giving four nodes and three antinodes in total, as shown at the right in Figure 2-21: we get three loops. In fact, we may also choose any more distant node to fix the string, creating a longer series of standing wave loops, as in Figure 1-5.

As we saw in Section 2.7, two neighboring nodes are exactly one half of the wavelength apart. Therefore, we can say: **to create a standing wave on a string with fixed ends, the length of the string must be a whole multiple of half the wavelength**; or equivalently: **the length of the string must fit a whole multiple of loops.**

What happens if we don't fix the other end of the string at a node, but fix the string between nodes? The short answer is that you get a messy wave that bounces back and forth between the two ends of the string and largely destroys itself as it crosses its own multiple reflections. You can see such messy waves in videos where people try to create standing waves by changing the frequency of shaking: this also changes the wavelength so it may not fit the distance between the ends of the string.[17]

You can try this yourself, as in that video: either use the online animation described in Section 1.3 or shake one end of a real string that is fixed at the other end. Either way, create a standing wave with a single loop by shaking it at just the right frequency; before you find that right frequency, the string will oscillate in a messy fashion. (We are cheating slightly here: the end of the string in your hand is moving as you shake, but that movement can be small, and it may look almost like an immobile node.)

If you do the experiment which I just described, you will notice an interesting behavior. When the wave you generate is messy, that string often opposes your hand's motion. But when you create a standing wave,

[17] See video "AP Physics 2: Waves 10: Resonance and Standing Waves on a String" by Yau-Jong Twu: https://www.youtube.com/watch?v=7xCmtYXewdk

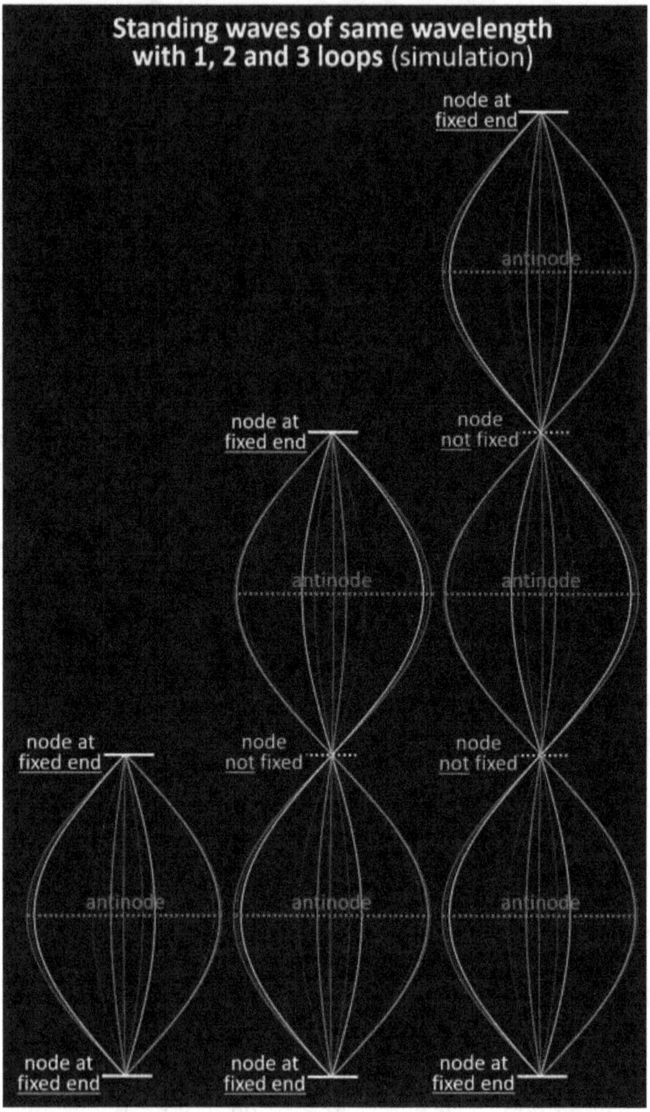

Figure 2-21: Three standing waves are shown as snapshots on a hanging string with fixed ends, as in Figure 2-1 but with equal wavelengths. They have one, two or three loops (at the left, middle, and right, respectively). Each wave has a node at each end, and zero, one or two nodes in between. An antinode exists between each pair of neighboring nodes.

the wave follows your hand's motion without opposing it; moreover, the standing wave then rapidly increases its amplitude, even with very little effort on your part.

As we have already seen, this behavior is very common in physics and technology and therefore carries its own name: **resonance**. It fully deserves its own section: see Section 2.9.

What happens if we superpose multiple harmonics on a single string with fixed ends? This situation is very common and important: it is normal for a string instrument making music. Indeed, activating a string is usually done by striking it (with a hammer as in a piano), by plucking it (as with a guitar) or by bowing it (as with a violin): this creates not only the fundamental frequency (the first harmonic), but also many higher harmonics, as we will describe in more detail in Section 5.5.

Therefore, multiple harmonics will be oscillating at the same time, superposed on each other. Figure 2-22 shows how such superposition takes place, one harmonic at a time. At bottom is drawn the first harmonic standing wave, which is the fundamental wave (thin black wave): if only this standing wave existed on the string, it would look

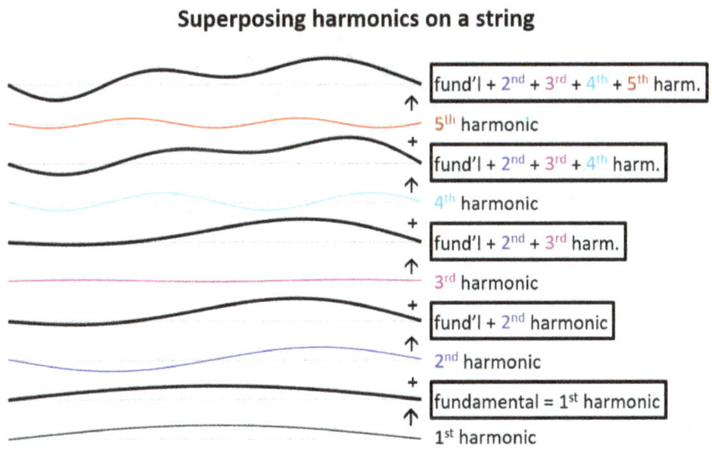

Figure 2-22: Snapshot of simulated harmonics (standing waves) superposed onto each other, one at a time from bottom to top.

exactly like this wave (it is copied as the lowest thick black wave, showing the actual string shape). The second harmonic (dark blue wave) has half the wavelength of the first: it is simply added on top of the fundamental wave and gives the second thick black wave, which combines the first and second harmonics. Adding the third harmonic (purple wave) to the fundamental and the second harmonic gives the third thick black wave. This is repeated by adding the fourth harmonic (light blue wave) and the fifth harmonic (red wave). The result of combining these five harmonics is the top black wave: that snapshot is what the string would look like to an observer. Additional higher harmonics could be superposed on top of the first five: because of their shorter wavelengths, they would add sharper features (narrower peaks and valleys) to the combined black wave, making it more and more complex.

In Figure 2-22 we notice how different the resulting superposed wave is from the individual harmonics contained within it. As we will discuss further in Section 5.5, we can produce any desired wave shape by such superposition of harmonics. Remember also that the string in Figure 2-22 is vibrating: each harmonic is oscillating at its own different frequency, such that the combined wave changes its shape constantly. This is best seen in my Animation 2*11, which also includes sound.

ANIMATION 2*11 — See — and listen to — my video WA1 at time 12:15 in its chapter "**Harmonics**" under the title "**Harmonics on a string — *listen to sounds***". (See details in the section References and Resources below.)

You may wonder: ***What effect do "weightlessness" and "freefall" have on waves on strings and in musical instruments?*** This is a slightly more advanced topic that is also very interesting: see Box 2-3.

BOX 2-3 — WAVES ON STRINGS IN "ZERO GRAVITY": Imagine that you are playing the guitar in an elevator (lift) at the moment when the elevator falls freely.

(A falling elevator is a favorite situation that physicists love to imagine, but hate to try out themselves, and most likely never experienced in person! They have a special name for this kind of experiment: Gedankenexperiment, which is

German for "thought experiment", namely an experiment conducted entirely in one's mind; such an experiment obviously needs to be checked in reality before it can become believable!)

You probably know that you and the guitar would also fall freely with the elevator: you would "float" within the elevator without pressing against its floor or ceiling. This situation is often loosely, and incorrectly, called **zero gravity** or **weightlessness**.

This is also exactly the situation of astronauts in a space station circling the Earth: they are in continuous **freefall**, but they are traveling so fast that they continually "miss" hitting the round Earth; they always fall beyond the horizon. They also float within the space station, just as in a falling elevator. You — including the guitar, the elevator, the astronauts and the space station — all still have "weight" in all these cases of freefall because weight is the force of attraction from the Earth on each of those objects: the weight in fact causes the falling in freefall.

The reason I mention freefall and gravity is that they, perhaps surprisingly, have no importance for the vibration of strings that are fixed at both ends. That is because the **waves on such strings are powered not by gravity but by the tension in the string**. As a result, **a guitar will produce the same music in a falling elevator and a flying space station as on the firm Earth; it will also produce the same tones on the Moon and on Mars**, even though gravity and air pressure are weaker there compared to Earth. All that is needed is air or another substance that can carry the sound to our ears.

2.9 Resonance and Quantization

As mentioned in Sections 1.3 and 2.1, we often hear or see **resonances** in daily life. They include music and voices, oscillations of bridges, boards and strings, singing of water pipes, rattling of machines, and tuning of television channels, among many others.

Let's summarize a few important aspects of resonances that we discussed earlier:

- resonances are vibrations of objects, including strings, boards, bridges and air.
- standing waves are resonances.

- resonances of waves often occur in families with different frequencies: a resonance with a fundamental frequency, and a series of related resonances with higher harmonic frequencies; for example, strings with fixed ends have a first harmonic with 1 loop and higher harmonics with 2, 3, 4, 5, *etc.* loops (however, pendulums such as swings, clocks, arms and legs do not have higher harmonics).

One common feature of all those physical resonances is this: **there is no need to inject much energy to create a resonance**. The reason is that at each oscillation, a little bit of injected energy adds up gradually to a large motion. This is clear with the collapse of the Tacoma Narrows Bridge, due to increasing twisting in a strong wind.[18]

When you walk at a fixed speed, try swinging your legs half as fast or twice as fast as you normally do: that will require much more energy because your leg muscles are then either counteracting or supplementing gravity most of the time!

An important feature is that **resonances occur only at certain frequencies**, namely with the correct time interval. The swing with a child will only swing dangerously high if you push it at just the right frequency. You can only topple a tree if you push it at the right frequency. When walking, you swing your legs at the frequency with which they naturally swing when you let them hang freely (so gravity can do most of the work to bring your leg forward for the next step). Water pipes "sing" only at certain water speeds that create water vibrations of particular frequencies. Wind creates sound with a particular tone when it blows at a given speed: think of whistling. Strings will form standing wave loops when repeatedly pushed at the right frequency, which depends on the length, material and tension of the string.

We see that the frequency at which an object resonates is a property of that object, not of the external input of energy or that input's frequency. **The resonance frequency of an object is therefore also often called its natural frequency**. **Only when the driving frequency**

[18] See story and video of the "Tacoma Narrows Bridge (1940)": https://en.wikipedia.org/wiki/Tacoma_Narrows_Bridge_(1940)#Film_of_collapse

matches the natural frequency (or a whole multiple thereof for waves) does the object resonate.

Waves have a significant advantage over other kinds of vibration (such as swings and pendulums): they have a wavelength, which allows more resonance possibilities. In particular, by doubling the driving frequency, we can make a string oscillate with two loops instead of one loop because the wavelength can be halved; tripling the frequency gives three loops with a third of the original wavelength, *etc.* These multiple resonances lead to harmonics and overtones in music and speech, as we will see in Chapters 3 to 6.

Also important is this: **on a string, multiple resonances can be started at the same time**. Multiple resonances can co-exist simultaneously on the same string due to superposition, similar to the co-existence of various waves shown in Figure 2-13. For example, in a **piano**, a string is struck once by a hammer, not repeatedly: this single strike can start up many resonances simultaneously, namely those with whole multiples of the natural frequency of that string. The same is true of plucking a **guitar**: a single pluck can start up many resonances with different frequencies at the same time. Bowing a **violin** also creates many simultaneous resonances in its strings.

Thus, **standing waves are resonances; musical instruments and speech exploit such resonances to create harmonious sounds**, as described in Chapters 5 and 6.

You may wonder: ***Why does doubling the vibration frequency of a string halve the wavelength***? This is due to the surprising fact of nature that the speed of a wave along a string does not depend on its frequency (which in turn can be understood from the mechanics of wave propagation using Newton's laws of motion). As a result, when you double the frequency of shaking the string, you get twice as many oscillations in the same length of string, which means a halved wavelength.

You can also triple the original frequency, giving a wavelength that is one-third the size of the original wavelength. This produces another standing wave with two nodes equally spaced between the two fixed ends, and therefore results in three loops, as shown at the right in Figure 2-1. We can continue multiplying the frequency by whole numbers (4, 5, *etc.*) to produce standing waves with four, five, and more loops.

We have seen that **a single string with a single length, under a single tension and with a single material can be used to produce a family of related sounds: the original frequency which we call the fundamental frequency; and whole multiples of that fundamental frequency which we call harmonics or overtones.**

It is important to realize that no other resonating frequencies are possible on that string with fixed ends. Any other driving frequency will produce a wave with a wavelength that does not fit the length of the string: the multiple reflections of such a wave from both ends of the string will mess up the wave itself by self-interference; this can be seen with the "Wave on a String" online animation mentioned in Section 1.3.

Resonance is an example of a fundamental behavior of waves called **quantization**; it is indeed fundamental to **quantum mechanics**, which is also called **wave mechanics**, as we will see in Chapter 13. **Quantization results from the fact that only a certain set of waves can exist in a limited space**, in this case those waves with whole multiples of the string's fundamental frequency, but not other multiples (for example, not 1.5 or 0.624 times the fundamental frequency). The only allowed waves in a limited space (like a string with two fixed ends in our case here) are the standing waves that fit that space.

We will also encounter quantization of waves with other types of waves, in particular sound waves in air and on surfaces (such as on drumheads and on water surfaces), electromagnetic waves (such as in lasers), and quantum waves (causing atomic structure).

2.10 What have We Learned in this Chapter?

We have seen that we can easily create or simulate simple waves on a string and examine their behavior. An important special wave is the standing wave, which is fundamental to voice and music, by forming resonances. To understand how such waves travel along strings, we looked at wave pulses, wave packets and wave trains on a string.

We also considered what happens when two waves "collide" on the same string: we found that these two waves can "destroy" each other briefly, but then resume their original shape and travel direction as if nothing had happened. In particular, we focused on the behavior of waves as they reflect back from the end of a string and interfere with

themselves. These unique behaviors result from the very important concepts of wave superposition and wave interference; we will see in later chapters that such behavior remains valid for other kinds of waves, such as sound waves in air or surface waves on water.

We also saw how beats between waves with nearby frequencies can be used to tune musical instruments.

Our exploration led us to the idea of quantization of waves, which is also important for other kinds of waves, including quantum waves.

Since waves on strings are due to tension in their strings rather than gravity, string instruments produce the same music in a falling elevator, in a space station, on the Moon, on Mars or anywhere else (as long as there is air or another substance that can carry the sound to our ears).

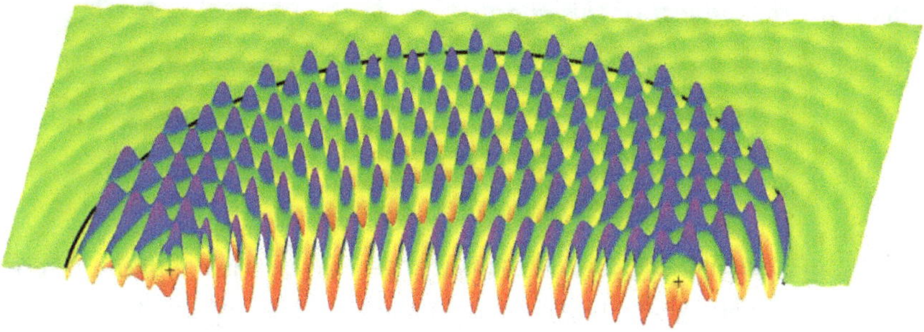

3

Waves in Air: Sound, Acoustics

We first show in this chapter that sound in air (and other gases) is made of waves, and that these waves are density and pressure waves. This is not obvious since waves in air are not visible. We discuss how such waves are created and heard. We find that such waves have much in common with waves on strings, especially if we consider sound in tubes as used in musical wind instruments: superposition and interference, standing waves, resonances, etc. also apply to waves in air.

New with waves in air is the three-dimensional space which allows them to travel in many directions, including especially in the shape of spherical waves, which at large distances from their source become plane waves. We also discuss waves in air confined within boxes and the mouth: we find that these waves can produce standing waves with many more frequencies and resonances than with one-dimensional strings. Also, a new possibility exists in three dimensions: waves can spontaneously turn around corners, which is called diffraction.

I recommend that you first watch my accompanying video to get a feeling for the subject of this chapter.[1]

3.1 Is Sound Made of Waves?

In Chapter 1, we illustrated waves on strings with a series of photographs (see Figure 1-1), while many online videos clearly exhibit their wavy shapes and motions. Can we do the same with sound? ***Do we actually have any visible evidence that sound in air is made of waves?*** Try to think of any situation where sound looked like a wave to your eyes!

Here is such a situation: watch a string vibrate, as in the two photographs of Figure 1-1b. In those photographs, the string vibrates too fast to see its exact shape. However, if you made that string vibrate more slowly (for example, with a heavier rope, like those used for rope skipping or in some online videos) or if you could photograph a faster snapshot of it, you could see its shape, similar to the flyfishing line in Figure 1-1c.

Now make your string vibrate faster: you will start to hear a sound as it swings back and forth through the air. At first, the sound will be very low-pitched, but as the string swings back and forth faster, the sound becomes higher-pitched. Somehow, the string's repeated motion has produced a sound that travelled from the string to your ear. That sound must have traveled through the air. And that sound depends strongly on the repeat frequency of the string's motion. These observations already suggest a wave of some sort, but we need more evidence to declare that sound is a wave.

With waves on strings, we noticed that such waves can equally well travel up or down a hanging string, as if gravity played no role. This is unlike objects with a mass (weight), which are pulled down by gravity on Earth: objects slow down after being thrown upward and speed up when falling downward. Sound waves in air travel up or down equally

[1] See my video "Waves in Air: *sound and musical instruments*" (WA2) on Everyday Physics by Michel A. Van Hove: https://www.youtube.com/watch?v=PMR49kj2PeY

well; in particular, if you shout upward, your shout will not "fall" back down.

Another important property of waves on strings was their "harmless collisions": we saw that they crossed each other as if nothing happened. Does the same happen with sound? Think of people speaking in a group, such as four people around a table: conversations between opposing people cross the same patch of air without changing each other. Colliding objects, however, usually change direction and may break up.

You probably have noticed echoes made by sounds: sounds can reflect off hard surfaces in ways that solid objects or liquids don't.

All the above are strong indications that sound in air is also a type of wave: sound exhibits many wave-like behaviors. **It has been verified over centuries of observation that sound indeed consists of waves**: in particular, music depends heavily on the wave character of sound. As we will see later (Chapter 9), sound also exists in dense materials (metals, glass, ceramics, living flesh, the Earth's crust, water, *etc.*).

3.2 What Kind of Wave could Sound be in Air?

Let's think of other ways to make sound besides shaking a rope. One way is to clap your hands or bump hard objects against each other: those collisions blow a pulse of air into all directions, which you can hear as a clap, but these claps don't have a musical character because there is no repetition in such collisions; we call them **noise**. The same happens when you pronounce the letter "t" alone, or when lightning causes thunder: you get noise. You can whistle by blowing air between your lips: then you may feel vibrations in your lips and corresponding musical sounds in your ear. Speaking and especially singing also give a musical impression because they cause repetitive vibrations in the air.

So, waves in air arise due to the motion of the air. Such motion causes local changes in the density of the air, which in turn causes local changes in the pressure of the air (you probably know that air is quite compressible, for example in balloons and in tires for cars and bicycles). When these changes of density and pressure vary regularly and rapidly, as in wave trains, we hear a musical sound.

Waves in air indeed consist of varying **compressions** and **decompressions** of the air itself; decompressions are also called **rarefactions**. Thus, **sound is a density wave, and at the same time, a pressure wave**. The study of the behavior of sound waves is called **acoustics**.

How different is the mechanism of waves in air compared to waves on strings? With strings, it is tension along the string that tries to straighten the string: this propels any disturbance, such as a pulse, along the length of the string. In air, density and pressure try to smoothen out over space to become as constant as possible (when air density and pressure are constant, we have silence): that smoothing action is the source of propulsion of sound waves in air. It is amazing that such very different mechanisms lead to the same phenomenon of waves. We will see the same thing happen again and again with other waves: on the surface of water, in solids, in space with electromagnetic waves or gravitational waves, *etc.* **Very different mechanisms produce waves with very similar behaviors.**

You may well ask: ***What is air and what are density and pressure of air***? You probably know that air is one example of a **gas**. Other examples are pure **oxygen** gas and pure **nitrogen** gas. **The air we breathe on Earth is actually a combination of gases**: mostly nitrogen (about 78%), a fair amount of oxygen (about 21%), and very small amounts of **argon**, **carbon dioxide**, **water**, *etc.* Oxygen, nitrogen, carbon dioxide and water are molecules that contain two or three atoms each, while argon is a single atom, not a molecule.

In air, the molecules and atoms all fly around at high speeds and collide against each other. They also collide with any surface, such as our skin or a balloon's skin. The fact that the molecules and atoms in air collide against each other and against surfaces puts **pressure** on them, just as a jet of water aimed at your hand puts pressure on your hand.

Sound is a rapid change in the air pressure that happens in less than about 1/20 of a second (this time corresponds to the lowest frequency of 20 hertz that humans can hear). In sound, the air molecules and atoms are briefly given slightly larger speeds in a particular direction. That creates a slightly larger pressure on neighboring air, which in turn puts additional pressure on air beyond that, *etc.*, thereby forming a wave that travels away.

How dense is air? Few people realize how dense air really is! We do notice it when high winds literally blow us off our feet, or when we put our hand outside the window of a speeding car. Near the Earth's surface, **one cubic meter (about 35 cubic feet) contains about 1.2 kilograms (or 2.6 pounds) of air**: that is the same weight (or mass) as 1.2 liters of water (or 1.3 US liquid quarts, a big bottle). For comparison, that density of air is close to one thousandth of the density of water (a cubic meter of water contains about a metric ton of water, which is the weight or mass of a small car).

What is the speed of sound in air? The speed of sound depends on the chemical composition of the air, as well as its density, pressure and temperature. In **"normal" atmospheric conditions** at the surface of the Earth (meaning average atmospheric pressure at a temperature of 20 degrees Celsius or 68 degrees Fahrenheit), **the speed of sound in air is**:

- **Mach** 1
- 343 meters/second (~1/3 kilometer/second)
- 1235 kilometers/hour (km/h)
- 1125 ft/second (375 yards/second)
- 767 miles per hour (US)

This speed is somewhat larger than that of passenger jet planes in cruise: these typically fly at Mach 0.75 to 0.85, which means 75% to 85% of the speed of sound.

Sound in air travels:

- 1 foot in 1/1125 second
- 1 meter in 1/343 second
- 10 meters (the size of a house) in 1/34 second
- 100 meters (the size of a sports hall) in 1/3.4 second
- 1 kilometer (the size of a small town) in 2.9 seconds
- 1 statute (US) mile in 4.7 seconds
- 10 kilometers (the size of a large city) in 29 seconds
- 100 kilometers (from city to city) in 5 minutes
- 1,000 kilometers (the size of a large country) in 48 minutes

You may know that it is easy to estimate your distance from the source of **lightning** or **fireworks**: the resulting **thunderclap** is sound, so it covers about 343 m (~1/3 km) or 1125 ft (~1/5 mile) each second following the lightning flash. (Light travels nearly a million times faster than sound; it covers about 300,000,000 m or 300,000 km each second, so we can ignore the tiny delay in seeing a lightning flash.) Thus, in easy round numbers: **count the number of seconds between a lightning or fireworks flash and its thunderclap, then divide that time by 3 to get its distance in km, or divide that time by 5 to get its distance in miles.**

How fast do molecules fly in air? Near the Earth's surface, **air molecules fly on average at a speed of around 450 meters per second, which is about 1,600 km/h or about 1,000 mph**. This speed is somewhat faster than the speed of sound. There is a simple reason for this: it is the flying molecules that carry the sound forward, from collision to collision, so the speed of sound comes from the speed of molecules. However, since the molecules fly in all directions, not only in the direction of the wave, their average forward speed is smaller and the wave travels more slowly than the molecules.

3.3 How is Sound Created?

We have seen in Section 3.2 that sound in air is due to the motion of the air, which causes density and pressure waves. Sudden claps, bumps, thunder or explosive letters like "t" and "p" create "non-musical" bursts of sounds, mostly called noise (or consonants in the case of letters!).

To make "musical" sound, including voices, a repetitive motion is needed. This can be achieved in several ways described next (we will further discuss the creation of music and voices in Chapters 5 and 6).

The most direct way to create repetitive sound is to use one or more **loudspeakers**, which act as pistons that rhythmically pump the air. A loudspeaker, of course, can produce noise as well as music and voice. The source of the vibrations of a loudspeaker is the electronics behind it: the electronics may **synthesize sound** (making sound by activating oscillations in electric circuits) or **play back** recorded sounds stored in digital or other form, which may have been recorded with one or more **microphones**.

Here is another way to produce repetitive sound, namely a sound wave: make a solid object vibrate so it acts like a loudspeaker and causes the surrounding air to vibrate like a wave. As we saw in Section 2.9, you can hit (or pluck or bow) a string under tension with fixed ends to make it **resonate** at its natural frequency: it will then drive the surrounding air at the same frequency. This is the approach used in **string instruments**, such as **pianos**, **guitars** and **violins**.

If you take a thin bar of metal or stone or wood, you can bend it and release it (for example, by hitting it). Because its bending is **elastic**, it will spontaneously bend back and forth repeatedly at a natural frequency (which depends on its size and material composition): it becomes a piston driving the air at that natural frequency. Good examples of this method to create sound are the tuning fork and the **xylophone**.

Oscillating air blown between your vibrating lips is used in musical **brass instruments**, such as **trumpets** and **trombones**.

Another source of vibrations to make music is, perhaps surprisingly, **turbulence** in the air. Turbulence consists of familiar strong irregularities in the flow of air (and other gases and liquids). A **flag** waves in the wind because the flagpole disrupts the regular flow of the wind, creating alternating motion to either side of the pole. When you **whistle** through your lips, similar turbulence occurs, causing a sound with frequencies that depend on the shapes and positions of your lips and on the speed of the air you blow out. Such turbulence is more visible in the water behind a moving ship, and in fast-flowing rivers with protruding rocks: you see eddies in the water that oscillate from side to side with characteristic frequencies. Thereby, a constant flow of air or water (with no apparent repetitive motion) turns into an oscillation (with a repetitive motion).

Turbulence is also caused when blowing over a hole in the side of a tube, thereby producing standing waves inside the tube, as in the **flute**. A very similar mechanism operates in the **recorder**, with a different mouthpiece.

A further approach uses a **reed**, which looks like a small length of drinking straw with one end almost flattened: blowing through it produces vibrations that feed into the air inside the tube. This method is used in **saxophones**, **clarinets**, *etc.*

Sounds for human communication, such as speaking and singing, are only useful in the range of frequencies that our ears can hear:

between about 20 hertz and about 20,000 hertz, as detailed in Section 3.4. However, our muscles can't even match the lower frequency of 20 hertz with rhythmic motions: we can't wave our hand 20 times per second or faster to produce an audible sound! For that reason, we need another source of sound to speak and sing: that source produces our voice.

The **voice** is based partly on vibrations of **vocal cords** located in the throat, both for **speaking** and for **singing**. Air from the lungs is pushed out between the two flexible vocal cords, which start to oscillate and cut the airflow into rhythmic puffs of air. This happens with a frequency that depends in large part on the size and tension of those cords. Therefore, the primary method of controlling the sound frequency of speech or singing is by changing the tension of the vocal cords by muscular action. However, the mouth above the vocal cords also affects the sound that is produced by favoring some frequencies over others: it is essentially a complex resonance chamber with a highly and quickly variable shape (including that of the tongue, cheeks and lips); the mouth is comparable, at least in function if not in tone, to the resonance chambers of guitars and violins.

3.4 Hearing Sounds

How do we hear sounds? Our primary organ for **hearing** sounds is, of course, the ear. Some sounds reach the ear by traveling through our bones, bypassing the ear canal. The ear is a very complex and sophisticated device that converts sound into electrical signals which are transmitted to the brain.[2] Here, I will only mention a few important aspects of hearing sounds.

One aspect of hearing is the neurological role of the **brain**, which is also very complex: **the brain can filter, modify and interpret the incoming sound in many ways.** The brain affects how we hear different frequencies, different sound intensities, and the pair of signals coming from our two ears; it interprets sounds as signals, words, sentences, messages, stories, and moods; it also interprets music as notes, chords, impressions, and moods, among other roles.

---

[2] See: https://en.wikipedia.org/wiki/Ear

Let's now consider the more physical side of hearing. A first question is: ***Can we hear all frequencies?***

If you listen as a string oscillates, you may hear its vibration as a musical note, for example on a guitar, violin or piano, or with a simple string that you stretch tightly. You can also do this with a stretched rubber band, although its musical quality will be poor.

However, you would not hear a musical sound from a string that you shake with your hand (you would at most hear a non-musical swishing noise of the string moving through the air): that is because you cannot shake a string fast enough to perceive it musically.

The human ear does not hear sounds with frequencies below about 15 to 20 hertz, or 15 to 20 repetitions per second, which would require that you shake the string back and forth at least 15 to 20 times in each second.

The human ear also does not hear "ultrasound" with frequencies above about 15,000 to 20,000 hertz. The audible frequency range is narrower for older people, especially at the high-frequency end: typically, people over 50 years old may not hear sounds above 11,000 hertz, while those over 70 may be deaf to sounds beyond 8,000 to 10,000 hertz.

The human ear is most sensitive to sounds between 1,000 and 4,000 hertz.

You may do a rough and free self-test of your hearing online to find out what the upper and lower limits of your own hearing range are.[3]

So, **our hearing is limited to frequencies in the range of about 15 to 20 hertz to about 15,000 to 20,000 hertz**. Other parts of our body can sense lower frequencies (sometimes called **infrasound**), mainly through the vibration (resonance) of certain body parts, such as inner organs or masses of flesh. Higher-frequency sounds, called **ultrasound**, cannot be heard but are much used for medical imaging of organs within the body: the higher frequencies (2 to 20 megahertz) correspond to smaller wavelengths, typically 0.1 to 1 millimeter, which allows the imaging of details on that scale.

---

[3] Use Online Tone Generator (be careful not to produce too loud sounds, as they may damage your hearing); just press PLAY and sweep the frequency with your mouse to find out what frequencies you can hear: https://www.szynalski.com/tone-generator/

Many animals can hear higher frequencies than humans: some dogs can hear up to 45,000 hertz, cats 65,000 hertz, rats 75,000 hertz, mice 90,000 hertz, bats, whales and porpoises over 100,000 hertz; other animals are more restricted in their hearing, such as the chicken (100 to 2,000 hertz), as well as other birds and fishes.[4]

The **loudness** (often called **volume**) of a sound comes from its wave **amplitude**. More precisely, the **intensity** of a sound is what matters physically because the intensity is the amount of energy contained in a wave; the intensity of a sound is the square of the wave amplitude.

The ear has to absorb this wave energy in the process of hearing. Too much energy will damage or destroy our ability to hear. But we often can't avoid large intensities inflicted on us by nature: a thunderclap, an explosion or the bang of large colliding objects could, literally, be deafening.

How does nature protect us from loud sounds? It seems that evolution has found an elegant trick to minimize the damage of loud sounds: **our sensitivity to sounds becomes progressively worse at higher loudness levels**. Basically, we hear progressively less of the sound's intensity as its loudness increases. Although we still perceive a higher intensity to be louder than a lower intensity, the perceived loudness is not simply proportional to intensity: doubling the intensity does not double the perceived loudness. Instead, perceived loudness grows much more slowly than intensity, especially for higher intensities.

Let's illustrate this with a couple of examples. In human hearing, we can compare the high sound level that starts to hurt (it is called the **threshold of pain**) with the much weaker sound of a mosquito flying near our ear: humans feel that the threshold of pain is about 14 times the perceived sound level of that mosquito. However, the actual intensity of the sound at our threshold of pain is about 10,000,000,000,000 times larger than the intensity of that mosquito's sound! Indeed, that is ten trillion (or 10 million million) times more energy, but only 14 times more perceived loudness!

Here is another more common example: compare the sound of traffic in a busy street with the background noise in the same street

[4] See "How Well Do Dogs and Other Animals Hear?" at https://www.lsu.edu/deafness/HearingRange.html

when it is quiet, such as at night. The perceived sound level in the busy street is almost double that of the quiet street. However, the sound intensity in the busy street is about 1,000 times higher than in the quiet street. We see that our hearing system "dampens" loud sounds extremely effectively!

The fact that our hearing system "dampens" loud sounds so strongly leads to a very special way to measure the perceived loudness of sounds, as follows.

The perceived loudness is commonly expressed in **decibels** (abbreviated as **dB**), a quantity that we often see in connection with sounds and noise. The decibel is a relative quantity that compares the perceived sound level to the weakest sounds that humans normally can hear, which is called the **threshold of hearing**, or **auditory threshold**. The threshold of hearing is at 0 dB, by definition. Box 3-1 gives a **table** of common sounds and their sound levels, both in dB and as intensities (the intensities in that table are also relative to the threshold of hearing). On this scale, **the threshold of hearing is around 0 dB, while the threshold of pain is around 130 dB**.

Loudness as detected by a human being is clearly a rather subjective quantity, as are categories like "very loud" and situations like "busy street". Different people have different thresholds of hearing and thresholds of pain, and these also depend on how long they listen to these sounds (a sudden loud explosion hurts less than spending whole evenings in loud concerts).

BOX 3-1 — PERCEIVED SOUND LEVELS *VERSUS* SOUND INTENSITIES.

The following table gives a rough idea of how humans perceive sound loudness, as compiled from a wide variety of sources. The left two columns are only suggestive and very approximate; an important factor that this table ignores is the duration of the sound: a long-lasting sound is perceived as louder than a brief sound. The dB level in the third column is tied mathematically to the intensity levels in the fourth column (through the logarithmic function), and is therefore a measurable quantity, once the intensity at the threshold of hearing is known. On average among people, and under normal atmospheric conditions, the intensity of the threshold of hearing is generally about 1 trillionth of a watt per square meter, so the threshold of pain is near 1 watt per square meter.

Apparent loudness	Type of sound	Sound level — in dB	Intensity relative to threshold of hearing
Deafening	Cannon firing nearby	220	10,000,000,000,000,000,000,000 $= 10^{22}$ = 10 sextillion
	Rocket launch	180	1,000,000,000,000,000,000 $= 10^{18}$ = 1 quintillion
	Jet engine nearby; balloon bursting	160	10,000,000,000,000,000 $= 10^{16}$ = 10 quadrillion
	Fighter jet takeoff	150	1,000,000,000,000,000 $= 10^{15}$ = 1 quadrillion
	Gunshot	140	100,000,000,000,000 $= 10^{14}$ = 100 trillion
	Rock band; jackhammer; jet plane takeoff	130	10,000,000,000,000 $= 10^{13}$ = 10 trillion
Very loud	Thunder; loud concert; siren	120	1,000,000,000,000 $= 10^{12}$ = 1 trillion
	Electric guitar at max. volume; chainsaw; leaf blower	110	100,000,000,000 $= 10^{11}$ = 100 billion
	Metro/subway; motorcycle; baby crying; hand dryer; industrial noise	100	10,000,000,000 $= 10^{10}$ = 10 billion
	Heavy truck traffic; lawn mower; hairdryer; welding	90	1,000,000,000 $= 10^{9}$ = 1 billion

(*Continued*)

Apparent loudness	Type of sound	Sound level — in dB	Intensity relative to threshold of hearing
Loud	Acoustic guitar; noisy restaurant; vacuum cleaner	80	100,000,000 $= 10^8 =$ 100 million
	Busy street; noisy office	70	10,000,000 $= 10^7 =$ 10 million
Moderate	Normal conversation; background music	60	1,000,000 $= 10^6 =$ 1 million
	Quiet office	50	100,000 $= 10^5 =$ 100 thousand
	Quiet conversation; typical room; library; quiet street; light rain	40	10,000 $= 10^4 =$ 10 thousand
Faint	Recording studio; whispering	30	1,000 $= 10^3 =$ 1 thousand
	Ticking watch; quiet countryside	20	100 $= 10^2 =$ 1 hundred
Very faint	Insect noises	10	10
	Threshold of hearing	0	1

How loud is loud? We can compare the energy in sound with other situations. Let's take the threshold of pain for sound as a reference: imagine standing in front of a big boombox of 1 x 1 meter blaring sound that starts to hurt your ears; this amounts to about 1 watt per square meter.[5]

How does this compare with other familiar situations? Sunlight hitting the surface of the Earth brings about 160 watts of power onto every square meter — 160 times more than the sound's 1 watt per square meter. This means that someone who is sunbathing receives 160 times more energy from sunlight than from a sound like thunder.

Our entire body constantly produces about 100 watts of power on average (mostly in the form of heat to keep our organs functioning, but also as movement, *etc.*); this is similar to a typical old incandescent lamp of 100 watts, or a dozen modern LED lamps. So, sound at the threshold of pain has about 1% of the power produced by the human body, or about 10% of the power of a modern LED lamp.

We see that the sound which we hear is normally quite weak compared to other familiar situations. This also shows that our hearing system is highly sensitive.

Above, we discussed the loudness and energy of sound waves. We can also ask: ***What are typical amplitudes of sound waves?*** We can distinguish between the amplitude of the motion of the air and the amplitude of changes in the air density and pressure.

As mentioned in Section 3.2, sound waves are due to back-and-forth motion of the air: ***How large is such motion?*** You probably have noticed vibrations on loudspeakers, especially the larger loudspeakers called **woofers** or **subwoofers** that produce low-frequency tones. You may also have felt the vibrations of the sides of the box in which such loudspeakers are held. Those vibrations are normally too rapid to follow, but you can see or feel their fast motion as a blur, similar to the blur of rapidly vibrating strings photographed in Figure 1-1b. The amplitude of such blurred vibrations is transmitted to the air outside the loudspeaker and becomes the amplitude of the back-and-forth motion of that air.

[5] Charles Taylor, *"Exploring Music: The Science and Technology of Tones and Tunes"*, Institute of Physics Publishing, 1992, page 21.

This amplitude of wave motion can be several millimeters in size at maximum intensity.

However, the resulting sound wave will radiate in all directions, and its amplitude will become smaller as it fans out and spreads its energy around. By the time this wave has expanded to become barely audible (near the threshold of hearing), its amplitude of wave motion will be about 1 million times smaller: that is around a millionth of a millimeter, or about a thousandth of a **micrometer**, or about 1 **nanometer**, the size of small molecules! **Amazingly, those of us who have excellent hearing are just able to hear air motion on the scale of molecules, which are totally invisible even with optical microscopes.**

How about the amplitude of the pressure (and density), namely the relative change in pressure and density? That is a rather different matter: it is extremely small even at the painful level (we will discuss only the pressure amplitude here since the situation with the density amplitude is very similar).

Surprisingly, **even near the painful level, sound only represents a pressure variation of about 1 part in 5,000**, relative to the normal air pressure on Earth. Indeed, such painful sounds only increase and decrease the normal atmospheric pressure by two hundredths of 1%, or 0.02%! **The weakest sounds we can hear give pressure differences that are about 1 million times smaller than that: 1 part in 5 billion.**[6] To appreciate how astonishing this is, let's compare such a sound wave in air to a water wave on an ocean with a depth of 10 kilometers (the largest known ocean depth is about 11 kilometers): 1 part in 5,000 for the painful sound corresponds to 2 meters of wave height on an ocean with a depth of 10 kilometers. It is amazing that a tiny "ripple" of 1 part in 5,000 in the air can be so painful to the ears: they are indeed hypersensitive! And 1 part in 5 billion for the weakest detectable sound represents 2 thousandths of a millimeter, or 2 micrometers, relative to 10 kilometers. So, **the weakest sound that some of us can hear is like an invisible ripple of about 2% of a hair's diameter on an ocean with a depth of 10 kilometers.**

[6] These numbers are derived from: Charles Taylor, *"Exploring Music: The Science and Technology of Tones and Tunes"*, Institute of Physics Publishing, 1992, pages 19–22.

For comparison, the typical daily variation in atmospheric pressure, between day and night, is about 0.1%, 50 times stronger than a painful sound. Of course, that variation (due in large part to temperature changes) happens very slowly, over hours, so we don't notice it. Much more noticeable is the pressure difference when we take an elevator: for example, rising 100 meters reduces the atmospheric pressure by about 1%; this happens in about a minute, and we certainly feel it in our ears, although it is far too slow to produce sound (we would have to go up and down at least 20 times per second to hear that pressure variation!).

A very valuable aspect of our hearing is the fact that we have two ears that independently measure sound, which is called **binaural hearing. This is particularly useful to detect from which direction sound comes.** Compare this with our eyes: a single eye sees fine details of the scene in front of us, whereas a single ear can hardly distinguish directions. The reason for this difference is the wavelength of light compared to the wavelength of audible sound: the very small wavelength of light (far shorter than the size of the eye's lens) allows the eye to pinpoint direction very accurately, while the long wavelength of sound (much larger than the ear) prevents sensing where it came from. All we can do with one ear is turn our heads to sense whether the sound seems louder in some general direction since the sound will cast a "shadow" behind our heads.

With two ears, we can overcome this weakness to some degree, allowing us to tell whether a sound came from the front left or the front right, for example. This is possible mainly because **there is a time difference for a wave arriving at the two ears**, and this time difference depends on the direction from which the wave comes. With two ears, the sound's shadow behind the head will also be more evident.

3.5 Visualizing Waves in Air

To better understand the behavior of waves in air, let's make them more visual.

In quiet air at constant altitude and temperature on Earth, the air density and its pressure are the same everywhere because any disturbance will even itself out over space; the same flattening out of disturbances occurs on strings and on the surface of water, as well as for all other kinds of waves that we will discuss. Let's see how this process creates the motion of waves in air.

In Figure 3-1 we imagine a series of claps in quiet air, forming a wave train: after the first clap, suddenly the air is denser in a small region in the center, so that region has both a higher density and a higher pressure, shown as the blue-purple peak. That pulse of "compression" pushes against the surrounding air, which has a lower density and a lower pressure. So, air moves from the denser center into the less dense surrounding air, which now becomes denser and gets compressed. In

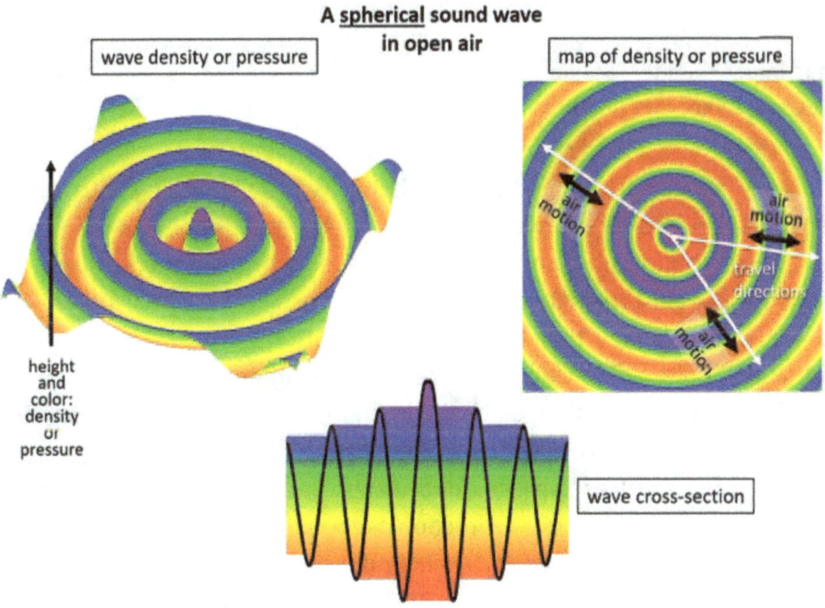

Figure 3-1: A repeated sound wave in air is sketched as radiating outward. These graphs show the density or pressure in a two-dimensional horizontal slice of air (almost the same wave shape occurs in non-horizontal directions thanks to very slow changes in density and pressure with altitude). In the left graph, "higher" in the vertical direction means a higher density and a higher pressure. The right graph is a 2D map of the height of the left graph: it also shows density and pressure. The middle graph shows a cross-section of the wave. The colors range from red (low density and low pressure), *via* yellow and green (medium density and medium pressure), to purple (high density and high pressure). For better visibility, this wave weakens slower than the inverse distance squared, and the central spike is limited in height.

turn, this pulse of denser air compresses the next ring of air, and so on: the compression thereby travels outward. The next clap repeats the same process a moment later, creating a second pulse of compressed air. Each successive clap adds a new pulse that follows the previous rings, forming a wave train that radiates from the center in all directions.

Between two claps, the air is decompressed in the center since air has been pushed outward; like the compressed pulses, those pulses of decompressed air also travel outward in between the pulses of compressed air. **This forms a wave train of repeating compressions and decompressions radiating outward**, as sketched in Figure 3-1. You may view this motion in my Animation 3*1.

ANIMATION 3*1 — See my video WA2 at time 2:56 in its section "**Waves in air**" under the title "**Sound waves in air — spherical waves**". (See details in the section References and Resources below.)

We see clearly in Figure 3-1 that **a wave expanding into open space forms spherical waves**. These waves can start from a small source of sound and spread out like an ever-growing spherical balloon that expands at the same speed in all directions. These **spherical waves are the normal way that sound reaches our ears**: the sound may start from a source like a clap, or a musical instrument, or a mouth, or a loudspeaker; it then spreads in all directions like a growing balloon; we can hear these spherical waves as they pass by our ears.

An important aspect of sound waves in air is that the air moves (back and forth) in the same direction as the wave travels: it is therefore called a **longitudinal wave**. In the example of Figure 3-1, the air vibrates away from and toward the center. This contrasts with a transverse wave on a string, where each piece of the string vibrates perpendicular to the string and therefore perpendicular to the direction in which the wave travels: see Section 2.3, and Figure 2-7 in particular.

We notice in the cross-section of Figure 3-1 that **the amplitude of the radiating wave decreases as it travels farther from the center**. That simply reflects our experience that sound weakens with distance: the wave expands through all available space and thus spreads out its energy. (To be more precise, the amplitude weakens as the inverse of the distance squared for three-dimensional radiation.)

If we follow a spherical wave far from its source in any direction, its spherical shape becomes less obvious: the wave crests and valleys then look straighter. As an analogy, think of standing on the surface of the near-spherical Earth, which is so flat (especially on the oceans) that we basically can't see its spherical shape. A spherical wave then looks more like the wave in Figure 3-2. We call this a **plane wave** because its wave crests and troughs are flat planes (perpendicular to the direction of travel of the wave). When the plane wave has a limited width, as is the case in Figure 3-2, we can call it a **beam**, such as a beam of sound or a beam of light. You may view this motion in my Animation 3*2.

ANIMATION 3*2 — See my video WA2 at time 5:29 in its section "**Properties of waves in space**" under the title "**Sound waves in air — plane waves and beams**". (See details in the section References and Resources below.)

Even though plane waves are special cases, there are real advantages to thinking about them because they have a very simple

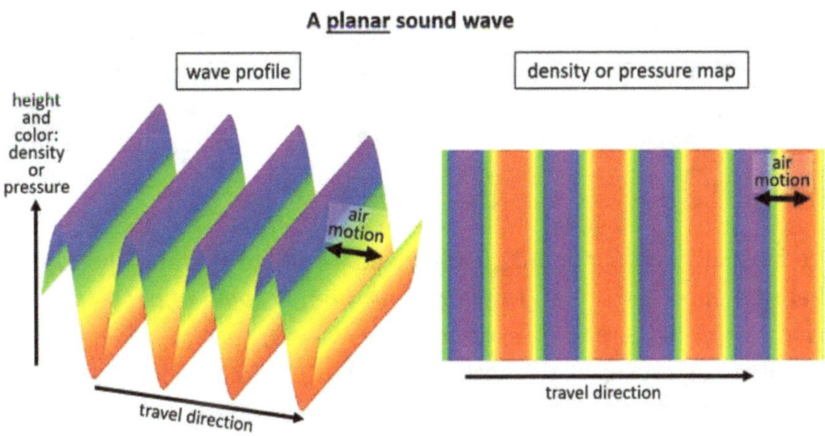

Figure 3-2: Four pulses of a planar sound wave train are sketched moving to the right in open air. The presentation is similar to Figure 3-1, showing air density and pressure (red is low, purple is high). The left graph shows a profile of the wave's density or pressure; the right graph shows a map of the density or pressure.

shape. One major advantage of a plane wave is that it is totally uniform, with the same amplitude and the same travel direction everywhere and at every moment.

How special are plane waves? A good way to decide whether we can think of a sound wave as a plane wave is to compare the distance between you and its source (where it was a spherical wave) to the size of your head. If your distance from the source of the wave is many times larger than your head, say tens of meters or more, then we can safely treat that wave as a plane wave in the region around your head.

There is another important situation where waves can be thought of as plane waves: inside musical wind instruments or more generally in long tubes, as we will discuss in Section 3.10.

Figure 3-1 represents a spherical wave in three dimensions (3D). But it can equally well represent a **circular wave** in two dimensions (2D), especially as I have only drawn a 2D slice of a 3D spherical wave: they look very much alike. Circular waves in air are not common, however, they would have to be sandwiched between two huge parallel walls or plates to prevent them from escaping into the third dimension. They are common on drumheads and cymbals, but we can't see them easily. On water, however, they are very common and very visible when waterdrops or stones fall on the water surface, as we will see in Section 11.1.

Because of the close similarity between 2D circular waves and 3D spherical waves, we will treat them on an equal footing and interchangeably in this chapter.

3.6 Sonic Booms and the Doppler Effect

We have all heard, or heard of, **sonic booms** due to aircraft flying faster than the speed of sound. ***How do sonic booms come about?*** We also have heard the sirens of emergency vehicles (police cars, firetrucks, ambulances), as well as the horns of normal vehicles, change their pitch as they pass by us: this is the famous **Doppler effect**. ***How does the Doppler effect come about?***

The common aspect of sonic booms and the Doppler effect is that the source of the sound is moving relative to the listener. Let's consider that situation in Figure 3-3. First, at top left, we see a motionless source

Spherical waves from sources at rest, at subsonic speed, at transonic speed, and at supersonic speed – Doppler effect and sonic boom

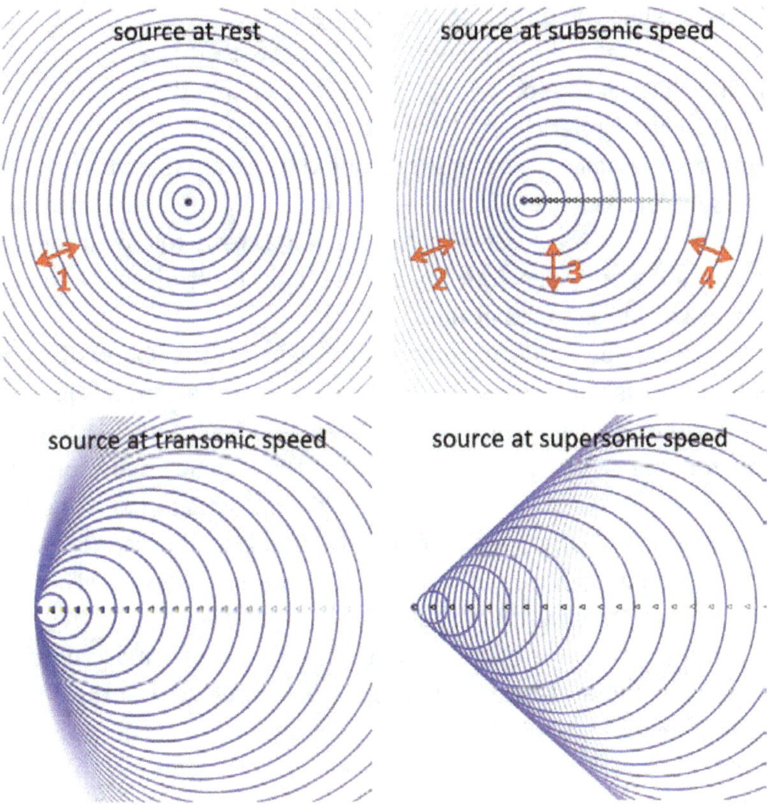

Figure 3-3: Spherical sound waves are sketched radiating from moving sources (such as an airplane, siren or horn): the sources are marked as little triangles pointing to the left, showing their movement. Only the spherical wave crests are drawn. At the top left, the source does not move; at the top right, the source is moving to the left slower than the speed of sound in a subsonic situation; at the bottom left, the source is moving to the left at the speed of sound in a transonic situation; at the bottom right, the source is moving to the left faster than the speed of sound in a supersonic situation. A motionless observer stands where the red arrows are placed: each double-headed arrow shows the distance covered by the wave crests in the same time, for example in one second.

of sound (such as a motionless airplane, siren or horn): it produces spherical sound waves, just like those we saw already in Figure 3-1, but now we only draw the wave crests. The wave crests are all circular (spherical in 3D) and travel outward at the speed of sound. Notice the wavelength, which is the crest-to-crest distance: the wavelength is the same everywhere when the source is at rest.

Now we make the source of sound move to the left, but slower than the speed of sound: we call this a **subsonic speed**. The wave crests remain circular (spherical in 3D) and still travel outward at the speed of sound from the point where they originated (not from the latest position of the source). As shown at the top right in Figure 3-3, this motion of the source causes the successive circular wave crests to crowd together toward the left and space out toward the right: this causes the Doppler effect, as we will see further below.

When the speed of the source is the same as the speed of sound, the source stays with its own sound, as we see at bottom left in Figure 3-3: the circular (or spherical) waves pile up on top of each other right where the source is. This is an example of **transonic speed**, or speed near the speed of sound. Transonic speeds are defined to be between 95% and 120% of the speed of sound. "Breaking the sound barrier" means speeding up through the speed of sound; only in 1947 did an airplane safely fly faster than sound.

You probably have heard of **Mach** speed, and you may know that **the speed of sound is often called Mach 1**. Therefore, transonic speed lies between Mach 0.95 and Mach 1.2. In air under normal conditions (such as a temperature of 20 degrees Celsius or 68 degrees Fahrenheit), the speed of sound is about 343 meters per second (m/s), which is 1235 km/h, or 767 mph. At this speed, you can cover a kilometer in 2.9 seconds or a statute mile (US mile) in 4.7 seconds. This speed of sound varies little with air pressure but slows down somewhat in cooler temperatures, as we find at higher altitudes above the Earth's surface. As a result, the speed of sound drops from about 343 m/s at sea level to about 295 m/s (1062 km/h, 660 mph) at normal cruising altitudes of passenger jets (around 10,000 to 13,000 meters or 30,000 to 40,000 feet): that is only a 14% drop in speed.

The speed of sound is very important for airplanes (and any other object traveling through air). In the transonic range, the flight of airplanes is more complicated than at lower speeds. In particular, there is much stronger air resistance (**aerodynamic resistance** or **drag**) at transonic speeds than below Mach 0.95 and above Mach 1.2. Airplanes need additional power to break the sound barrier; for example, the famed former supersonic passenger airliner Concorde needed to use its "afterburners" (called "reheats" in Britain) to accelerate from Mach 0.95 to Mach 1.7 (its constant cruising speed was around Mach 2.00 to 2.02 depending on load; and its maximum allowed speed was Mach 2.04 to avoid damage such as overheating due to air friction).

We next give the source a speed larger than the speed of sound, as shown at the lower right in Figure 3-3: we call this a **supersonic speed**. Currently, no passenger airplanes fly over the speed of sound, but many military aircraft do, some up to about Mach 3. Many missiles (military rockets carrying bombs) travel above the speed of sound. Rockets carrying satellites, goods and people up to orbit around Earth, or beyond, reach Mach 1 within minutes after they take off. (To orbit Earth safely above the atmosphere, a speed of about 27,000 km/h or 17,000 mph is needed; this would be something like Mach 25, except that Mach speed has no meaning outside the atmosphere, where no sound is possible!)

A moving object creates sound simply by disturbing the air, just like a boat creates waves on water as it moves. At supersonic speed, the source catches up with the wave crests and finds itself ahead of all the sound that it created before. In Figure 3-3, we see very clearly how the wave crests pile up on top of each other along a V-shaped wave front (in three dimensions, this V-shape is actually a cone), instead of right at the source as at Mach 1. This **piling up of sound implies a very strong sound front traveling at the speed of sound in directions away from the airplane: this shockwave is the famous sonic boom**.

Because of its conical rather than spherical shape, the sonic boom can travel much farther than normal sounds. For this reason, the supersonic passenger jet Concorde was banned from flying at supersonic speeds above populated areas since it could destroy windows, *etc.*, in addition to causing much noise: that considerably reduced its commercial value, together with its excessive fuel consumption. Supersonic military jets are also limited in where they can fly at supersonic speeds. At sufficient speed, a boat creates a V-shaped bow wave that is very similar to a sonic boom, as we will discuss in Section 11.10.

When the source's speed exceeds Mach 5 (five times the speed of sound), we speak of **hypersonic speeds**. Above Mach 10, speeds are called **high-hypersonic speeds**, and above Mach 25 they are called **re-entry speeds** because they are common during the re-entry of space vehicles returning to the Earth's surface from outer space. The bottom right sketch of Figure 3-3 still shows these cases: the V-shape simply becomes narrower and more pointed as the speed increases.

Hypersonic speeds in the Earth's atmosphere are reached by some rockets and many missiles: at such speeds, more complex physical effects play a growing role, in addition to increasing heat due to air friction. There have been many proposed designs for hypersonic vehicles, some of which are being tested; one type, called **waverider**, literally rides on its own shockwave; the compressed air in that shockwave can even become so hot that it ignites fuel and propels the vehicle, although at slower speeds other propulsion methods, like **rockets**, **ramjets** and **scramjets**, are still needed.

(Rockets mix two onboard fuels to create a chemical reaction, basically a controlled gradual explosion, that propels the vehicle; they don't require air and therefore can operate in outer space. Ramjets and scramjets mix one onboard fuel with outside air to achieve the same outcome: ramjets and scramjets are jet engines that use their own speed to compress the outside air, instead of using compressor blades; inside ramjets, the fuel-air combustion takes place at subsonic speeds of the air, while inside scramjets, this happens at supersonic speeds of the air.)

At high hypersonic speeds, managing temperature becomes particularly critical: materials reach their strength limits at the high temperatures caused by the high speeds. "Re-entry" speeds refer to the entry of space vehicles into the atmosphere of a planet like Earth: these speeds are typically around 28,000 km/h or 17,500 mph, or more.

How about the Doppler effect? To discuss the **Doppler effect**, first imagine yourself in the situation with a motionless source at the top left in Figure 3-3: you stand at the position of the double-headed red arrow "1", listening to the siren of an immobile police car at the center. Imagine counting how many wave crests pass by you in one second: that gives you the frequency that you hear (the frequency is the number of repetitions of the wave per second). The length of the red arrow in Figure 3-3 represents the distance covered by a crest in one second: here I count about four crests in the length of the arrow, so we have a frequency of about 4 hertz (real-world frequencies would be much larger than this but would be difficult to draw).

Now the police car moves toward you at a subsonic speed, as at the top right in Figure 3-3: you are in position "2", and again, you count how many wave crests pass by you in one second. The length of the red arrow tells you it is about eight crests: this means a frequency of about 8 hertz, about twice the original 4 hertz before the police car moved. Here, it is important to realize that all the wave crests still travel at the same speed (the speed of sound), independent of the speed of the source, so all the red arrows have the same length.

When the police car is near you, your position is "3": the car moves almost perpendicular to the wave crests that it sends toward you. Now you count about 4 crests per second, so that you again hear the original frequency of about 4 hertz.

We can repeat this exercise after the police car has passed and while it goes away from you toward the left: that is the situation "4" in Figure 3-3. Now the crests are spaced farther apart. Counting how many crests pass by you in a second (shown again by the length of the arrow) gives about 2.5: this means you now hear a frequency of about 2.5 hertz, lower than the original siren's frequency of 4 hertz.

In this example, you would hear a drop in the apparent frequency of the siren from about 8 hertz through 4 hertz to about 2.5 hertz, as the police car approaches you and then goes away. In more realistic situations, the Doppler effect is not quite so strong. For example, with a car driving at 70 km/h and using a sound speed of 1200 km/h, the Doppler effect changes the apparent frequency by about 5.5% (upward when the car approaches and downward when it leaves): that is an 11% drop in frequency as the car passes you. That change in frequency is easily heard[7]: it is about 1/9th of an octave, about one white key over on the piano keyboard.

(We should note: the Doppler effect is <u>not</u> related to the change in sound <u>intensity</u> which also happens at the same time: the sound becomes louder as the car approaches you and then becomes weaker as

[7] You can hear the Doppler effect online at https://en.wikipedia.org/wiki/Doppler_effect

it goes away from you, simply because of the weakening with distance of all spherical waves.)

The Doppler effect exists not only with sounds but also with electromagnetic waves (especially radars) and other waves. It is used to measure the speeds and direction of travel of vehicles, liquids (such as blood flowing in veins), stars, and wind speeds in rainstorms.

3.7 Wave Interference in Space

In Section 2.5, we discussed that two waves on a string are superposed on each other, causing **constructive** and **destructive interferences**. The same is true for waves in air.

For example, in Section 2.6, we mentioned the interference effect called **beats** on strings. Beats also exist in air, giving a "wha-wha-wha-wha-…" sound: all you need are two sources of sound waves with slightly different frequencies, such as two loudspeakers. We showed in Section 2.6 how to use online sound sources with your computer to produce beats that you can hear; that section also explained the principle of beats, so we don't repeat that story here. You can also see and hear such beats in my Animation 2*7, mentioned in Section 2.6.

A well-known use of the interference of sound waves is **noise-canceling headphones**, also called **sound-canceling headphones**. (Noise is an undesired sound.) The idea is simple: **add to noise or sound its "opposite", so they cancel each other by destructive interference**. In Section 2.5, we saw how a pulse that is in "antiphase" to another identical pulse can destroy it (at least temporarily): see the green snapshot in the righthand sketch of Figure 2-10.

We can permanently destroy a more complex wave (such as noise) as follows: let's take as an example the topmost wave in Figure 2-13 (the thick black wave); turn it upside down (meaning: make all its peaks into dips and all its dips into peaks); add the upside-down sound to the original sound, so they add up to zero; that is silence, and we have canceled the noise.

You may object that this method also cancels any desired voices and music: indeed, this approach does not only suppress noise but also suppresses all desired sounds. You may wish to remove noise from a beautiful song, but sound cancellation removes both. Such selective cancellation is strictly impossible. One imperfect way around that is to suppress only certain frequency ranges, such as high and low frequencies, leaving untouched (or stronger) the middle frequencies that are more common in voices and music than in noise. All of this requires electronics to invert the incoming "noise", and add them together, perhaps also filtering out some frequencies.

We now switch to a different situation: ***What happens when two plane waves cross each other?*** The top part of Figure 3-4 shows two identical plane waves in the shape of two beams crossing in the middle.

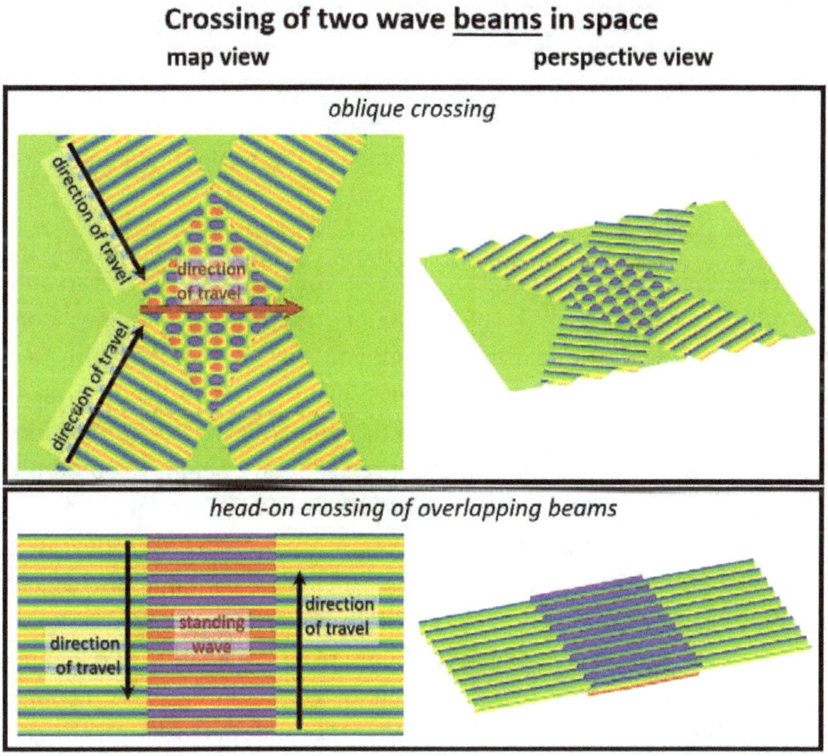

Figure 3-4: Two identical plane waves, forming two beams, interfere with each other by simple superposition. *At the left* are map views, *at the right* perspective views. The presentation is similar to Figure 3-1, showing air density and pressure (red is low, purple is high). *Upper graphs*: the two beams cross obliquely; interference in the overlap region causes a "checkerboard" or "egg-crate" pattern; this pattern moves rigidly to the right, not up or down or obliquely. *Lower graphs*: the two beams travel in opposite directions, overlapping in the middle; now a standing wave is formed in the overlap region.

(This image visually resembles Figure 2-11 but is quite different: we now look at one snapshot of two waves in 2D rather than a sequence of snapshots of two waves in 1D.)

This wave crossing leads to remarkable behavior: when the beams cross at oblique angles (top pair of graphs), we see a new interference pattern that looks like a checkerboard or an egg crate. There are peaks (purple) and dips (red) separated by green lines. As seen in my Animation 3*3, this checkerboard or egg-crate pattern slides rigidly to the right (as indicated by the red arrow in Figure 3-4), not in the direction of either beam. Nevertheless, in any fixed position, the wave oscillates as it passes through.

ANIMATION 3*3 — See my video WA2 at time 6:00 in its section "**Properties of waves in space**" under the title "**Crossing plane waves — checkerboard/egg-crate interference pattern**". (See details in the section References and Resources below.)

Here is one way that you can convince yourself that this remarkable pattern is reasonable: in Figure 3-4, extend the purple/blue lines from the two beams into the overlap region; you will see that they intersect where the purple peaks are located. Similarly, the red/orange lines intersect at the red dips. To reinforce this point, Figure 3-5 adds a third crossing beam: you can imagine the three separate beams piled on top of each other by following their wave fronts shown as straight lines.

As with waves crossing on strings, the two plane-wave beams in Figure 3-4 cross without changing each other: they continue unchanged after crossing, unlike objects that collide.

Another remarkable behavior takes place when the two beams collide head-on, as in the bottom pair of graphs in Figure 3-4. Here, the two beams, one coming from the top and the other from the bottom, overlap in the middle. The result is the **standing wave** shown in the overlap region in the middle: indeed, as shown in my Animation 3*4, this standing wave has no motion to the right or left, or up or down in the figure. The green lines are immobile: they are nodes. The purple ridges and red troughs oscillate between purple (high)

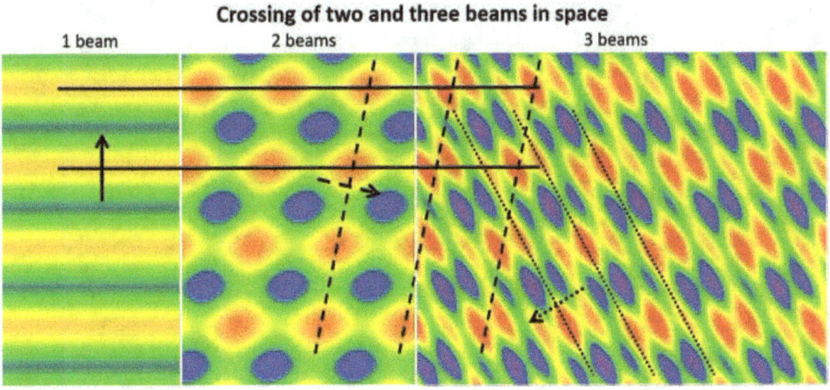

Figure 3-5: Three identical plane waves, forming three beams, interfere with each other by simple superposition in map views; arrows show the directions of travel of each beam. *At the left* is a single beam. *In the middle*, a second beam crosses the first. *At the right*, a third beam crosses the previous two. The presentation is similar to Figures 3-1 and 3-4, showing air density and pressure (red is low, purple is high).

and red (low) without moving. In fact, this pattern is a special case of the checkerboard/egg-crate pattern of the oblique crossing (at top in Figure 3-4): the small "cells" (purple or red) are simply stretched horizontally to infinity.

ANIMATION 3*4 — See my video WA2 at time 6:36 in its section "**Properties of waves in space**" under the title "**Opposing plane waves — standing wave pattern**". (See details in the section References and Resources below.)

What happens when we mix spherical waves in three-dimensional space? Figure 3-6 shows the superposition of two identical spherical waves: each one is like the spherical wave sketched in Figure 3-1, as reproduced in the left graph of Figure 3-6. The single spherical wave expands uniformly in all directions. The two identical superposed spherical waves also expand in all directions, but they interfere: they cancel each other out by destructive interference in certain directions, shown by the green streaks radiating outward; in other directions, the waves interfere constructively and thus concentrate

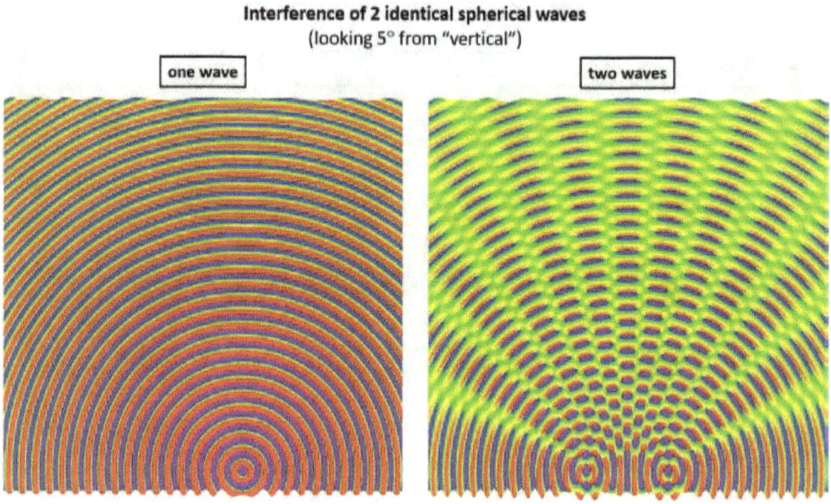

Figure 3-6: Two identical spherical waves interfere with each other by simple superposition. *At the left,* a single spherical wave is shown, like that of Figure 3-1; *at the right,* two identical spherical waves interfere. The sources of the spherical waves are near the bottom in both graphs; they could be your computer's loudspeakers. The presentation is similar to Figure 3-1, showing air density and pressure (red is low, purple is high). The view is slightly tilted to expose the wavy profiles along the bottom and top edges. Also see the blow-up in Figure 3-7.

intensity, shown by the small blue-red "ribs" radiating outward. You may recognize the checkerboard or egg-crate pattern of Figure 3-4: it shows up here in very distorted form, but the mechanism of its formation is identical.

How does the interference pattern of the two spherical waves in Figure 3-6 ***move?*** This pattern has an interesting combination of different motions. On the line between the two sources (near the bottom of the figure, which is shown magnified in Figure 3-7), we have essentially a standing wave: two waves move in opposite directions "through" each other, as along a string (see the wave trains forming standing waves in Figure 2-20 and at bottom in Figure 3-4); the only difference compared to strings is that the spherical waves weaken with distance, so they do not have exactly the same amplitude and can only perfectly cancel each other (by destructive interference) at the mid-point between the two sources.

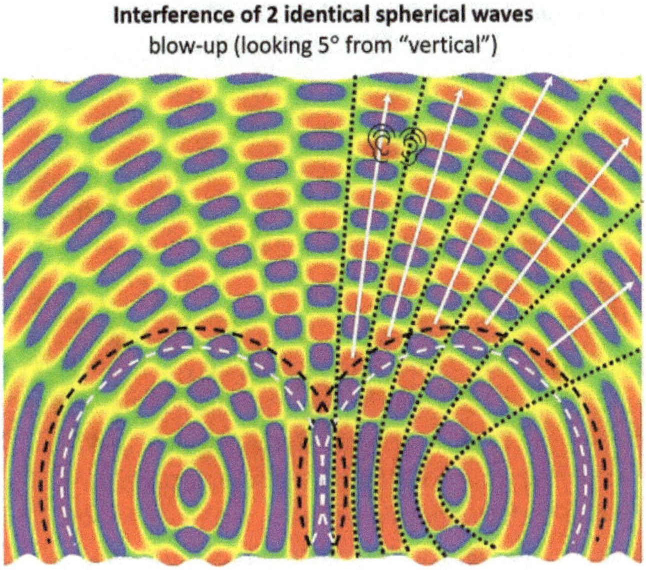

Figure 3-7: Blow-up of the right graph of Figure 3-6. The dashed lines follow spherical ridges (white) and troughs (black). The dotted black lines are curved nodal planes, where the combined wave gives near-destructive interference: here, the right ear is located in a nodal position, while the left ear is placed in an antinodal position, where strong oscillations due to constructive interference take place. The white arrows show the travel directions of the combined wave.

Further away from the two sources, the wave pattern radiates outward: the "green" streaks that we see spreading out in Figures 3-6 and 3-7 (where they are shown as dotted lines) are nodes where there is little wave action, so they are "silent" avenues. Between those "green" streaks, we see a pattern of peaks and troughs that move outward like a spherical wave (just like the wave shown at the left in Figure 3-6): the arrows in Figure 3-7 show the travel directions of these waves. We thus see "beams" of sound in several directions. It is very helpful to view this motion in my Animation 3*5.

ANIMATION 3*5 — See my video WA2 at times 3:50 and 4:22 in its section "**Properties of waves in space"** under the title "**Crossing spherical waves in air — radiating interference pattern (1 of 2)**" and "**... (2 of 2)**". (See details in the section References and Resources below.)

Can you hear constructive and destructive interferences? Indeed, you can: you may experience these variations in sound in my Animation 3*6. You may also use, for example, the webpage in this footnote.[8]

> ANIMATION 3*6 — Listen to my video WA2 at time 4:53 in its section "**Properties of waves in space**" under the title "**Interfering spherical waves — an important signature of waves**". (See details in the section References and Resources below.)

In either case, move your head around: do you hear clear variations in the sound intensity? Do your two ears hear different variations? You may cover one ear to increase the effect (two ears may hear different things and confuse you); or you may also approach a nearby wall or other large flat surface and hear similar variations in sound intensity. These variations are wave interferences. They produce peaks and dips in intensity where constructive and destructive interferences take place, respectively.

One way to confirm that these are wave effects, and not just a consequence of the shape and layout of your room or computer hardware, is simply to change the frequency of the sound. On that same webpage, repeat what you did but now with a higher frequency, say 1200 hertz, and then with a lower frequency, say 300 hertz. Do you notice a difference in the intensity variations (besides the difference in tone)? You should be able to hear the variations in intensity with shorter motions of your head at higher frequencies, and *vice versa*. The reason for this difference is the change in wavelength: at higher frequencies, the wavelength of sound is shorter, so the peaks and dips in intensity are located closer together; we thus can confirm that this is a wave effect and that it is not due to your room or hardware.

---

[8] Use Online Tone Generator, The 432Hz Frequency (be careful not to produce too loud sounds, as they may damage your hearing): https://onlinetonegenerator.com/432Hz.html. You should enter a relatively high frequency, such as 800 hertz (to create wavelengths comparable to the size of your head), and press Play. This method and the similar method in my Animation 3*6 use your computer's two loudspeakers, which act as the two sources of identical spherical waves. If your computer has only one loudspeaker, a wall can serve as mirror to create a second identical wave, as we will see in Section 3.8.

Specifically, peaks and dips are located about one quarter of a wavelength from each other. The wavelength in normal air ranges from about 17 meters (~56 feet) at 20 hertz, *via* about 1 meter (~3 feet) at 300 hertz, ~38 centimeters (~1.3 feet) at 800 hertz, and ~25 centimeters (~1 foot) at 1200 hertz, to ~17 millimeters (~0.67 inches) at 20,000 hertz. Therefore, the distance between peaks and dips varies from ~4 meters (~14 feet) at 20 hertz, *via* ~25 centimeters (~1 foot) at 300 hertz, ~10 centimeters (~4 inches) at 800 hertz and ~6 centimeters (~2 inches) at 1200 hertz, to ~4 millimeters (~0.14 inches) at 20,000 hertz: you should hear the strongest intensity variations when moving your head by such distances.

The wavelength also explains that your two ears often hear different intensities in the experiment above: one ear may be located at a peak while the other may be at a dip of the interference. This is illustrated with two ears in Figure 3-7 (which is a blow-up of Figure 3-6): the right ear is drawn at a "quiet" node (on a dotted line where there are only small pressure changes), while the left ear is at a "loud" antinode (where there are larger pressure changes) between two nodes.

You can also produce such waves and interference on the surface of water by rhythmically poking a pair of fingers into the water, for example.

The radiating interference pattern of Figure 3-7, often called two-point-source interference pattern, is very famous in physics. The reason is that it provides a very distinctive signature of wave behavior: only waves can produce such a pattern. It is therefore an important test for other waves as well, such as waves on water, in solids, electromagnetic waves, quantum waves, *etc*. It is closely connected with the Young's double-slit experiment, which we will discuss in Section 12.9 and which was central in developing the wave theory of light as well as the quantum theory of matter.

How does the two-point-source interference pattern of Figures 3-6 and 3-7 arise? The mechanism causing this interference pattern is better seen in the blow-up of Figure 3-7. Here you can recognize the individual spherical waves created at each of the two sources. Near their source, the waves dominate like a volcano over the waves coming from the other source. As they radiate out, the waves weaken and become more comparable to each other: then the interference becomes more "competitive" in the form of peaks and dips. For example, follow the ridge of a wave leaving the left source (marked as a white dashed circle): where it crosses the similar white-dashed ridge from the other source, we see multiple peaks due to constructive interference (ridge + ridge = higher peak). Likewise, follow troughs (black dashed circles): where they cross, we see multiple dips (trough + trough = deeper dip).

On the other hand, where ridges meet troughs, we find yellow/green areas due to destructive interference (ridge + trough = destruction): these are **nodal planes**, some of which are marked as dotted black lines in Figure 3-7. Nodal planes are similar to nodes on a string; these correspond to the nodes ("nodal points") that we discussed many times on strings. In two dimensions, the nodes become lines; in three dimensions, the nodes become planes (the nodal lines need not be straight, and the nodal planes need not be flat). Note that we rarely get perfect destruction: the reason is that the heights of the ridges and depths of the troughs are rarely equal (due to decreasing amplitudes away from the sources and different distances from the two sources), so they rarely add up to exactly zero.

What happens if we add more sources? Instead of two identical wave sources, as used in Figures 3-6 and 3-7, let's use seven aligned identical sources with equal spacing, as is done at the left in Figure 3-8. The result is roughly similar to the case of two sources (Figure 3-6), but exhibits some new features marked by arrows. These new features are almost straight wave fronts that look like three "beams" pointing along those arrows: they are indeed almost plane waves (described in Section 3.5, see Figure 3-2). **So, we have here a method of forming strong beams of sound in particular directions.**

The reason for one beam going straight up in Figure 3-8 is the following: in that direction, the crests of all seven spherical waves closely coincide (and their troughs also closely coincide), so the seven waves reinforce each other. This goes at the cost of other directions where peaks and troughs destroy each other: these are the greener directions, showing weak amplitudes there. However, we see exceptions: the two other arrows, to the right and left, also mark plane waves; how do they come about? This is an interesting situation: in those two other directions, the peaks coming from neighboring sources again coincide, but they have slipped by a full cycle! Namely, one source's crest

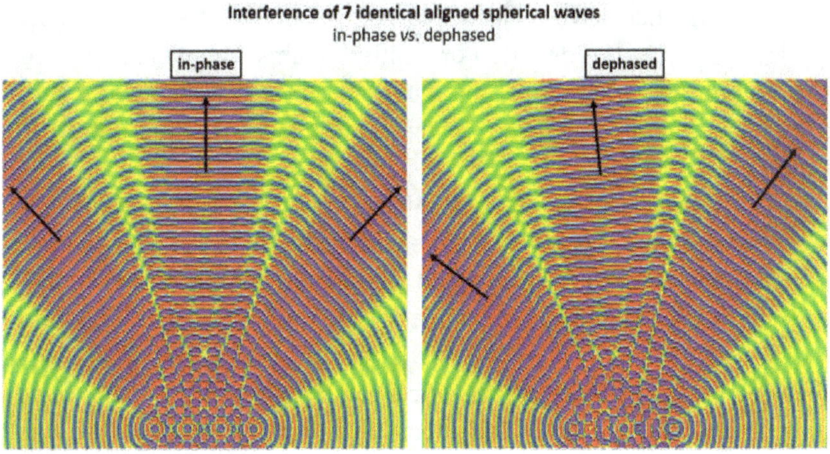

Figure 3-8: Seven identical spherical waves come from seven aligned, identical sources (located near the bottom of each graph). The presentation is similar to Figure 3-6, but viewed "vertically", like a map. The right-hand graph shows the effect of a small time delay in successive sources: sources to the right "fire" a bit earlier than sources to the left, so they are dephased, which rotates the "beams" slightly to the left as they radiate out.

coincides with the following crest of the neighboring source. The result is the same: constructive interference.

Next, we play another trick: ***What happens if we have delays between the waves leaving the different sources?*** Suppose that the rightmost source "fires" its wave a bit before the next source to its left; the following source is also delayed relative to the second, and so on down the line of seven sources (you often see such delayed firing in fireworks, where rockets follow each other after short delays). Therefore, waves coming from the right-hand sources are a bit ahead of those to their left. This gives the result seen at the right in Figure 3-8: **the interference pattern has rotated** toward the left! The three beams marked with arrows have also rotated toward the left. We can increase the rotation angle by simply increasing the delay time in the firing of the sources.

The technical term for such a system of delayed sources is a **phased array**.[9] It allows turning the direction of emitted beams without rotating

[9] See: https://en.wikipedia.org/wiki/Phased_array

the emitters themselves: electronic control of these emitters allows quickly modifying the direction of emitted beams. The words "phased" and "dephased" refer to the time delays: a delay in firing a wave shifts it from being in-phase toward being in antiphase and then onward to being in-phase again.

An important application of this idea is medical **ultrasound imaging**: it uses sounds with such high frequencies that we don't hear them and allows rapid and wide scanning of organs within the body. Another application is **phased-array radar**, mainly used by the military. It operates not with sound but with electromagnetic waves, using exactly the same principle: instead of rotating the whole radar antenna round and round, as is the normal method of using radars, the radar beams can be swept left and right extremely rapidly by electronic control, without rotating the radar antenna at all. Such radars can even be permanently built into a wall instead of standing on a separate tower, and this can even be done on a ship, despite its variable orientation as it turns or swings on waves.

Can we shape sound differently with more sources? Indeed, using more loudspeakers, we can produce many wave patterns, somewhat similar to the **surround sound** used in movie theaters, rooms, cars, *etc.* Figure 3-9 shows a few simple examples using a ring of 36 loudspeakers. The strength of each loudspeaker is varied, as well as its dephasing relative to its neighbors (similar to the dephasing used at the right in Figure 3-8). It is possible to focus sound in one small spot (left graphs), or to create a near-silent zone (right graphs), whether in the center of the ring (top graphs) or away from the center (bottom graphs).

The focusing in one spot is spectacularly demonstrated with converging water waves in a video online.[10] Again, wave superposition is used, here with water surface waves: a circle of many "paddles" around a water tank sends little waves simultaneously to the center of the tank, creating a water fountain almost 30 meters high. Delaying the motion of successive paddles and changing their relative strengths also make it possible to shape almost any desired water wave pattern in the tank.

---

[10] See "90 ft. Vertical Spike Wave in Slow Mo" by The Slow Mo Guys: https://www.youtube.com/watch?v=iWKFPTgkpXo&t=3s

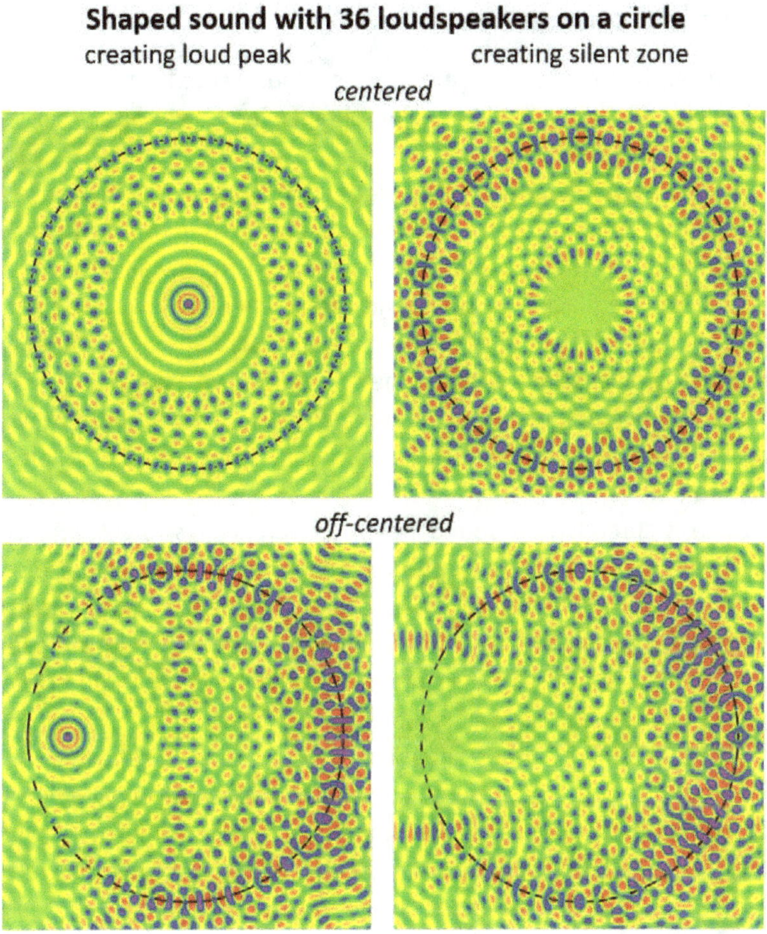

Figure 3-9: A ring of 36 loudspeakers (equally spaced along the black circle), each producing spherical waves. Depending on the strength and phase of each loudspeaker, interference creates different wave patterns. *Top left*: a strong peak is created at the center of the ring. *Top right*: a silent zone is created near the center. *Bottom left*: a strong peak is created near the edge of the ring. *Bottom right*: a near-silent zone is created across the edge of the ring.

Instead of placing loudspeakers in a circle, they could also be arranged in any other shape, such as along the sides of a room. By adjusting their phases and strengths, any wave pattern can still be generated. However, walls and objects will add reflections that can change the resulting pattern; in principle, they can be compensated for, although they do complicate the design.

3.8 Reflection of Sound Waves

We have all heard sounds (voices, music, noise, *etc.*) being **reflected** from walls, windows, buildings, mountains, *etc.*: we often call the reflected sounds **echoes**.

How do sounds travel from room to room inside a house or building? For comparison, try to throw a ball from one room to another room, with open doors: a ball bounces around from wall to wall, but only goes to one place at a time, and rarely where you were hoping it would go! Sound is also reflected by the walls, but it spreads around and goes to many places and rooms at the same time: we have seen (Section 3.5) that sound radiates in all directions, unlike a ball. Thereby, your voice can reach anyone in the house, wherever they may be, as long as doors are open.

Once it spreads out, sound travels generally in straight lines (but we will see an interesting exception in Section 3.9: sound can also turn around corners!). When sound reaches an obstacle like a wall or other solid or liquid substance, it will be reflected. (We know and will see in Chapter 9 that sound can also penetrate and travel through solid or liquid substances, such as walls, although less easily.)

Can you think of other situations where sound is reflected? One example is our ear: sound is reflected by the external part of our ear and channeled into the inner ear. Another example is a horn, in which sound is channeled from a loudspeaker through a tube into a preferred direction. Sound is also reflected inside musical instruments and inside our mouths, as we will discuss in Chapters 5 and 6, respectively.

You may know that a flat "planar" sound wave, like the wave shown in Figure 3-2, is reflected in a single direction when it hits a flat surface, and that this direction depends simply on the incoming direction. Figure 3-10 illustrates this. The situation is very much like that of light reflected by a mirror. We call this **specular reflection**: the reflected light or sound makes the same angle with the mirror or wall as the incoming light or sound (we will explain the reason for this same angle further below).

In Figure 3-10, in the upper left graph, the mirror (or wall) lies horizontally: see the gray line. We look horizontally along that mirror as a wave beam approaches from top left. It is reflected at the same

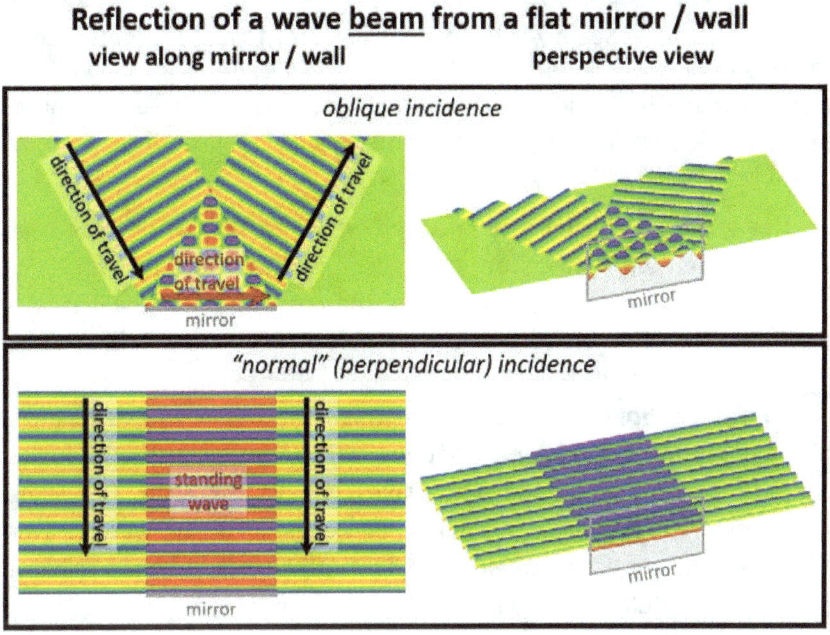

Figure 3-10: A beam coming from above is reflected into a beam going back up, by a mirror (gray) at the bottom of each sketch. At the left are "map" views looking along the mirror; at the right are perspective views looking through the mirror, like a transparent window. *Upper graphs*: the wave arrives in an oblique direction; the reflected beam causes a "checkerboard" pattern due to interference between the incoming and reflected waves; while the beam edges are fixed in space, the pattern moves continuously to the right along the mirror, as indicated by the red arrow. *Lower graphs*: the beam arrives perpendicularly to the mirror; the reflected beam goes back in the opposite direction, and the interference with the incoming wave creates a standing wave pattern, which does not move. Note how similar these waves are to those in Figure 3-4.

angle relative to the mirror. In the upper-right graph, we tilt that image backward, so we look at the wave and mirror from behind the mirror. We make the mirror transparent (like a window which is not transparent to the sound), to better see the beam approaching it. If the beam arrives perpendicularly to the mirror (see the bottom pair of graphs), the beam also leaves the mirror perpendicularly. You may view this wave motion in my Animation 3*7, which also illustrates a different mirror geometry.

ANIMATION 3*7 — See my video WA2 at times 7:24 and 7:55 in its section "**Reflection of waves in air**" under the titles "**Reflection of plane waves — virtual source behind wall/mirror**" and "**Reflection of plane waves — another example**". (See details in the section References and Resources below.)

Note the great similarity between Figure 3-10 and Figure 3-4: as we have seen with waves on strings (see, for example, Figure 2-18), reflection of a wave is very much like the crossing of two waves. We again see the checkerboard/egg-crate pattern and the standing wave pattern in the case of reflection (Figure 3-4).

However, how do we know all this? Of course, we can simply observe nature and accept what we see without further thought. But let's try to understand why waves behave this way. The same explanation will be valid for all kinds of waves, including water waves and electromagnetic waves such as light, *etc.*, so it is worth thinking about this carefully, step by step.

To show that the result is not limited to narrow beams and to make it more intuitive, let's start now with a short wave packet (instead of a beam), as shown in Figure 3-11. The wave packet has four ups and downs (just like the wave shown in Figure 3-2). Since our mirror is transparent, we see the "footprint" of the wave through the mirror from below: the footprint is the black wavy curve in the right-hand graphs (its black "wavelength" is different from the wavelength of the wave away from the mirror, shown with white arrows). The footprint slides continuously to the right along the mirror as the incoming wave advances.

The central argument of our explanation is that the footprint of the incoming wave must also be the footprint of the outgoing (reflected) wave. The reason is that the incoming wave is in fact the source of the reflected wave, and this is true at every point along the mirror's surface. Therefore, the outgoing (reflected) wave must have exactly the same footprint as the incoming wave. In other words, the reflected wave must match the same black wavy curve. This is shown in the middle of Figure 3-11, where only the reflected wave is drawn, but it is "attached" to the same footprint serving as its source.

There is one more important point: the wavelength after the reflection must be the same as before the reflection because the

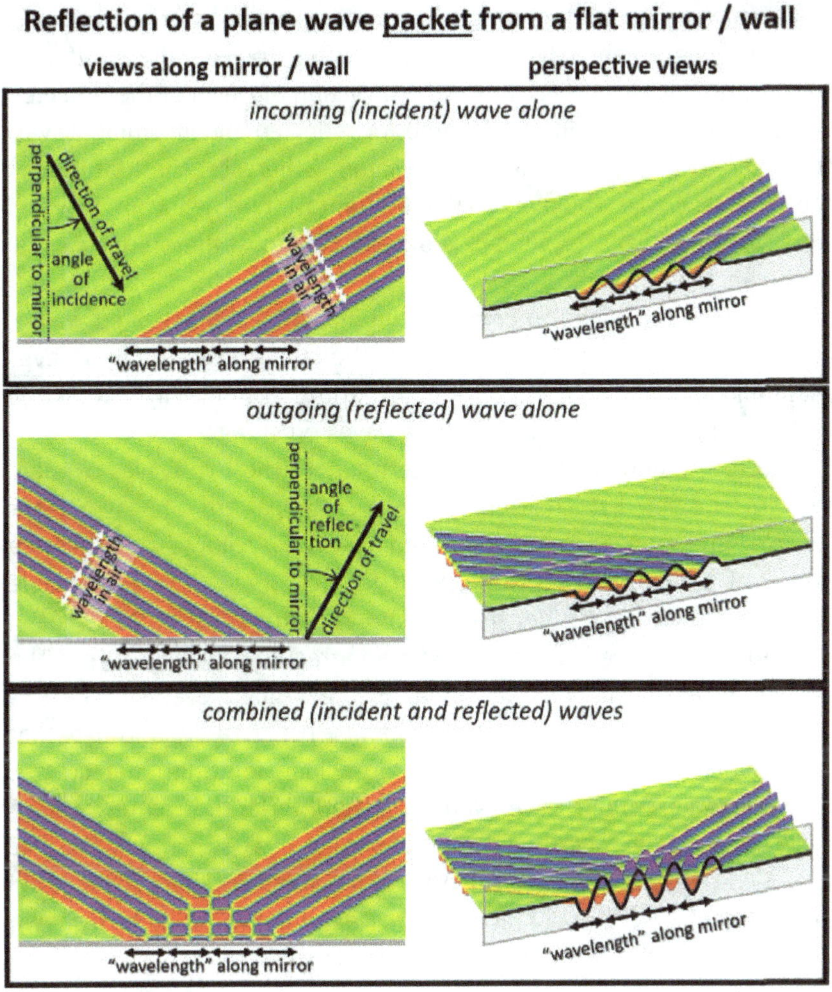

Figure 3-11: A plane wave packet coming from the upper left is reflected by a mirror (gray) at the bottom of each graph. At the left are "map" views looking along the mirror; at the right are perspective views looking through the mirror, like a transparent window. The packet has four up-and-down oscillations shown in purple-to-red; the packet is extended as weak yellow-green waves for comparison with the full wave train shown in Figure 3-12. The small double-headed arrows show the sound's wavelength in air (white) and along the mirror (black). *Upper graphs*: only the incoming packet is drawn; the black wavy line at the right shows the packet's "footprint" on the mirror. *Middle graphs*: only the reflected packet is drawn, ignoring the incoming packet; it has the same "footprint" as the incoming packet. *Lower graphs*: both the incoming and reflected packets are drawn, including their interference in their region of overlap; they still have the same "footprint" on the mirror, but have doubled in amplitude.

wave still travels in the same environment (such as air for sound); the wavelength can only change if the wave travels into a different substance. This requirement fixes the direction of travel of the reflected wave: since the footprint is the same for reflected and incoming waves, the angle of travel must also be the same. Otherwise, the footprint of the reflected wave would not match the footprint of the incoming wave. As drawn in Figure 3-11, the angle of reflection must therefore be equal to the angle of incidence.[11]

Now we only need to combine the incoming and reflected waves, as is done at the bottom of Figure 3-11. We "superpose" the two waves by simple addition, as we have done in various examples in Section 3.5, and in Chapter 2 for the reflection of waves on strings.

Note the new interference pattern where the two waves overlap. We get the familiar checkerboard/egg-crate pattern of wave peaks (purple) and troughs (red), separated by green/yellow strips that are nodes of zero oscillation. Before we further discuss this remarkable wave pattern, let's consider the reflection of a full wave train.

How is a wave train reflected compared to a wave packet? The answer is again simple: the same, just extended everywhere. This is shown in Figure 3-12: here, compared to Figure 3-11, the packet is extended throughout space, forming a long footprint that creates an extended reflected wave. Naturally, the interference pattern is also extended (see the lowest pair of graphs in Figure 3-12): in fact, it exists everywhere, so we can't see the original incoming wave or the reflected wave separately (you may recognize the extended wave pattern shown very weakly in Figure 3-11 for direct comparison).

The egg-crate-like or checkerboard-like wave pattern at the bottom of Figure 3-12 is the characteristic and universal interference pattern of two identical waves traveling in two different directions. Changing the angle between those two directions only changes the side

---

[11] More precisely: the wavelength in air, shown as short white double-headed arrows in Figure 3-11, must be the same in the reflected wave as in the incoming wave; but the "projected wavelength" shown as small black arrows along the footprint must also be the same for both waves. These two requirements uniquely determine the reflection angle to be identical to the incidence angle: no other reflection angle will satisfy those requirements.

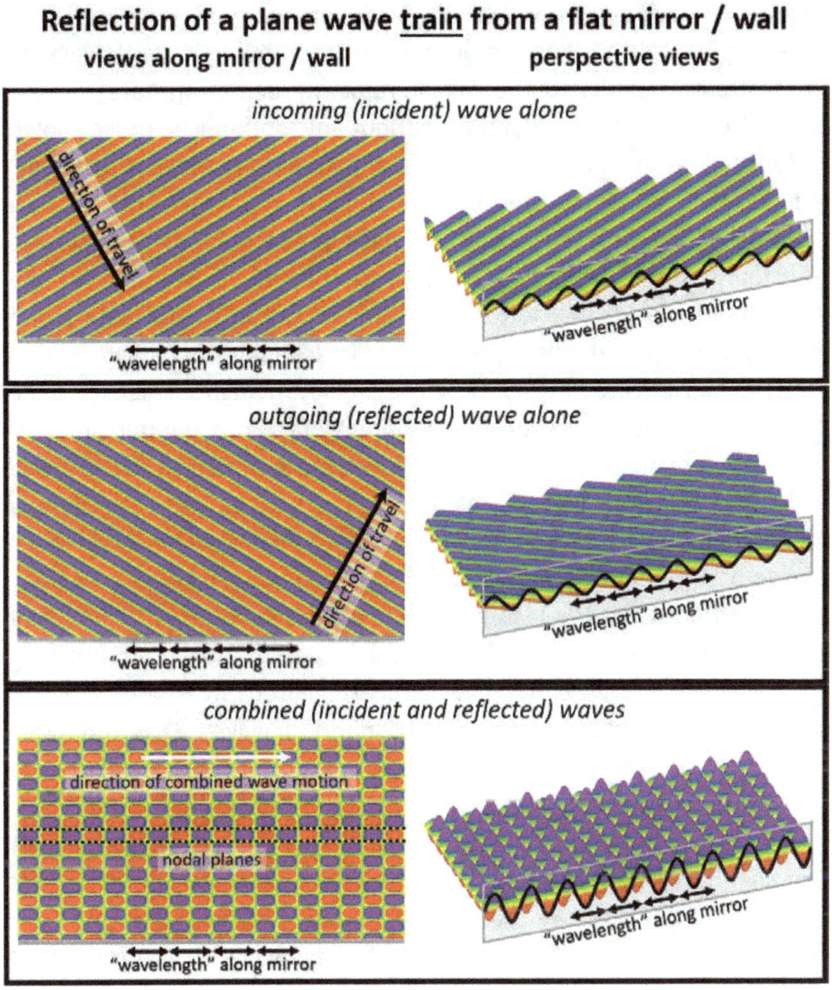

Figure 3-12: As Figure 3-11, but for a plane wave train instead of a plane wave packet: the packet of Figure 3-11 is here extended everywhere. All views here are identical to those in Figure 3-11, for direct comparison.

lengths of its rectangular "cells" (compare Figures 3-10 and 3-12, for example).

How does this interference pattern move? The answer is also very simple: the wave patterns at the bottom of Figures 3-11 and 3-12 move rigidly to the right as the wave passes by. You can literally take the image and slide it at constant speed to the right along the mirror: that is what the wave pattern will look like at every moment. It looks exactly like an egg crate that slides smoothly in one direction. In particular, the green streaks parallel to the mirror remain green (some are marked with dotted lines): they are nodes without any motion, like those shown as dotted lines in Figure 3-7. The peaks and valleys between the nodal lines move rigidly along the mirror, like a string wave steadily moving along the string. This is the pattern that we observed in the case of a reflected beam in Figure 3-10: within the beam, we have exactly the same situation as at the bottom of Figure 3-12. And the formation of a standing wave with the perpendicular beam (see bottom of Figure 3-10) is just a special case of this: now the small rectangular "cells" of the checkerboard pattern are stretched infinitely far along the mirror.

I am sure that you have never consciously noticed such a sound wave pattern in your life: the main reason is that our ears are not good at detecting sound patterns in space. However, you may have seen such a wave pattern along a wall in water: a ship passing by creates a bow-wave that looks like the incoming wave packet in Figure 3-11 and that is reflected from a flat wall to produce a similar interference pattern.

You may recognize this checkerboard pattern in Figures 3-6 and 3-7, due to the interference of two spherical waves. It is the same effect, only curved instead of rectangular because of the circular shape of spherical waves.

Another comparison is very illuminating: we can look back at the standing wave formed at the free end of a string, as seen in Figures 2-18 and 2-20. Exactly the same standing wave pattern is present in our mirrored sound wave, in the direction perpendicular to the mirror. To see this, fix your eyes on a spot, say in the middle of the lower graph of Figure 3-11 or 3-12. As the sound wave slides along the mirror across the point you are watching, it has nodes in all the green stripes parallel to the mirror (it does not oscillate there). But it has antinodes midway between those green stripes (it oscillates most there, as shown by the purple peaks and red dips), like a traveling wave; in particular, the footprint travels continuously to the right. We may conclude: **Wave reflection from a mirror is a combination of a standing wave perpendicular to the mirror and a traveling wave along the mirror.**

What if the plane wave hits the mirror straight on? Now the wave fronts are parallel to the mirror. This situation is very much like the case of a wave on a string hitting the free end of the string, just extended laterally along the mirror. The result is now therefore a standing wave in front of the mirror: it is the standing wave seen at the bottom of Figure 3-10, but extended far along the mirror. Seen along the mirror, the wave profile is then just like the string shape shown in Figures 2-18 and 2-20 for the free string end. The standing wave now has long nodal and antinodal lines parallel to the mirror: there is no motion parallel to the mirror and no motion perpendicular to the mirror. We will encounter this situation again in tubes and boxes in Sections 3.10 to 3.13, where we will see its importance for making music and voices.

What happens when the mirror is rough? Suppose we scratch the mirror or roughen it up in other ways. For example, we could reflect sound from a rough stone wall instead of a flat wall. Then the footprint on the rough mirror becomes very complicated and irregular, as it is no longer two-dimensional: that complex footprint serves as the source of the reflected wave, which then also becomes very complicated and irregular; now, the reflected wave no longer needs to match a regular wave-like footprint. The result is scattering of the wave in all directions: we call this **diffuse scattering** of waves, or **diffuse reflection**, similar to light reflected from paper that is matte (non-glossy).

What matters here is the size of the roughness compared to the wavelength of the sound: if the roughness is larger than the sound's wavelength, the wave is scattered in all directions. The same is true for light: matte paper is rough compared to the wavelength of light (which is a fraction of a micrometer and therefore too small for our eyes to see); glossy paper is smooth on that scale and therefore reflects light specularly, giving shiny reflections.

What happens when the mirror or wall is "leaky", meaning partly transparent? This happens a lot when sound hits doors and walls: part of the sound penetrates the door or wall, with the rest being reflected back into the air. We will discuss this interesting situation in Chapter 10, since we first need to learn a bit more about how sound behaves in solid substances.

How is a spherical wave reflected by a flat wall or mirror? We can use our knowledge of the plane-wave reflection to understand what happens with a spherical wave, as illustrated in Figure 3-13. If we focus

Reflection of a spherical wave from a flat mirror / wall

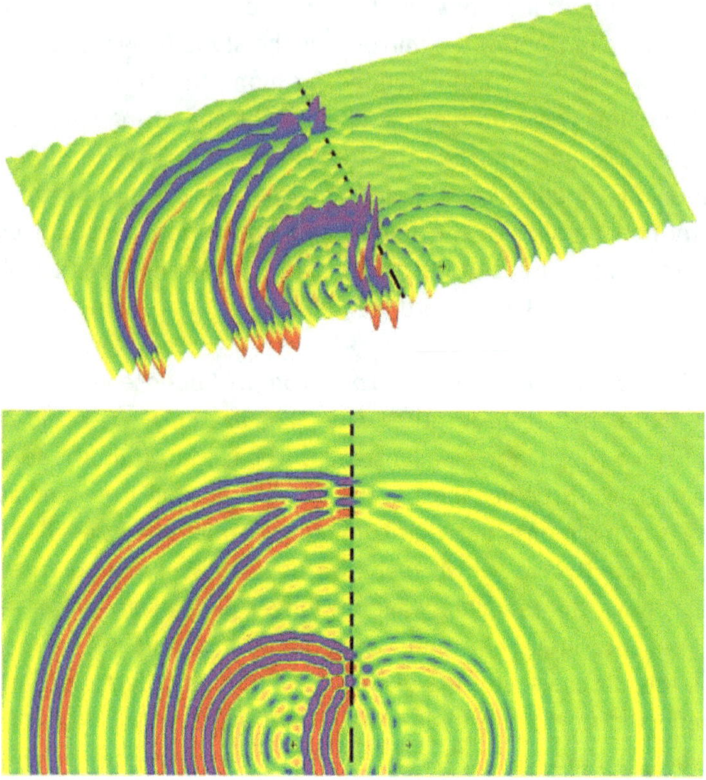

Figure 3-13: Two spherical wave packets starting from just left of the mirror (shown as a black line) are reflected by that mirror, in a perspective view (top graph) and a "map" view (bottom graph). The reflected waves are also spherical and behave as if they started behind the mirror from a "virtual" source (at the righthand + sign). The waves outside the packets and to the right of the mirror are shown weakened, for direct comparison with the two-source system shown in Figure 3-6, which gives the resulting wave for a full wave train. (The "wiggles" on the wave pulses are due to the contribution of these other weak waves.)

on every point along the mirror (shown as a black line), a spherical wave looks locally like a plane wave; it is then reflected "specularly", meaning with a reflection angle equal to the incident angle, just like a plane wave. The total effect of this reflection is remarkable: the reflected wave is also a spherical wave, except that its source is behind the mirror! We have hereby recreated the situation shown at the right in Figure 3-6:

we have two identical sources, now one real (in front of the mirror) and the other virtual (behind the mirror), which together produce the same resulting wave pattern as that shown in Figure 3-6. Thus, by viewing my Animation 3*5 (mentioned above), you will see exactly how a spherical wave is reflected by a flat wall or plane: in the animation the mirror runs up/down through the middle of the image (as it does in Figure 3-13), the right side being entirely virtual.

We are actually very familiar with this situation: a light source (like a lamp) in front of a mirror is itself mirrored as a "virtual" source behind the mirror, so that we can see two sources of light. The reflected spherical wave is very real, even though it comes from a "virtual" source. We can say the same about the reflected plane wave that we discussed above: it also looks like it came from behind the mirror. Similarly, the string waves that we discussed in Section 2.7 are reflected from the end of a string as if they came from a virtual source beyond the end of the string.

Thus: **a flat mirror reflecting a wave creates a "virtual" source mirrored behind the mirror.**

Next, we may wonder about mirrors/walls that are curved. We are familiar with parabolic mirrors used in spotlights, flashlights and searchlights to create a narrow beam of light, and with parabolic antennas used to capture weak television signals. Let's see how they operate.

How is a spherical wave reflected by a parabolic reflector? Figure 3-14 illustrates this case. Let's start a spherical wave containing two pulses at the so-called "focus" of the parabola (a special point near its tip). In the left graphs, we see how this 2-pulse wave packet can travel freely to the right as spherical waves. We also see how this wave packet is reflected from the parabolic mirror: close to each point along the curved parabolic mirror, the spherical wave looks planar, and the mirror also looks straight, so the reflection is "specular" as before. While the free spherical wave spreads out, it weakens with distance. By contrast, the reflected wave constantly gains width as more reflection feeds into it, and being planar instead of spreading, it keeps a constant amplitude. Thus, gradually the weakening spherical wave is converted to a steady plane wave packet traveling to the right. As a result, a full wave train (in the right graphs of Figure 3-14) can gradually build up a wave that looks

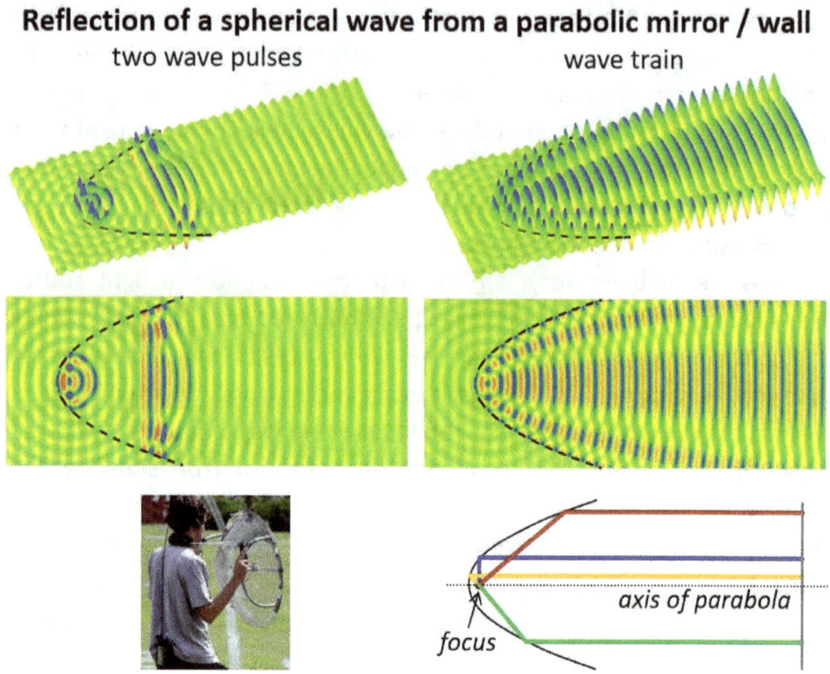

Figure 3-14: A parabolic mirror is shown as a black line in a perspective view (*top*) and a "map" view (*middle*); its "axis" points straight to the right from its "focus", marked with a + sign in the map views, where spherical waves are centered. *Left graphs*: Two spherical wave packets have started from the focus of the parabola; they are reflected specularly from the parabolic mirror. The reflected waves become "planar" (with straight wave crests and troughs) as if they had started as plane waves far behind the mirror at the left. *Right graphs*: As at the left, but for a full wave train that comes from the same point source. The waves outside the packets and to the left of the mirror are shown weakened as if the mirror were slightly transparent, as a guide to the eye. *Bottom left*: Parabolic microphone. (*Source*: J. Glover, under CC BY-SA 3.0, extracted from https://commons.wikimedia.org/wiki/File:ParabolicMicrophone.jpg.) *Bottom right*: Wave paths from the focus point at left; the paths of equal lengths end on the same vertical line (gray).

increasingly planar as it travels farther from the source. You may view the motion of such a wave in my Animation 3*8.

ANIMATION 3*8 — See my video WA2 at time 8:31 in its section "**Reflection of waves in air**" under the title "**Reflection of spherical waves — parabolic mirror**". (See details in the section References and Resources below.)

The critical aspect of the shape of the parabolic reflector is illustrated at the bottom right in Figure 3-14. Let's follow the colored wave paths from the focus point as they reflect specularly from the mirror. These paths all have the same length when they reach any straight line (such as the gray line) perpendicular to the axis of the parabola. This means that wave crests will arrive all along that straight line at the same time, thus forming a plane wave.

What happens when we reverse the situation and send a plane wave into a parabolic reflector? Figure 3-14 is also valid for this situation: just imagine the plane wave coming from right to left into the parabolic reflector along its axis. The plane wave is reflected (specularly) from the parabolic mirror, causing a spherical wave that converges onto the "focus" of the parabola, with all wave crests arriving in the focus at the same time. With a perfectly shaped parabolic mirror, an infinitely wide wave can in principle be concentrated into an infinitely small focus, giving infinite amplification; in practice, parabolic mirrors are limited in size, while the focus has the size of the wavelength, so the concentration of energy is limited. Thereby a weak distant sound coming from a particular direction can be concentrated at the focus of the parabola, while avoiding other sounds (that may be stronger) which come from other directions and are reflected in other directions.

This capability of parabolic mirrors is used in directional microphones, for example to pick out one person's voice in a noisy environment. A parabolic microphone is illustrated in Figure 3-14: these are often used to pick out the voice of a sports player (or coach) far away. Large parabolic mirrors were also used along the coast of southern England before the invention of the radar to detect the sound of distant warplanes coming from continental Europe.[12] Parabolic antennas are widely used with electromagnetic waves, for TV transmission on Earth, for communications with distant satellites and space probes, as well as for the observation of stars and galaxies with optical and radio telescopes.

How is a spherical wave reflected by an elliptical reflector? An ellipse is a special kind of oval: a stretched circle with two "centers" called foci (a parabola is a special kind of ellipse, in which one "center"

[12] See: https://en.wikipedia.org/wiki/Acoustic_mirror

is moved infinitely far away). Figure 3-15 illustrates this case. We see how 2-pulse wave packets coming from one focus of the ellipse are (specularly) reflected by the elliptical mirror. The reflected waves converge onto the second focus of the ellipse. You can view the motion of waves inside an ellipse in my Animation 3*9.

ANIMATION 3*9 — See my video WA2 at time 9:02 in its section "**Reflection of waves in air**" under the title "**Reflection of spherical waves — elliptical mirror**". (See details in the section References and Resources below.)

The reason for this behavior is very similar to that for the parabolic mirror, as illustrated at the bottom of Figure 3-15: all wave paths that go from one focus to the other *via* a specular reflection by the elliptical mirror have the same length, so that all wave crests arrive simultaneously at the other focus.[13] We can also reverse the situation: a spherical wave starting at the second focus will be reflected to the first focus; Figure 3-15 is valid for this case as well. A wave train (shown in the right graphs) behaves the same way as wave pulses and wave packets. In fact, the wave train pattern in Figure 3-15 is very similar to that in Figures 3-6 and 3-7 for two sources. However, there is one important difference: with two sources, as in Figure 3-6, the combined wave radiates out in all directions (as can be seen in my Animation 3*5); by contrast, with one source at the focus of an ellipse, the spherical wave travels toward and converges on the other focus. You may view this "inside" motion in my Animation 3*9.

An elliptical mirror allows a person at one focus to speak normally and be heard by someone else at the other focus as if they were standing mouth-to-ear, even if there are other sounds inside the elliptical room.

There is an interesting twist to the elliptical mirror: we can ask what happens to the sound that has been focused on the other focus point.

[13] You may know the following trick for drawing an ellipse, based on the same principle: take a string of fixed length, attach its two ends to two points on a page, for example with two pins stuck in the page (these will become the focus points of the ellipse); with a pencil, stretch the string to form a triangle; pull the pencil around the two focus points, keeping the string stretched; the pencil will then draw an ellipse (and the string will form a specular reflection path from that ellipse).

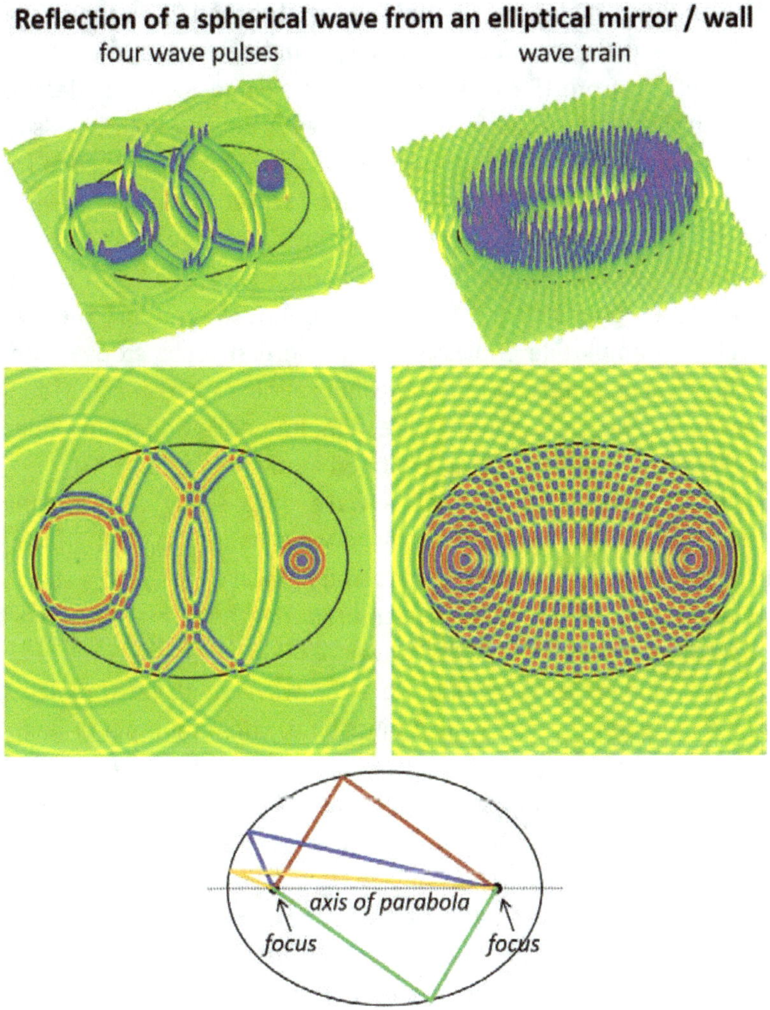

Figure 3-15: An elliptical mirror is shown as a black line in a perspective view (*top*) and a "map" view (*middle*). *Left graphs*: Four spherical wave packets start at the left from one "focus" of the ellipse (one of the packets has gone beyond the ellipse at the far right: we see its reflection closing in on the right focus). Specular reflection takes place at the mirror. The reflected waves are also spherical but converge onto the second "focus" of the ellipse, at right. *Right graphs*: same for a full wave train. The waves outside the packets are shown weakened, as are all waves outside the mirror, as if the mirror were slightly transparent: this serves as a guide to the eye and for direct comparison with the two-source system shown in Figure 3-6. *Bottom*: wave paths from one focus point to the other; the paths of different colors have equal lengths.

If the sound arriving at the second focus is absorbed, it disappears, and if the source at the first focus continues emitting, we get a continuous wave traveling from the first focus to the second focus, as described above. But if nothing absorbs the sound at the second focus, the sound passes through that point and radiates out again from the second focus as a new outgoing spherical wave. It is then focused again by the elliptical mirror, but back onto the starting point, giving a clear echo there! And this will repeat itself: the sound can bounce back and forth between the two focus points. If the first focus continues emitting, the waves will build up amplitude endlessly. But if the first focus stops emitting, the whole wave pattern within the elliptical mirror becomes a standing wave: there is no motion in any direction; immobile nodal lines (planes in 3D) "freeze" that pattern, as exhibited in my Animation 3*9!

This situation reminds us of a wave on a string with two fixed ends: as we saw in Section 2.8, the wave's bouncing back and forth between the two ends creates a standing wave, assuming that the wavelength and string's length match properly (the string's length must be a whole multiple of half the wavelength). The elliptical mirror is a 2D or 3D analogue of the string. Here the wavelength must match the distance between the foci (which must be a whole multiple of half the wavelength), so that the reflected waves do not cancel out the initial wave. When this condition is satisfied, we can get a standing wave in 2D or 3D, with immobile nodes. You may also view this immobile situation in my Animation 3*9.

(The above discussion assumes a 3D room that is elliptical not only in the horizontal direction, but also in the vertical direction, forming an ellipsoid, similar to the shape of an American football but less pointed; an elliptical room with vertical walls, such as the Oval Office in the US White House, will also form echoes, but they will be weaker and they will decay over time faster because the "focus" of the reflected wave will spread out vertically and partly cancel itself out.)

3.9 Can Waves Turn Around Corners? Diffraction

Imagine standing near the corner of a building. Your friend is around the corner. ***Can you throw objects at your friend around the corner?*** **No**, unless there are other walls or obstacles from which you could bounce those objects, or unless you throw objects with special aerodynamic effects such as paper planes, frisbees, spinning balls, boomerangs, motorized helicopters, *etc.* Normal objects do not turn horizontally (but they do curve down due to gravity), so they can't turn horizontally around obstacles by themselves.

Can you talk to your friend around the corner? **Yes!** And you can also talk to someone who is hiding behind a thick tree, well protected from thrown objects. These facts alone prove that **sound waves can turn around obstacles**. **We call this capability diffraction**. **This capability of sound waves exists also for all other waves**. It is another very important general property of waves of all sorts, such as waves on water, sound in solids and liquids, electromagnetic waves, *etc.*

We may describe diffraction simply as the spreading out of a wave. Familiar situations are sound coming out of our mouth, or out of a room, or out of a wind instrument: when sound finds new directions to expand in, it will do so. An analogy is the expansion of a gas into new space: this can be viewed as the expansion of a wave pulse, which indeed can turn around corners.

Figure 3-16 illustrates how a sound wave turns around a corner. We consider a plane wave (which is a spherical wave extremely far from its source). It travels from left to right along a wall and approaches a corner in that wall: for better visibility, we make the two walls transparent like windows (see bottom sketch in Figure 3-16). Let's think of what happens at the corner. There, as the wave comes from the left, it can suddenly "escape" around the corner: the pressure that builds up at the corner can indeed push air around the corner, starting a "new" wave that radiates away into the open space around the corner. (Not shown in Figure 3-16 is that the corner also causes ripples in the wave that continues straight through: the spherical wave coming from the corner spreads out in all possible directions. We will see this more clearly and in more detail in Section 12.9 with electromagnetic waves.)

The situation is very much like that near the source of the spherical wave in Figure 3-1: the wave will try to go in all available directions. We can say that **the corner point acts like the source of a new wave: this is again the Huygens principle** that we mentioned in Section 2.2. Moreover, according to that principle, all points in the plane wave away from the corner also act as sources of new waves.

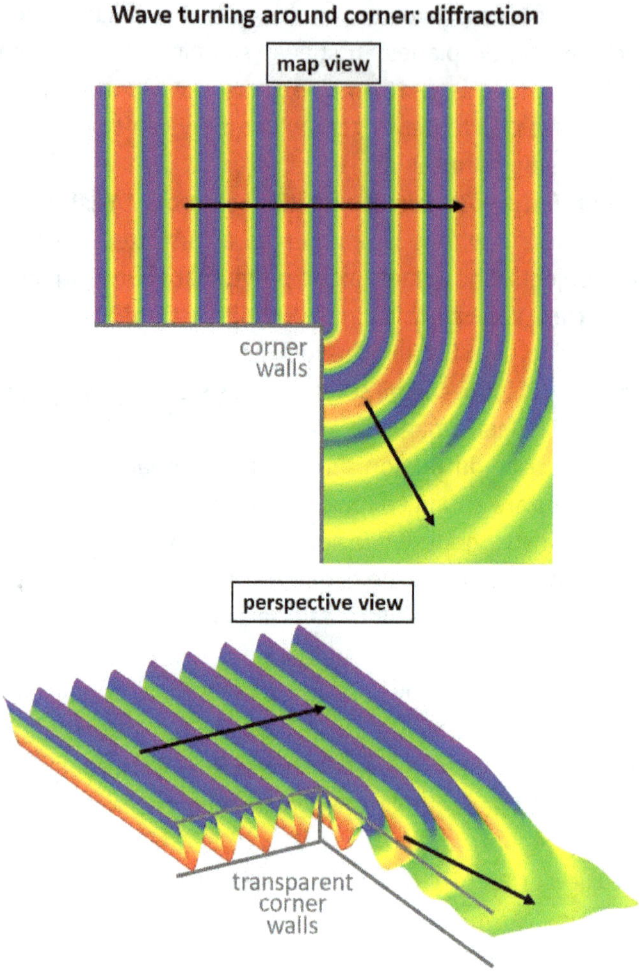

Figure 3-16: A plane wave curves around a transparent solid corner, like a corner made of two windows. The corner acts like a source of spherical waves in the "shadow" region. At the top is a map view, and at the bottom is a perspective view, displayed as in Figure 3-1.

The wave around the corner has spherical wave fronts, but not spherical amplitudes. The wave fronts indeed form spheres around the corner point because the wavelength and speed of the wave remain the same as for the plane wave. But the amplitude of these wave fronts is more complex. The reason for this is simple: it is not only the corner point that feeds energy into this spherical wave, but also the plane wave that continues straight past the corner; the plane wave continues to feed energy into the spherical wave. In addition, all spherical waves weaken as they spread out. As a result, we get a spherical wave that weakens as it travels farther from the plane wave: see Figure 3-16. Your friend behind the corner will hear you less loudly than someone straight ahead!

Turning around corners, or diffraction, is a very general, characteristic and important behavior of waves. Diffraction indeed happens whenever a wave encounters a disturbance. A common example is an obstacle in the wave's path, such as a tree or other solid object: the wave can then spread out in all available directions, because the obstacle itself serves as a new source of the continuing wave, as we saw here with a corner.

The situations depicted in Figures 3-6 to 3-8 can also be viewed as diffraction, if we assume that the "sources" drawn there are actually little objects that scatter a sound wave which approaches from below the image: in the spirit of the Huygens principle, the objects then become sources of the "new" spherical waves that are drawn in those figures. Similarly, the two sources in Figures 3-6 and 3-7 can be viewed as two small holes in a screen that block a plane wave coming from below; we will describe this situation in Section 12.9 for the case of light waves.

3.10 Waves Inside Long Tubes

Musical instruments using waves in air — called **wind instruments** — mostly contain the air in **tubes**. Examples are trumpets, horns, recorders, flutes, clarinets, and saxophones. The tubes may be curved to save space, but they can be viewed as being straight from the point of view of sound waves traveling through them. Many such tubes have a constant cross-section, like a straight cylinder: we will thus assume a straight cylindrical shape here because it already exhibits all the important properties of sound waves in wind instruments, while being simpler to think about than curved or conical tubes.

How do waves travel in long tubes of air? Figure 3-2 shows a plane wave in free space. That shape of wave fits perfectly in the air of a straight cylindrical tube, often also called an **air column**, as long as the wave's travel direction is parallel to the tube's direction. We can create a wave in a tube by using a simple piston that pushes air a bit along the inside of the tube: that compresses the air in front of the piston, which causes a pulse of compressed air to travel down the tube. If the piston pulls back, a pulse of decompression is created, which also travels along the tube, followed by another compressed pulse when the piston pushes forward again. We see that the motion of the air is along the tube direction, which is also along the direction of travel of the wave, as shown in Figure 3-2: we indeed create a **longitudinal wave**.

The mechanism of this movement is exactly the same as with the spherical wave shown in Figure 3-1; the only difference is that now the wave cannot spread out but is limited to the interior of the straight tube, much like a wave on a string is limited to staying on the string: we can therefore speak of a 1D sound wave in air. As a result, **the wave simply travels along the tube with constant amplitude** (assuming no losses due to friction or sound production, for example), as sketched in Figure 3-2.

This feature has been extensively used to carry voices over long distances, for example on ships to send instructions from the bridge to the machine room: before the telephone, many captains would tell the machine room to change the power of the engines by talking through a long **speaking tube** or **voice pipe**; many such tubes are still in place on ships, especially military ships, as a backup or as a more secure method of communication.

3.11 Waves Inside Tubes with Open and Closed Ends

Tubes in musical wind instruments have a limited length: their ends can be open or closed; this corresponds to the free or fixed ends of strings, respectively. As we saw for strings in Chapter 2, reflection of waves from the ends can cause standing waves: these are central to making music.

The same ideas apply also to sound waves in tubes of limited length. But there is one important difference: in string instruments, the strings are normally fixed at both ends, while in wind instruments, open ends are common. Wind instruments thus offer more variety: a tube can be open at both ends, or closed at both ends, or open at one end and closed at the other end. And that does not count the option of making holes in the side of the tube, a very common case that offers even more possibilities because openings act like open ends, even when small!

In Section 2.7, we saw how waves are reflected from the ends of strings with free and fixed ends. Reflection from a free end caused a standing wave with an antinode at the free end, while a fixed end caused a standing wave with a node at the fixed end: see Figure 2-20.

How are waves in air tubes reflected from their open and closed ends? The case of the closed end is straightforward: a compressed wave pulse approaching a closed end (a wall blocking its passage) will simply build up air pressure against that wall. That overpressure has only one way to go — back along the tube in the opposite direction — and overpressure implies a compressed pulse that will travel backward into the tube. This is similar to a pulse on a string with a free end (we called its reflected pulse "in phase").

A moment later, the overpressure at the closed end of the tube becomes underpressure, so we have an antinode in terms of pressure (and density): maximum variation. However, in terms of air motion, the closed end blocks motion along the direction of the tube, which means a node in motion. **In general, with air, we get nodes of motion at antinodes of density or pressure, and *vice versa*.**

If, however, the pulse reaches an open end of the tube, it can fan out in all directions into the surrounding air. Sound waves have a high speed (the speed of sound, under typical conditions on Earth, is around 343 meters/second, or 1235 km/h, or 767 mph, which is also called Mach 1). Therefore, a pulse of compressed air coming out of a tube will push its way into the surrounding open air and leave a decompression where it left the tube. This decompression will be filled mainly by air from inside the tube, thereby starting a decompression pulse going backward into the tube: thus, **a compressed pulse is reflected back into the tube as a decompression pulse** (while a part of the pulse escapes into the surrounding air). The same is true of a decompression pulse reaching the open end of the tube: **a decompressed pulse will be partly reflected as a compression pulse back into the tube**.

We therefore have the opposite behavior compared to waves on strings: on strings, the fixed end reversed the sense of the wave (we called it "in antiphase"), while the free end kept it ("in phase"), see Figure 2-16.

A slight complication with open ends is that the point where reflection occurs is not simply the center point of the tube's opening (unlike the end of a string which is sharply defined): because of the complex interaction between the surrounding air and the pulse leaving the tube through its end, it turns out that the reflection point is farther out in the external air, namely about one quarter to one half (depending on other geometrical factors) of the tube's diameter outside the tube's end. It is as if the tube was actually that much longer!

It is very surprising and interesting that the reflection of a wave back into the open end of a tube is very strong (at least for tubes that are narrower than the sound's wavelength): only a small part of the pulse or wave is lost into the outside air, even when the tube is wide open! The outside air acts like a very bouncy cushion. This implies that a wave traveling inside a tube with one or two open ends can be reflected back and forth multiple times and build up a strong standing wave between its two ends: that is crucially important for producing music. We will therefore discuss standing waves in tubes next.

3.12 Standing Waves in Tubes

We have observed much similarity between waves on strings and waves in tubes, but also some differences in how they behave when reflected by the ends of tubes *versus* strings (see the last section). This suggests that **standing waves** in tubes will also show similarities and differences compared to strings.

As mentioned in Section 3.11, tubes can have either open or closed ends (we will not discuss the additional possibility of holes in the sides of tubes, like those in flutes and other wind instruments; their role, in simple terms, is to shorten the tubes, thereby favoring shorter wavelengths and thus higher frequencies). In string instruments, normally only fixed ends exist. New options exist in air: tubes with two open ends, tubes with one open end and one closed end, and tubes with two closed ends.

In Section 2.8, we saw that the fixed ends of strings imply nodes at those ends: this means that the strings cannot move at those ends. In tubes, the situation is a bit more varied, as we saw in Section 3.11: at closed ends, we have nodes of air motion but antinodes of density and pressure. At open ends of tubes, however, we saw that we must have antinodes of motion, but nodes of density and pressure, since the air inside and outside the tube must equalize the density and pressure there.

We now have limitations on the motion of waves at both ends of a tube: either nodes (at closed ends) or antinodes (at open ends). These limitations restrict how a wave can fit within the length of a tube: Figure 3-17 shows several resulting possibilities. **It is interesting to see the variety of situations that can arise with waves in tubes, even without considering other holes in the tubes!** More details of those possibilities are explained in Box 3-2. An online video[14] also explains and illustrates (very spectacularly with a burning **Rubens tube**) how such standing waves arise. You may also view my Animation 3*10 of Figure 3-17, and read more details on Figure 3-17 in Box 3-2.

ANIMATION 3*10 — See my video WA2 at time 10:11 in its section "**Standing waves in tubes and boxes**" under the title "**Standing waves in tubes — comparison with waves on strings**". (See details in the section References and Resources below.)

BOX 3-2 — FURTHER EXPLANATIONS OF FIGURE 3-17: The upper row of sketches shows a tube with two open ends: waves are fit within that tube in such a way that they have a node of density and motion at both ends (shown by green/yellow colors), starting from the longest wavelength possible. This can only be done when the tube length is equal to a whole multiple of half a wavelength; so, at the left, the tube length is half a wavelength (shown by a light blue curve), while the sketch to its right has a tube length equal to two halves of a wavelength, which is one wavelength; the next sketches have tube lengths equal to three and four half wavelengths, respectively.

The lower row of sketches in Figure 3-17 shows a tube with two closed ends: here, waves are fit within that tube such that they have an antinode

(Continued)

[14] See video "A better description of resonance" by Steve Mould: https://www.youtube.com/watch?v=dihQuwrf9yQ; the Rubens tube has holes along its side, but they are made small and dense enough that they do not significantly change the behavior of the standing waves inside the tube, unlike, for example, the holes in flutes and other wind instruments.

(*Continued*)

in density and pressure at both ends (shown in purple for crests and red for troughs). This again can only be achieved if the tube length is equal to a whole multiple of half a wavelength, as with the open ends. Nodes and antinodes are interchanged between the cases with open ends *versus* closed ends.

However, the mixed case of both open and closed ends is a bit more complex, as shown in the middle row of Figure 3-17. Now we must have a node at one end and an antinode at the other end. This leads to the rule that the tube length must be an odd number of quarter wavelengths. Thus, at the left, we have a quarter (1/4) wavelength fitting inside the tube. The three situations to its right have 3/4, 5/4 and 7/4 wavelengths, respectively, within the tube.

One aspect that we mentioned previously changes the above results slightly: the effective length of a tube with an open end is somewhat larger than the actual length of the tube itself. This means that in the above comparisons between tube lengths and wavelengths, we should artificially increase the tube length by about one quarter to one half of the tubes' diameter for proper fitting of the waves within the tubes.

You can also make such standing waves on the surface of water by yourself in a dish, sink or bathtub, all of which have closed ends: you may use a flat object like your hand, a dinner plate, or a spatula to rhythmically push water back and forth; when you find the right frequency, a standing wave should build up and can easily spill out!

How does air move in a standing wave? This question can be easily answered by considering that **air moves from regions of high density and pressure to regions of low density and pressure, simply driven by the pressure difference**. That is also the principle for wind blowing in the atmosphere: a large atmospheric pressure difference causes a strong wind in the direction from high to low pressure areas.

The air flows that result from density and pressure differences are sketched as arrows in Figure 3-18, where snapshots taken at three successive times are shown from left to right for tubes with both ends open (upper graphs) and tubes with both ends closed (lower graphs). We start with the left snapshots: in the regions with the highest and lowest densities and pressures ("highs" and "lows" in weather language, shown again in colors from purple to red), there is no air motion. But air is moving from highs to lows (from purple toward red). Because of that

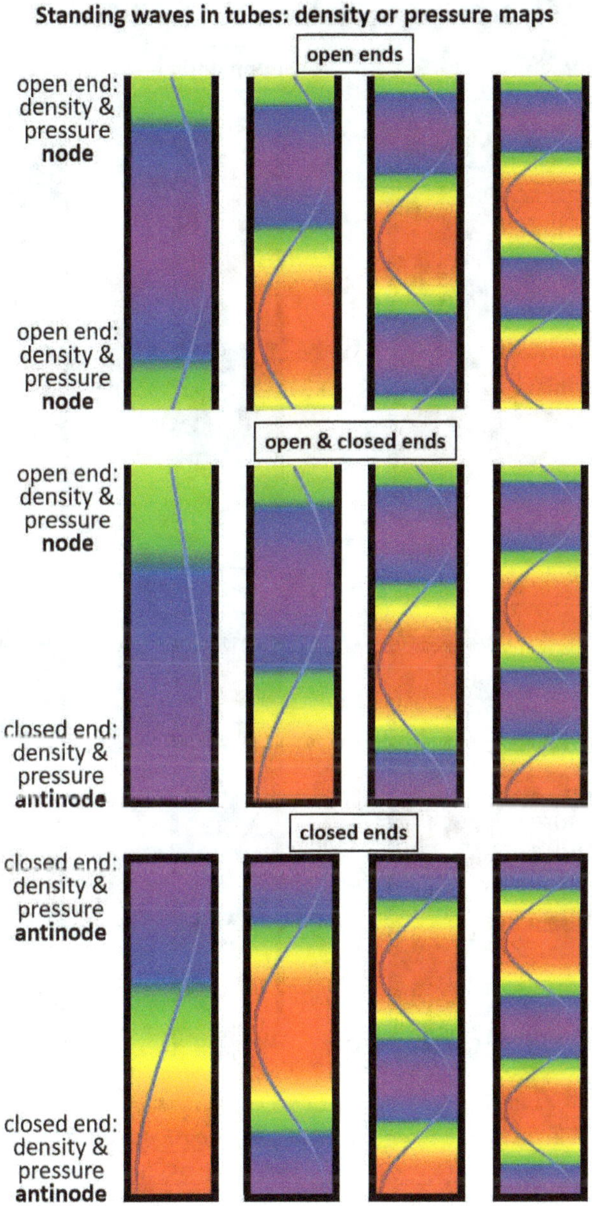

Figure 3-17: Some possible standing waves are shown in tubes that have both ends open (*top*), one end open and the other closed (*middle*), or both ends closed (*bottom*). The heavy black lines show the walls and closed ends of the vertical tubes. The same color scheme is used as in Figures 3-1 and 3-2: purple for high density or pressure, red for low density or pressure. The thin blue wavy lines show the density or pressure profiles along the length of the tubes. From left to right are standing waves with progressively shorter wavelengths. See Box 3-2 for more details.

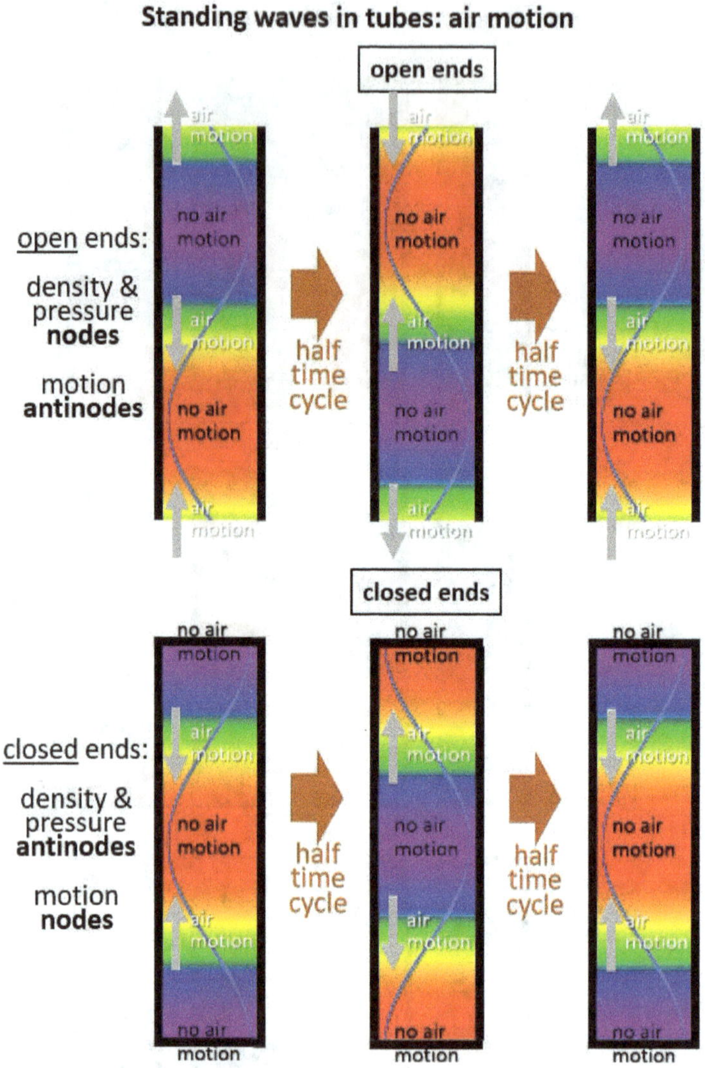

Figure 3-18: The time evolution (over one full cycle from left to right) is shown for two standing waves in tubes that have both ends open (*upper* graphs), or both ends closed (*lower* graphs), using the same design as in Figure 3-17; the colors here still represent air pressure or density. The motion of air is shown with gray arrows: air moves from regions of high density and pressure (purple) towards regions of low density and pressure (red). The case with one open end and one closed end is not shown: it is a simple combination of the two cases drawn here.

flow of air, a half cycle later (see middle sketches), the highs and lows are exchanged: highs become lows, and lows become highs. Now the direction of air flow reverses, so that another half cycle later (rightmost sketches), we recover the initial situation of the left sketches. The cycle repeats itself from here on.

Perhaps the most notable feature of this air flow is what happens to the nodes and antinodes. **The nodes in the density and pressure (where density and pressure do not change) are the antinodes of air motion**: there, the air flows the most! And *vice versa*: **the antinodes in the density and pressure (where density and pressure change most) are the nodes of air motion**: there the air does not flow.

3.13 Waves Inside Boxes and Mouths

In Sections 3.10 to 3.12, we looked at sound waves in tubes. These are essentially one-dimensional systems, so we could focus on waves that travel along the length of the tubes and ignore waves traveling across the tubes. However, at very high frequencies, meaning short wavelengths that are comparable to or smaller than the diameter of the tube, waves can also travel across the tube: the situation then becomes more three-dimensional. Waves in a cylindrical tube are rather more complex to discuss, so we will focus on boxes with rectangular shapes: they already exhibit the important properties of more general shapes.

How about standing waves in three-dimensional boxes? If we choose three-dimensional boxes that are rectangular, the situation becomes relatively simple again. In particular, in that case, we can think of standing waves in each of the three dimensions separately and use our knowledge of waves in "one-dimensional" tubes.

For simplicity, let's limit ourselves to completely closed boxes (even though closed, you could still hear the sound inside by drilling a small hole in the box). Open box sides let waves escape before building a standing wave, so we will ignore open boxes here. We do require that the side walls of the box be very stiff (as are tubes), so the sound waves cannot make the side walls vibrate and also cannot escape through the tube walls.

Let's look at some examples of standing waves in a rectangular box in Figure 3-19. Since it is difficult to draw three-dimensional waves, Figure 3-19 only shows them in two dimensions: we plot only the part of the wave that touches the bottom of the box. The other parts of the standing waves would look very similar in other slices through the box. You may also view my Animation 3*11 of Figure 3-19.

ANIMATION 3*11 — See my video WA2 at time 11:22 in its section "**Standing waves in tubes and boxes**" under the title "**Standing waves in boxes — resonance chambers, drums and bells**". (See details in the section References and Resources below.)

What we see is a landscape of peaks and dips alternating like a checkerboard: each peak or dip is equivalent to a loop on a string (see Figure 2-1 or 2-21). As in a loop, over time, a peak collapses into a dip and then grows back into a peak, repeatedly and without moving away, thus deserving the name standing wave. Similar to the situation shown in Figure 3-18, air flows from peaks toward dips, but now in varying directions.

Are there nodes in three- or two-dimensional standing waves? Indeed! Drawn in Figure 3-19 are **nodal lines** (black dotted lines): these correspond to the nodes ("nodal points") that we discussed many times before. In two dimensions, the nodes become lines; in this simple case of a rectangular box, the nodal lines are straight. In three dimensions (for example, inside our box), the nodes and nodal lines become **nodal planes**. As with nodes, both the nodal lines and the nodal planes are locations where the wave does not change; more precisely, at nodal lines and planes, a standing sound wave has constant density and pressure. This implies that **if you put your ear at a nodal line or plane, you would hear nothing**!

We notice in Figure 3-19 that there are no nodal lines along the edges; also, the edges go over peaks and through dips. The reason is that we chose the edges to be closed walls, so antinodes of pressure and density exist there.

How do these standing waves relate to vibration frequencies and resonances? For standing waves on a string of a given length, we saw in Section 2.9 that each standing wave has its own frequency:

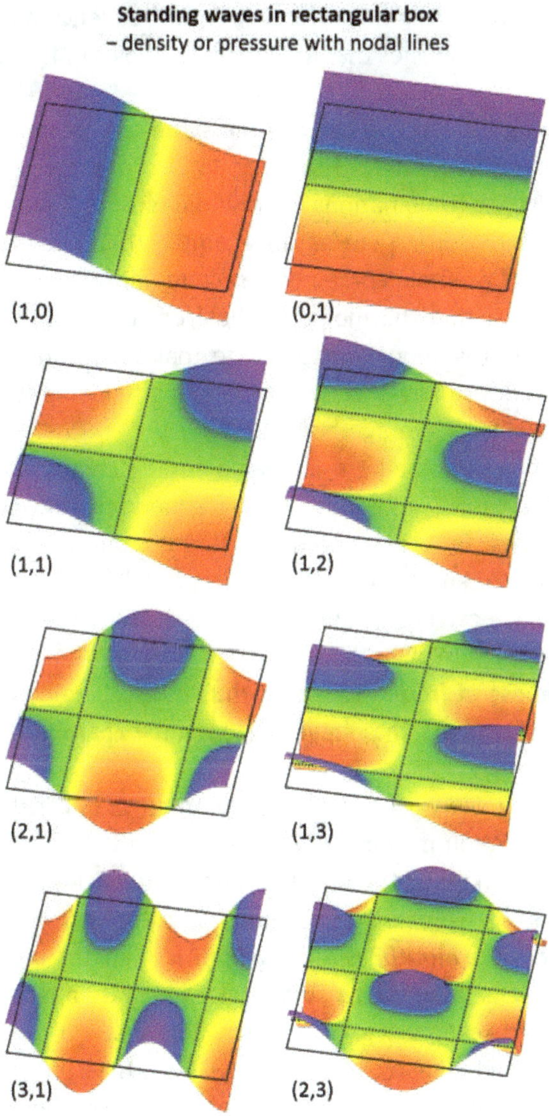

Figure 3-19: Eight examples of two-dimensional standing waves on the 2D bottom of a 3D rectangular closed box, outlined by the black lines, in a perspective view. Shown in color is the density or pressure (not motion). We use the same design as the left sketches in Figures 3-1 and 3-2. The dotted lines are straight nodal lines, where the density or pressure does not change over time. The first number in each label counts the number of nodal lines from left to right, while the second number counts the number of nodal lines from front to back; these numbers also indicate the number of half wavelengths that fit in the box and correspond to the number of loops shown on strings in Figures 2-1 and 2-21.

more precisely, the string has a fundamental frequency and a series of harmonics with frequencies that are whole multiples of the fundamental frequency.

In a rectangular box, we can use the same reasoning in each of the three dimensions separately. However, in two and three dimensions, we additionally get combinations of "one-dimensional" waves that exist in different dimensions. Take as an example the lower right case in Figure 3-19: that example combines a second harmonic in the left-right dimension with a third harmonic in the front-back dimension, hence its label is (2,3). It is important to realize that the frequency of such a combination is not simply a whole multiple of a fundamental frequency: instead, it is a kind of average.[15]

The important point here is that a three-dimensional case adds many standing waves, and thus many frequencies, and consequently many resonances. By using a shape different from a rectangular box, such as a sphere or an odd-shaped room, especially with other reflecting objects inside, we can create an even greater variety of standing waves, frequencies and resonances.

This richness of frequencies and thus musical sounds is exploited in musical instruments that have a resonance chamber, such as the guitar and violin.

You may ask: ***Why are resonance chambers in musical instruments not rectangular?*** You may already have guessed the answer: the more complicated the shape of the box, the more frequencies it can generate. We will discuss the role of these chambers further in Section 3.14.

How about the mouth? Is it like a box? Indeed, **the mouth is a very complex box: its shape and its soft sides can create a large number of frequencies and resonances**. In addition, the shape of the mouth can be rapidly changed in many different ways by muscles, further enriching and varying the sounds it produces. The nose cavity can also be included to form sounds. We will discuss this in Chapter 6.

---

[15] More precisely: for 2D rectangular boxes, the frequency of such a combination is the square root of the sum of the squares of those two frequencies — or of three frequencies in three dimensions, such as inside 3D rectangular boxes. Note that in Figure 3-19 we only draw the pressure wave (or density wave) along the 2D bottom of the 3D box.

3.14 The Role of Resonance Chambers

The strength of the sound coming from a string is not sufficient to fill a large space like a concert hall, so **amplification** is needed: a string is so thin that it does not push much air around. That is one reason for having several violins in one orchestra. But there is also a method to increase the sound of each violin, as follows:

A number of musical instruments have 3D **resonance chambers**, such as the violin and the guitar, or have 2D **resonance plates**, such as the piano. **Their main function is to take the weak sound from the strings and to make larger plates vibrate**: these plates in turn push against a much larger area of air and thereby send much stronger sounds toward listeners.

We have seen in Section 3.13 how air vibrates in boxes with very solid walls. If boxes have flexible walls, these act as plates that can make the surrounding air vibrate. We will see how such plates vibrate in Section 4.2.

As their name indicates, resonance chambers utilize resonance as the method of amplification. The idea is that **the (weak) sound from a string will activate resonances of the nearby resonance chamber**. This is most effective when the chamber has a resonance frequency that is very close to the main frequency coming from the string. Mismatched frequencies dampen out quickly and transfer energy to the matched frequencies, further strengthening the desired frequency. It follows from this that the resonance chamber should have many different resonance frequencies: **chambers and plates with complex shapes provide this richness of resonance frequencies**. The resonance chamber of the violin is a perfect example of this complexity.

The increased energy and power of amplification come partly from suppressing or weakening other frequencies, and partly from gradually building up the resonance vibrations (similar to amplifying a child's motion on a swing by pushing it repeatedly).

3.15 What have We Learned in this Chapter?

This chapter has taken many concepts from the preceding chapter about waves on strings and applied them to waves in air, showing the

value of such wave concepts even when the underlying physics changes significantly: air is quite different from strings, so the mechanism of wave motion is also quite different. A surprising aspect of waves in air is their weakness and our ears' high sensitivity: even sounds that hurt our ears are amazingly weak in terms of amplitudes and energy!

The concepts of wave superposition, interference and standing waves remain valid in air: we applied them here to waves in tubes (for music) and boxes, including the human mouth. Standing waves play a special role in resonance chambers and resonance plates that amplify weak sounds coming from vibrating strings.

The three-dimensional space in which sound travels creates new opportunities since now waves can travel in many directions: spherical waves are a good example. Another novelty is that waves in 3D can turn around corners, unlike normal objects; this is called diffraction. Sound also gives rise to sonic booms and the Doppler effect. These properties will also be valid for other kinds of waves, for example electromagnetic waves, including light.

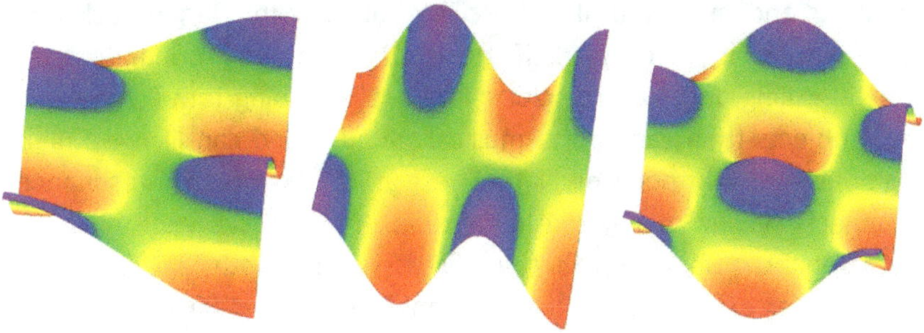

4

Waves on Sticks, Plates, Drums and Flags

This chapter focuses on vibrating objects that are two-dimensional. To better understand the case of 2D plates (such as cymbals and metallic drums), we first consider 1D flexible sticks, such as poles, trees, musical triangles and tuning forks. Although waves on sticks operate differently from strings, they also exhibit wave behavior, including interference, reflection at the ends, and standing waves (resonances). As with air in boxes compared to tubes, we find that waves on plates, drums and flags also behave very similarly to waves on strings.

4.1 Waves on Sticks

If you bend and then let go of a flexible stick like a pole, a small tree or a branch, it will sway back and forth at a frequency that it chooses by itself. You can do the same with a plastic comb, knife or ruler if you hold one of its ends fixed (but it will oscillate faster than a larger stick). The same happens if you drop a knife or metal bar on the ground, or tap on a metal railing, musical triangle, xylophone or tuning fork (Figure 4-1): these objects then can emit a musical sound. Wind can likewise make flag poles and light poles oscillate. These are all examples of resonances on objects having mainly one dimension: they behave very much like the resonances on strings that we discussed in Chapter 2. I will call all these harder objects **sticks**: they are long, narrow and flexible.

The mechanism of wave motion on such sticks is quite different from the case of **strings**. In strings, it is the **tension** within the string that is responsible for opposing any disturbance of the string and for moving a wave along the string, through elastic stretching. **In sticks, it is the elastic bending of the metal, wood or other material that is responsible for wave motion**: when we bend a stick, it resists that deformation and tries to straighten itself out, thereby pushing the deformation along in the form of a wave; no tension is needed.

Despite the different mechanisms, waves generally behave similarly on sticks as they do on strings: in particular, waves also superpose with "harmless" interference, waves are reflected at the ends, and waves resonate in the form of standing waves. Sticks therefore can exhibit standing waves like those illustrated in Figure 2-21, for example.

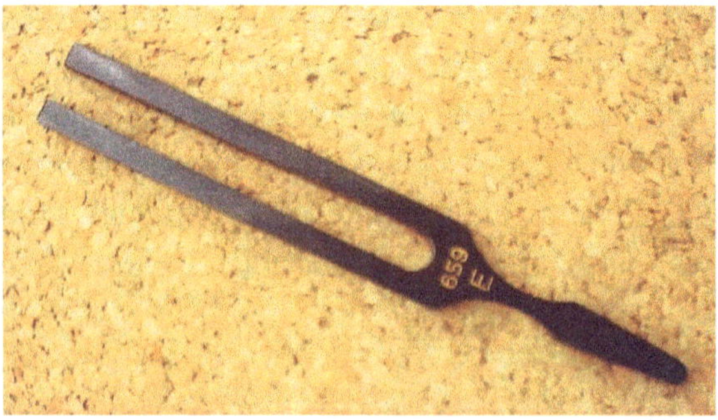

Figure 4-1: Tuning fork marked to play note E at frequency 659 Hz (hertz). (*Source*: John Walker, in the public domain, https://commons.wikimedia.org/wiki/File:TuningFork659Hz.jpg.)

Nevertheless, there is a minor difference between strings and sticks in their behavior at their free ends. For strings, we saw an S-shape at the free end (see Figure 2-20 and the leftmost blue sketch in Figure 4-2), with the model used there. For sticks, that S-shape is replaced by a simpler J-shape, as shown in the blue middle sketch of Figure 4-2: it is the shape that you obtain, for example by pushing against the free end of a comb, knife or ruler; you simply bend the tip to one side or the other.

Another more substantial difference occurs at the fixed ends of sticks, illustrated in red in Figure 4-2. There are now two options, depending on how the end of the stick is held fixed: either the end can still freely rotate (as is also the case with the fixed end of a string), or the end may be held straight by clamping it between two flat platelets, so the end cannot rotate; these are called "rotatable" and "non-rotatable" in Figure 4-2. These motions can be seen animated online.[1]

The same situation may occur at a node. On a string, a node is immobile, but the string rotates around that node. However, on a stick, a node could either rotate or be fixed straight, like the fixed end. This also creates more options for types of standing waves on sticks: we will see an analogous situation for plates in Section 4.2.

As we have seen before with strings and tubes, the behavior at the ends has a strong influence on the standing waves. We see here that the standing waves on sticks can be different from those on strings or in tubes, simply because of the different structure of the ends and nodes.

Simple sticks, such as those used in the **xylophone**, can be used to make music: see Figure 4-3. The bars of the xylophone have free ends. They are typically supported at two points that force rotatable nodes on the resonances in those two points. Each bar has a size that makes it resonate at a desired frequency when hit with a mallet: thereby many notes can be generated by hitting different bars in the right sequence.

The **triangle** is another musical instrument in the shape of a stick (since it is not a closed triangle, it has two free ends). Its two curved corners and two free ends give it a richer set of resonances than would a straight stick; its sharp corners are relatively stiff, causing nodes nearby.

---

[1] See "Flexural Bending Mode Shapes and Boundary Conditions" by Daniel A. Russell: https://www.acs.psu.edu/drussell/Demos/Flexural/bending.html.

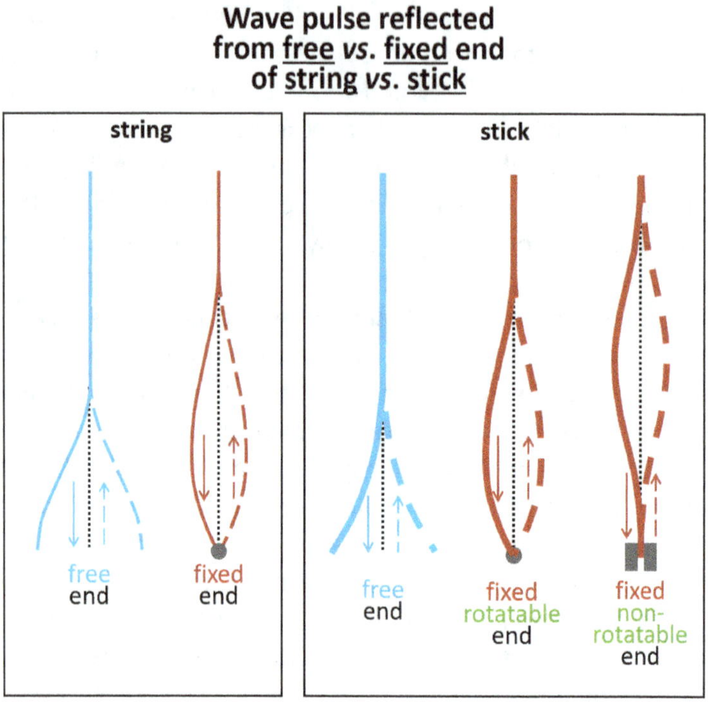

Figure 4-2: Sketches of the behavior of a wave pulse being reflected from the end of a string *versus* a stick: the pulse approaches from the top (on the left side of the dashed lines) and is reflected back up (on the right side of the dashed lines). The deviation from the straight string or stick at rest (dotted lines) is exaggerated for better visibility.

Figure 4-3: A xylophone. (*Source*: cropped from Ferbr1, under CC BY-SA 3.0, https://commons.wikimedia.org/wiki/File:Tres_xil%C3%B3fonos.JPG.)

4.2 Waves on Plates

Every home has many **plate**-like surfaces that vibrate when you hit them: windows, doors, cupboards, tables, boxes, pans, dishes, wine glasses, *etc.* Some of them resonate at particular frequencies. You may notice that large and heavy surfaces (doors and tables) vibrate at lower frequencies, while small and stiff surfaces (dishes and drinking glasses) resonate at higher frequencies. But you are rarely able to see those vibrations: they usually oscillate too fast for your eyes to follow their rapid back-and-forth motion. You can feel such vibrations with your fingers, but you may also see them indirectly after sprinkling powder on those surfaces.

Figure 4-4 shows six beautiful standing waves on a square metal plate. Such patterns were studied and popularized by the German physicist **Ernst Chladni** in the 18th century and carry his name: **Chladni figures** on **Chladni plates**. His approach, still frequently used today in spectacular demonstrations, is illustrated in Figure 4-5: a metal plate is clamped down in its center but free along its edges, while a bow causes it to vibrate up and down, much like a bow is used to play a violin. Nowadays, Chladni plates are mostly activated electronically, like loudspeakers which they resemble very much.

The pattern of white lines is made with white powder on the plate: the powder bounces away from locations with stronger vibrations (these are **antinodes** and their surroundings) toward locations with the least vibrations (these are **nodes** and **nodal lines**), where the powder accumulates into white lines. **The patterns thus show standing waves of the plate.**

The patterns that are formed depend on the shape of the plate. You can see online[2] other patterns on plates with a variety of shapes (rectangular, triangular, pentagonal, septagonal and star-shaped); see also the collection of short videos at the website in that footnote, nicely showing the effect of the bowing action.

---

[2] See the collection of Chladni figures and videos in "Ernst Chladni" (note: septagonal means seven-cornered like a heptagon, but with unequal sides): https://en.wikipedia.org/wiki/Ernst_Chladni

Figure 4-4: (left): Six Chladni patterns of white powder accumulating near nodal lines on a square vibrating plate. (*Source*: http://dataphys.org/list/images/uploads/2017/09/chladni-plates-wide-wallpaper-1280x800.jpg, under CC-BY-SA).

Figure 4-5: (right): Chladni's method of creating Chladni figures. (*Source*: Wikipedia, in the public domain, https://en.wikipedia.org/wiki/Ernst_Chladni#/media/File:Bowing_chladni_plate.png).

These Chladni patterns also depend very much on the frequency of vibration of the standing wave: changing the speed of bowing changes that frequency and makes the powder jump from one pattern to another. It is fascinating to see the evolution of the Chladni patterns as the frequency gradually increases, as beautifully exhibited online.[3] This is a great illustration of **quantization**: **the plates jump suddenly between different vibration modes, despite a smoothly varying driving frequency**.

We see clear patterns forming only at certain frequencies. Nevertheless, there are many such frequencies: many more than we

[3] See video "Resonance Experiment!" by brusspup: https://www.youtube.com/watch?v=1yaqUI4b974

would expect from a string or stick. The reason is the two-dimensional nature of a plate (we already mentioned this multitude of frequencies in connection with rectangular boxes in Section 3.13). We also see clearly that the nodal lines become tighter as the frequency increases: that is because the wavelength shrinks with increasing frequency; parallel nodal lines are about half a wavelength apart. Thus, we can tell that in Figure 4-4 the upper left and middle right patterns have lower frequencies (longer wavelengths), while the middle left and upper right patterns have higher frequencies (shorter wavelengths), and the two lower patterns are in between.

Another factor influencing the Chladni figures is the method of clamping the plate. In the above examples, the plates are all clamped at the center in such a way that the center is a node (see the powder accumulating there); in addition, the plate at that node cannot rotate out of the plane of the plate. This is analogous to the "fixed non-rotatable end" of the stick in Figure 4-2 but applied to a node at the center of the plate instead of at the side of the plate. In fact, this clamping method is an important reason why the white nodal lines on Chladni plates are usually curved, instead of being straight as with air in the rectangular box of Figure 3-19.

The Chladni figures have a strong resemblance to the standing waves which we saw at the bottom of a box containing air: see Figure 3-19. The white nodal lines of the Chladni figures correspond to the black nodal lines of the box with air. And the antinodal regions (from where the powder bounces away) correspond to the purple peaks and red valleys of the box with air.

Chladni plates are not practical for making musical tones, compared to the violin for example, despite the similar bowing technique. Also, their very metallic sounds are not as pleasant.

Cymbals are much more popular instruments made from metallic plates. Their circular shape (and any other shapes they may have) produces even more frequencies simultaneously when hit by a mallet or when two plates hit each other. You can see the resulting movements in a slow-motion video online.[4]

---

[4] See video "Vibration. See the unseen: Cymbal at 1,000 frames per second", by Fluke Corporation: https://www.youtube.com/watch?v=kpoanOlb3-w

The player has little control over which frequencies are produced: in fact, cymbals radiate a very wild variety of sound frequencies (as is illustrated in the spectrogram and sonogram at the top of Figure 5-17 in Chapter 5). The reason for this is that the circular shape does not produce harmonics: instead of generating whole multiples of the fundamental vibration frequency (as on strings and square plates, for example), circular plates create many frequencies that are not harmonically related to the fundamental frequency. Cymbals therefore give a more violent sound, especially when it is loud.

The **bell** is another musical instrument that uses the vibrations of a plate: in this case the plate is strongly curved into the traditional bell shape and is relatively thick. Even though we are tempted to call a bell three-dimensional, its metal has the form of a bent two-dimensional plate.

Unlike cymbals, however, each bell is tailored to have a dominant resonance frequency despite its more complex shape; this and many other frequencies are activated by a clapper. Bells with matched frequencies are therefore assembled in a carillon that allows playing musical compositions. Heavy bells are very loud and are therefore common in church towers, from which they can be heard far and wide.

A **wine glass** is similar to a bell and, as you surely know, can easily be made to "sing". The favorite way to make a wine glass sing is to rub its edge with a wet finger, as shown online.[5] You probably also know that a wine glass can be broken by making it resonate too much, as demonstrated online[6]: this may be done with a loudspeaker but also with a very strong human voice,[7] if it produces the right frequency to activate the vibration.

---

[5] See video "Resonance standing wave wine glass", by science with bobert: https://www.youtube.com/watch?v=AxWzVPdubjs (note in this video the "spectrum" at top right: it shows a major peak at the dominant frequency emitted by the glass).

[6] See video "Breaking a wine glass using resonance", by iflamenko: https://www.youtube.com/watch?v=17tqXgvCN0E (note that this video shows the vibration in slow motion, something that nobody can see with eyes).

[7] See video "How I broke a wine glass with my VOICE (using science!)", by Physics Girl: https://www.youtube.com/watch?v=Oc27GxSD_bI

4.3 Waves on Drums and Flags

We can use what we learned in the preceding chapters and sections to better understand vibrations on drums and flags. ***Despite clear similarities between drums and flags, flags are not used to make music: why not?***

A drumhead is made of a membrane or skin that is tightly stretched around a circular frame. A flag is made of light cloth that is typically attached only at two neighboring corners.

Thus, a **drumhead** is under external tension, pulled from all sides, much like a string under tension with fixed ends: the drumhead is indeed a 2D analogue of a 1D string. The drumhead can vibrate like a flexible 2D plate.

Most drums consist of two drumheads facing each other. The air box between them transmits the vibrations of one drumhead to the other and helps amplify the sound: it forms a resonance chamber. The shell connecting the two drumheads also transmits vibrations.

The circular shape of a drumhead gives rise to many frequencies, just like the cymbals and bells that we discussed in Section 4.2. You can watch the vibrations of a drumhead online[8] (interestingly, that video shows a loudspeaker oscillating in slow motion, clearly exhibiting an amplitude of motion around several millimeters). The drumhead's vibration frequencies are high enough to be heard and used in music. Since each drum is made to produce mainly one fundamental frequency (one tone), multiple drums are needed to produce multiple tones.

Many other forms of drums are possible: for example, the Caribbean **steel drum** or **steelpan** is based on the shape of oil drums. In this case, the drumhead is itself made of stiff metal and is nearly spherical, but it contains various flatter portions that act like small metallic drumheads with different resonance frequencies. Different notes can thus be produced by hitting different parts of the steelpan.

Flags (as in Figure 1-1f) are also two-dimensional. Their three free sides are similar to the Chladni plates of Section 4.2; the fourth side, between the fixed corners, is under more tension than the rest of the

[8] See video "Circular Membrane (drum head) Vibration", by Dan Russell: https://www.youtube.com/watch?v=v4ELxKKT5Rw

flag: that side can be viewed as a string with fixed ends, but it plays a minor role in the flag's vibrations.

The driving force of the flag's vibrations is the wind. The wind firstly stretches the flag: the friction of the air along the cloth puts tension in the flag in the direction of the wind flow (thus mostly horizontally). This is similar to a hanging string, in which gravity causes internal tension.

The wind also causes the side-to-side flapping of the flag through air **turbulence**: we discussed this mechanism in Section 3.3. In the case of flags, the turbulence is mainly caused by the flow of air around the flagpole; however, even without flagpole there would be some turbulence, as you can experience by holding a piece of cloth or paper in the wind. The frequency of such turbulence depends on the wind speed but is generally below our hearing range; we can hear the flapping, but that is like hearing successive claps or bangs, which are not musical.

4.4 What have We Learned in this Chapter?

We have extended our understanding of waves from strings, air in tubes, and air in boxes to flexible elastic sticks, plates, drums, and flags. We find that many of the wave properties are closely similar in all these situations, despite different mechanisms (like tension in strings, drums and flags, compression/decompression in air, or elasticity in sticks and plates).

However, the structural details of the ends or edges have a strong influence on which standing waves can form and on the corresponding frequencies that can become sound. This is also true of the shapes of vibrating surfaces (square *versus* circular *versus* bell-shaped, *etc.*).

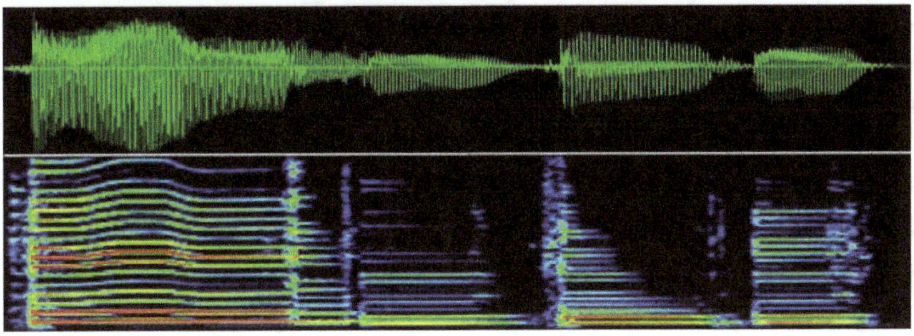

5

From Waves to Music

This chapter will link standing waves to the notes that a ***musical instrument*** *makes. Starting with a simplified model of a piano, we will discuss how string vibrations correspond to notes, pitches and frequencies, and how they are controlled to deliver* ***music****. We will also address what distinguishes the sound of different musical instruments, as well as the recording and playback of sounds. We will consider synthetic sounds as well.*

I recommend that you first watch my accompanying video to get a feeling for the subject of this chapter.[1] *See also an excellent other video.*[2]

[1] See my video "Waves in Air: *sound and musical instruments*" (WA2) on Everyday Physics by Michel A. Van Hove: https://www.youtube.com/watch?v=PMR49kj2PeY

[2] See video "Resonance and the Sounds of Music" by Walter Lewin: https://www.youtube.com/watch?v=f4M-6tWtkoA

Let's first summarize how **musical sounds** go from a **source** (such as a string or a tube, your mouth or a loudspeaker) to a detector (such as your ear or a microphone). Two typical cases, the sounds from a guitar and a flute, are illustrated in Figure 5-1. Many types of sources exist, such as: plucking a guitar string with a finger; hammering a piano string with a key; bowing a violin string; hitting a xylophone bar, a cymbal plate or a drumhead; using lips to make air vibrate in a flute; using vocal cords to speak or sing; and using electric currents to make a loudspeaker or earphone vibrate.

The result is the vibration of a **string**, a **plate** or an **air column**. The strings in guitars, violins and pianos produce a relatively weak sound: that sound needs amplification to be heard well; this is achieved with a **resonance chamber** (the "box" of a guitar and violin) or a **resonance plate** (the **soundboard** of a piano).

Then the sound can travel through the air to a listener, spreading out through space, reflecting off surfaces, weakening as it goes.

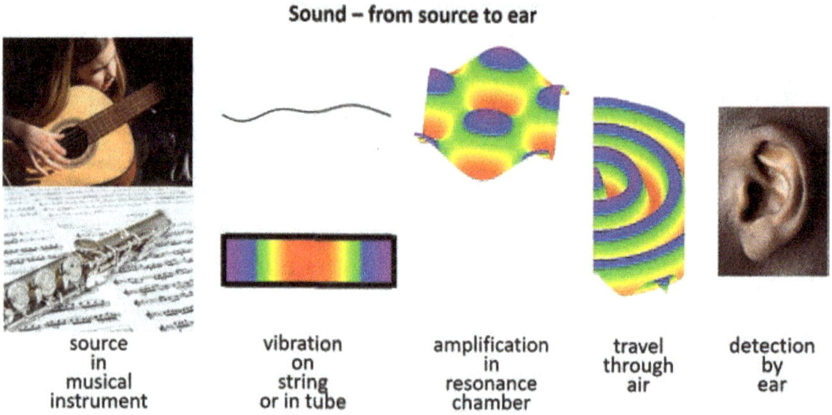

Figure 5-1: Tracing sound from a source (guitar or flute) to an ear. *From left to right*: The finger causes a vibration of a string of the guitar, while blowing through the lips causes a vibration of air in the flute's tube. Since the guitar string produces only weak sound in the air, it is first amplified in a box: a resonance chamber. The sound then radiates from the musical instrument to an ear. (*Image sources*: Takkk, Guitarist_girl.jpg, under CC BY-SA 3.0, https://commons.wikimedia.org/wiki/File:Guitarist_girl.jpg; Petar Milošević, Flute with musical notes, under CC BY-SA 4.0, https://commons.wikimedia.org/wiki/File:Flute_with_musicial_notes.jpg; Genusfotografen, Human right ear (here reversed), under CC BY-SA 4.0, https://commons.wikimedia.org/wiki/File:Human_right_ear_(cropped).jpg.)

Of course, the sound can also be recorded (on a vinyl record, on a magnetic tape, on a compact disk or in an electronic memory), for storage, copying, mixing, modification, amplification, transmission and rebroadcast.

We will discuss many of these aspects of musical sounds in this chapter. The fascinating case of voices will be the subject of Chapter 6.

5.1 A Simplified Piano

We start with a **piano**, pictured in Figure 5-2. Of particular interest is the fact that each white or black key has its own separate mechanism for hammering strings.

The keyboard is sketched in Figure 5-3. A normal piano has 88 keys (including the black keys) that produce 88 different notes (some modern pianos have 97 or 108 keys). In Western music, the notes and keys are usually grouped into **octaves**, numbered 1 to 7, while a few more notes and keys are at the left (in octave 0) and at the right (in octave 8).

Figure 5-2: A standard upright piano, opened to show the keyboard with white and black keys at the bottom, which are individually linked to the hammers (top white pads) and strings (seen above the hammers). (*Source*: Pko, under CC BY-SA 3.0, https://commons.wikimedia.org/wiki/File:Upright_piano_inside.jpg.)

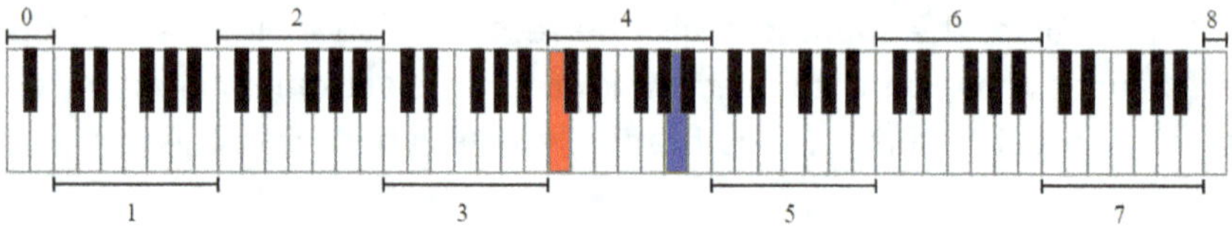

Figure 5-3: The keyboard of a standard piano, with octaves numbered. The red key is C_4, while the blue key is A_4. (*Source*: AlwaysAngry, re-colored under CC BY-SA 3.0, https://commons.wikimedia.org/wiki/File:Piano_Frequencies.svg.)

Each octave has 8 white keys with corresponding notes that are known familiarly as "do – re – mi – fa – sol – la – si/ti – do". In the American Standard system of pitch notation, the red-colored key is labeled as C_4 (it is a "do", as shown in Figure 5-4), while the blue-colored key is A_4 (it is a "la"). By worldwide convention, the note A_4 should have the frequency 440 hertz: all other string frequencies are tuned (meaning adjusted) to harmonize with that one note A_4.

In addition to the white keys, there are black keys that give intermediate notes, called sharp or flat notes: a "sharp" note, marked as ♯, has a slightly higher frequency or pitch than its neighboring "white" note, for example C_5♯ is slightly higher than C_5; a "flat" note, marked as ♭, has a slightly lower frequency or pitch, for example G_4♭ is slightly lower than G_4. The 7 white keys (ignoring the second "do") and 5 black keys in an octave form 12 notes in the Western musical scale.

Let's now look at my model of a simplified **piano** shown in Figure 5-4. It has a shortened **keyboard** with 14 colored **keys** of octaves 4 and 5, and one from octave 6, attached to 15 colored **strings** of different lengths (all the strings are otherwise identical here). The black keys are not used in this model.

What happens when you hit a key of a piano? In the model of Figure 5-4, a hammer strikes the string of the same color. As we have seen in Section 2.9, **each string can then resonate at its natural frequency and at its harmonic frequencies**. These frequencies correspond to standing waves, some of which are drawn in Figure 5-4: each colored solid line is the fundamental vibration with the string's natural frequency, while the dashed, dash-dotted and dotted lines are harmonics (with half, third and quarter wavelengths, respectively).

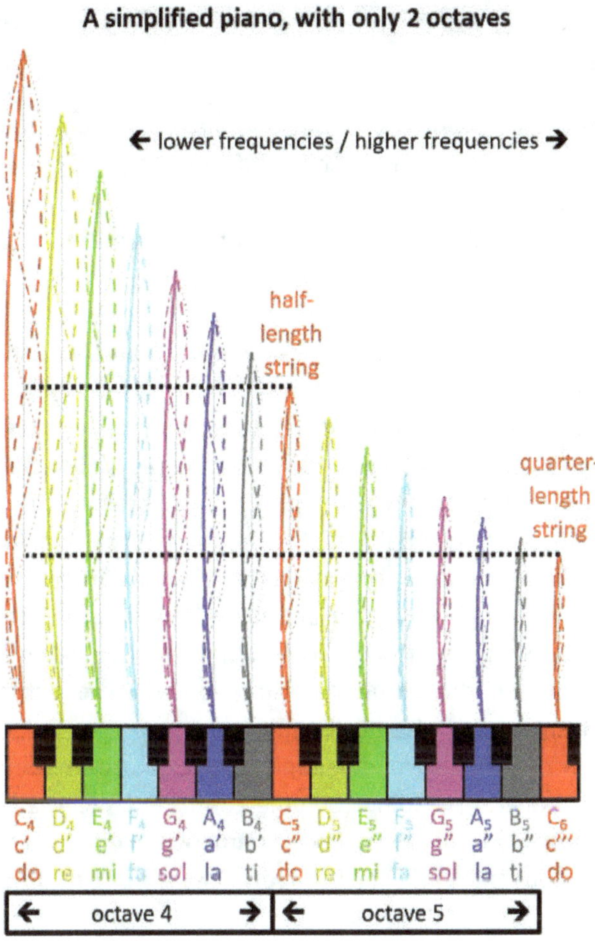

Figure 5-4: My model of a shortened and simplified piano with only 14 strings in 2 octaves and one more string in the next octave. A normal piano has 88 keys (including the black keys) covering 7 octaves. Here we include only 2 octaves near the center of the normal piano's full keyboard: from key C_4 to key C_5, and from key C_5 to key C_6 (shown as colored boxes) using the American Standard pitch notation (or from c' to c" to c''' in the Helmholtz notation; or from "do" to "do" to "do" in the "do – re – mi" notation; "ti" is common in America, "si" in Europe). Each colored key is connected through a hammer to a string of the same color, but with different lengths. The black keys in this model are not connected to strings but serve as a guide when looking at a real piano; in a real piano, they would be connected to additional strings that produce intermediate ("sharp" and "flat") notes. Four standing waves are drawn on each string, with 1, 2, 3 or 4 loops for each standing wave.

5.2 How do String Vibrations Correspond to Note and Pitch?

First, we must ask: ***What are frequency, pitch, note and tone***? These terms are closely connected and, unfortunately, sometimes used interchangeably.

We have defined frequency as the rate of repetition of a periodic event, such as the passage of wave peaks. This is a precise quantity that can be measured objectively with an instrument (such as a clock or oscilloscope). It is commonly expressed in hertz, namely as the number of repetitions per second.

When we listen to a sound that was generated at a certain frequency, our auditory system, including our brain, estimates that frequency. This is a subjective estimation and not a precise measurement: for example, it is found that our estimate of the frequency can depend on the loudness of the sound; also, our estimate deviates somewhat from the correct frequency for both higher and lower frequencies. Furthermore, when multiple frequencies occur in a sound, the estimated frequency can differ from all the component frequencies. **The estimated (or perceived) frequency is often called pitch**.

In practice, however, many people and texts use the word "pitch" to describe frequency, leading to some confusion! **We shall use pitch to mean estimated (or perceived) frequency, which is subjective.** Fortunately, **the pitch is usually very close to the fundamental frequency**.

Suppose that you produce a simple sound with a musical instrument or with your mouth, such as an A_4 (or a' or "la"). **The names such as A_4 or a' or "la" then identify the note that you are making**: so the note is primarily a label. That note could be the single frequency of 440 hertz, but more likely it contains also many higher harmonics (multiples) of that frequency.

Sound can be soft, warm, harsh, metallic, thunderous, *etc*. Such words describe the tone of a sound. The tone depends especially on the relative amplitudes of the different harmonic frequencies present in the sound.

We see that the fundamental vibration of a string, which has the string's natural frequency, gives the **note** of this string and of its key on the keyboard. This frequency is basically the **pitch** that you hear: a low note or pitch corresponds to low frequencies, while a high note or pitch corresponds to high frequencies (thus, shorter strings create higher notes and pitches). Figure 5-4 shows some of the common names of these notes, for example C_4 ("do") and A_4 ("la") (often the labels 4 and 5 are not written). When you sing do – re – mi – fa – sol – la – ti/si – do, you may be singing the notes $C_4 – D_4 – E_4 – F_4 – G_4 – A_4 – B_4 – C_5$, also written as c′ – d′ – e′ – f′ – g′ – a′ – b′ – c″ (but you may also be singing one or more octaves higher or lower).

Hitting a key thus creates several resonances on the corresponding string. The snapshot in Figure 5-5 shows a simulated example of resonances at five different frequencies: the fundamental frequency (thin black standing wave at left) and four higher harmonics (thin colored waves) to its right. These standing waves usually have different amplitudes, as seen in the colored curves, and add up, as shown by the thick black waves, which accumulate the harmonics: the rightmost black wave is the only wave seen on the string, as it contains all the 5 resonances created by hitting the key for this string.

The piano offers another interesting option through its rightmost pedal, called the **damper pedal**. Pressing that pedal removes dampers from all strings: as a result, all strings can resonate with the currently played strings for a longer time after the player has released the current key, producing a very rich, continuing sound.

The bottom of Figure 5-5 also shows the **spectrum** of the combination of five simple waves at the top of that figure. A spectrum displays all the frequencies present in the combined wave as individual peaks. The height of each peak shows the amplitude of that component. **Thus, the spectrum is a very convenient picture summarizing important information about the wave: a spectrum shows all frequencies that are present and their relative amplitudes.** We will use this concept of spectrum many times.

Note, however, that in practice, the peaks in a spectrum are often not as sharp as shown in Figure 5-5. This happens especially when there are many closely spaced frequencies, for example in human voices.

Figure 5-6 illustrates a real spectrum recorded from a piano playing the note C_4 ("do"); such a spectrum is also often called a **spectrogram**. This spectrum shows half the audible range, from 0 to 10,000 (= 10^4)

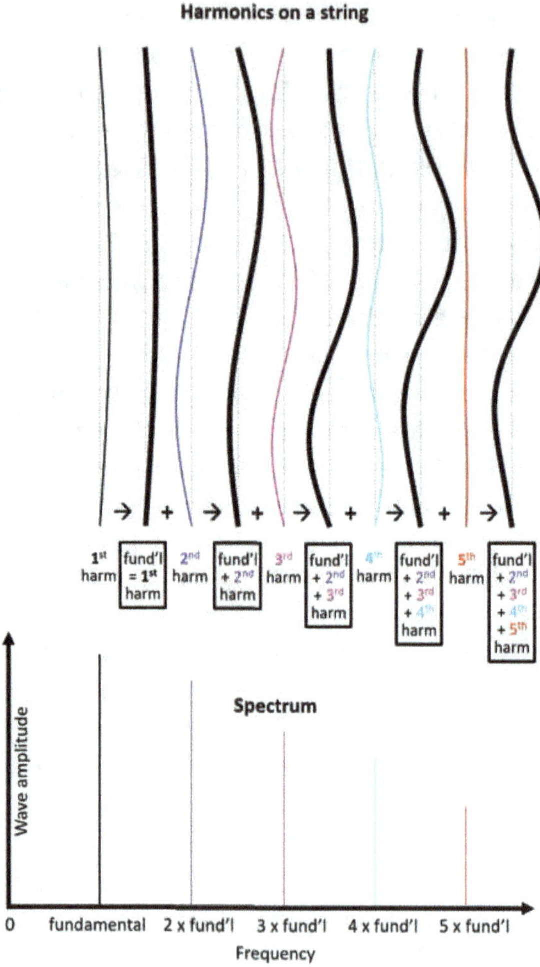

Figure 5-5: *At the top*: Simulated waves with harmonic frequencies are added on a string with two fixed ends (similar to the waves in Figures 5-4 and 2-22). In these snapshots, the fundamental frequency wave is shown at the left (black, thin); it is also called the first harmonic. The second harmonic (dark blue) has double the fundamental frequency: adding it to the fundamental gives the second thick black wave. Adding the third harmonic to the fundamental and the second harmonic gives the third thick black wave, *etc.* The rightmost black wave is the resulting shape of the string with these five harmonics.

At the bottom: The spectrum of the combined wave above shows the amplitude of each component harmonic separately; this spectrum thus has five peaks of decreasing height at the corresponding frequencies.

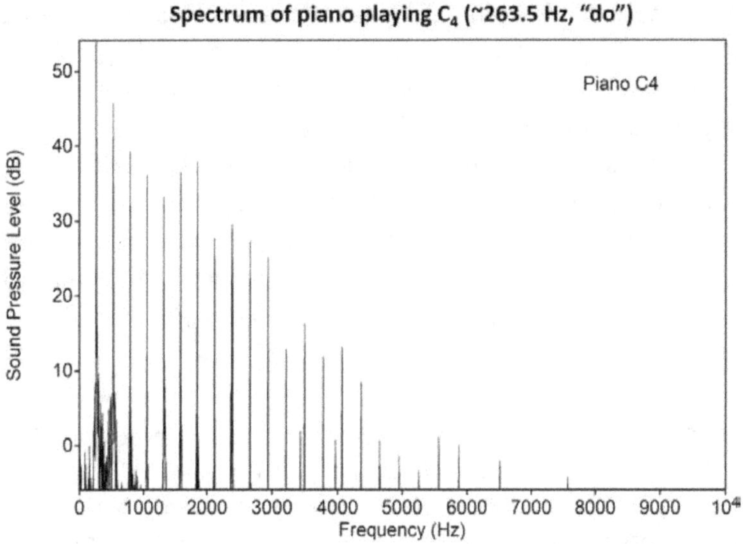

Figure 5-6: Real spectrum from a piano playing C_4 (do), covering half the entire audible range, from frequency 0 to 10,000 hertz (Hz). (*Source*: sound from University of Iowa Musical Instrument Samples (MIS); spectrum produced with Praat).

hertz; this piano plays almost no audible sounds above about 6,000 hertz (the level 0 dB is the threshold of hearing, as noted in Section 3.4).

The most obvious feature of this real piano spectrum is its multitude of peaks: I count over 20 peaks, compared with only 5 peaks in Figure 5-5. Most of these are harmonics of the fundamental frequency, which is the leftmost high peak. **The piano has many harmonics: this gives it a very rich sound. The spectrum can thus be used to show and explain the "character" of a musical instrument**, as we will do in Section 5.7.

The piano used for Figure 5-6 produced a C_4 frequency of about 263.5 hertz: you can see peaks at almost all whole multiples of this frequency up to nearly 6,000 hertz: 263.5 hertz (1 ×), 527 hertz (2 ×), 790.5 hertz (3 ×), 1,054 hertz (4 ×), *etc.*

Also visible in Figure 5-6 is that peaks in a spectrum generally become weaker toward higher frequencies, although a few can be stronger than lower-frequency peaks. **Often, the fundamental resonance at the natural frequency is the loudest**, as is the case here.

The smaller "non-harmonic" peaks at low frequencies are partly due to the way in which the sound starts: a single strike by a piano key can activate waves of many frequencies, also on other strings, but those waves that do not resonate die off quickly.

What does the composite sound wave look like in reality? We again take the sound from the piano mentioned above: Figure 5-7 plots the actual recorded **wave form** of the sound produced by that piano. You can also listen to the sound of Figure 5-7 in my Animation 5*1. The waveform is the pressure variation as time passes, as measured by a microphone.

ANIMATION 5*1 — Listen to my video WA2 at time 12:39 in its section "**Viewing sounds**" under the title "**Viewing the sound wave of a piano**". For comparison, that piano sound is also followed by similar sounds from a human voice, a guitar, a cymbal, a xylophone, a violin, a saxophone, and a trumpet (see details in the section References and Resources below.)

In Figure 5-7, we plot the wave against time, rather than against distance from the source (the piano), as in Figure 5-5: this corresponds to recording a wave as it passes a fixed point, such as your ear or a microphone. Waveforms plotted against time look essentially the same as waves plotted against position, as we have drawn so far in this book, so we don't have to worry about this distinction; in particular, a sine wave looks like a sine wave either way.

Two **stereo** channels were recorded simultaneously by different microphones and are drawn below each other in Figure 5-7. The two wave forms are clearly quite different from each other, which is probably mainly due to different placements of the two microphones: they may be at different distances; they may hear the piano from different directions; and they may record different reflections from walls, *etc.* (the two microphones themselves could also be different). Similar differences can occur between your two ears, giving you stereo hearing: stereo sound tells you approximately where a sound comes from, so it gives the realistic impression that sounds from different instruments originate in different locations.

Even though the two stereo wave forms look quite different, they do contain the same frequencies with very similar relative amplitudes, so they sound very similar.

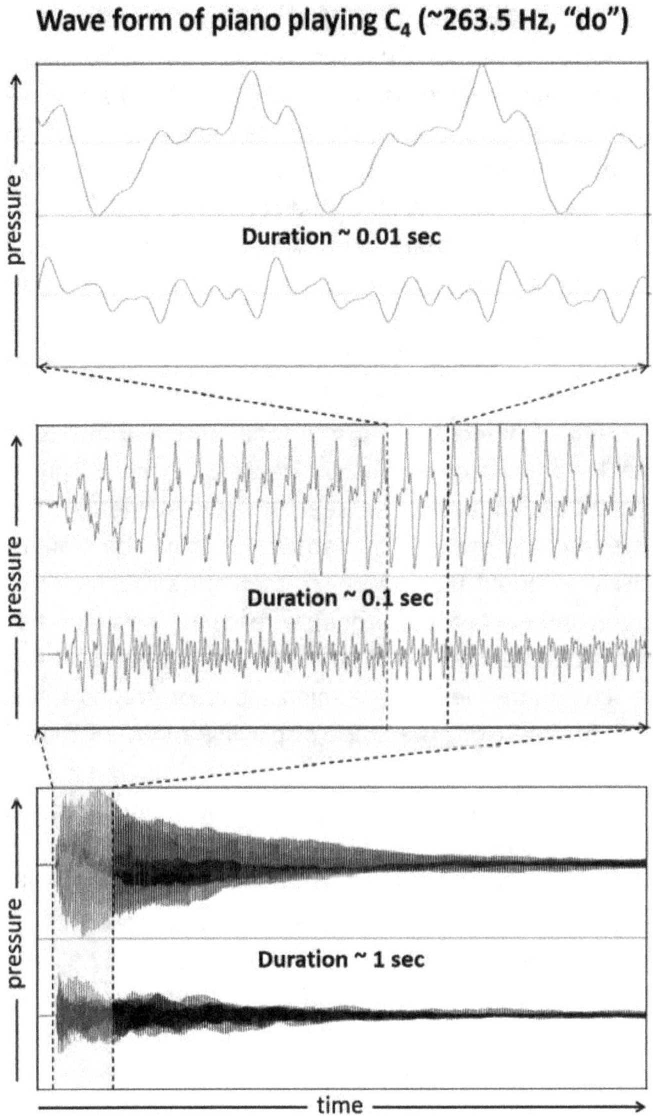

Figure 5-7: Waveform recorded from a piano playing C_4 ("do"), showing the wave's amplitude as measured by a microphone as a function of time: the vertical axis is the air pressure; the horizontal axis is the time. Each pair of graphs contains two stereo waveforms recorded simultaneously from that piano by two microphones in different locations. The lower pair shows about 1 second of sound, from silence at the left, through the key strike in the middle, to near-silence at the right. A tenth of a second (colored in pink) is blown up in the middle pair, including the key strike at left. One tenth of that waveform (colored in light blue) is further blown up in the upper pair, spanning about one hundredth of a second. (*Source*: sound from University of Iowa Musical Instrument Samples (MIS); spectrum produced with Praat.)

One way to understand this difference between stereo channels is to look at Figure 2-13. There, the sum of the colored waves (excepting the bottom dark-blue wave) is shown as the thick black wave at the top. But imagine shifting any of the colored lines to the left or to the right: such a shift would very much change the thick black curve. Such shifts are possible because each of the colored waves may come from different parts of the vibrating string in the piano, so they don't arrive at the same time at different microphones or ears.

The large difference we see between wave forms in the two stereo channels is a clear warning: **We cannot easily recognize the tone, the note, or even the instrument just by looking at the <u>wave form</u>. The <u>spectrum</u> is more characteristic of the tone, note and instrument.**

In the lower pair of graphs of Figure 5-7, the C_4 piano key is struck to start the sound, so we see the wave suddenly start at left. As time proceeds to the right for about 1 second, the sound initially grows quickly (in about 0.02 seconds), then stays roughly constant for about 0.1 second (in the pink zone); after that, it weakens gradually for 1 second (and beyond). If we blow up the wave form, we see the wave character more clearly: the middle pair of graphs shows about 0.1 seconds of time from the start, displaying about 25 repetitions of the wave motion in that time.

Another tenfold blow-up produces the upper pair of graphs in Figure 5-7: now we see only two full repetitions of the wave motion in the time span of about 0.01 seconds. A single repetition can be measured (with the software that produced these images) to take 0.0038 seconds: that period corresponds to a frequency of $1/0.0038 \sim 262.9$ hertz, which is close to the known frequency of 263.5 hertz (in fact, the frequency measured in this manner when the sound is fading away is closer to 266 hertz, so the frequency produced by the piano seems to increase slightly over time and 263.5 hertz is an average over the full duration of the sound).

Notice how different the two stereo sounds are in the top pair of graphs in Figure 5-7, even though they would sound very similar if you heard them separately: this shows how difficult it would be to identify the instrument just by looking at the wave form!

Does the spectrum change over time, as the wave strengthens and then weakens? Indeed, the answer is visible in Figure 5-8 for the

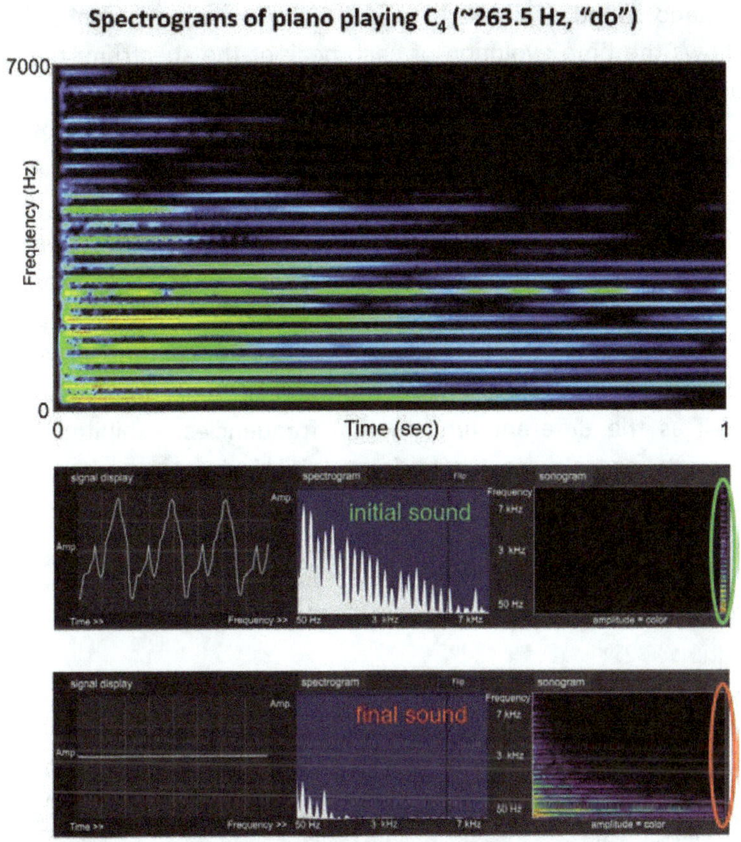

Figure 5-8: Spectrograms recorded from a piano playing C_4 ("do").

Top graph: The vertical axis is the frequency; the horizontal axis is time. This graph, from left to right, corresponds to the 1-second duration of the lower wave form in Figure 5-7. Each horizontal band corresponds to one peak in the spectrum of Figure 5-6. Its color shows the intensity level (red is the most intense, dropping through yellow, green and blue to black for the most silent). (*Source*: sound from University of Iowa Musical Instrument Samples (MIS); spectrum produced with WaveSurfer).

Bottom two graphs: The two strips show the initial and final sounds of the same piano recording shown at the top. *At the left*, the "signal display" shows the wave form as at the top of Figure 5-7, but averaged over the two stereo channels. In the center, the "spectrograms" are spectra similar to that in Figure 5-6. The two "sonograms" *at the right* are similar to the top graph in this figure, using the same colors to indicate amplitude (these two graphs span about 8 seconds instead of 1 second at top); the green and red ovals mark the times when the wave form and spectra at left and at center were recorded; the "initial" sound is much stronger than the "final sound". The "50 Hz" labels mark the left edge of the spectrograms, and the bottom edge of the sonograms. (*Source*: these graphs were produced with the app A.L.M.S. — Audio Literacy for Music Students; on a PC or smartphone, the graphs of this app are dynamic, showing the waves and spectra evolving in real time; here only snapshots are shown).

same piano sound of Figures 5-6 and 5-7. The upper graph in Figure 5-8 shows the time evolution of each peak of the spectrum: the color of each horizontal band shows the amplitude of the frequency peaks in Figure 5-6 as they change over time toward the right. As expected, each frequency gradually weakens after reaching a maximum early on. However, as we look closer, we see that some peaks grow again, even a few times, like multiple echoes, before fading away. **This time evolution of the spectrum will depend on the musical instrument and is another aspect that helps us identify what kind of instrument is being played.** Figure 5-9 illustrates this in a longer series of sounds, but now from a guitar: we clearly see the different intensities of the various frequencies (as well as the different fundamental frequencies, exhibited by the varying spacing between the horizontal "ribs" due to the harmonics). In Section 5.7, we will consider in more detail how musical instruments differ.

Figure 5-9 **also helps understand how much useful information is available in such graphs.** You can listen to the sound of Figure 5-9 in my Animation 5*2.

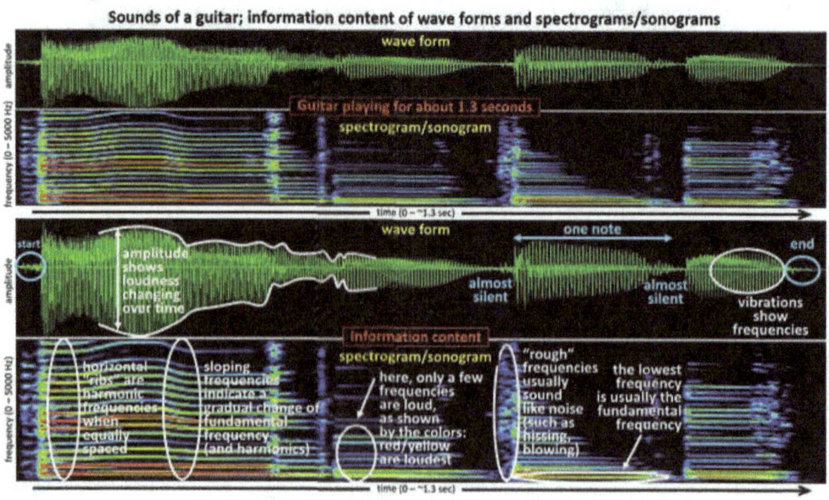

Figure 5-9: Waveform (*upper graph*) and spectrogram (*second graph*) recorded from a guitar playing for about 1.3 seconds. The two *lower graphs* show useful information that can be seen in the two *upper graphs*, and generally in all waveforms and spectrograms/sonograms. (*Source*: sound from Mixkit, in the public domain; plotted with WaveSurfer.)

ANIMATION 5*2 — Listen to the sound of a guitar in my video WA2 at time 14:06 in its section "**Viewing sounds**" under the title "**What do wave forms and spectrograms (also called sonograms) show us?**" (See details in the section References and Resources below.)

A note on vocabulary will be useful here, as different authors and software programs may use slightly different terms: **A spectrum displays amplitude (vertically) *versus* frequency (horizontally) at a fixed moment in time; a spectrum is also called a spectrogram. A spectrogram can instead display a frequency spectrum (vertically) as it evolves over time (horizontally); a spectrogram is also called a sonogram**. We will use both spectrogram and sonogram interchangeably, depending on the software used to produce them: no confusion should arise.

What does a human voice look like? For the human voice, due to the complex shape of the mouth, **we can expect a complex spectrum**. So let's start with the simplest possible sounds: a musical scale, as sung and displayed in Figure 5-10. Here we see the notes "do – re – mi – fa – sol – la – si/ti – do" climbing through one full octave, very much as they would in many musical instruments. The spectrum of these notes is still rather simple: the spectrum for each note consists of a fundamental frequency (shown as the rising bottom red dashed line) and many harmonics near whole multiples of that fundamental frequency; the upper red line has double the frequency of the lower red line. The red lines show that the fundamental frequency and the higher harmonics increase steadily from one note to the next. The horizontal blue dotted line shows that after climbing through one complete octave, the frequency has doubled: it reaches the second harmonic of the initial "do", which has double the fundamental frequency. The lower graph extends the frequency range to 5,000 Hz from the 1,000 Hz limit in the middle graph: it shows that the harmonics (the horizontal "ribs") continue to high frequencies. The frequencies wiggle rapidly up and down: these wiggling frequencies are an artistic feature of singing, whereby the frequency itself oscillates (it is often called **vibrato**: see Figures 6-4 and 6-20, as well as Sections 6.9 and 6.11): the wiggles are amplified in the higher harmonics.

Now let's look at real words: Figure 5-11 shows a recording of the words "YES! Yes? Noo…". The spectrogram is overwritten with the

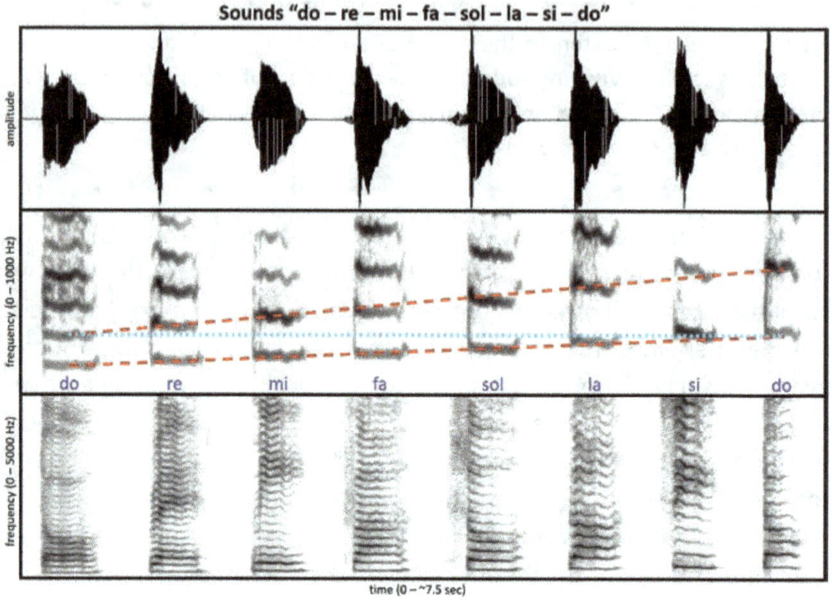

Figure 5-10: Waveform (*at the top*) and spectrograms (*at the center* and *bottom*) recorded from a woman singing the notes "do – re – mi – fa – sol – la – si/ti – do" over one full octave. The central spectrogram zooms in on smaller frequencies than the bottom spectrogram: its range is 0 to 1,000 *versus* 0 to 5,000 hertz. Each note takes about 0.5 seconds. The consonants (like the "d" of "do" and "r" of "re") are very brief: we see mainly the vowels (like the "o" of "do" and "e" of "re"). (*Source*: recorded with a smartphone; displayed with Praat).

letters at the time they start being pronounced. We can recognize the individual letters by their different wave forms. Also, the 's' gives a high-frequency hissing noise high up in the spectrum, from about 2,500 hertz to beyond 5,000 hertz (above the top of the middle spectrogram); the 'o' gives a low-frequency sound below 4,000 hertz; the brief 'y' has both low and high frequencies.

Can you see those tonal changes in the middle spectrogram? Notice the rib-like stripes near the bottom following the ups and downs of the intonation! Those ribs are sets of harmonics that drop or rise together: we see nicely how the second harmonic frequency always remains 2 times the fundamental frequency; similarly, the third harmonic always has a frequency 3 times that of the fundamental, and so on for higher harmonics.

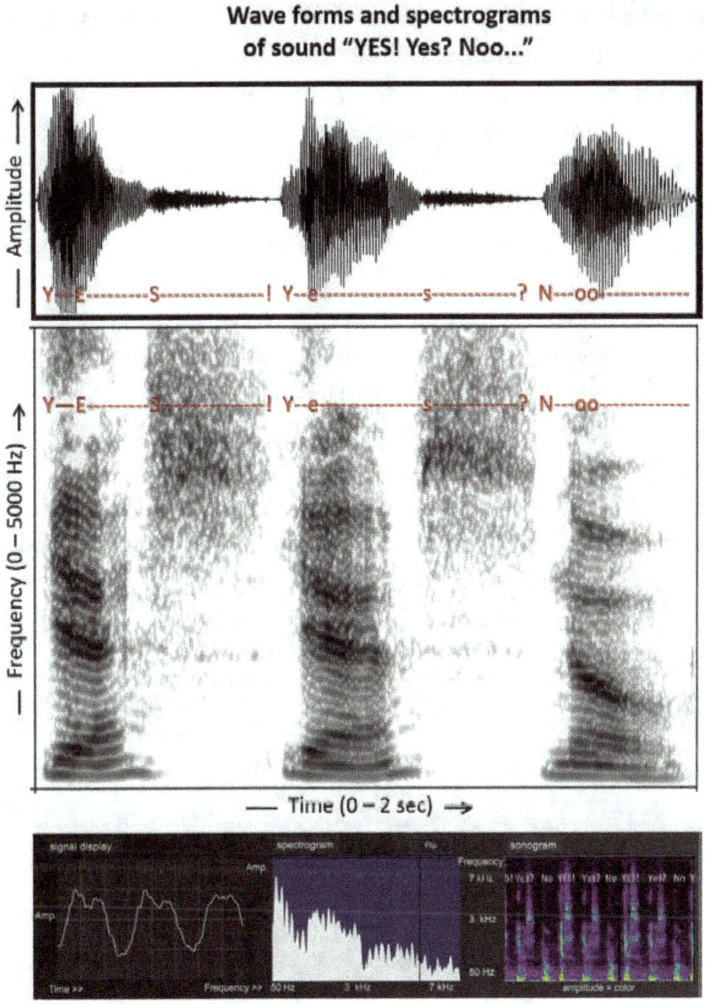

Figure 5-11: Waveforms and spectrograms recorded from my voice saying "YES! Yes? Noo...", for about 2 seconds.

At the top: Waveform (mono, not stereo) with the three words marked (letters are placed where their sounds start, while dashes show how long the letters continue, as in 'Y---E--------S---'), displayed in Figure 5-7. (*Source*: recorded with smartphone; displayed with Praat.)

At the center: Spectrogram of the wave form at top with the three words marked, displayed at the top in Figure 5-8. The frequency range is from 0 to 5,000 hertz. Black is the most intense.

At the bottom: Graphs for the same sound as at the top, displayed at bottom in Figure 5-8; the signal display and spectrogram show the sound of the letter "Y" in "YES!", while the sonogram repeats the "YES! Yes? Noo..." several times. (*Source*: recorded by a smartphone; displayed with Praat for the top two graphs, with A.L.M.S. — Audio Literacy for Music Students — for the bottom graph.)

We thus can "see" the intonations in the spectrograms, expressing moods in addition to words.

The top half of Figure 5-12 shows the same voice pronouncing the same words in the same way as in Figure 5-11, but with colors instead of grays. In addition, this adult male's voice is compared with that of a 7-year-old girl speaking the words "yes" and "no" (bottom half of Figure 5-12). Young girls speak with a much higher pitch (frequency) than adult males, and this is visible in the figure: notice how the green waves oscillate faster (more tightly in the wave form) for the girl than for the male; also notice that the "ribs" in the spectrograms are spaced apart much more for the girl than the male, due to the girl's higher frequencies. We will return to the interesting topic of the voice in Chapter 6.

What does "noise" look like? Let's contrast **noise** with the sounds from a piano and the voices shown above. There are many kinds of noise: basically, all undesirable or unpleasant sounds are noise, while

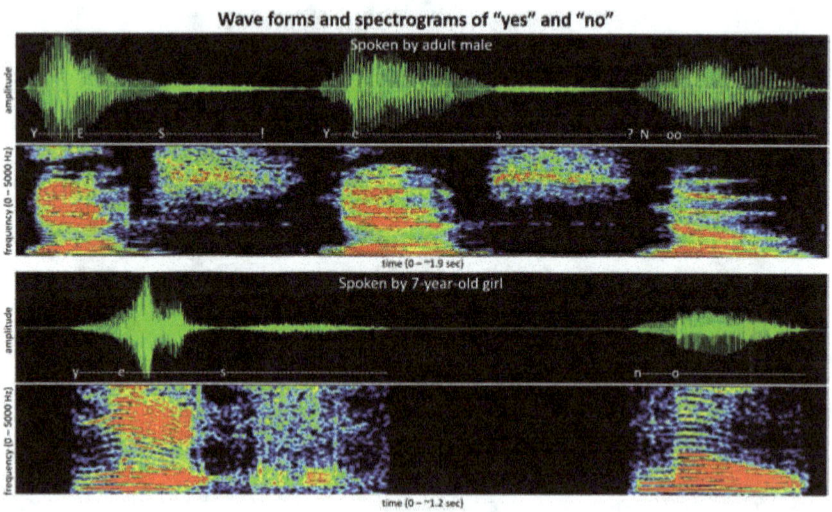

Figure 5-12: Waveforms and spectrograms recorded from my voice as an adult male (top pair of graphs) saying "YES! Yes? Noo..." and a 7-year-old girl (bottom pair of graphs) saying "Yes" and "No". The adult male sound is identical to that in Figure 5-11, but shown in color here (red/yellow are most intense). (*Source*: recorded with a smartphone; plotted with WaveSurfer.)

a more restrictive definition of noise could be "non-musical" sounds. Figure 5-13 shows sounds that I recorded in a noisy restaurant, sounds of crumpling pages of paper in a quiet room, and sounds from a rushing mountain stream. **We see some obvious features of noise: wildly irregular wave forms (signal display), dense spectrum with no harmonics (spectrogram), and wildly varying spectrum (sonogram). We prefer sounds with more regularity and more harmonics that gives them a more musical character!**

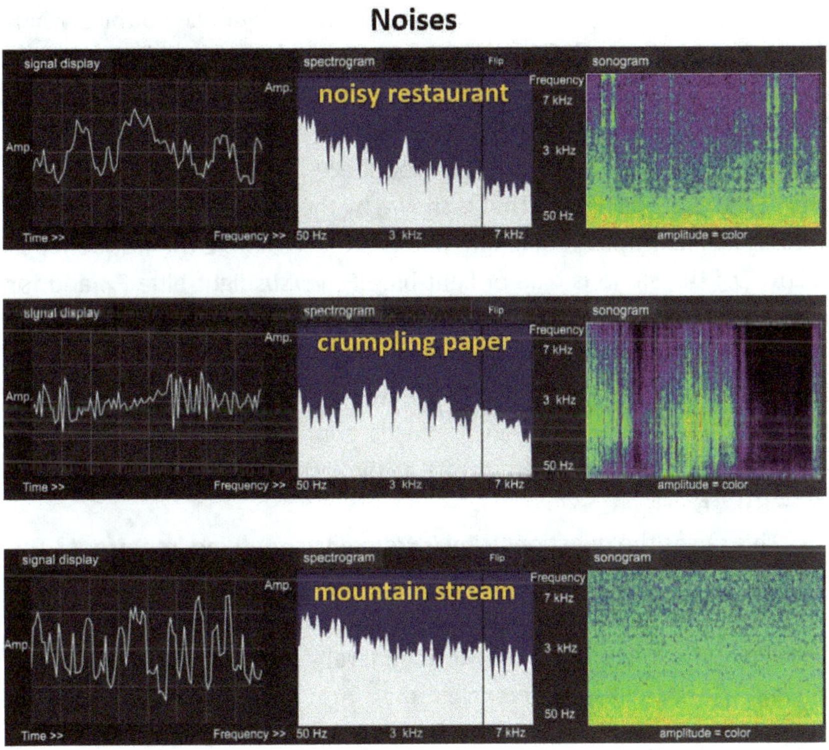

Figure 5-13: *Top*: Sounds recorded in a very noisy restaurant. *Middle*: Sounds due to crumpling paper. *Bottom*: Sound from a mountain stream. Displayed as in Figure 5-8; the sonograms again cover about 8 seconds. The signal displays at the left cover about 0.014 seconds, but only show the sound wave about every 0.0002 seconds, making the signal more angular than in reality because the highest frequencies are not included (see the discussion about digital sound sampling in Section 5.8). (*Source*: recorded with smartphone; the graphs were produced with A.L.M.S. — Audio Literacy for Music Students).

5.3 The Role of Halving or Doubling the Length of Strings

To understand music, it is important to ask: ***What is the role of halving or doubling the length of strings?*** Look more closely at the three red strings in Figure 5-4. The middle one (C_5) has half the length of the leftmost string (C_4), while C_5's length is double that of the rightmost string (C_6). This halving and doubling is chosen on purpose and is very important. Since the strings are identical to each other (apart from their length, but including equal tension), **the frequency doubles when the string is shortened by half.** Therefore, the note C_5 has double the frequency of C_4, and is one octave above C_4; and C_5 has half the frequency of C_6, and is one octave below C_6.

The same length halving and frequency doubling exists for all the other strings in Figure 5-4. That is shown by the strings' colors: the yellow note D_5 has half the string length and therefore double the frequency as yellow D_4; the same is true of light-blue F_5 *versus* light-blue F_4, and for dark-blue A_5 *versus* dark-blue A_4, *etc.* Now you can see why the lettering "C – D – E – F – G – A – B – C" is repeated (with different numbering 4, 5, *etc.*) and also why the notes "do – re – mi – fa – sol – la – ti/si – do" are repeated: each repetition produces another octave. Octaves on a normal piano are numbered from 1 at the left (A_1 – B_1) to 7 at the right (A_7 – B_7): see Figure 5-3.

This raises the question: ***Why is frequency doubling important?*** The answer appears to be related to how the human brain functions. Our brain feels a close similarity between two tones related by a doubled frequency, such as C_4 and C_5. More generally, **we hear harmony when mixing sounds with frequencies that are simply related**, such as frequencies in simple ratios like 1:2, 2:3, 3:4 or 2:5 that involve small whole numbers. And that remains true of any other notes besides C_4 and C_5: it includes the "black" notes ("sharp" and "flat") in between the notes shown in Figures 5-3 and 5-4.

In fact, the desire for "harmony" is so strong that, for most people, it is more important than playing the correct frequency for C_4, say: we tolerate incorrect frequencies, but dislike inaccurate frequency doubling! We generally prefer simple frequency relations like 1:2, 2:3, 3:4 or 2:5 over more complex relations like 7:9, 5:11 or 3:8, for example. You can test your own preferences online by mixing two sounds of different frequencies.[3]

However, there is a conflict between the desire for simple harmonious relations between frequencies (like 1:2, 2:3, *etc.*) and the desire to have, say, 12 equally spaced frequencies in an octave (so that we can provide a continuous set of frequencies with the same relation between each successive pair of frequencies). The currently accepted solution is the **equal-tempered scale** with 12 equally spaced frequencies in an octave. On this scale, the relations between frequencies are not as simple as 2:3, *etc.*, but they are close enough to such simple relations for most purposes.

It should be noted that "harmony" depends to a certain extent on culture (such as Western *versus* Asian). The perception of harmony also depends on the kinds of music people were exposed to when they were young.

5.4 What Controls the Sound Frequency Produced by a String?

A wave's natural oscillation frequency, and therefore its note, depends on several factors: the frequency is increased if the string is made shorter or lighter, or put under greater tension; conversely, the frequency is decreased if the string is made longer or heavier, or put under less tension.

So you can change a string's frequency by changing its length (by changing the distance between its ends) or its tension (by pulling more or less on both ends); you can also change its mass (by switching to a different string).

---

[3] Online Tone Generator, The 432Hz Frequency (be careful not to produce too loud sounds, as they may damage your hearing). If you open the following link twice as separate pages in your browser, you can let one page make the 432 hertz sound, while on the other page you can easily sweep through other frequencies by placing your cursor in the frequency box and pressing your keyboard's up and down arrows, so you hear both frequencies at the same time: https://onlinetonegenerator.com/432Hz.html

You can do the same thing with another Online Tone Generator; it allows you to sweep frequencies with your mouse: https://www.szynalski.com/tone-generator/

In musical string instruments, this **tuning of the frequency is accomplished by a human tuner or player who adjusts the string's tension to obtain the desired frequency**, for example, as we saw in Section 2.6 with beats.

5.5 How Can We Make a String Oscillate with Two Fixed Ends?

If both ends of a string are fixed, you can no longer shake an end to start a wave, but you can still shake any other piece of the string to make the whole string oscillate. Repetitive shaking with your hand is, however, not a convenient way to produce music, because our hands cannot shake the string fast enough to generate the desired frequencies of music: try to shake your fingers back and forth ten times per second; you need much faster shaking than that to produce audible sounds!

Other ways to cause waves on a string that is fixed at both ends are to **strike**, **pluck** or **bow** it. Striking means rapidly hitting a small piece of the string, as in a **piano**. Plucking is done by pulling a small piece of the string sideways and then releasing it suddenly, as is often done on a **guitar**, **mandolin**, **lute**, **pizzicato violin**, *etc*. Bowing means sliding a sticky **bow** across the string: by repeated sticking and slipping, this bow causes waves in the **violin's** string.

It is easy to strike, pluck or bow a string, but none of these actions favor a particular frequency of vibration: these actions have no clear repetition and thus no clear frequency. As a result, striking, plucking and bowing create many waves at once on that one string: the result is a mix of waves of different frequencies and a corresponding complex tone. However, we have seen that the string, due to its tension, will favor one dominant frequency to form a note: this will filter out all frequencies except the fundamental frequency of the string and its harmonics.

The result will look like Figure 5-5: the string will vibrate with several waves, the fundamental (which is the first harmonic), the second harmonic, ..., the fifth, as well as more harmonics, namely the sixth, ..., the twentieth, and so on without end. By superposition, all these harmonic waves add up to a combined wave that corresponds to the rightmost black wave in Figure 5-5.

We can ask: ***How strong are the harmonics that are generated when we strike, pluck or bow a string?*** This question is important because different musical instruments can be distinguished by the sound they generate, which is due to the relative strengths of the harmonics they produce, as we will discuss in Section 5.7. Let's consider here the plucking of a string, as it is the easiest case to describe.

When you pluck a string (for example, on a guitar), you pull the string sideways, so it forms a triangle; an example is drawn as a gray line in Figure 5-14. Before you release the string, let's assume it is briefly motionless: the string has no movement. The tip of the triangle is bent by being pulled sideways. The tension from the two straight sections tries to pull the bent tip back toward the relaxed position of the string. However, the two straight sections themselves are not pulled toward their relaxed positions because a straight string only has forces pulling parallel to the string itself (we can ignore the relatively weak force of gravity here).

Now release the string: the bent tip will start moving toward its relaxed position, dragging with it the neighboring pieces of the straight

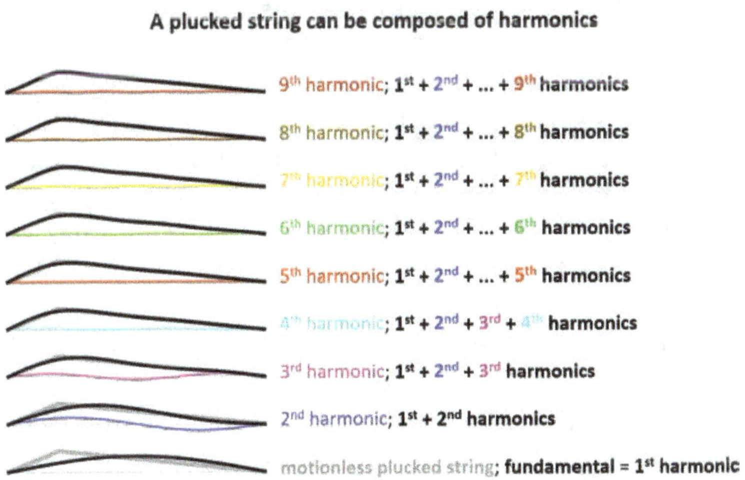

Figure 5-14: A triangular-shaped plucked string with fixed ends (drawn as a gray line) can be represented as a combination of harmonics (lowest black and colored waves). More harmonics (only nine are shown here) allow a more exact representation, especially of the sharp tip of the triangle.

sections. This is similar to starting a wave pulse on a string, as we discussed in detail in Section 2.2 (see Figure 2-6), except that now we create two wave pulses which go toward opposite ends of the string. These wave pulses will be reflected by the two ends of the string, coming back to interfere with themselves and each other. This seems to become rather complex and messy! Fortunately, there is a more accurate way to describe this process, using the harmonic standing waves of Figure 5-5.

The general idea of the simpler description is this: 1) we represent the stationary plucked string as a combination of all possible harmonics on that string (this already gives the relative strengths of the harmonics); 2) after releasing the string, we let each harmonic oscillate as it does naturally, with its own frequency and unperturbed by the other oscillating harmonics; 3) at each moment in the future, we sum up all those harmonics to produce the wave shape at that moment. Let's discuss these three steps next.

Step 1) Figure 5-14 shows how we can combine harmonics to form the stationary triangular wave. At bottom, we start with a simple fundamental standing wave with one loop: the black sine wave is very roughly similar to the gray triangular wave, but it undershoots at the left, overshoots at the right and misses the tip entirely. The second harmonic (shown in dark blue) is added to try to correct the errors of the fundamental wave: this increases the left part and decreases the right part, so that the result (next black line, labeled "1st + 2nd harmonics") looks more like the desired triangle (gray). There still are clear differences from the triangle. The third harmonic (purple) is then added to try to correct those differences: the result (next black line, labeled "1st + 2nd + 3rd harmonics") indeed resembles the gray line more closely.

This process of adding higher harmonics is continued: going up in Figure 5-14, we see the tip in the black combined wave slowly becoming sharper, while the two sides slowly become straighter, looking more like the perfect gray triangle. The figure stops at the ninth harmonic (dark red at the top), but could continue forever (or at least until the tip is sufficiently sharp).

What we have done here is a remarkable and extremely important feat of physics: we have represented one shape (the triangular string) by a combination of simple shapes (the harmonic waves that fit the

length of that string). Most remarkable is that this can be done for any possible shape of the string, not only triangles; it can even be done for the square wave and the sawtooth wave shown in Figure 2-8, despite their very sharp corners (these corners, however, require many very high harmonics). The importance of this representation by simpler waves is that we can now quite readily and accurately predict the future behavior of the plucked string after we release it in step 2.[4] See Box 5-1 for a more general discussion of the importance of such approaches in physics.

BOX 5-1 – THE POWER OF MATHEMATICS IN PHYSICS: Figure 5-14 illustrates a remarkably successful technique used by physicists and engineers to describe complex situations in many areas of science and technology, from musical strings and telecommunications to medical imaging: this particular technique is called Fourier expansion or Fourier transform, whereby a shape that can be very complicated (like the shape of a string) is represented by a combination of simple shapes (in our case, the harmonic waves with sine shapes). Since we know how the individual simple shapes behave (they are oscillating standing waves), we can thereby predict how their combination will behave in the future. Thus, a single technique can solve a wide variety of problems, thereby avoiding trying to solve each case separately.

This and other powerful techniques (such as the theories of gravity, electromagnetism and quantum mechanics) are based on the fascinating observation that nature behaves just like mathematics. It is indeed far from

(Continued)

[4] You may wonder, and rightly so, how we know the amplitudes of the different harmonics that add up to the triangular (or any other) wave shape. These amplitudes are given by the so-called Fourier coefficients, which are mathematical formulas that can be found on the web, for example at Wolfram MathWorld: https://mathworld.wolfram.com/FourierSeriesTriangleWave.html. In brief: any wave can be replaced by a Fourier series, which is the collection of harmonics in our case; the series can have an infinite number of harmonics. Each harmonic can contain a sine wave and a cosine wave (which is just a shifted sine wave). For waves with fixed ends, only sine waves are needed, with amplitudes given by the last formulas at the above link. Figure 5-14 shows the case of m = 5, for which the tip of the triangle is at 1/5th of the length of the string); the nine first amplitudes are: 0.7444, 0.3011, 0.1338, 0.0465, 0, –0.0207, –0.0246, –0.0188, –0.0092 (for a tip height of 1). Notice how these successive amplitudes generally shrink, as plotted in Figure 5-14.

(*Continued*)

obvious that nature should follow the same principles as the mathematics invented (or discovered) by humans.

As a simple example, when we put together 5 balls and 2 balls, we count 7 balls: this observation agrees with the algebra which says that $5 + 2 = 7$. This example seems very obvious, but think of the price of balls: the price of 7 balls may not be equal to the price of 5 balls plus the price of 2 balls, if we get a volume discount (such as a 50% discount for balls beyond the 5th); algebra does not predict this because algebra does not include discounts! Humans invented discounts (which therefore vary in unpredictable ways), while the rest of nature does not offer discounts.

Thus, we are lucky that nature follows the strict logic of our mathematics. This amazing fact allows us to model nature in formulas, equations, laws, *etc.*, that can be extremely accurate and reliable. We can thereby predict with superb accuracy when the next lunar eclipse will occur and exactly where on Earth it will be visible; similarly, we can aim a spacecraft to land on another planet; we can even send a spacecraft to land on a tiny, distant and fast-moving comet or asteroid. We can also accurately predict the physical properties of new composite materials that don't even exist in nature before we manufacture them. This predictive power of physics explains why humans could design and optimize machines that were not present in nature, from steam engines and lasers to smartphones.

All this is due to the possibility of mimicking nature in mathematical equations: solving the mathematical equations then tells us what nature would do under new circumstances. However, physics at present does not have all the necessary equations for all circumstances. For example, you may have heard of dark matter and dark energy in the cosmos: the word "dark" reflects our ignorance about them. Someday, physicists will hopefully understand dark matter and dark energy, and thereby advance science further, just as relativity and quantum mechanics were major advances in the last century.

In this book, we avoid mathematics and focus on the physical forces and processes. Physics can indeed be viewed on two different levels. On one level, we can understand the forces and processes with no mathematics at all: we can understand why and how waves travel along strings, or how forces keep the Sun, planets and spacecraft together. On the mathematical level, we can make detailed predictions of what those forces and processes will cause in the future, such as the precise timing of eclipses and accurate spacecraft trajectories.

Step 2) We know how individual standing waves (the harmonics) behave: they oscillate across the string without any motion along the string; see Sections 2.1 and 2.8. We also know that higher harmonics oscillate at higher frequencies (double the fundamental frequency for two loops, triple for three loops, *etc.*); see Section 2.9. So releasing the plucked string means that we can simply release the different harmonics simultaneously: each will start oscillating at its own frequency. For example, the fundamental wave (with one loop) will have a certain frequency of oscillation; the second harmonic (with two loops) will oscillate twice as fast, *etc.* It is very important to realize that these harmonics will oscillate totally independently of each other, so the shape of the combined wave can be predicted from then on, as we do in step 3.

Step 3) Figure 5-15 shows a computer simulation of how the plucked string behaves after its release. It only shows the time evolution

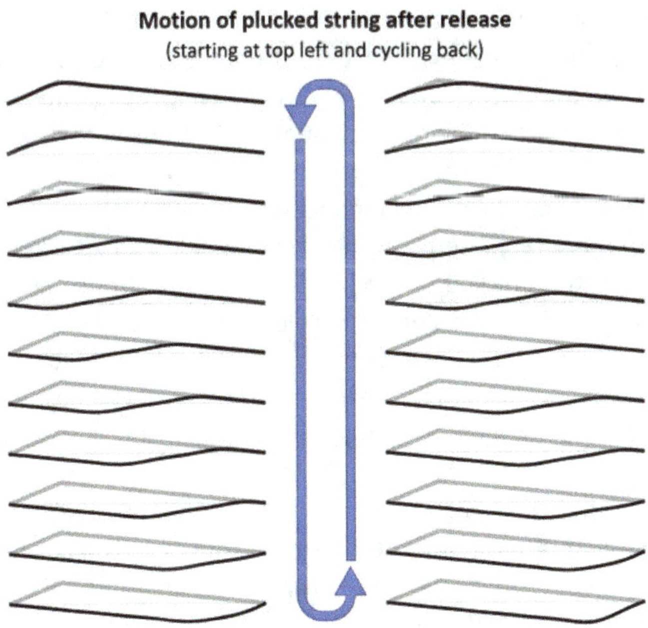

Figure 5-15: A triangular-shaped plucked string with fixed ends (drawn as a gray line) is released at the top left and allowed to evolve over time (following the blue arrows). It cycles back to the starting triangular shape and repeats the cycle endlessly.

of the combined top black wave of Figure 5-14; it includes the first nine harmonics and closely resembles the triangular plucked string (shown again in gray). Figure 5-15 follows the wave during one whole cycle (which is the time it takes the fundamental wave to oscillate once): after this time, the cycle repeats itself identically again and again (we ignore losses due to friction or emission of sound, which would gradually stop the oscillations). You may also view this motion in my Animation 5*3.

ANIMATION 5*3 — See the behavior of a plucked guitar string in my video WA1 at time 15:38 in its section "**Harmonics**" under the title "**String instruments: oscillation of a guitar string**". (See details in the section References and Resources below.)

Starting at the top left in Figure 5-15, the tip of the triangle is pulled down by the nearby straight string segments, which in turn are pulled down by the tip. This sends two pulses to the left and right along the string (as we imagined at the beginning of this section). The simulation follows the pulses as they reflect from both ends and interfere with themselves and with each other. Notice that, while the harmonics do not move along the string, their combination does! **We see that combining standing waves can produce traveling waves, just as combining traveling waves can produce standing waves** (see Section 2.7).

We observe very interesting behavior: at half the cycle (see bottom right frame), the plucked string has moved its tip from upper left to lower right: the tip has somehow traveled from one end of the string to the other end, and from one side (above) to the other side (below). And, after another half-cycle, the tip has returned to its initial position. From then on, the motion repeats itself again and again. You can watch online very similar motion on a bowed violin string.[5]

The central conclusion of this section is: **Plucking a string creates a large number of harmonics. The same is true of striking or bowing a string.** (The latter two cases are slightly more complicated than for the plucking because of additional initial motion.)

[5] See video "Bowed violin string in slow motion" by ViolinBOW: https://www.youtube.com/watch?v=6JeyiM0YNo4

5.6 How Many Strings does a Musical Instrument Need?

Today's standard **piano** produces 88 notes with about 230 strings (most notes use three identical strings, while some notes use two or one); a single octave on a piano contains 12 notes (including the white and black keys). At the other extreme, the Vietnamese **đàn bầu** and the related Chinese **duxianqin** have a single string; the Chinese **erhu** has two strings; a **violin** has four strings; and a **guitar** typically has six strings. ***Even a one-string instrument can produce multiple notes: how so?***

Remember that a single string can produce different frequencies by varying its length, or its tension, or its mass. The question is whether it is practical to do so while performing music. The length of a string is easiest to change rapidly: **pressing a finger at the right position on the string fixes it there, so its vibrating length is shortened, which produces a higher frequency of vibration. This is the system used in violins, guitars, *etc.*** In violins, the player can place a finger anywhere along the **fingerboard**; in particular, the frequency can be smoothly changed by sliding the finger along the string. In guitars, as shown in Figure 5-16, metallic strips called **fretbars** are used: they predefine certain frequencies. Fretbars limit the number of possible string lengths but permit easier control of the frequencies.

Figure 5-16: The metallic fretbars across the fingerboard of a guitar are used to change the vibrating length of strings. (*Source*: by Tomgally, in the public domain, https://commons.wikimedia.org/wiki/File:Frets,_guitar_neck,_C-major_chord.jpg.)

A further option with violins, guitars, *etc.* is to pull the string sideways (for example across the fingerboard or along a fretbar), thereby increasing the length and tension of the string, and thus changing the frequency continuously.

We see that a few strings can already produce many notes: the violin with its four strings is a perfect example of that capability. Nevertheless, having more strings does allow more combinations of simultaneous notes (such combinations are called **chords**): on a piano, you can place up to ten fingers on at least ten different keys at the same time (not to mention arms, elbows, nose, or even feet), allowing a very rich range of combinations of notes.

5.7 Distinguishing the Sounds from Different Instruments

Can you recognize a piano, guitar or violin when they play the same note? Of course you can hear the difference, but why do piano, guitar, violin, etc. sound so different? According to our exploration so far, all of these instruments can produce the same fundamental natural frequency as well as the same harmonic frequencies. And this will remain true also for wind instruments and other **musical instruments**; nevertheless, these other instruments sound even more different than the string instruments.

Figure 5-17 compares sounds from seven different instruments. It is visually clear that there are major differences in the character of their sounds, especially when looking at their frequencies: see the central spectrograms. You can compare the sounds of several instruments in my Animation 5*4.

ANIMATION 5*4 — Listen to my video WA2 at time 15:00 in its section "**Sounds of musical instruments**" under the titles "**A noisy instrument: the cymbal**", "**A contrasting percussion instrument: the xylophone**", "**String instruments: piano *vs.* violin**", and "**Wind instruments: saxophone *vs.* trumpet**". (See details in the section References and Resources below.)

However, we have to be careful with such comparisons for several reasons: First, as we have discussed in Section 5.2 in connection with

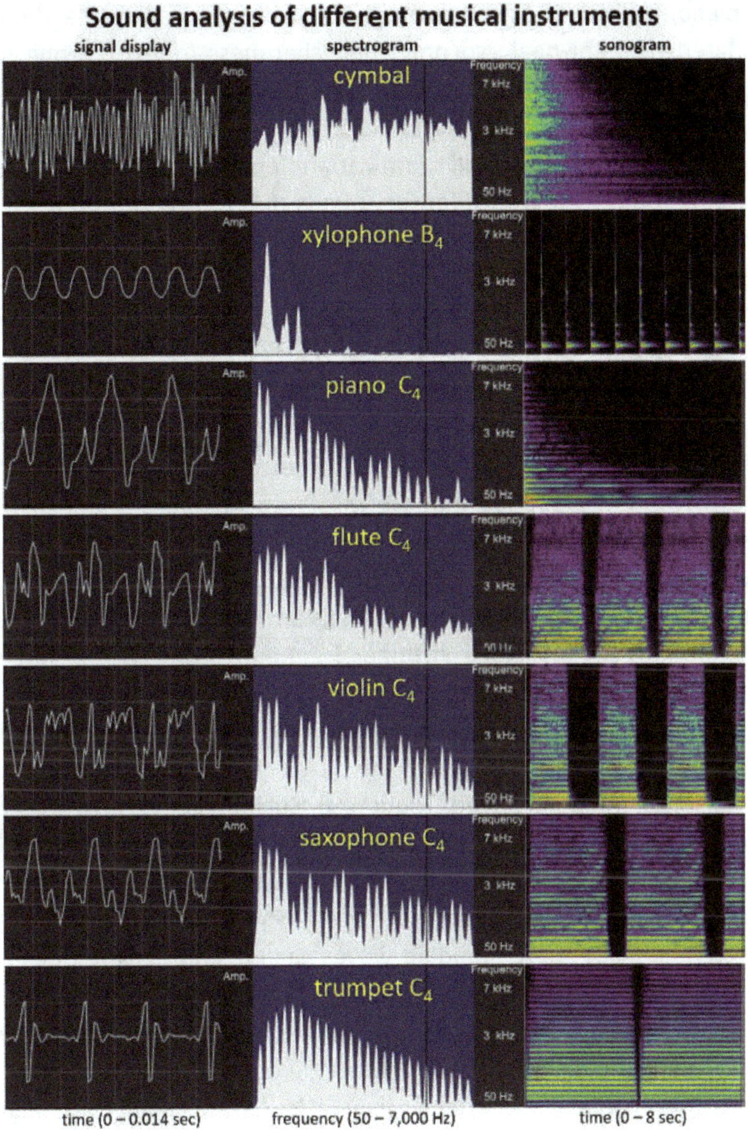

Figure 5-17: The sounds of different musical instruments are analyzed as in the bottom of Figure 5-8. The instruments played C_4 (about 262 hertz, except the xylophone, which played B_4, about 494 hertz, and the cymbal, which has no dominant frequency or pitch). The signal displays at the left cover about 0.014 seconds, while the sonograms at the right cover about 8 seconds. For the cymbal and piano, the sound of a single hit is shown in the sonogram. For the other instruments, the sound recording lasted less than 8 seconds and is therefore repeated until 8 seconds have passed. The spectrograms were recorded near the moment of maximum intensity, marked by red/yellow colors in the sonograms. (*Source*: sounds from University of Iowa Musical Instrument Samples (MIS); these graphs were produced with the app A.L.M.S. — Audio Literacy for Music Students.)

the piano, individual signal displays (wave forms) cannot be directly compared peak-for-peak, so only their sharpness (due to strong high frequencies) and regularity (due to the dominance of harmonics) are really meaningful. Second, the same instrument recorded in different environments (such as a small room with reflecting walls *versus* a sound-absorbing room) can give different sounds due to **reverberations** from wall to wall. Third, the placement of the microphone(s) also affects the recorded sound. Fourth, individual instruments of the same type also differ in their sounds. Fifth, the way the player handles the instrument affects its sound.

Let us start with the sound of a **cymbal** (at the top in Figure 5-17). This sound results from striking the cymbal once with a mallet. The spectrogram and sonogram show a wild set of frequencies, similar to some of the noise sounds that we displayed in Figure 5-13. There is no single dominant frequency, and there are no obvious harmonics. However, compared to the noises in Figure 5-13, we do see stronger frequencies standing out: in the sonogram for the cymbal, we see horizontal streaks indicating stronger frequencies, but we do not see such horizontal streaks in the noises of Figure 5-13. It is also notable that the amplitudes of middle and higher frequencies are often higher than those of lower frequencies: this gives the cymbal sound its high metallic pitch (high perceived frequency). Such a variety of frequencies is not surprising, because the cymbal indeed sounds like noise; that sound is even richer in frequencies than a simple metal plate dropped on a hard floor because the cymbal has a much more complicated shape that creates many more standing waves than a simple plate does. We also see in the sonogram how the frequencies die out over time: high frequencies die out faster than low frequencies, but we also see that a few low frequencies die out more slowly than others nearby. So, after a few seconds, the cymbal gives a lower pitch (lower perceived frequency).

Next, we look at a **xylophone**. We notice immediately: its signal looks close to a simple sine wave; its spectrum has few strong frequencies, one of which dominates (in this case B_4 at about 494 hertz), while most of the others are simple harmonics of B_4; and it has very few higher frequencies. We also see in the sonogram that the sound of the xylophone weakens very fast: it dies out within half a second or so, compared with about 8 seconds for the cymbal (in the sonogram, the xylophone's signal is repeated identically about every second). We can understand this behavior from the simple shape of a xylophone bar (see Section 4.1 and Figure 4-3): a rectangular box shape creates relatively few standing waves; also, such a bar rapidly radiates its energy away in all directions, while its support may quickly dampen the vibrations.

The **piano** (already discussed in some detail in Section 5.2) has a richer spectrum dominated by many harmonics of the fundamental frequency C_4 (about 262 hertz). As a result, its signal is complex but very repetitive. The higher frequencies are weak: there are almost no significant frequencies above about 5,000 hertz (this is partly due to the **soundboard**, which is designed to absorb higher frequencies). We note again in the sonogram how the amplitudes of different frequencies vary with time in a complex manner (see also Figure 5-8): apparently, beats occur (with repeat times of about 1 to 2 seconds) at some frequencies, and these beats are different from each other; such beats may be due to vibrations on different strings that are very close in frequency. It turns out that humans are quite sensitive to such time-dependent behavior: this particular pattern is characteristic of the piano and helps distinguish it from other instruments.

Turning to the **flute**, also playing C_4, we see a similarly rich spectrum as for the piano, but there are more non-harmonic frequencies between the harmonics, especially in the middle frequencies. At low frequencies, the fundamental C_4 is not the strongest frequency: nearby harmonics are stronger, giving a more complex sound. There are additional frequencies above 5,000 hertz. The sonogram looks very different because of the way the flute is played: unlike the piano, xylophone or cymbal, which are activated by a sudden strike, the flute is here activated by continuous blowing. In Figure 5-17, the note C_4 is kept going for about 3 seconds at constant loudness (and the sonogram repeats that note identically three more times). Within the 3 seconds of the note, there are variations in the amplitudes of the frequencies, which may be related to changes in the blowing by the player, such as changes in the shape of the lips or the strength of the air stream.

Next, we consider the **violin**, again playing C_4. The violin spectrum shown in Figure 5-17 shares features with the piano or the flute: it has few non-harmonics like the piano, higher frequencies like the flute, *etc.* But the higher frequencies are stronger, giving the violin a more metallic sound. One notable difference visible in the sonogram is that the frequencies themselves vary somewhat: within the 2 seconds of the tone, we see the frequencies dip and then rise slightly, in a mild U shape (see the higher harmonics which magnify this effect). This suggests a change in the vibrating length of the string, possibly due to moving the finger on the fingerboard (see Section 5.6). This possibility of the violin helps distinguish it from other instruments.

The spectrum of the **saxophone** strongly resembles that of the violin in this example. One difference, however, is the constant frequencies seen in the 3 seconds of playing the note C_4, but this may be affected by the way the player blows over time. The rising intensities toward the end of the 3 seconds of sound may be due to a damper being removed from the saxophone, which can also change the perceived pitch.

Lastly, the **trumpet** shows a very distinctive spectrum with a quite smooth sequence of frequency amplitudes, stretching into higher frequencies; the amplitudes peak around the fifth harmonic, leaving the fundamental C_4 and lowest harmonics relatively weak. Furthermore, the amplitudes stay very steady over time in this recording.

Are the differences which we have found in the sounds of seven musical instruments sufficient to distinguish the instruments? The differences we see between the flute, the violin and the saxophone, for example, are rather subtle: inexperienced ears may not pick them up. In fact, it has been found that the steady sounds from such instruments may indeed be difficult to tell apart. Another significant factor has been discovered: the initial start-up of the sound, called the **starting transient**; this factor is very important in identifying the instrument, even though it all happens within a fraction of a second.[6] Basically, different frequencies build up at different times and at different speeds, and this behavior is recognized by the brain and thereby provides another characteristic signature of the instrument. Let us therefore look more closely at the start-up of the sounds of the same instruments: see the blowups in Figure 5-18.

[6] Charles Taylor, "*Exploring Music: The Science and Technology of Tones and Tunes*", Institute of Physics Publishing, 1992.

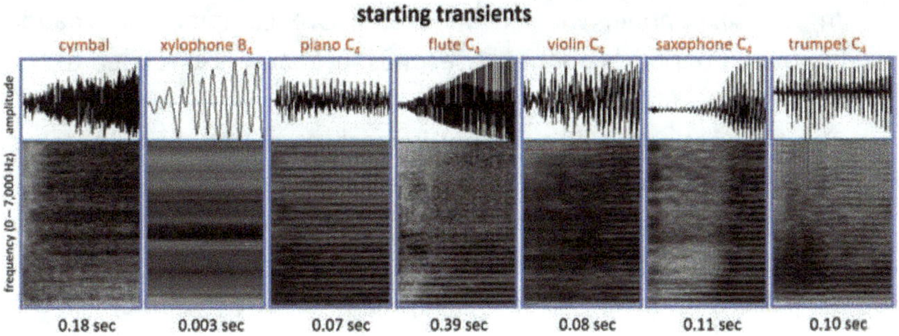

Figure 5-18: The initial sounds (starting transients) are shown for the same musical instruments as in Figure 5-17. For each instrument, the top panel displays the wave form (signal amplitude), while the bottom panel displays a sonogram, as in Figure 5-17; both cover the time indicated below. (*Source*: sounds from University of Iowa Musical Instrument Samples (MIS); analyzed with Praat.)

We first look again at the cymbal (at left in Figure 5-18). Its amplitude of vibration rises in about 0.2 seconds, but the higher frequencies are slightly delayed, as we can see at the top of the sonogram. This indicates that the vibrations need some time to spread out over the area of the cymbal and settle into standing waves.

The situation is totally different with the xylophone: here the wave grows within about 0.001 seconds, and almost immediately settles into its stable spectrum (this happens so fast that the mathematical analysis of the spectrum didn't even separate out the initial growth). The stiff, compact xylophone bar very quickly settles into its few standing waves.

For the piano, the initial transient takes about 0.01 seconds: during that time, we see complex variations in the frequency spectrum. This results from the spreading of the initial string vibrations to other strings and to the soundboard, where standing waves also take time to build up.

The flute is slower to start, within about 0.4 seconds in our example, probably due to the player building up their air stream, which itself appears to change considerably, especially at the higher frequencies.

The violin follows the piano's initial transient, but takes more time to develop standing waves in its resonance chamber in this example. This may be due not only to the particularly complex shape of the resonance chamber but also to the irregular beginning of the bowing action.

In the saxophone, we see an even more complex and chaotic buildup over a period of about 0.1 seconds, probably also related to the player's actions.

The trumpet's starting transient is more organized, but also takes about 0.1 seconds to stabilize. Again, the player has a big role here.

Now how do we distinguish different musical instruments? One important characteristic of a musical instrument is its spectrum: it can contain a single dominant frequency (xylophone), or two or more strong frequencies (piano, flute, *etc.*), together with some of their harmonics; or the spectrum can contain a dense multitude of unrelated (non-harmonic) frequencies (cymbal). So, **different instruments play, for the same note, different mixes of frequencies: this gives instruments their different tone quality, acoustic color or character, often called timbre** (timbre is a French word, which sounds like "tam-br" in English). If you played the fundamental frequency alone (which is possible with an electronic synthesizer or online), you could not tell which instrument produced it. In addition, however, we have seen how the time evolution of the spectrum plays a significant role in characterizing an instrument. In particular, **the initial buildup, the starting transient, provides a very important component of the character of a musical instrument.**

5.8 Recording and Playback: Analog and Digitized Sounds

Today, most recording of sounds is done electronically by digitizing the sound waves and storing them as digits in electronic memory for later playback. Nevertheless, analog recording and playback still exist, for example, with vinyl "records" using phonographs. Also very common until the 1980s was magnetic tape recording, especially with audio cassettes.

Analog recording and playback rely on duplicating the sound waves in a different medium for permanent storage. The **phonograph**, invented by Thomas Edison in 1877, was the earliest such approach; it evolved toward today's **vinyl disc records**. The idea is to let a sharp needle "write" a sound wave (detected by a microphone) directly into a soft recording material, leaving a permanent trace of the sound wave. Indeed, if you look closely at a vinyl record (vinyl is a kind of plastic), you

Figure 5-19: Photograph of grooves in a vinyl record, including white specks of dust. (*Source*: Shane Garwin, under CC BY-SA 2.0, extracted from https://commons.wikimedia.org/wiki/File:Vinyl_groove_macro.jpg.)

will see a snaky groove that goes round and round in a spiral, as seen in Figure 5-19; you may also notice that the two sides of the same groove are often different because each side is used for one of the two channels of a stereo recording. That recording can be copied for distribution to many buyers.

For **playback**, a needle in the buyer's phonograph follows the same groove, including the stereo effect, and converts it to an electrical wave that activates one or more loudspeakers (usually through an amplifier to make the sound stronger). Damage, especially scratches across several loops of the spiral, is a disadvantage of vinyl records. Also, it is not practical to record a new signal over an old recording.

Magnetic analog recording started in the 1930s in Germany, using **magnetic tapes**. The principle here is to copy a sound wave into a permanent wave-like variation of the magnetism in a metallic tape. This is achieved with an electromagnetic "head" which converts the sound wave into a strong, varying magnetic field that modifies the local magnetism in the tape as it moves along the head. Stereo sounds can easily be recorded on two parallel tracks; furthermore, multiple tracks can record sounds heard from different angles to produce "surround sound", for example for the reproduction of a concert or conversation through multiple loudspeakers placed around a room.

Such a magnetic recording is faithfully copied onto many identical tapes for distribution to buyers. Magnetic recording allows overwriting an earlier recording, making it particularly attractive for home and office use. The compact cassette became especially popular thanks to its ease of use, leading to the pocket-sized Sony Walkman and its successors.

Playback is the reverse process of recording: a magnetic head senses the varying magnetism in the tape and converts it into an electrical signal that activates loudspeakers (after amplification).

Digital recording and **playback** have largely replaced analog recording and playback, starting in the 1970s, in parallel with digital photography. Compared to analog information, digital information is easily modified (for example, to reduce noise or to mix different sounds), easily stored and copied (in electronic files), largely permanent (with very little damage or loss of information over time), and very compact (electronic storage takes little space).

The principle of digital recording is to chop up a wave into very brief "samples", just like digital photography chops up an image into small "pixels" (picture elements). This process is called **sampling**. The wave's variation in time is replaced by a series of successive samples of that sound, as shown in Figure 5-20: the blue wave is replaced by a list of the heights of the red bars; the remainder of the blue wave between the red bars is simply ignored. Each sample is thus converted to a number which tells the air pressure at that moment in time. Sampling reduces the amount of information contained in a wave, allowing more compact storage and faster transmission.

However, such sampling loses parts of the original sound, especially the fine details. In order to keep the fine details of the wave, it is important to use a sufficiently dense sampling, called **sampling rate**. In Figure 5-20, the example using a low sampling rate misses many rapidly varying details of the wave: such details are high-frequency parts of the sound. Therefore, a high sampling rate is needed to represent high frequencies correctly. In practice, a sampling rate that is double the highest desired frequency is suitable. Since humans hear up to about 20,000 hertz, it is necessary to use a sampling rate of about 40,000 hertz (meaning 40,000 samples per second, or 1 sample every 1/40,000 seconds). Typical sampling rates used in digital audio are 32,000, 44,100, 48,000 and 96,000 hertz, depending on the application.

In digital recording, the amount of information stored and transmitted is further reduced, in addition to sampling, by "rounding off" the sample heights (pressure values) in a process called **quantization**. There is no need to list the sample heights with very high precision if we cut out significant parts of the wave anyway during sampling. The idea is this:

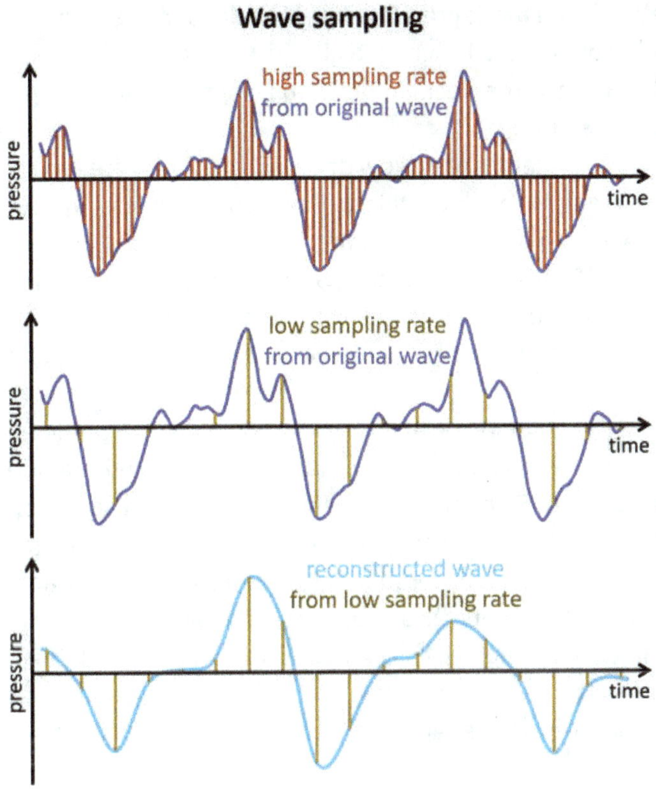

Figure 5-20: The digital sampling of a wave replaces the original continuous wave shape (dark blue wave) with a list of pressure heights at sampling times (vertical bars) that are regularly spaced. With a high sampling rate, a dense set of samples is recorded: red bars. With a low sampling rate, only a few samples are recorded: yellow bars. The yellow bars ignore the details of the original sound, namely the high frequencies: they can only produce a smoothed low-frequency sound, shown in light blue. The dark blue wave is from a piano playing C_4 for about 0.01 seconds. (*Source*: sound from University of Iowa Musical Instrument Samples (MIS).)

we rarely write numbers with ultra-high precision like 73.59320176, but we round them off to, for example, 73.6 when a precision of three digits is sufficient; this saves a lot of storage space, with no loss of relevant information. The same is done in sound quantization. Quantization is based on the binary counting system common in computers (which uses only the binary digits 0 and 1 instead of the decimal digits 0, 1, 2, 3, 4, 5, 6, 7, 8, and 9 of the familiar decimal system). A typical level of quantization is the **bit depth** 16: a sample height will be rounded off to a binary value of at most 16 binary digits, just as we rounded off a decimal value to at most three decimal digits (bit depth is also known as **word length** and **resolution**).

Figure 5-21 sketches the flow of sound from a source (a mouth) to a detector (an ear) through digital recording and playback. The digital part includes two main devices in a computer: an analog-to-digital converter ("A to D" or ADC); and the reverse, a digital-to-analog converter ("D to A" or DAC). The computer can also process the digital signal in various ways, while storage devices can serve as destinations or as sources of digital sounds.

Digital magnetic recording is similar to analog magnetic recording, but it only records a list of numbers, namely the sampled pressure values mentioned above. Since numbers can be written as a series of 0's and 1's (in the binary counting system), only two degrees of magnetization are needed: one for the number 0 ("non-magnetic") and another for the number 1 ("magnetic"). By contrast, in analog magnetic recording, the magnetization must be continuous to represent all the possible pressure levels between the crest and valley of the sound waves: this requires magnetic materials that are much more precisely tailored. Digital magnetic recording is usually done on disks, such as floppy disks

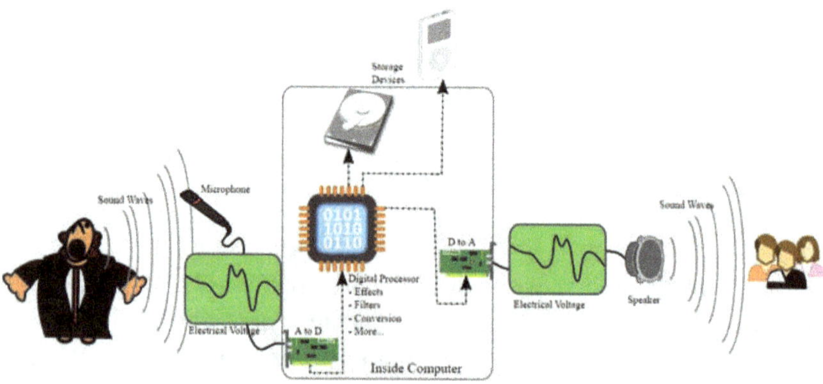

Figure 5-21: A flow of audio from sound waves through a microphone to an analog electrical voltage, A to D converter, computer, optional storage device, D to A converter, analog electrical voltage, speaker and finally as sound waves again. (*Source*: Teeks99, under CC BY-SA 3.0, https://commons.wikimedia.org/wiki/File:A-D-A_Flow.svg.)

or hard disk drives, or on USB flash drives, which allow higher densities of information than do magnetic tapes.

More successful than magnetic digital recording has been **optical digital recording**, as witnessed by today's proliferation of CDs (compact discs) and related products. The basic idea is very simple: each number 1 of a digital recording becomes a small pit in a surface, while each number 0 does not. A light ray (actually a sharp laser beam) is reflected from that surface and reveals the succession of 0's and 1's that describe the sound. These 0's and 1's are converted using a digital-to-analog converter to an electrical signal that is sent to loudspeakers.

5.9 Synthetic Sounds and Music

The electronic age has allowed creating new sounds, called **synthesizing**, such as **electronic music** (**synthetic music**) and **electronic speech** (**synthetic speech**); the abbreviation **synth** is also often used. In addition, electronics allow changing existing sounds in new ways, including combining separate recorded sounds and adding effects like echoes and many others.

Electronic music has given rise to many new styles of music.[7,8] We also hear electronic speech every day in public announcement broadcasts about transportation, in weather information, in telephone answering trees that guide us to desired information or real persons, in navigation systems that tell us how to travel to a destination, *etc.* Many written texts can be listened to because of speech synthesis (for example, spoken books, ebooks, PDF files, and modern word processors). We will discuss speech synthesis more specifically in Section 6.11 and concentrate on the synthesis of general sounds in this section.

Can we combine different sounds, for example, from a piano and a singer? To address this question, we look back at a very important aspect of sound synthesis: the **superposition principle** which we discussed in

[7] See the long "List of electronic music genres" at https://en.wikipedia.org/wiki/List_of_electronic_music_genres

[8] See, for example, the videos "Famous synth sounds of the 80's and 90's" by Estuera at https://www.youtube.com/watch?v=Lsz0gGRWyvM and https://www.youtube.com/watch?v=2Z6peX7fZ5c

Section 2.5. This principle allows us to combine different sounds in such a way that they are not changed. Combining sounds is called **mixing**. The nice thing about the superposition principle is that it allows us to add any two sounds to each other. This is also what happens when you hear two separate sounds, for example, two voices coming from two people: the sounds simply add up in your ear. (The brain is often able to distinguish between the two sounds, but that is a completely different topic that we cannot get into here.)

Thus, we can combine a piano tune with a human song simply by adding the wave coming from the piano to the wave coming from the singer, even if they were recorded at different times or different locations. It's as if they were recorded together: listeners will not be able to tell whether they are hearing a single joint recording or the mix of two separate recordings. We can also literally produce a virtual concert by recording all the orchestra's instruments separately and later combining their sounds electronically. This is routinely done by placing multiple microphones near different musicians or groups of musicians in an orchestra or band and recording their separate sounds. Later, those separate sounds can be mixed with adjusted loudness to produce the best combination (other aspects of the sounds can also be changed, as we will discuss below): the result is a virtual "improved" concert.

We can illustrate the mixing of sounds by using simple waves, like the sine waves of Figure 5-5. As shown in that figure, we start with the leftmost wave, the first harmonic or fundamental wave. To this first wave, we then add the second harmonic (the dark blue wave in Figure 5-5), producing the combination of the first and second harmonics shown as the fourth line from the left. Next, we add a third harmonic (purple), and then a fourth harmonic (light blue) and finally a fifth harmonic (red). The result of those additions is the rightmost wave in Figure 5-5; we could add higher harmonics as well. The resulting wave could be the sound of a flute, or a singer, or anything else.

You can see an example of such a combined wave in Figure 5-19, which shows the soundtrack on a vinyl record recorded by a phonograph: the groove faithfully traces such a wave. The phonograph can play back this recorded wave to reproduce the original sound with great fidelity: it does so by converting the oscillations of the groove into electronic oscillations that are sent to a loudspeaker.

In electronic synthesis, we can mimic the phonograph entirely electronically: we can mix waves as shown in Figure 5-5 or from any other source (voices, musical instruments, machinery, noises, *etc.*) and produce electronic waves that look exactly like those on the vinyl record (they could even be recorded on a vinyl record if we wanted to). These synthesized waves can then go directly to a loudspeaker, or, more likely, they are stored digitally for later playback on a loudspeaker located anywhere. Such electronic synthesis can be achieved with specialized electronic devices called **synthesizers** or with general-purpose computers.

Since we can produce any desired sound by combining sine waves like those of Figure 5-5, we already have a method to electronically produce any sound. In particular, we can synthesize sounds that no musical instrument or voice can produce: with electronics, an infinity of entirely new sounds are possible. Two examples of electronic music that are hard to produce with standard musical instruments are shown in Figures 5-22 and 5-23. You can hear these synthetic sounds in my Animation 5*5. Much of the music we hear today is either produced entirely electronically, or at least manipulated electronically.

ANIMATION 5*5 — Listen to my video WA2 at times 17:22 and 18:31 in its section "**Synthetic music**" under the titles "**Synthetic music: 'Space soundscape'**", and "**Synthetic music: 'Musical electronic glitch'**". (See details in the section References and Resources below.)

How can we change existing sound? A different approach to synthesizing sound is to use original recorded sounds and modify them in any of multiple ways: this is often called **signal processing**. Examples of processing are filters, delayed echoes, reverb (reverberations), chorus, phasing, flanging, pitch shifts, amplitude modulation (AM), frequency modulation (FM or pitch modulations), and various other distortions. Such processing is widely used in popular music.[9]

[9] See "Effects: All You Need To Know... And A Little Bit More" at https://www.soundonsound.com/techniques/effects-all-you-need-know-and-little-bit-more

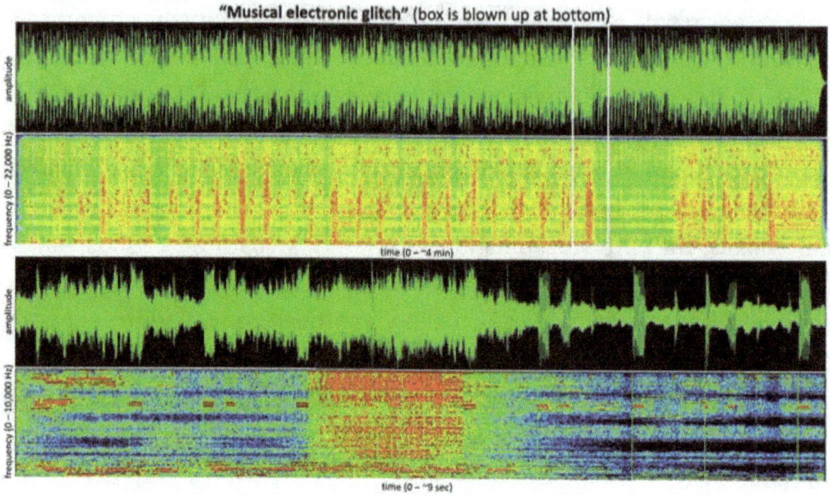

Figure 5-22: Synthetic music entitled "Musical electronic glitch" shown as wave forms and spectrograms. The upper pair of graphs shows about 4 minutes of sound, which goes up to a frequency around 22,000 hertz. The sound in the white box is blown up in the lower pair of graphs, which covers about 9 seconds and goes up to 10,000 hertz. It is most unusual for musical instruments (and human voices) to reach 22,000 hertz (the normal limit of human hearing) with similar loudness as at low frequencies. Also notable is the large number of perfectly horizontal red/yellow lines, indicating constant frequencies but varying intensities of those frequencies (this is an example of amplitude modulation, described below); in addition, there are no separate notes (but a bang every 8 seconds or so), and very few of the frequencies are harmonics. (*Source*: sound file "Musical electronic glitch" at Mixkit in the public domain; analyzed by WaveSurfer.)

Let's illustrate a few of these techniques for modifying existing sound. Some of them can also be used in the process of creating new sounds.

Filtering is probably the most familiar technique for modifying sound; it is very similar to light filtering, in which some colors of light are removed or reduced. The idea in the case of sound is also to remove or reduce the intensity of some range of frequencies. If we used a sharp cutoff frequency, the result would be too abrupt for most applications: instead, a gradual cutoff is used, so the boundary is imperceptible.

For example, **low-pass filtering** means letting only low frequencies pass by removing or reducing high frequencies. **High-pass filtering** is the opposite: letting only high frequencies pass by removing or reducing

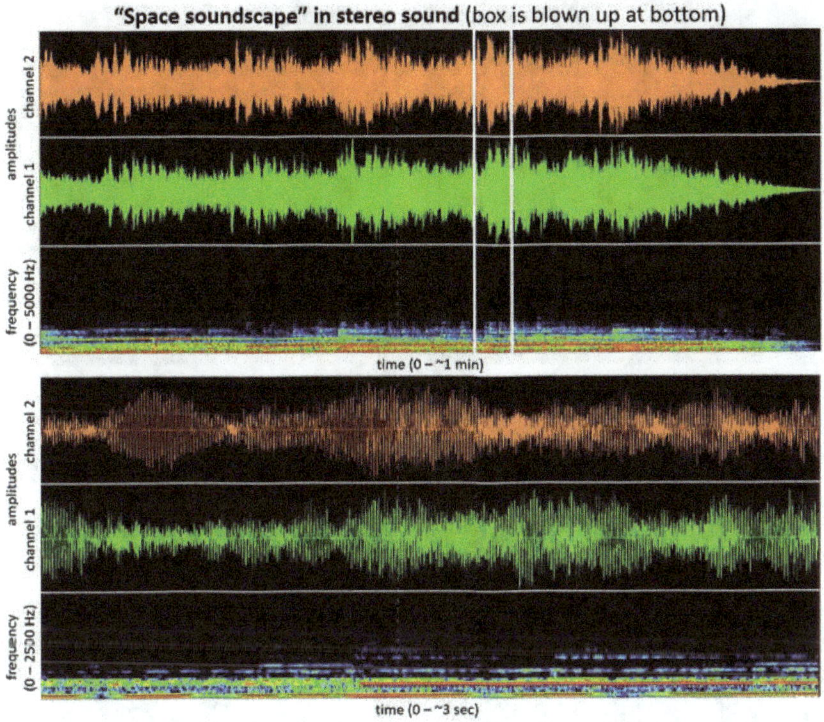

Figure 5-23: Synthetic music entitled "Space soundscape" shown as stereo wave forms and spectrograms. The upper three graphs show about 1 minute of sound, which goes up to 5,000 hertz. The sound in the white box is blown up in the lower three graphs, which covers about 3 seconds and goes up to 2,500 hertz. This sound is limited in frequency to only about 1,000 hertz and is seen to include very few harmonics. Its most notable aspect is the large difference between the two stereo channels (meant for the left ear and right ear; they are shown as separate wave forms here, colored orange *versus* green): this gives a sound that seems to drift from left to right and back, so it gives a feeling of floating in space. As in Figure 5-22, the frequencies are constant but vary in intensity, and there are no separate notes. (*Source*: sound file "Space soundscape" at Mixkit in the public domain; analyzed by WaveSurfer.)

low frequencies. Other examples are **bandpass** and **band stop**, as well as **peaknotch** (often written as **peak/notch**) and **bass amplification**: bandpass suppresses both low and high frequencies, while band stop removes a limited range of frequencies; peaknotch amplifies a limited range of frequencies, and bass amplification only amplifies low frequencies. Several of these are illustrated in Figure 5-24.

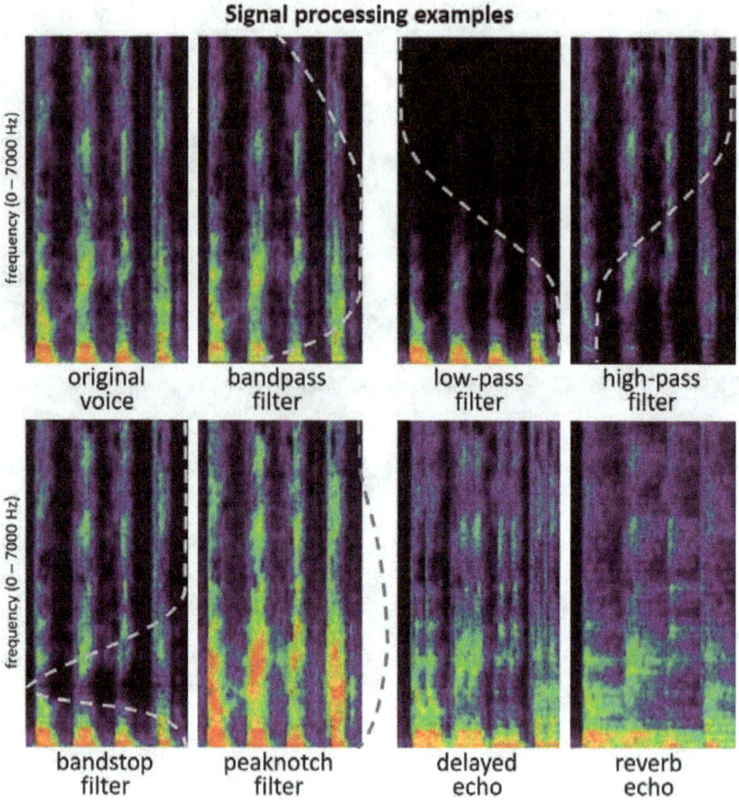

Figure 5-24: Examples of signal processing: in each image, time flows to the right for about 3.5 seconds and frequency rises upward from 0 to 7,000 hertz. The original sound, shown at top left, is the words "do re mi fa" recorded from an adult male. The blue-to-red colors represent intensity, and red is the most intense. The light-gray dashed curves indicate the "transparency" of the filters: toward the right is more transparent, toward the left is more opaque. In this example, the bandpass filter broadly weakens both low and high frequencies. The low-pass filter weakens only high frequencies. The high-pass filter weakens only low frequencies. The bandstop filter weakens a narrow range of frequencies. The peaknotch (peak/notch) filter strengthens (amplifies) a wider range of frequencies, indicated by a dark-gray curve. The delayed echo delays a copy of the original sound by a fraction of a second. The reverb echo smears out and weakens the original sound over time. (*Source*: recorded by smartphone; analyzed by A.L.M.S. — Audio Literacy for Music Students.)

Other techniques illustrated in Figure 5-24 are **delays** and **reverb**: they are different forms of **echoes**. A delayed sound is simply a copy of the original sound that is sent out a bit later, like an echo that has bounced off a distant wall (it may be weakened like a real echo would be). Reverb (short for reverberations) also copies the original sound but sends it out continuously while weakening it, like multiple echoes that have bounced off very many walls at various distances: this simply smears out the sound over time.

Amplitude modulation and **frequency modulation** are two related techniques: they take a simple wave called a carrier wave that has a single high frequency called the carrier frequency; they then modify that carrier wave so it "carries" another lower-frequency wave, which can be a music or voice wave and is often called a signal. We will see a nice example in Section 6.6: the rolling of the sound r (in "brrrrr"). In amplitude modulation (**AM** for short), the signal wave is simply added to the carrier wave, while in frequency modulation (abbreviated to **FM**), it is the frequency (rather than the amplitude) of the carrier wave that is changed. Vibrato and whistling with varying pitch are good examples of FM, as we will see in Section 6.11.

The AM and FM techniques are compared in Figure 5-25. Here the same signal (like a sound wave shown as the white curve) is added to the same carrier wave of high frequency (the rapidly oscillating green curve). AM modulates the amplitude of the carrier wave: the frequency of the wave remains constant, but the height of the wave varies irregularly as shown; that height variation (the white curve) contains the useful information, namely the sound. By contrast, FM varies the frequency, which variation contains the same sound (derived from the same white curve), while it produces no height variation.

AM and FM are best known in the transmission of radio and television waves: these are not sound waves but electromagnetic waves (which we will discuss in Chapter 12). However, a radio or television receiver converts that electromagnetic wave into an electronic wave that can be sent to a loudspeaker to produce sound. The carrier wave has an extremely high frequency far beyond our hearing range, so we cannot hear that carrier frequency: the electronics in our radio or TV set remove the carrier wave, leaving only the variations in amplitude or frequency, thus giving us the desired sound signal. Because AM is more

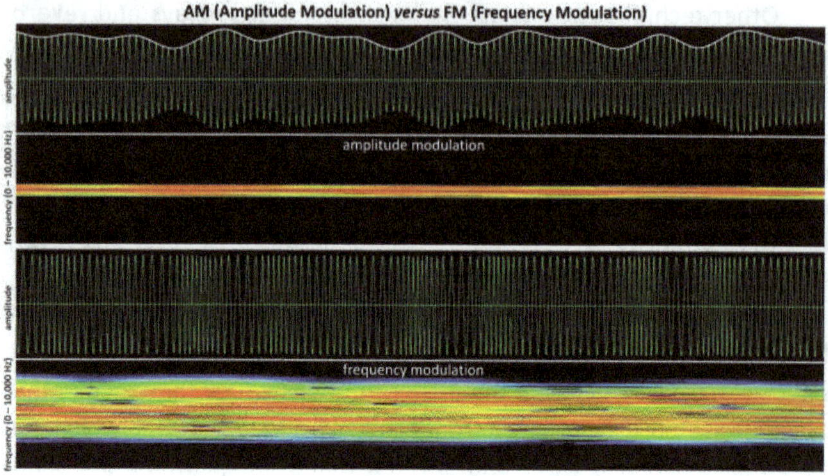

Figure 5-25: *Top pair of graphs*: Amplitude modulation, showing varying amplitude but near-constant frequency (the amplitude modulation creates slightly changed frequencies called sidebands). The white curve is the original sound wave that was added to the carrier wave; it modulates the carrier wave. *Bottom pair of graphs*: Frequency modulation for the same original sound wave (white curve above), showing fixed amplitude but a broad range of varying frequencies; notice how the waves tighten and widen in this case, unlike for amplitude modulation at the top. These two examples use the same carrier frequency of 10,000 hertz, which is much smaller than in AM and FM radio or TV transmission. (*Source*: designed with Praat; displayed with WaveSurfer.)

sensitive to noise (coming from electrical machines, lightning, *etc.*), FM is favored for radio and television transmission.

Other examples of frequency modulation are the Doppler effect (in which the frequency is changed due to the relative speed of the source and the ear; see Section 3.6), many sirens and the vibrato effect (see Sections 6.9 and 6.11).

5.10 What have We Learned in this Chapter?

We have been able to link the standing waves on a string with the musical notes produced by a piano, violin or guitar, for example: each string can resonate at its natural (or fundamental) frequency and at its harmonic frequencies. The natural frequency usually is the resulting

note. These frequencies depend on the length, tension and weight of the string.

Our ear-brain system hears an estimated or perceived frequency, often called the pitch of the sound. The perceived character of the sound (soft, warm, harsh, thunderous, *etc.*, as determined by harmonic frequencies) is the tone of the sound.

The spectrum of a sound is a convenient visual method to display the character of a sound by exposing its frequencies as well as their individual amplitudes. A spectrogram or sonogram can also display the evolution of the sound over time; this evolution differs among different musical instruments, helping us to distinguish them.

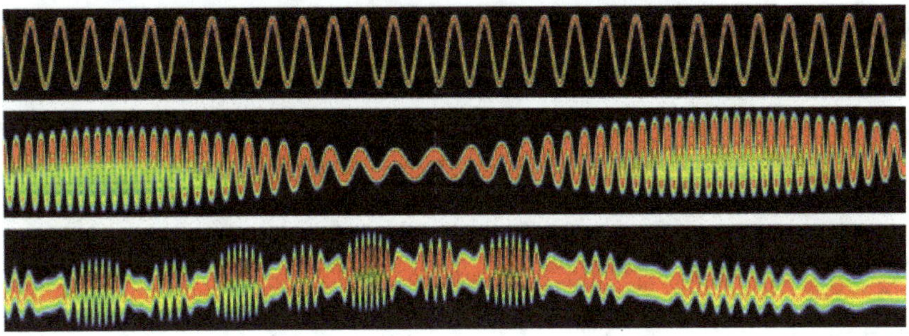

6

From Waves to Human and Animal Voices

We probably use our voice more than any other method to communicate our thoughts to other people. It is fascinating to realize that our mouth makes all sorts of hisses, pops, groans and vibrations to express simple and abstract ideas. And let's not forget the pleasure of singing and whistling!

In this chapter, we explore how we go from those hisses, pops, groans and vibrations to words. Some of the very interesting questions we will discuss are: How do the sounds that we make become the letters and syllables of words? How different are the voices of males, females and children? How different are the accents, dialects ("varieties") and languages that humans use? And how are these connected with the sounds that our mouths make? How different is singing from speaking? What are whistling, yodeling, whispering and humming? How can voices be synthesized? How different are animal sounds from human voices?

I recommend that you first watch my accompanying video to get a feeling for the subject of this chapter.[1]

6.1 Overview

Let me first outline how we will proceed with this fascinating topic of voices. We know that, in many languages including **English**, the individual sounds used in speech are either vowels (such as 'a', 'e', 'i', 'o', and 'u') or consonants (such as 'b', 'c', 'd', 'l', 'm', 'n', 'p', 's', 't', and 'z'). The **vowels** generally result from resonances (standing waves) in the mouth of vibrating sound waves coming from the vocal cords: the vowels use vibrations with harmonic frequencies, like the violin or piano (see Figure 5-17). By contrast, the harder **consonants** are either brief explosions that generate many unrelated frequencies (such as 'p', 't', 'k', 'b', 'd' and 'g'), like the cymbal (see Figure 5-17), or are "liquid" or hissing sounds (such as 'l', 'f', 's' and 'z'), which also contain unrelated frequencies. The sound of consonants is mainly due to air streaming fast through tight openings between lips, teeth and tongue, as with whistling; resonances in the small space between the lips and the teeth contribute high-pitched sounds like those of 's' and 'z'.

The vowels depend more on the internal shape of the mouth than do the consonants, since vowels are strongly influenced by resonances in the cavity of the mouth. The vowels are more affected by the tongue, the lips, the opening of the mouth (the dropping of the jaw) and the nasal cavity (by access to the open space within the nose), all of which can be varied by consciously activating muscles. Consequently, the greater variability of vowels plays a larger role in defining **accents** than do consonants.

Therefore, we will first consider how the mouth influences sounds, in particular vowels. Then we will discuss consonants, followed by accents, dialects and languages.

[1] See my video "Waves and Voices: human and animal" (WA3) on Everyday Physics by Michel A. Van Hove: https://www.youtube.com/watch?v=VwIwJ5Jtiks

Accents used in different regions or social classes are different ways of pronouncing letters or syllables, often due to different positions of the tongue, lips, *etc*. Accents are often parts of dialects (dialects are sometimes also called varieties), but dialects also contain different words and often different grammar: dialects are not just variations of pronunciation. Normally, dialects are mutually understandable: a person speaking one dialect can understand a person speaking a related dialect without having learned it. By contrast, different languages are not mutually understandable: they have too many differences in words and grammatical rules.

We will consider the interesting case of Chinese because it is a so-called tonal language, where a vowel can have different variations of tone that give it different meanings.

We will also discuss special forms of the voice: singing, whistling, yodeling, whispering and humming. In addition, we will address a wide range of animal voices: they can have clear similarities but also large differences with human voices. We will also consider synthetic sounds and, in particular, synthetic voices, since we hear such sounds every day.

The voice is clearly a complex topic: to keep the story reasonably simple, I will focus here on some of its main physical aspects. More details and broader discussions can be found in the literature.[2]

We will use the following notation: single quotes, as in 't' and 'th', denote a sound and its approximate pronunciation, while double quotes, as in "to" and "then", denote the spelling of a word or part of a word without indicating its pronunciation. Various other notations are also used in different publications.

6.2 Structure and Movements of the Mouth

It is very instructive to watch how the **mouth** moves while it speaks or sings: X-ray movies show that much motion takes place inside the

[2] P. Ladefoged and K. Johnson, *"A Course in Phonetics"*, Wadsworth Cengage Learning, Boston, 2014; E. Sapir, *"Language: An Introduction to the Study of Speech"*, Harcourt, Brace, New York, 1921; F. de Saussure, edited by C. Bally, A. Sechehaye and A. Riedlinger, translated and annotated by R. Harris, *"Course in General Linguistics"*, Duckworth, London, 1983.

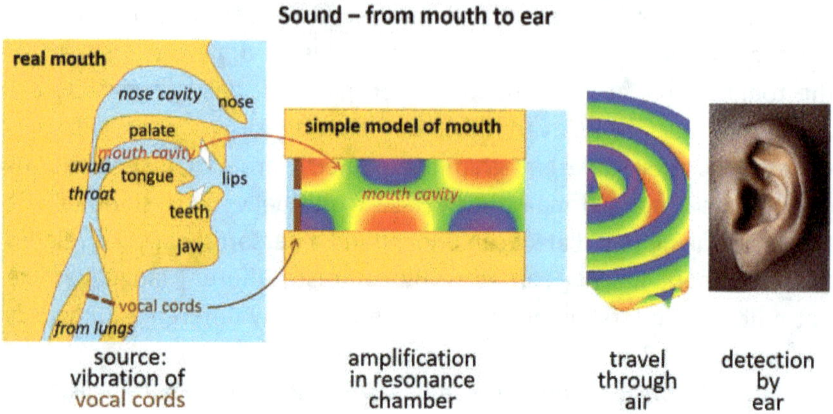

Figure 6-1: Tracing sound from mouth to ear (compare with Figure 5-1 for the case of musical instruments). At the left, a sketch of the mouth structure shows major organs involved in making sounds. The vocal cords, when tightened, create vibration in the air passing through them. The mouth cavity, sketched as a simple model to its right, functions as a resonance chamber: it amplifies certain frequencies and weakens others. From the mouth, sound radiates outward, reaching the ear. (*Photo source*: Genusfotografen, Human right ear (here reversed), under CC BY-SA 4.0, https://commons.wikimedia.org/wiki/File:Human_right_ear_(cropped).jpg.)

mouth.[3] Such motions change the shape of the mouth cavity in order to produce each letter or syllable. Several movable organs are particularly important for speech; they are visible in the videos and are sketched at the left in Figure 6-1: **lips**, **jaw**, front tip of the **tongue**, top of the tongue, back of the tongue, and **uvula** (a "door" which controls the passage of air to and from the nose). You can also see and hear this in my Animation 6*1.

ANIMATION 6*1 — See my video WA3 at time 2:15 in its section "**Waves in the mouth**" under the title "**Sources of the human voice: from vocal cords to pitch/tone**". (See details in the section References and Resources below.)

[3] See "Live video of movements during speech production" (by magnetic resonance imaging) by Medgaget: https://www.youtube.com/watch?v=uTOhDqhCKQs; "Echtzeit-MRT-Film: Sprechen" (in German) by MaxPlanckSociety: https://www.youtube.com/watch?v=6dAEE7FYQfc; "Singing in the MRI with Tyley Ross — Making the Voice Visible" by Tyley Ross: https://www.youtube.com/watch?v=J3TwTb-T044

In brief, human sounds are created as follows: Our lungs push (or pull) air through our vocal cords. If the vocal cords are wide open, the air can rush through the mouth and nose, as in breathing. If we restrict the resulting airflow with tight lips, teeth, or tongue, we can create hissing sounds like the letters 'f', 's' and 'z', or whistling. The sudden opening of the lips or tongue creates "explosive" sounds like 'k', 'p' and 't'. All these letters are called consonants. If, however, the vocal cords are tightened, the airflow will make them vibrate. We can control the frequency of this vibration by changing the tightness of the vocal cords: this gives the pitch of our voice, like do-re-mi-.... Vibration of the vocal cords causes vibration of the air in the mouth and nose (see middle of Figure 6-1). These act as resonance chambers, amplifying some frequencies and weakening others. Movements of the tongue change the shape of the mouth cavity: this produces the different vowels, like 'a', 'e', 'i', 'o' and 'u'.

Thus, **to produce consonants** (such as 'b', 'c', 'd', 'l', 'm', 'n', 'p', 's', 't', and 'z'), **the lungs push air through narrow openings due to the lips, teeth and tongue; the vocal cords are then wide open to let the air flow freely.**

For vowels ('a', 'e', 'i', 'o', 'u', also called "voiced sounds"), **the vocal cords and tongue play a more important role**. To understand how we produce different vowels, let us start with a simple model of the mouth: the rectangular box (which we discussed in Section 3.13) drawn in Figure 6-1.

What is the role of the mouth? The mouth has two central tasks, shown in Figure 6-1. **The first task of the mouth is to activate a vibrating sound** at one end, which is done in a real mouth inside the larynx with the **vocal cords** (often called **vocal folds**). **By tensing the vocal cords with our muscles and blowing (or inhaling) air between them with our lungs, we can create a sound in a wide range of frequencies.** A higher tension of the vocal cords produces a higher frequency of vibration in the air passing through to the mouth cavity.

The frequency due to the vocal cords is usually called f1 or F0: it will be the **fundamental frequency** of a vowel, giving it its **pitch**. For example, f1 = F0 can be the C_4 frequency (about 262 hertz) that makes the 'o' sound in "do – re – mi – ...". In addition, the vocal cords will also create **harmonics**, for example C_5, C_6, *etc.*, at frequencies that are whole multiples of f1: in particular, C_5 with f2 = 2 × f1 ~ 524 hertz, C_6 with f3 = 3 × f1 ~ 786 hertz, *etc.* (in this example, these are the "do" notes in higher octaves, also shown in Figure 5-4). These harmonic frequencies are shown at left in Figure 6-2: they are the spectrum of the vocal cords.

The second task of the mouth is to amplify some frequencies and weaken other frequencies, which is also the purpose of the resonance chamber of a guitar or violin; the cavity of the mouth (and sometimes also the nose) is the mouth's resonance chamber. This task relies on **standing waves** inside the mouth, also sketched at center in Figure 6-1 (which is analogous to Figure 5-1 for musical instruments). Figure 3-19 represents such standing waves for rectangular boxes.

The waves with the lowest resonance frequencies dominate vowels: they are normally labelled F0, F1, F2, F3, *etc.* and shown in blue in Figure 6-2. The frequency F0 = f1 is the lowest frequency of the vocal cords, while F1 is the lowest resonance frequency of the mouth cavity when it has the shape needed to make a particular vowel. The frequency F1 depends mainly on the length of the tube, which in a real mouth could

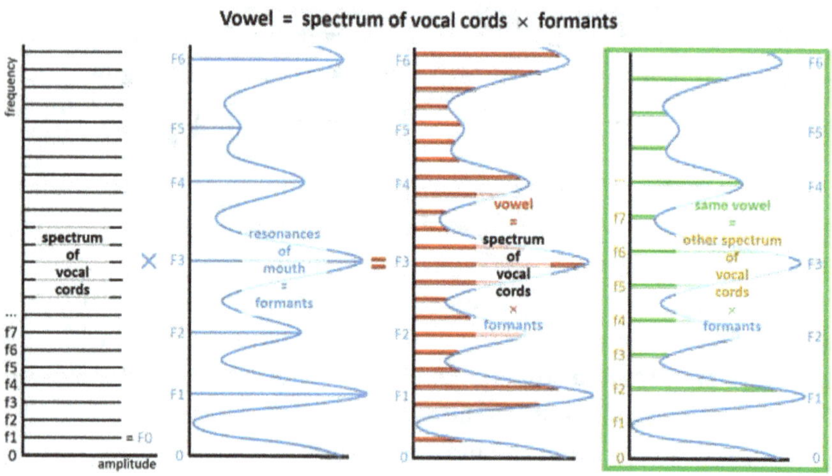

Figure 6-2: This sketch shows how sound from the vocal cords (black, at the left) is strengthened or weakened by the resonances and formants of the mouth (blue, at the center left) to produce a vowel (red, at the center right). Changing the fundamental frequency f1 = F0 still produces essentially the same vowel (green, at right) because the formants (blue) are not changed.

be the distance from the vocal cords to the teeth. In an ideal simple tube, F2, F3, *etc.* are whole multiples (harmonics) of F1, just as f2, f3, *etc.* are whole multiples of f1. However, in a real mouth, F2, F3, *etc.* are not whole multiples of F1; we will discuss the case of the real mouth in the next section.

How does a resonance chamber, such as the mouth, work? It functions as follows: waves with frequencies near the chamber's fundamental and harmonic frequencies are amplified, while other frequencies are weakened. This is shown in Figure 6-2 by the blue wavy line: that line has peaks near F1, F2, F3, *etc.*, indicating that waves with frequencies which fall near such a peak will be favored. We can call this blue wavy line an "amplification factor", or a "transmission coefficient", or a "transparency factor", or a "filter", or an "envelope". The peaks of this wavy blue line are called the **formants** because they "form" the sounds of the different vowels: formants are thus "peaks of amplification".

6.3 Vibrations in Real Mouths: Vowels

The shape of a real mouth is much more complex than a tube or rectangular box, as are most real resonance chambers in musical instruments. Consequently, its resonances are less regular: the resonances are not simply whole multiples of F1. That is why the peaks in the blue wavy line of Figure 6-2 are not equally spaced. Also, the peak heights can differ considerably: as we will see soon, this will be very important for making distinct vowels.

The result of the amplification or weakening by the mouth's resonance chamber is shown in red at the center right in Figure 6-2. Each wave from the vocal cords (black frequencies f1, f2, f3, *etc.*) is "multiplied" by the "amplification factor" (blue wavy line) to give the final red frequency amplitudes. We notice that the red frequencies are large near the peaks of the blue wavy curve: thereby, the resonances of the mouth have "formed" the final vowel coming out of the mouth by favoring some frequencies over others. This gives the formants of the final vowel.

What happens if we change the vocal cords' fundamental frequency f1 = F0? This is shown (in green) at the right in Figure 6-2:

f1 has been roughly doubled in that example, compared to the black spectrum at the left; now only the green frequencies are produced by the vocal cords. However, the mouth cavity, if its shape has not changed, still has the same resonance frequencies F1, F2, F3, *etc.* (shown in blue). Now the resonance frequencies still "form" the strengths of the final frequencies, shown in green. Therefore, the same vowel is still pronounced, even though the vocal cords produce a different pitch f1. This is very important: different people naturally produce different vocal cord frequencies f1; in particular, children produce a high pitch f1, while women produce a lower pitch and men produce an even lower pitch. Nonetheless, the formants F1, F2, *etc.* can still produce the same vowels, as long as the mouth takes the right shape for that vowel. **In other words, all of us can produce the same vowels, despite having different vocal cords, by controlling the shape of the mouth; thus, men, women and children can understand each other!**

How does all this look in reality? Figure 6-3 shows actual vowels pronounced by a female and a male: they are the vowel sounds 'a', 'e', 'i', 'o', 'u' in the English words "had", "head", "he", "go", and "boot", respectively. In the traces of the wave amplitude (shown in green), you may notice that the female wave forms oscillate faster than the male wave forms due to the higher frequency of the female voice. We also see that the loudness declines during the half-second or so of pronouncing the vowels, that is despite a conscious effort to keep a constant loudness. For singing, better control of the loudness would be necessary! You can also see and hear these vowels in my Animation 6*2.

> ANIMATION 6*2 — See my video WA3 at time 5:20 in its section "**Consonants *vs.* vowels**" under the title "**Consonants *vs.* vowels: synthetic *vs.* real vowels**". (See details in the section References and Resources below.)

The spectrograms are more instructive than the waveforms, as shown in Figure 5-9. We will present many spectrograms in this chapter, as they allow us to "see" the structure of a sound more clearly than when we hear it. **Spectrograms are rich in information: they show not only the loudness and pitch (perceived frequency) of a sound but also its frequency composition (similar to the spectral decomposition**

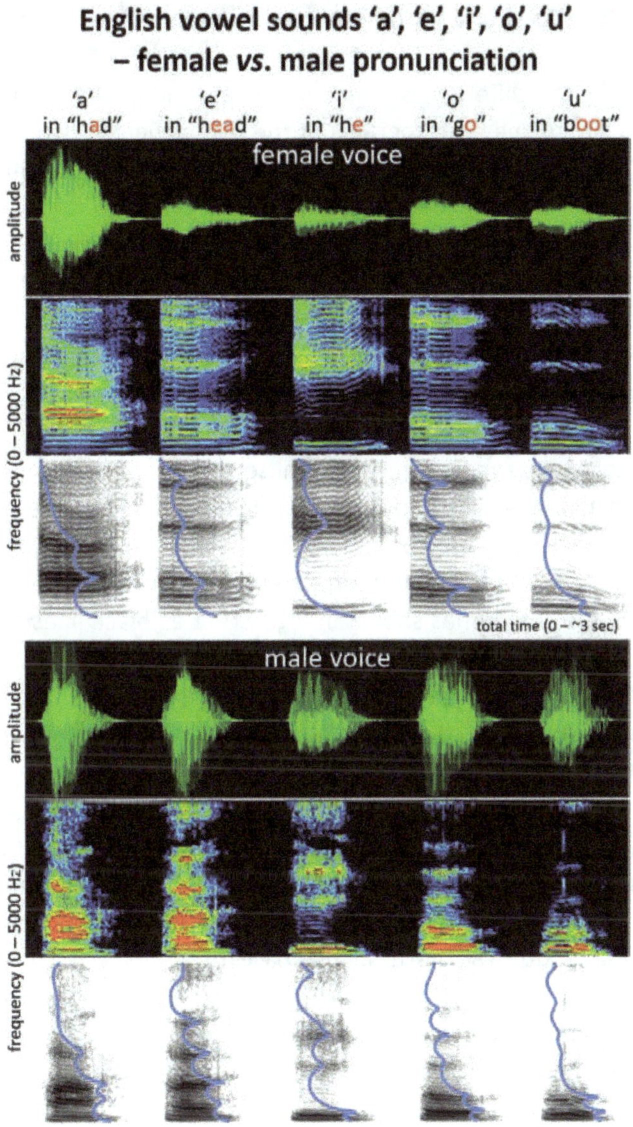

Figure 6-3: Waveforms and spectrograms for the vowels 'a', 'e', 'i', 'o', and 'u' pronounced in English (as in the words "had", "head", "he", "go", and "boot", but without the consonants "h", "g", "b", "d" or "t") by a female (*top*) and a male (*bottom*). The spectrograms are duplicated in black and white *versus* color to highlight different aspects (dark gray and red/yellow show higher intensities). The formants show up as brighter red/green bands in the color spectrograms, and as broad horizontal dark bands in the black and white spectrograms. They are also shown as sharp peaks in the blue curves, pointing to the right, like the blue peaks in Figure 6-2. Each recording takes about 3 seconds for 5 vowels. (*Source*: recorded with Praat, displayed with Praat (black and white) and WaveSurfer (color).)

of light into colors) and its time evolution with as much detail as we want.

To make the differences between vowels and gender more visible, Figure 6-3 shows the spectrograms of the five vowels in two equivalent fashions: in color, where red is loudest, and in black and white, where dark is loud.

One general difference between the female and male voices in Figure 6-3 is the "cleaner" shapes in the female spectrograms, compared with the "messier" or "noisier" shapes in the male spectrograms. This may explain why this particular female's voice is in reality distinctly clearer than this male's voice: this female's voice indeed can be understood much farther away than this male's voice.

The second clear feature of all these vowel spectrograms is their "ribbed" structure (also seen for a guitar in Figure 5-9): we can easily see 20 or so stronger horizontal lines for each vowel, especially in the female case. These are the **harmonic frequencies** of the vocal cords: f1, f2, f3, *etc.*, the same ones shown in Figure 6-2 as black lines at the left and as red or green lines at the right. These lines are actually not quite horizontal, indicating that the fundamental frequency changed during the pronunciation, which lasted about half a second for each vowel: it is challenging to maintain a perfectly constant pitch (and a constant loudness) even in such a short time!

We also see from the tightness of the "ribs" that the male frequencies are lower than the female frequencies. Indeed, measurement for the vowel 'a' shows an f1 of about 130 hertz for the male voice, while the female f1 is about 180 hertz in the case of Figure 6-3. (In addition, remember that the higher frequencies are whole multiples of f1 in ideal situations; that seems to be at least approximately correct here.)

Another important observation is that the pitch f1 is about the same across all vowels for the same person, as we discussed in Section 6.2. When speaking, we unconsciously always select the same "normal" pitch f1 when pronouncing vowels, but if we are excited, we use a higher pitch, and when we are in a somber mood, we use a lower pitch. The "normal" pitch is higher for adult females (typically 165 to 255 hertz) than for adult males (85 to 155 hertz), and even higher for children (250 to 300 hertz). We see additional higher harmonic frequencies in the spectrograms for 'e' and 'i': that is why the 'e' and 'i' sounds are actually higher pitched than the other vowels 'a', 'o' and 'u'.

6.4 How are Vowels Formed and Distinguished? Formants

Let's now compare the vowels with each other in Figure 6-3. The most obvious differences between vowels are the strong, broad horizontal bands: they are red/green in the color spectrograms, and dark gray in the black/white spectrograms. We also see that these bands are about the same, although not quite identical, when comparing the female and male voices: that similarity makes the vowels pronounced by a female sound like the same vowels pronounced by a male, so females and males can understand each other.

Furthermore, the bands usually remain horizontal even when the f1 frequencies change: in other words, the bands do not depend on the pitch chosen by the speaker. This helps make the vowels "universal" in the sense that you can always hear the same vowels spoken by different speakers, regardless of the speaker's normal pitch or mood at the moment.

These bands are the **formants** F1, F2, F3, *etc.* that we saw in Figure 6-2 (the blue wavy lines there). Notice the similarity between the red peaks in Figure 6-2 (third graph from the left) and the strong bands in Figure 6-3: each band in Figure 6-3 contains several "ribs", just as each blue peak in Figure 6-2 contains several frequencies (horizontal red lines). The blue wavy lines in Figure 6-3 show the intensity and width of each band. These blue wavy lines are calculated from the actual voice by a mathematical procedure (in the software Praat); they correspond to the blue wavy lines in Figure 6-2. Since the blue wavy lines are due to the shape of the resonating mouth, we see how strongly the formants are deformed by the complex shape of the mouth, and how they influence which harmonics from the vocal cords are amplified.

These strong bands or formants are the signatures of the different vowels: they allow us to distinguish one vowel from another.

Let's compare the bands or formants of the two vowels 'o' and 'u' in Figure 6-3 (they are the fourth and fifth from the left, under "go" and "boot"): they look quite similar. Now pronounce these two vowels yourself and feel what is different between them (but make sure you don't pronounce the 'u' as the full words "you" or "yew" but as in "go" and "boot"!). You should notice that the main difference is the shape of your lips, with a slight change in the jaw's position, which also changes the tongue's position; on the other hand, remember that the external lips have less influence on the formants than the internal tongue because the standing waves (the resonances) depend mostly on the internal shape of the mouth. Also, you can notice that, in these examples, pronouncing 'u' increases a blowing sound between the lips, which does not change the formants: indeed, if you pronounce an 'o' and add to it only some blowing through tighter lips, you get essentially the 'u' sound with practically the same mouth shape (and thus the same formants) as for 'o'!

This simple comparison between 'o' and 'u' starts to give us some insight into the creation of vowels. Let's explore further.

Does the fundamental frequency f1 affect the identity of the vowels? Suppose that you sing "a" (as in "had") with an increasing frequency f1: "a – a – a – a – a". Will the vowel 'a' remain recognizable as 'a', without shifting to 'e', 'i', 'o', 'u' or anything else? The answer is yes. The spectrograms of Figure 6-4 show an example of this: we recognize the formants (bands) of the vowel 'a' visible in Figure 6-3, which show up here as perfectly horizontal bands (dark bands in the black and white graphs, and bright bands in the color graphs): the lowest band is outlined with red and white boxes. By contrast, the frequencies of the "ribbed" lines are not horizontal. Apart from rapid ups and downs, the ribbed lines generally climb upwards to the right, as the frequency f1 increases: you can follow a "rib" all the way from left to right in the speaking sound and notice how it jumps up a bit 4 times, especially the last time. The ribbed lines climb because they are harmonics and thus whole multiples of the fundamental frequency f1, which indeed climbs as well. This upward trend is most visible in the singing spectrograms, particularly as growing spacing between the lines. The ups and downs are oscillations of the frequency f1 itself, giving a "vibrato" effect that is much used in singing (we will see it again in Sections 6.9 and 6.11). You can also see and hear this in my Animation 6*3.

ANIMATION 6*3 — See my video WA3 at time 6:36 in its chapter "**Singing**" under the title "**Singing: singing the vowel A**". (See details in the section References and Resources below.)

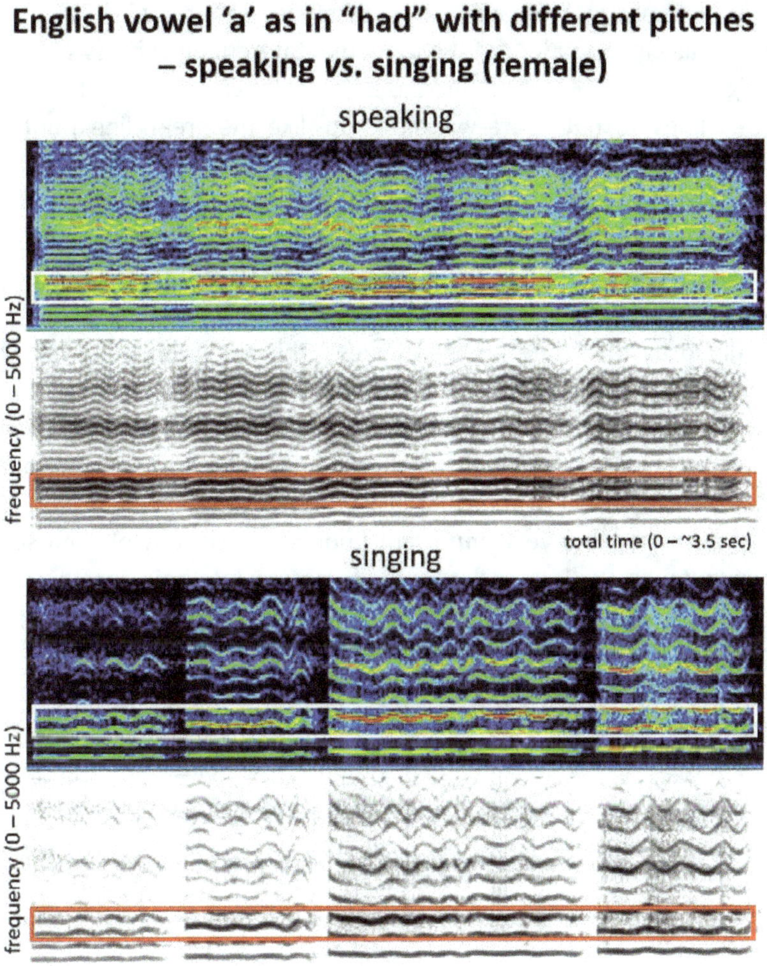

Figure 6-4: Spectrograms showing the vowel 'a' as in "had", being spoken (top two graphs) and sung (bottom two graphs) by the same female with 5 increasing pitches from left to right. The white and red boxes indicate the lowest band: formant F1. The lowest line (gray in the black and white graphs, green/yellow in the color graphs) has frequency f1, giving the pitch of the sound. From left to right, f1 here rises from 190 to 294 hertz in speaking, and from 333 to 534 hertz in singing. The frequencies oscillate up and down, especially in singing, to give a musical "vibrato" effect. Each recording covers about 3.5 seconds. (*Source*: recorded with Praat, displayed with Praat (black and white) and WaveSurfer (color).)

To summarize, we have seen in Section 6.3 that a vowel's identity is essentially independent of the fundamental frequency f1, as shown at the right in Figure 6-2: as f1 increases (and all harmonics of f1 as well), the red lines in that figure will move up (like the green lines) but will be "capped" by the blue wavy line, which does not change (because the mouth shape does not change). So, the formants (the peaks in the blue curve) remain the same, and the vowel remains the same. This is also very clear in Figure 6-4: there the pitch f1 increases toward the right, but the formant bands do not move up or down (the formant F1 is outlined by the white and red boxes). This shows that the mouth's shape remained constant, and the vowel also remained the same. The bottom spectrograms of Figure 6-4 show the effect of singing with higher pitches f1: the vowel remains the same because the formants remain the same. We will further discuss singing in Section 6.9.

We now have a very important understanding: **Vowels and some consonants as well, are determined mostly by their formants F1, F2, *etc.* and less by their pitch or fundamental frequency F0 = f1.** Importantly, the formants F1, F2, *etc.* are also fairly constant between people who speak the same language, dialect and accent: this enables them to understand each other well.

Another very important observation is the following: among the formants F1, F2, F3, F4, *etc.*, the first two, F1 and F2, dominate to define the vowel. The higher-frequency formants, F3, F4, *etc.*, are usually not needed to recognize that vowel; however, they are more characteristic of the individual person who is speaking. **Thus, formants F1 and F2 allow us to recognize the vowel, while formants F3, F4, *etc.* allow us to recognize who is speaking**. (Of course, our brain automatically does all the work necessary to recognize vowels and speakers, so we are not aware of this analysis going on all the time in a conversation.)

A very convenient consequence is that we can focus on just the formants F1 and F2 when discussing vowels in a language, dialect or accent, as we will do in the next section.

Why is speaking in helium so funny? If we inhale helium gas before speaking, we discover that our voice seems to become strangely high-pitched, quacking like a duck. If instead we first inhale the gas sulfur hexafluoride, the opposite happens: our voice seems to become oddly low-pitched. We can easily explain what happens by using Figure 6-2:

the resonances of the mouth, shown as blue wavy lines, are changing with the different gases. The reason is that the speed of sound depends on the gas, changing the sound's wavelength, so that the resonances in the mouth (the peaks F1, F2, *etc.* in the blue curves of Figure 6-2) shift to other frequencies. If the speed of sound increases threefold, as it does in helium compared to air, the resonance frequencies also increase threefold; but if the speed of sound decreases threefold, as it does in sulfur hexafluoride, the resonance frequencies decrease threefold. This effect is similar to making your mouth three times smaller or three times larger, respectively, while still using air. Notice, however, that the spectrum of the vocal cords (the black lines in Figure 6-2) does not change with the gas because those frequencies depend on the structure of the vocal cords and not on the gas. So, the fundamental frequency f1 does not change, but the different gases amplify different frequencies. You may listen to speech sounds in air *versus* other gases on the web.[4]

6.5 Where do the Formants F1 and F2 Physically Come from? Vowel Charts

It has been found that the formant F1 is related to how close the tongue gets to the ceiling of the mouth (the palate shown in Figure 6-1), while F2 is related to the position of the tongue in the forward/backward direction (towards the teeth *versus* towards the throat). Indeed, most vowels can be produced by independently controlling only those two positions of the tongue: vertical *versus* horizontal.

Since we have two variable quantities (F1 and F2) that define vowels, we can conveniently draw a 2D **vowel chart**, also called a **vowel map**, as is done in Figure 6-5: from top to bottom, the mouth is opened wider, thereby dropping the tongue lower; from left to right, the tongue is pulled to the back of the mouth.

You can pronounce the words shown on the chart one after the other and feel whether your jaw and tongue move as indicated when

---

[4] See "What's Really Happening When You Inhale Helium" by Science Insider: https://www.youtube.com/watch?v=CWdjOpU-p6Y. Or download and play the wav sound files for "Ordinary Speech" and "Helium Speech" from https://www.animations.physics.unsw.edu.au/jw/speech.html#helium

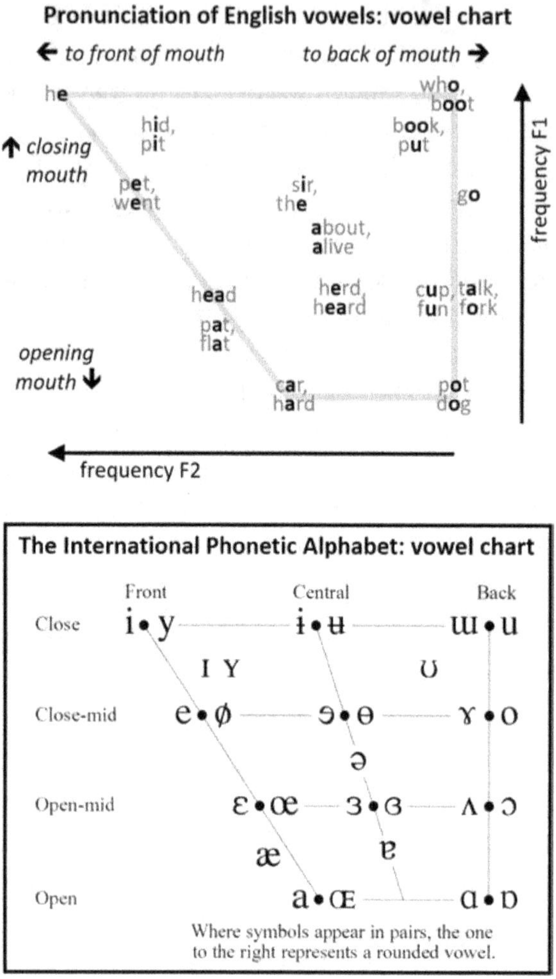

Figure 6-5: The top chart compares the main English vowels (shown in bold inside words) in terms of the formant frequencies F1 and F2, which are related to the position of the tongue (from a closed to an open mouth for F1, and from front to back of the mouth for F2). The bottom chart is the official vowel chart of the International Phonetic Association: it shows standard symbols for a variety of pronunciations of vowels in various languages; these symbols are used in many dictionaries to indicate proper pronunciation, but they require hearing spoken sounds to correctly pronounce them. (*Source*: the top chart is the author's adaptation of the bottom chart reproduced from the International Phonetic Alphabet chart of 2015, under CC-BY-SA 3.0, http://www.internationalphoneticassociation.org/content/ipa-chart.)

you pronounce the bold vowels. For example, say repeatedly "he – ha – he – ha ...": do you notice how your jaw goes up-down-up-down-..., so that your tongue also goes up-down-up-down-...? When your tongue is up, it forms a smaller cavity above it and therefore a higher-pitched sound. Or say "he – ho – he – ho...": your tongue now should move backward-forward-backward-forward... (in addition, your lips probably move, but that is not necessary). When the tongue is pushed forward, the smaller cavity in front of it favors higher frequencies.

Normally, we are not conscious of the movements within the mouth when we talk: take the time now to feel those movements as you slide your voice from one vowel to another through the vowel chart, both between nearby vowels and more distant vowels. Such sliding between two vowels is called a **diphthong**. Examples are particularly common in English:

- "day" and "eight": here "ay" and "eigh" both slide from 'ea' as in "head" to 'ee' as in "he"; thus "day" sounds like 'd – a – ee' and "bake" sounds like 'b – a – ee – k' .
- "I"/"hi"/"high"/"bye": these words slide from 'a' as in "pat" to 'ee' as in "he" .
- "cow"/"loud": both "ow" and "ou" slide from 'a' as in "pat" to 'oo' as in "boot", which is similar to the 'u' discussed above .
- "toe"/"tow": these slide from 'o' as in "go" to 'oo' as in "boot".

You can use these examples to practice sliding slowly around the vowel chart and to feel the movements within your mouth.

The gray outline in Figure 6-5 shows the practical limits that our mouth can reach (some animals can go far beyond those limits, as we will see in Section 6.12). This means that any language must have a set of vowels that fits within these limits of the human mouth's capabilities. To create a language that is rich in sounds and thus rich in meanings, we would choose as many vowels as possible within these limits. However, if the vowels get very close together, the mouth has difficulties generating them distinctly, and our ears and brain have difficulties distinguishing them. We should also remember that different people have different mouths, which will inevitably create somewhat different sounds: to avoid misunderstandings, these sounds should not overlap from one person to another. Such factors limit how close the vowels can be to each other. **In practice, the number of distinguishable vowels is limited to around 20**: in vowel charts of various languages, we will indeed see fewer than 20 vowels (Figure 6-15 in Section 6.8).

How can we pronounce a "foreign" vowel? Before leaving the vowels for the consonants, let's think about a vowel that is rather common in European languages but non-existent in English: the 'u' which is used, for example, in **French** and **Dutch**, and the identical 'ü' in **German** (the German letter 'ü' is called "u umlaut"; it is also often spelled "ue" in German text and German names such as Mueller, which is pronounced exactly like Müller; in particular, internet addresses normally use "ue" rather than "ü", since the "ü" is hard to type outside Germany).

This 'u' is not pronounced as the English 'oo' or 'you'! You can listen to this 'u' sound online.[5] In English, this 'u' is borrowed in the French word "résumé", although often mispronounced. You may be familiar with the pronunciation of 'ü' in the city name München (which is German for Munich, sometimes written as Muenchen in German), and in the city name Nürnberg (sometimes written as Nuernberg in German, or Nuremberg in English). This German 'ü' and French 'u' are illustrated in Figure 6-6 within different words (this figure also illustrates other consonants, which we will discuss further in Section 6.6).

In the IPA vowel chart at the bottom of Figure 6-5, this 'ü' or 'u' is marked as [y] near the top left corner: this suggests that it is close to the 'e' of the English word "he" (symbol [i] in the chart), namely close to 'ee' as in "he" and "bee"; the difference is that the German/French 'ü'/'u' is spoken with very tight, rounded lips. Therefore, one way to pronounce this 'u' is to start saying 'ee' (which opens the lips wide) and then almost close the lips, leaving only a small round opening; but don't go so far as to whistle!

How about nasal vowels? **French** has several **nasal vowels** that are not used in English. For example, nasal sounds are used in the French

---

[5] Listen to the audio sample at https://en.wikipedia.org/wiki/Close_front_rounded_vowel. And see the video "How To Pronounce U in French" by Learn French With Frencheezi: https://www.youtube.com/watch?v=xRetc-sfn8o

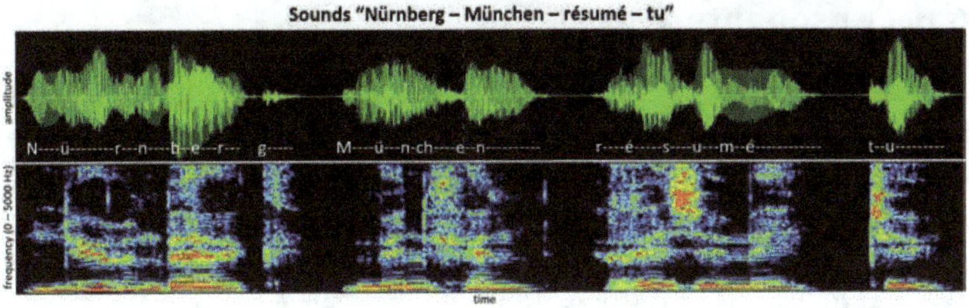

Figure 6-6: Waveform and spectrogram of the German names "Nürnberg" and "München", as well as the French words "résumé" and "tu" (meaning "you" in English). Each marked letter shows the start of the corresponding sound, which continues as marked by dashes (such as 'ü--------'). Note the brief silence before the explosive 'g' of "Nürnberg", pronounced as a weak 'k'. The recording covers about 3.6 seconds. (*Source*: recorded with a smartphone, displayed with WaveSurfer.)

words "an" (meaning "year") and the equally pronounced "en" (meaning "in"), as well as in the differently pronounced "vin" (meaning "wine"), "on" (meaning "we") and "un" (meaning "a" or "one"); important is that the final consonants such as "n" are not pronounced separately. Nasal vowels are produced by the air passing through the mouth and nose at the same time: resonances within the nose cavity give them their nasal character.

These nasal vowels are different from the English sound 'ng' in "ang", "eng", "ing", "ong" and "ung": the English 'ng' is produced entirely in the nose, whereas the nasal vowels combine the mouth cavity with the nose; 'ng' is categorized as a consonant (see Section 6.6).

Incidentally, the French sound 'an' also occurs with many other French spellings, such as the final sound in "blanc" (English "white"), "grand" ("large"), "rang" ("rank"), "dans" ("in"), "tant" ("many"), "en" ("in"), "temps" ("time"), *etc.* This nasal sound is somewhat related to the 'a' sound in the English "at": one way to pronounce it is to say 'a' as in the English "at", but to open the passage to the nose, causing a stronger resonance in the nose. The other nasal sounds can be produced in the same manner. For the French 'in', you can start with the 'i' in the English "in" and make it more nasal. For the French 'on', you can start with the 'o' in the English "on". And for the French 'un', you can start with the 'u' in the English "bun".

6.6 Consonants

We have focused so far on vowels to show how the vocal cords and the internal resonance cavity of the mouth can be controlled to produce different sounds. By contrast, **consonants** are produced by blocking — or at least restricting — the flow of air through the mouth or nose. Blocking (with the tongue or lips) and then releasing the airflow through the mouth produces the letters 't', 'd', 'p', 'b', 'k' and 'g': they are brief silences followed by brief **explosions**; 't', 'p' and 'k' are stronger explosions than 'd', 'b' and 'g', respectively. There is no "consonant map" to compare with the "vowel map" of Section 6.5. A detailed description of consonants used in many languages is available online.[6]

Figure 6-7 shows the sound "pssst" pronounced by two males: there is a silence at left before a very brief (almost invisible) 'p'; another silence separates the longer "sss" from the 't', to build up the pressure for the explosive 't'. The 't' itself is followed by an involuntary and strong 's' representing the longer outflow of accumulated air. Try out all these sounds and feel what your mouth is doing for each sound! The spectrogram shows that many frequencies are produced that look more like noise than music: there are no harmonics. Not shown is that the frequencies extend all the way up to about 17,000 hertz (compare with other noise in Figure 5-13). You can also see and hear the sound "pssst" in my Animation 6*4.

ANIMATION 6*4 — See my video WA3 at time 5:56 in its section "**Consonants *vs.* vowels**" under the title "**Consonants *vs.* vowels: consonants**". (See details in the section References and Resources below.)

The explosive letters 't', 'p' and 'k' can be softened, as in "doe", "though", "thorough", "bone", "phone", "go": such sounds are illustrated in Figure 6-8. In these cases, air is allowed to leak out so that the airflow is not totally blocked and there is no explosion. This outflow of air gives a hard 's' sound or softer 'z' sound, also as shown in Figure 6-8.

[6] See, for example: http://dialectblog.com/the-international-phonetic-alphabet/ipa-consonants/

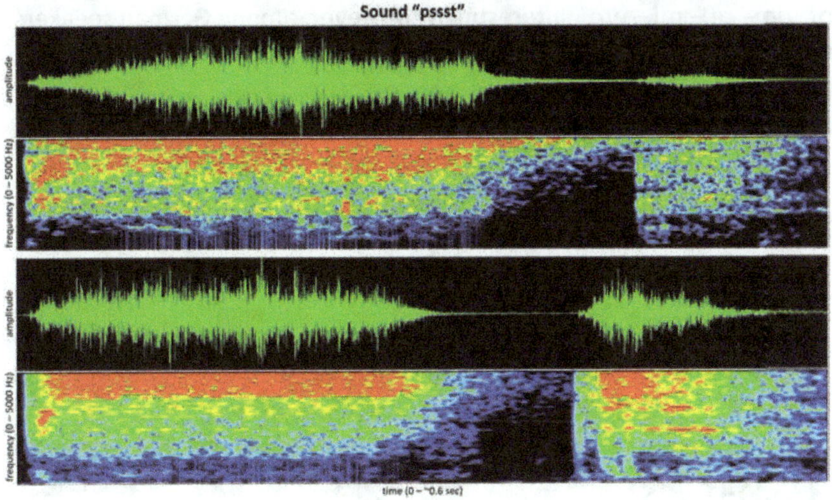

Figure 6-7: Waveform and spectrogram of the sound "pssst" pronounced by two males (top and bottom). The 'psss' and the 't' are separated by a near silence. The upper and lower recordings cover about 0.8 and 0.5 seconds respectively. (*Source*: recorded with smartphones, displayed with WaveSurfer.)

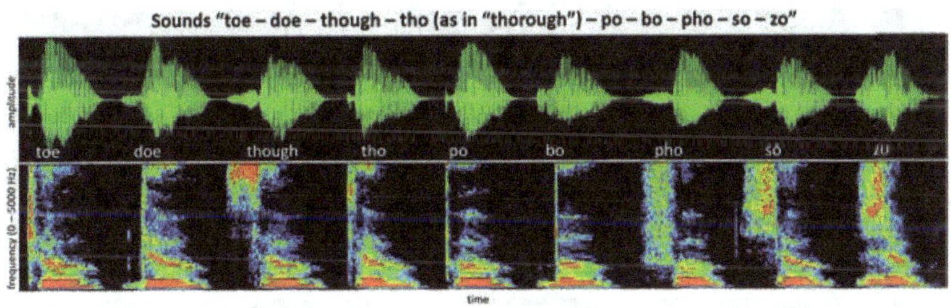

Figure 6-8: Waveform and spectrogram of "toe", "doe", "though", "tho" (as in "thorough"), "po", "bo", "pho", "so" and "zo", all with the same 'o' as in "go", pronounced by a male. The 'o' is visible as horizontal "ribs" in the spectrogram; the consonants do not normally have such "ribs". The recording covers about 4.8 seconds. (*Source*: recorded with a smartphone, displayed with WaveSurfer.)

What is the mechanism for producing soft consonants? These sounds result from turbulence in the air as it races through narrow spaces, similar to wind shrieking through small openings or around wires. Each consonant produces a wide range of frequencies, including many high frequencies when the airflow is fast. So, consonants don't have a fundamental frequency or pitch: you cannot sing them on a scale of "do – re – mi – *etc.*" Interestingly, however, among German speakers, the Bavarians and Austrians make no distinction between 't' and 'd', and between 'p' and 'b'.

Figure 6-9 illustrates the verse "Happy birthday to you!" It is sung by an adult female (including the **vibrato** effect mentioned in Section 6.4), spoken by a 7-year-old girl (her higher pitch is seen as larger spacings between "ribs") and synthesized by Microsoft Word (speech synthesis is further discussed in Section 6.11). This figure combines consonants and vowels and shows many of the effects that we have discussed so far.

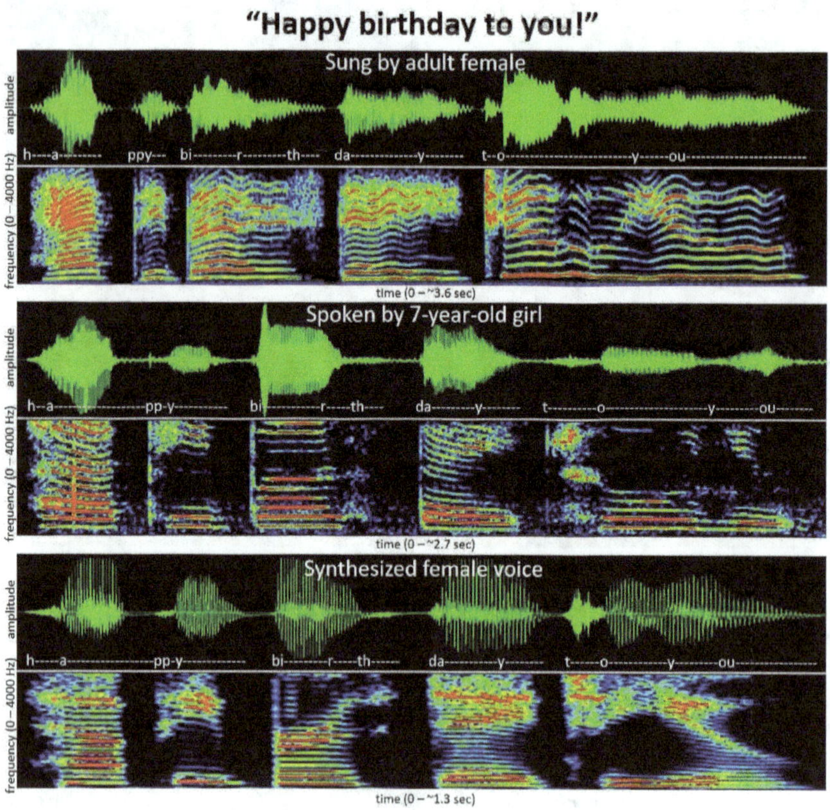

Figure 6-9: Waveform and spectrogram of one verse of "Happy birthday to you!" sung by an adult female (*top*), spoken by a 7-year-old girl (*middle*) and synthesized as a brisk adult female voice by Microsoft Word (*bottom*). The three recordings cover about 3.6, 2.7 and 1.3 seconds respectively. (*Source*: recorded with Praat, a smartphone and Microsoft Voice Recorder, respectively; displayed with WaveSurfer.)

Very visible is the up-and-down "vibrato" variation in frequency (pitch) during singing, which gives a richer and more pleasing melodious sound than in speaking. You can also see and hear singing in my Animation 6*5.

ANIMATION 6*5 — See my video WA3 at times 8:22 and 8:49 in its section "**Singing**" under the titles "**Singing: singing *vs.* speaking (1 of 2)**" and "**... (2 of 2)**". (See details in the section References and Resources below.)

Further consonants in English include:

- 'm' and 'n' (as well as 'ng') are nasal sounds in which the airflow goes entirely through the nose; their "nasal" sound is due to resonance in the series of folds within the nose cavity; they may end with a small explosion due to reopening the mouth to pronounce the next letter, as seen especially after the first 'n' in "München" in Figure 6-6.
- 'v' lets air flow between upper teeth and lower lip.
- 'l' as in "look" allows air to escape around the sides of the tongue, giving a "liquid" sound.
- 'y' as in "you" makes air flow over the middle of the tongue; it is called a semiconsonant because it includes some vowel-like vibration.
- 'h' as in "home" makes air flow over the back of the tongue.
- 'w' as in "we" is produced by blowing gently with rounded lips; it is also a semiconsonant.

What about the consonants 'r' and 'g'? They deserve more discussion, as they can pose challenges to English speakers, especially when they learn other European languages.

How is 'r' pronounced? The 'r' is often ignored after a vowel in British English and in some parts of the US's New England and South, but not elsewhere in the US. However, some British speakers, as well as opera singers, like to emphasize a strongly **rolling** 'r', particularly at the beginning of a word, for example when saying the word "rolling". Everyone tries to roll the 'r' when pronouncing "brrrrr" to express the feeling of shivering in the cold! How is this rolling sound produced? Try saying "brrrrr" and keeping the 'r' rolling as long as you can: do you feel

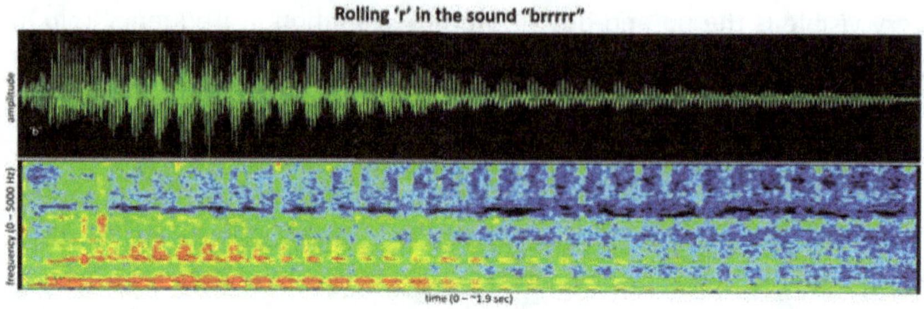

Figure 6-10: Waveform and spectrogram of the sound "brrrrr". The initial 'b' takes about 1/20 of a second, after which the rolling of 'r' repeats about every 1/20 of a second. The recording covers about 1.9 seconds. (*Source*: recorded with a smartphone, displayed with WaveSurfer.)

the vibration in your mouth? What vibrates? It's your tongue, but what part of the tongue? It is probably the top of your tongue that vibrates if you are an English speaker.

There are also other ways of pronouncing the rolling 'r': by using the front or the back of your tongue. Speakers of **Spanish**, **Italian** and **Dutch** often roll the 'r' with the tip of the tongue, close to the teeth: this produces a very strong rolling effect, illustrated in Figure 6-10. French speakers have three types of 'r'; one type rolls the 'r' with the back of the tongue, almost in the throat: it feels and sounds almost like snoring or gargling (similar to the **Scots** 'ch' in "loch" or the Dutch 'g', 'gh' or 'ch' that we will discuss next). By contrast, the German 'r' is quite soft.[7]

What does "rolling" mean when pronouncing 'r'? It is a repeated variation of the loudness of the sound, as seen in Figure 6-10: the sound amplitude oscillates with a repetition time of roughly 1/20 of a second (meaning at a frequency of about 20 hertz). Rolling the 'r' is thus an example of amplitude modulation (see Section 5.9). Our ears can hear such a low frequency quite easily because it causes an oscillation of the amplitude of a wave that already has a much higher audible frequency produced by our vocal cords (which in our example is about 125 hertz and its harmonics, as shown by the very rapid oscillations of the wave

---

[7] The pronunciation of 'r' is nicely explained in the video "Dutch pronunciation: the letter R" by Bart de Pau: https://www.youtube.com/watch?v=7C8iwl2pNlQ

and the tight horizontal "ribs" in Figure 6-10). This rolling effect is very similar to the "beats" that we discussed in Section 2.6 and illustrated in Figure 2-15. However, the rolling effect is different from the vibrato effect, where it is the frequency itself that varies repeatedly, not the amplitude (see Figure 6-4).

How is 'g' pronounced? The letter 'g' (and the frequently equivalent 'gh') has several possible pronunciations in English. It is a relatively hard, explosive sound, somewhat similar to 'k', in the words "go", "ground" and "ghost". It is a softer, 'dj'-like sound in "page" and "gem". It is an even softer, 'zj'-like sound borrowed from **French** in "massage" and "triage", as well as the second 'g' in "garage" (although some people pronounce that "g" like 'dj' when emphasizing "garage" rather than "garage"). In some words, the "g" may even be unspoken, as in "design" and "foreign".

Dutch (spoken in the Netherlands) uses a much harder guttural sound 'g' (and the equivalent 'gh' and 'ch') than in English. The Dutch 'g', 'gh' and 'ch' sound a bit like snoring or gargling, or even like clearing one's throat[8]: imagine trying to spit out a mosquito that accidentally entered your throat! Most English speakers can hardly pronounce this sound, used twice in the famous painter's name "Van Gogh"; **Scots** use it in words like "loch". The neighboring **Germans** have a similar but somewhat gentler sound 'ch' as in "Bach" and "hoch" ("high"). They also have an even softer sound 'g' in words like "fertig" (at least in some regions, meaning "ready") and the similar sound 'ch' in "ich" ("I") and "München". The Germans otherwise pronounce 'g' much like the English 'g' (a soft 'k'), for example in "Garten" (meaning "garden") and "Hamburg" (in fact, the German "rg" often sounds close to 'rk'). The sound 'g' in the ending "burg" is the same as in "Nürnberg" illustrated in Figure 6-6. Nevertheless, Germans have great difficulty pronouncing the Dutch 'g', 'gh' and 'ch'; this is even more surprising as the Dutch 'g', 'gh' and 'ch' sound becomes softer as you go southward in the Netherlands and into Belgium (in the closely related **Flemish** language), resembling more the softer German 'g' and 'ch'. Spanish also has a

---

[8] See video "Learn how to pronounce G and H in Dutch" by The Language Academy: https://www.youtube.com/watch?v=7982L-EWnaA

guttural letter 'j'[9], while **Arabic** has a guttural 'gh' (as in the letters خ and غ)[10]: these sounds are usually a bit softer than the Dutch 'g', 'gh' and 'ch', especially in Spanish. We will further discuss regional variations in Section 6.8 in connection with accents.

6.7 Tonal Languages: Mandarin Chinese and Cantonese Chinese

A few languages are **tonal**, such as **Chinese**, **Vietnamese**, **Thai**, **Cherokee**, and some African languages. "Tone" here refers to the pitch, and more precisely to a variation of the pitch within a single vowel, such as a climbing pitch or dropping pitch, or also different pitches for the same vowel: these are called **contour tones**. We can say that a vowel pronounced with such different tones becomes multiple different vowels, with separate meanings in words. Tonal languages thus use the tone to change the meaning of a word rather than to emphasize a word.

Let's start with a **non-tonal language** such as English. We can say "This house is blue." and emphasize separately each of the words in this sentence by increasing its pitch (frequency) and/or its intensity (the underlining shows the emphasis):

- "This house is blue." adds the information "not those other houses".
- "This house is blue." adds the information "not that car parked near the house".
- "This house is blue." adds the information "not another color".

The emphasis does not change the meaning of the words "this", "house" or "blue" but draws attention to them and thereby adds precision.

What is a tonal language? In a **tonal language**, the tone does change the meaning of a word. A well-known example is the sound 'ma' in **Mandarin Chinese** (which is spoken in most of China, where

[9] See video 'How to Say "J" | Spanish Lessons' by Howcast: https://www.youtube.com/watch?v=pSzbu1kjnAI

[10] See video "How to pronounce غ & خ like a real Arab — The proper technique and muscle training — Lesson 4", by Arabic 101: https://www.youtube.com/watch?v=at0bJhzsBpI

this language is called **Putonghua**), illustrated in Figure 6-11: depending on the **tone**, 'ma' as a word can have four totally different meanings, as described in Table 6-1. **Cantonese Chinese** (which is spoken mainly in the southern Chinese province of Guangdong, including the city of Guangzhou/Canton, and in Hong Kong and Macau, as well as in older Chinese communities overseas) uses nine different tones; these are illustrated in Figure 6-12 and described in Table 6-2.

We see that the Chinese tones can completely change the meaning of the sounds: one basic sound, like 'ma' in Mandarin, or 'si' in Cantonese, can have up to 4 or 9 different meanings, respectively.

In Mandarin, one tone is flat (constant pitch), while one tone increases its pitch, another decreases and the fourth dips and rises. In Cantonese, two tones rise and one tone dips; the six other tones are flat, but may have a high, medium or low pitch; three of these six are short versions of the other three, giving nine different meanings.

The representation of the tones as simple straight lines or straight arrows (as done in Tables 6-1 and 6-2) is an oversimplification: the tones evolve over time in a more complex fashion, as the spectrograms in Figures 6-11 and 6-12 make clear. In particular, each tone tends to

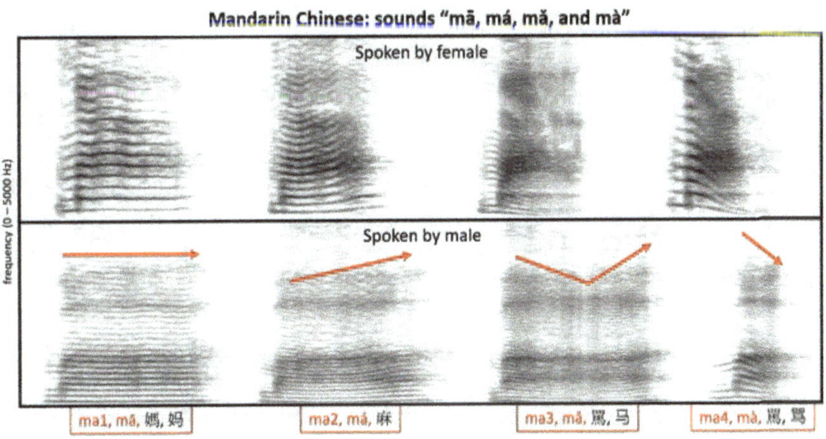

Figure 6-11: Spectrograms of Mandarin Chinese pronunciation of the tonal sounds in "ma1 – mā", "ma2 = má", "ma3 = mǎ", and "ma4 = mà", by a female (*top*) and male (*bottom*). The red arrows indicate the direction of the tonal variation, which is also sketched in Table 6-1. Both the traditional and simplified Chinese characters are shown; they are equal for "ma2". (*Source*: recorded with a smartphone, displayed with Praat.)

Table 6-1: The four tones in Mandarin Chinese. (See notes below the table.)

pinyin	numbered examples	tone variation	traditional Chinese character	simplified Chinese character	English meaning
mā	ma1, ma55	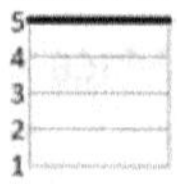	媽	妈	mother
má	ma2, ma35	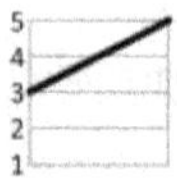	麻	麻	hemp
mǎ	ma3, ma213		馬	马	horse
mà	ma4, ma51	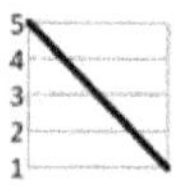	罵	骂	scold

Notes on the standard notation used in this table: **Pinyin** is a Romanized version of Chinese characters that includes an accentuation mark over the letter to indicate one of the four tones (Pinyin is used mainly for teaching Chinese, for spelling Chinese names in English, and for typing into computers in Mainland China; Pinyin also often appears on shops in Mainland China, but the accentuation mark is mostly omitted, leading to confusion since the tone and meaning are then lost, except through the context). The accents in ā, á, ǎ, and à indicate the direction of change of the tone, also sketched in the third column. The form 'ma1', *etc.* is equivalent to 'mā', *etc.* The form 'ma35' explicitly indicates the variation in tone, as sketched in the third column, showing the tone variation, which can range between 1 (lowest frequency) and 5 (highest frequency), 3 being neutral (in these sketches, you can imagine the tones 1-2-3-4-5 to be the tones do-re-mi-fa-sol): thus, 'ma35' indicates that the tone varies from level 3 to level 5 (from mi to sol). Each Chinese character is normally equivalent to a single English word; but each Chinese character sounds like a single English syllable, such as 'i' (meaning "one" in English), 'er' ("two"), 'san' ("three"), 'liou' ("six") or 'wang' ("king"). The **traditional Chinese characters** are used mainly in Taiwan, Hong Kong, Macau and overseas. The **simplified Chinese characters** are used on the Chinese Mainland, in Malaysia and in Singapore. A fifth or zeroth or neutral tone also exists in Mandarin Chinese: its pronunciation depends on the preceding syllable.

start at the "default" or "neutral" pitch of a voice: the vocal cords then quickly adjust up or down before producing the main dipping or rising or high or low sound of the tone (you can see this from the "ribs" in Figure 6-11, as the rib separation is initially about the same for all four tones). You can also see and hear Chinese pronunciation in my Animation 6*6.

ANIMATION 6*6 — See my video WA3 at times 9:27, 9:58 and 10:28 in its section "**Tonal languages: Chinese**" under the titles "**Tonal languages: Mandarin Chinese**", "**Tonal languages: Cantonese Chinese**" and "**Tonal languages: Mandarin *vs.* Cantonese Chinese**". (See details in the section References and Resources below.)

Most speakers of non-tonal languages, including English, have difficulties speaking and even hearing the varying tones of tonal languages. Their ears and brains are not accustomed to distinguishing such tones.

Another aspect of tonal languages is the difficulty of expressing emphasis of the type I wrote as "This house is blue", since the tone is used for a different purpose than to emphasize. Instead, in a tonal language like Chinese, an emphasis is usually given as an additional word, for example: "This specific house is blue" meaning "This house is blue".

Cantonese Chinese, like all other forms of the Chinese language, is usually written with the same characters as Mandarin Chinese (although a separate Cantonese character set exists). However, many of these characters are pronounced quite differently. For example, numbers are written identically in Mandarin Chinese and Cantonese Chinese,

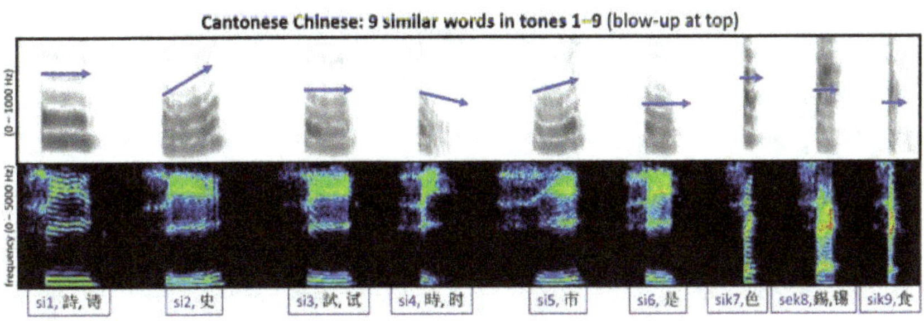

Figure 6-12: Spectrograms of Cantonese Chinese pronunciation of the nine tonal sounds in similar words. The frequency scale in the bottom spectrogram is blown up five-fold in the top spectrogram, making the ribs look very broad there while emphasizing the tonal variations marked by blue arrows, as also sketched in Table 6-2. In Cantonese, the final consonant is often very weak or silent, such as the 'k' here in 'sik' and 'sek'. (*Source*: recorded with a smartphone, displayed with Praat (*top*) and WaveSurfer (*bottom*).)

Table 6-2: The nine tones 1–9 in Cantonese Chinese. The ten digits 0, 1, 2, ..., 9 and the set of words from Figure 6-12 are used as examples. (See notes below the table.)

tone number	numbered examples	tone variation	Chinese character (trad., simpl.)	English meaning
1	saam1		三	three
	jat1		一	one
	si1		詩，诗	poem
2	gau2	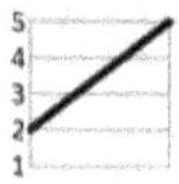	九	nine
	si2		史	history
3	sei3	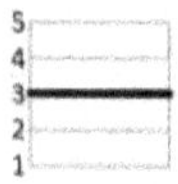	四	four
	si3		試，试	examination
4	ling4	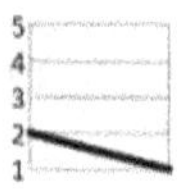	零	zero
	si4		時，时	time
5	ng5	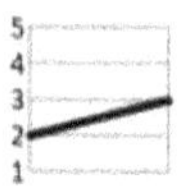	五	five
	si5		市	market
6	yi6	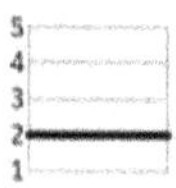	二	two
	luk6		六	six
	si6		是	is
7	chat7	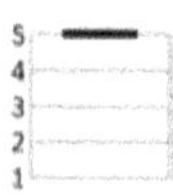	七	seven
	sik7		色	color
8	baat8	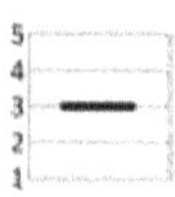	八	eight
	sek8		錫，锡	kiss
9	luk9	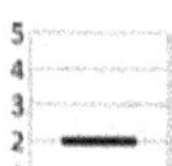	六	six
	sik9		食	eat

Notes: The last three tones (7, 8 and 9) are essentially short versions of tones 1, 3 and 6; therefore, some linguists prefer to say that Cantonese has six tones instead of nine. The third column graphically shows the variation in tone, between 1 (lowest) and 5 (highest), 3 being neutral, as in Mandarin Chinese. The nine tone numbers (given in the first column) are often attached to the romanized form of a word to indicate the tone, as illustrated in the second column, such as 'saam1' meaning "three pronounced in tone 1", and thus 'saam' spoken with a steady high pitch. In Cantonese, the numbers 3-9-4-0-5-2-7-8-6 (namely 'saam-gau-sei-ling-ng-yi-chat-baat-luk') happen to have the tone numbers 1 to 9, in that order. Non-Cantonese speakers may find it difficult to pronounce the 'ng' for "五" (meaning 5); it appears in many words, including at the beginning of a word like 'ngan' (meaning "small"): one way to pronounce 'ng' is to start by saying the English word "wing", then removing the 'w' to say 'ing', and finally shortening the 'i' until only 'ng' is left.

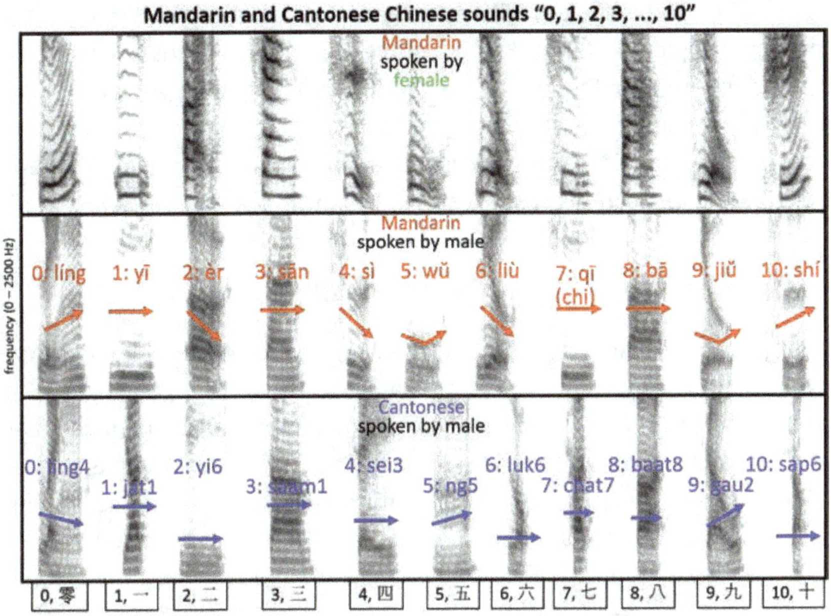

Figure 6-13: Spectrograms of counting from 0 to 10 in Mandarin (top two graphs) and Cantonese Chinese (bottom graph). The top case is spoken by a female; the lower two are by a male. The tones are indicated by accents in Mandarin, by tone numbers in Cantonese, and by arrows in both languages, as in Tables 6-1 and 6-2. (*Source*: recorded with a smartphone, displayed with Praat.)

but are partly pronounced very differently, as shown in Figure 6-13 (see especially the numbers 1 and 2); there also are differences in both vocabulary (characters) and grammar. As a result, even though they share the same written form, spoken Cantonese and Mandarin Chinese are not mutually understandable and thus can count as separate languages (as we will discuss in Section 6.8).

The pronunciation of Chinese tones is different from that of western accentuation marks, such as the **German** "ü" *versus* "u", or the **French** accents in "é", "ê", and "è" *versus* "e". These French accents can change the meaning of words, but do not always change their pronunciation.[11] These accents normally have a flat tone (similar to tones 1, 3 and 6 in

[11] Examples of changes in meaning of French words without change in pronunciation include: "a" (English "has") *versus* "à" ("to/at"); "la" ("the") *versus* "là" (there); "ou" ("or") *versus* "où" ("where").

Table 6-2), can be spoken with a constant pitch, but can differ in higher formant frequencies. The "accent grave" as in "è" means a low pitch, not a descending pitch, while the "accent aigu" as in "é" means a high pitch, not a rising pitch.[12] By contrast, several of the Chinese tones must rise and/or drop in frequency level to produce the correct different meanings, as shown in Tables 6-1 and 6-2.

6.8 Accents, Dialects/Varieties and Languages

With modern transportation and telecommunications, we are increasingly exposed to **accents**, **dialects** (which are **varieties of languages** as well as dialects understood as ancient official languages) and **languages** that are different from our own. This happens through travel, online videos, movies, exchange visits, international marriages, and migrations, among many ways to mix and compare cultures.

It is fascinating to explore how language varies in space and time. An obvious source of variation is geography, but there are also other reasons for linguistic variations, such as social groupings, professional settings and family relations: these different groups of people may use different pronunciations of the same words, partly by tradition, or by habit, or to "be different", or even for fun. We will explore further below how accents, dialects and languages can evolve over time, but let's first define these concepts more carefully.

What do we mean by "accent", "dialect", "variety" and "language"?

"Accent" can have the simple meaning of emphasis of a word or syllable, such as when we emphasize "old" in "old house" to stress its age. In this book, we are more interested in the more linguistic meaning of "accent", as in the "New York accent" being a variation of English.

Accent, in the following discussion, refers to the way words are pronounced in different regions or social classes, but excludes differences in words and grammar. For example, a New York accent and

[12] The "accent circonflexe" as in "ê" is pronounced the same way as in "è"; it usually means that an ancient Latin letter is not pronounced and has been dropped, such as the letter "s" in "hâte" and "fête", meaning "haste" and "feast" in English, respectively; nonetheless, the "s" in "festival" is still written and pronounced in French.

a Boston accent clearly have distinct pronunciations; some words and expressions may differ between the two regions, but these are a minor aspect of the difference in accents. In extreme cases, two accents can sound so different as to make them mutually unintelligible. However, in normal conversation, most of what is said in a different accent can be understood.[13]

Dialect (a variety of language) refers mainly to the use of different words and grammar, in addition to the possibility of different sounds, as in "accents". Typical examples of related dialects are British English, American English, Australian English, Indian English, *etc.* (these dialects are collectively called **Englishes**). In extreme cases, "dialects" may be mutually unintelligible, even with identical accents, because the words and grammar differ significantly. As one example among many, most Germans don't understand Swiss German because it is so different from Standard **German**. (Actually, Swiss German is a family of dialects and subdialects that can vary from one valley to another or even from one village to the next, which mostly understand each other. This is not unique to Switzerland, but perhaps especially clear in that multi-lingual country with isolated valleys.)

In practice, "accents" and "dialects" are usually mutually intelligible because we can normally grasp the meaning of what is being said through the context of the spoken conversation or written text: we "fill in" or "guess" from the rest of the story the information that we did not understand.

Why use the word "variety" instead of "dialect"? In some parts of the world, a "dialect" is considered to be a lower-level variety of the "proper" or "correct" or "official" language: a dialect may be viewed as a regional or lower-class or outdated variety of the "official" language. This

[13] See the videos "1 Language, 3 Accents! UK vs. USA vs. AUS English Pronunciation!": https://www.youtube.com/watch?v=pOHTr-MxtGM; "English Accents Ranked from Easy to Hard to Understand" from around the world: https://www.youtube.com/watch?v=aIDfw3wrK7M; "British Accents Ranked from Easiest to Hardest" from around the United Kingdom: https://www.youtube.com/watch?v=oV8_rdJok38; "Accent Expert Gives a Tour of U.S. Accents — (Part One)": https://www.youtube.com/watch?v=H1KP4ztKK0A; "Accent Expert Gives a Tour of U.S. Accents — (Part 2)": https://www.youtube.com/watch?v=IsE_8j5RL3k; "Accent Expert Gives a Tour of North American Accents — (Part 3)": https://www.youtube.com/watch?v=Sw7pL7OkKEE

is often the result of one region or class in a country dominating other regions or classes and imposing its variety of the language. Examples are the **Queen's English** or **King's English** in the United Kingdom (also called the BBC pronunciation or the Received Pronunciation), or the Parisian variety of **French** dominating the rest of France. Linguistically, such downplaying of dialects is not relevant: the official dominant language is equally a dialect like any other dialect. All dialects are linguistically equivalent, being simply different ways of saying and writing the same thing. **Linguists therefore prefer to use the word "variety" over "dialect" to avoid favoring one dialect over others**; in the following, we will use the more familiar term "dialect".

When dialects differ so much as to be largely unintelligible, we speak of languages. For example, the English language evolved by mixing languages spoken on the nearby European mainland (including German and French words in particular); English, German and French are now clearly mutually unintelligible.

It may be difficult to decide whether two language varieties are "accents" or "dialects" or "languages" relative to each other, partly because differences in pronunciation, vocabulary and grammar all occur at the same time. There is also no clear boundary separating these concepts, especially as accents, dialects and languages vary so much within themselves from person to person, as we will discuss next. Nevertheless, there exist language atlases and many language maps, including online.[14]

How do accents show up in spoken language? Let's start our discussion of accents by looking at the sounds in Figure 6-14: they are the same five English words spoken by an American male and an Italian male. We see clear differences between the two pronunciations: some are individual differences that can exist between any two persons, while others are more systematic differences of foreign speech and are therefore often called "foreign accent". The individual *versus* foreign differences are difficult to distinguish when comparing the speech of only two people. By comparing the speeches of many people, the

[14] See, for example, the many highly-detailed regional maps of languages and dialects at http://www.muturzikin.com/countries.htm

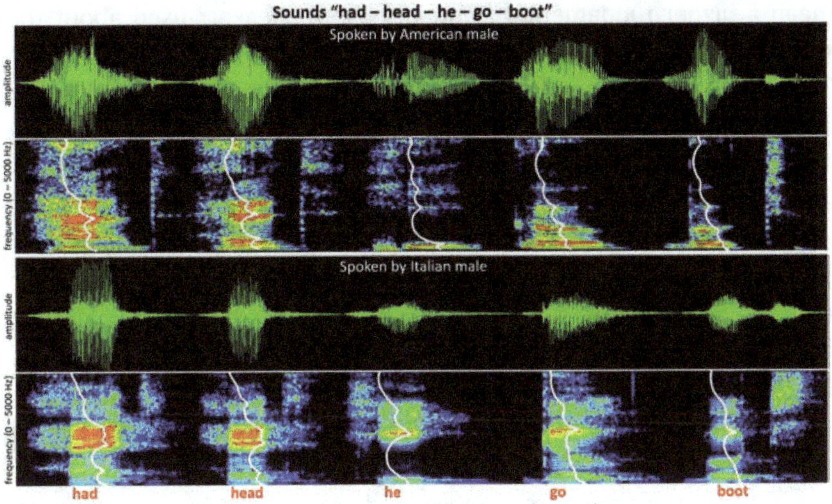

Figure 6-14: The words "had", "head", "he", "go" and "boot", spoken by an American male (top pair of graphs) and an Italian male (bottom pair of graphs). The peaks pointing to the right in the white curves show the formants of the five vowels, as presented in Figure 6-3 (in Figure 6-3, only the vowels were spoken and presented, while here the consonants are included). The upper and lower recordings cover about 2.8 and 3.3 seconds respectively. (*Source*: recorded with smartphones, displayed with WaveSurfer.)

"foreign accent" can be better identified. Nevertheless, let's look more closely at the two samples shown in Figure 6-14.

The white wavy curves in Figure 6-14 show the formants of the vowels (they are equivalent to the blue wavy lines drawn in Figure 6-3): their peaks pointing toward the right represent strong frequencies. As we saw in Section 6.4 and Figure 6-3, the formants dominate and distinguish the vowels. In this American *versus* Italian example, the formants are recognizable from one person to the other, so they are indeed pronouncing the same vowels.

However, there are also significant differences. The formants are not quite identical in frequency or strengths between the two voices: some of these differences may be universal among American *versus* Italian speakers, and therefore can define different accents. Another very visible difference is the higher strength of the middle frequencies in the Italian voice of Figure 6-14, which show up as red regions: in the American voice, the lower frequencies are strongest. This does not mean a higher fundamental frequency (both speakers used about the same fundamental frequency of 120 to 130 hertz), but more strength in higher formants of the Italian voice: this corresponds to a higher average frequency of the Italian voice. One more difference is the more abrupt termination in the American voice of the 'd' in "had" and "head", and of the 't' in "boot": this may also be a difference between accents.

How many accents exist? We have seen that, strictly speaking, each person has a unique voice and therefore a unique accent different from that of every other person. An accent then is a personal combination of ways to pronounce various letters and words: for example, each person uses their own individual vowel map. **We can estimate that it takes only a few dozen changes in pronunciation to create individual accents for everyone**, by using the following simple reasoning.

Let's consider speakers of English and imagine drawing a geographical map showing the regions where people drop the letter 'r' (saying 'ba' for "bar", for instance). Now we have 2 regions using different accents, one dropping the 'r' and one keeping the 'r'. Next, on the same map we can also draw the regions where people say 'd' instead of 't'. These regions probably are not the same as the regions dropping or keeping the 'r'. This splits each of the 2 earlier regions in 2: we get 4 regions with 4 accents (one accent uses both 'r' and 'd', another 'r' and 't', yet another 'd' but not 'r', and the fourth neither 'r' nor 'd'). If we go on to draw the regions where people pronounce "I like" as 'aa laak', we split up each of the 4 regions into 2, resulting in 8 regions with 8 accents. We thus find 8 regions with different combinations of 3 changed sounds (this 8 is $2 \times 2 \times 2 = 2^3$, or 2 to the power 3), causing 8 different accents. If we add a 4th changed sound, we generate 16 different sound combinations (2^4), meaning 16 different accents. Each additional changed sound doubles the number of possible accents: with 26 different changed sounds (one for each letter of the alphabet), we can create $2^{26} = 67{,}108{,}864$ or nearly 70 million different accents. This number suffices to give each Briton a different accent; a few more changes of sounds will give all the other English speakers around the world their own personal accents!

Thus, there is no clear answer to the question of how many accents there are. Nevertheless, let's consider "national accents", since we often think in those terms.

What distinguishes "national accents", for example, the "British English accent" from the "American English accent"? English speakers can fairly easily distinguish the British from the American accent, even though there are many large variations within the UK and also within the US. When we compare such "national accents", we focus on just a few obvious and systematic differences between the dominant accents of the different countries, while ignoring many other local variations. If you ask on the internet what the differences are between the British and American accents, many webpages focus on just the letter 'r'. Wikipedia, for example, answers: "British English and American sound noticeably different. The most obvious difference is **the way the letter r is pronounced**. In British English, when r comes after a vowel in the same syllable (as in car, hard, or market), the r is not pronounced. In American English, the r is pronounced". If you look a bit harder, you will find other differences, such as the 't' pronounced by Americans as 'd' in "water". We can thus say that a "national accent" is a very simplified "basket" or "stereotype" of just a few characteristic pronunciations that ignores a wide variety of other differences between accents.[15]

How do accents originate and evolve over time? It is easy to imagine that a small group of people could someday change their pronunciation of a particular vowel: that single changed vowel starts a new accent, which may die out soon or spread further among other people. This can happen in a classroom of children, or in a family, or in a professional setting, or in a gang, *etc*. The reason for such a change in pronunciation may be for fun, or for identification as a group, or through new interactions with another group, *etc*.

If a changed vowel spreads out to other people, it can expand geographically or within certain social circles, such as families or professions or the fans of a singer. Another changed vowel or consonant can originate anywhere and spread in similar ways, usually to different groups. Such spreading of accents resembles the spreading of viruses and, more generally, the natural evolution of species. The speed at which languages evolve is rather slow: even though new words and

[15] An amusing personal experience is that I learned English partly in the USA and partly in the UK; at one time people told me that my English accent was "mid-Atlantic", mixing British and American pronunciations.

expressions are added every year to a language, its pronunciation only changes noticeably over many decades or even centuries. After all, young and old people still need to understand each other, and they like to read old literature.

If the spoken language evolves, why does the written language not follow exactly the spoken language? It is interesting how slowly the written language adjusts to changes in the spoken language. The permanence of text written on paper helps to perpetuate it, so the mismatch between written and spoken language tends to grow over time. Even though the spelling of some languages is sometimes "officially" updated, such as in **German**, **Dutch** and **French**, such changes are not always accepted by the general population. The written and spoken forms of both English and French are especially mismatched; this is amazing, given the lack of official control of the English language and the strict (but not very effective) official control of the French language.[16]

We can imagine individual accent changes occurring anywhere and anytime, spreading and overlapping with each other, as we described above when answering "How many accents exist?". As a result, each location and social group will have a somewhat different accent, or different combination of accents: in fact, each individual person will have a different accent, composed of some combination of modified pronunciations within their social groups.

---

[16] I once tried to count in how many ways the sound "o" can be written in French. (You can do the same in English or another language for different vowel sounds, but beware of diphthongs in English: the 'o' in "so" is not the same as the 'ow' in "sow"!) I found in French 14 different ways to spell a short 'o' and 43 ways to write a long 'o', but such numbers depend on regional varieties and definitions. Examples for the short 'o' include the following (the 'o' sound is underlined): Paul, homme, bol, album. Examples for the long 'o' include: au, chaud, saut, eau, eaux, Peugeot, haut, home, hôtel, kilo, dos, pot, tôt. Similarly, I found 47 ways to spell the French sound 'è' as in "très" and "paix", and 26 ways to spell the French 'é' as in "né" and "clef" (also spelled "clé" with the same meaning). In English, I found 13 ways to spell the English short 'o' as in "good" or "put", 15 ways to spell the English long 'oo' as in "do" or "true", and 13 ways to spell the English 'yoo' as in "you", "few" or "use".

The creation of accents happens especially when a group of people migrates to an isolated region, such as an island: accent changes will then be contained within their isolated region and will not be "corrected" or affected by people elsewhere. This has happened with the French language in Quebec and Louisiana, with the German language in Pennsylvania among the Amish, and with the Dutch language in South Africa, for example. Over time, larger changes can also occur: new words or meanings of words may be created, and grammatical rules may change. New words and grammar will result in new **dialects**.

You can easily think of examples of such language evolutions around you and around the world. In the case of English, we can mention that there are close to 40 dialects in the British Isles! There are also many English dialects around the world: about eight in Canada, around 30 in the US and others in Australia, New Zealand, India, South Africa, Hong Kong, Singapore, *etc*. These varieties of English are often called **Englishes**.

If a new dialect has changed so much as to become unintelligible to people speaking other related dialects, it becomes a new language. We already mentioned that English has become unintelligible to German and French speakers, and *vice versa*. Another example is the group of Latin-based languages (the so-called Romance languages): **Italian**, **French**, **Spanish**, **Portuguese**, **Romanian**, *etc*.

Can languages be distinguished by their vowels? Let's think a bit more about languages and the **vowel charts** of Figure 6 5. As we mentioned, a language can have at most around 20 distinguishable vowels to allow enough separation between its vowels. So, if we imagine shifting any one vowel in the chart of Figure 6-5, neighboring vowels will shift as well to keep them distinguishable. We therefore can expect vowels in different languages to form a different "vowel map" on the chart of Figure 6-5, and that is exactly what happens: another language can have a very different vowel map, as shown in Figure 6-15. Nonetheless, some vowels tend to be easier to pronounce (such as the 'i' in English "he") and therefore remain almost the same in different languages. It is interesting to note in Figure 6-15 that some languages have only a few distinct vowels, such as **Spanish** and **Italian**, making them easier to pronounce, especially for foreigners.

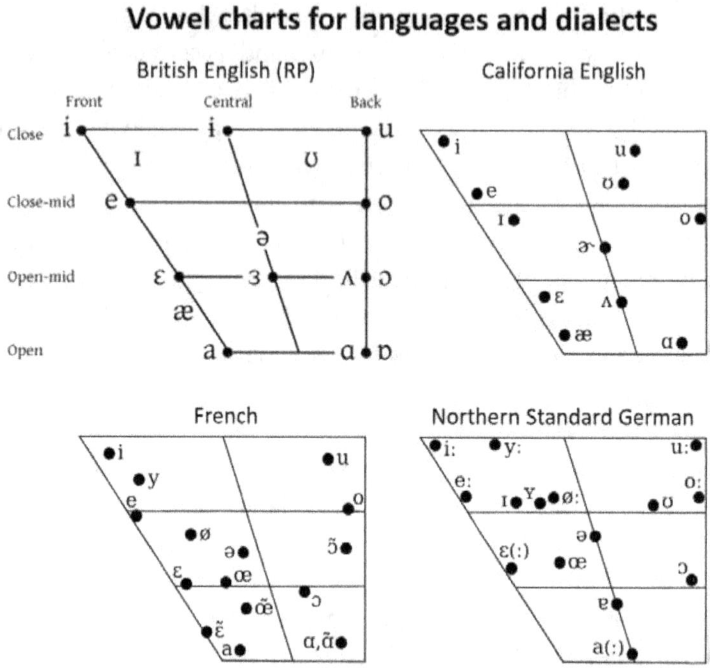

Figure 6-15: These charts compare the pronunciation of vowels in several languages and two dialects (the "Queen's English" called "Received Pronunciation", and Californian American English), using the same chart format as in Figure 6-5. The symbols in these graphs are used by linguists to recognize more precisely the sound of each vowel. For English-language non-linguists, it is easier to use the top chart of Figure 6-5 to imagine the sounds: for example, the vowel 'œ' in some of the charts shown here is located slightly to the left of the center of the chart; looking at that location in Figure 6-5 then tells us that 'œ' has a sound between 'ea' in English "head" and 'a' in English "about". The symbol ":" means a long vowel. For those interested in writing these symbols in Microsoft Word, PowerPoint, *etc.*, most of them are available in font Arial Unicode MS in the subset IPA Extensions; see also http://www.internationalphoneticassociation.org/content/ipa-chart.

(*Sources*: these charts come from Wikipedia, under CC-BY-SA 3.0 or 4.0, https://commons.wikimedia.org/wiki/File:English_vowel_chart.png, https://commons.wikimedia.org/wiki/File:California_English_vowel_chart.svg,
https://commons.wikimedia.org/wiki/File:French_vowel_chart.svg,
https://commons.wikimedia.org/wiki/File:Northern_Standard_German_vowel_chart.svg,
https://commons.wikimedia.org/wiki/File:Spanish_vowel_chart.svg,
https://commons.wikimedia.org/wiki/File:Italian_vowel_chart.svg,
https://upload.wikimedia.org/wikipedia/commons/d/d3/Beijing_Mandarin_monophthongs_chart.svg,
https://commons.wikimedia.org/wiki/File:Cantonese_vowel_chart.svg.)

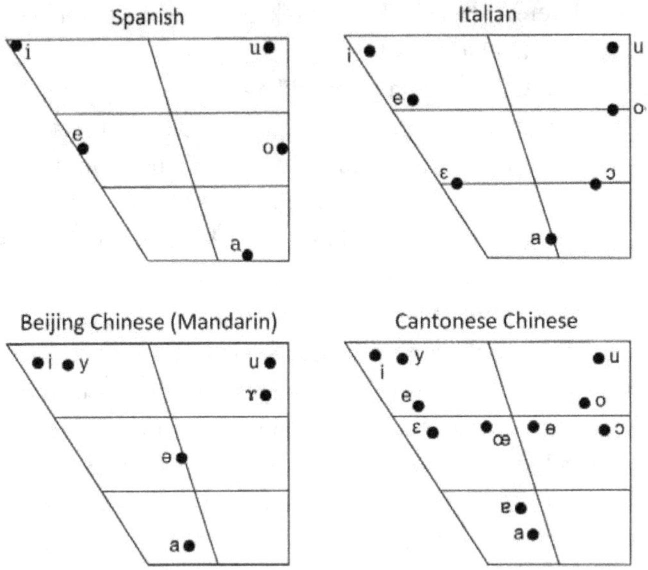

Figure 6-15: *(Continued)*

Polyglots, people who can speak multiple languages, may have two or more vowel maps stored in their brain. Remember that a vowel map tells the shapes of the mouth needed to produce the vowels of a given language. This means that polyglots can unconsciously and automatically switch from one language's vowels to another language's vowels (and similarly for consonants), without "foreign accent". This happens mainly to people who learn more than one language when young, typically before their mid-teens. They often remember vowel maps all their life, even if they forget words; fortunately, words are easier to re-learn than vowel maps.

What are "foreign accents"? People who learn another language at a later age have much trouble generating that new language's vowel map in their brains: this appears to be mainly due to the fact that the muscle movements that control sounds, for example to make formants, tend to "freeze in" and prevent other sounds from being made. This causes "foreign accents": in a **foreign accent**, one language's sounds are used to pronounce another language's words. Thus, a German who learns English later in life will likely speak English with a German accent, meaning with German sounds.[17]

Another interesting aspect of multi-lingual speaking is the following: due to migrations, mixed marriages, frequent and rapid travel, electronic communications, *etc.*, many people nowadays speak more than one language (they are polyglots) or more than one dialect. Just think of the number of languages used in "global" cities like London and New York City: apparently, in London, over 250 languages are spoken[18], and in New York perhaps more than 600![19] (Of course, we have to be careful with such numbers: "speaking a language" is not the same as speaking it perfectly.) **All these coexisting languages will locally influence each other by creating new local accents and perhaps dialects. Even on the level of accents, any person may use several accents in daily life**: one accent at home, another at work, a different accent in a sports club, yet another accent in a job interview, *etc.*; such variations are also called **language registers**, and are used in different social situations. We can then also expect that one person may mix his or her own languages or dialects or accents, as very young children often do.

In brief: languages, dialects/varieties and accents are fascinating, evolving and interconnected structures, especially in the modern world which sees so many migrations, so much travel and such easy communications.

---

[17] Here is a specific example of an accent being "frozen in". In Section 6.6, we described that, while the English 'g' is quite soft, the Dutch 'g' and the equivalent Dutch 'gh' and 'ch' are a much stronger guttural sound, which the neighboring Germans have great difficulty pronouncing. This fact was exploited by the Dutch during World War II to detect occupying Germans who were trying to blend in by speaking Dutch in the Netherlands: the Dutch would ask a stranger to pronounce the name of the Dutch city Scheveningen; if the stranger pronounced the "Schev" as in the English "shave", following standard German pronunciation instead of using the Dutch guttural 'ch', he or she was clearly German. Such examples illustrate how strongly accents can be frozen in, because the Germans in World War II were fully aware that they could face this linguistic but dangerous challenge.

[18] See "Languages Spoken in London": http://projectbritain.com/regions/languages.htm

[19] See "New map shows over 600 languages spoken in NYC": https://www.6sqft.com/new-map-shows-over-600-languages-spoken-in-nyc/

6.9 Singing *Versus* Speaking

We have already visually compared singing with speaking in Figures 6-4 and 6-9. We saw that variations in tone or pitch gives a melodious tone to singing, including the vibrato effect.

How different is singing from speaking? Members of a choir often say that singing is all automatic: this indicates that singing is a relatively easy extension of speaking. One difference is that the fundamental frequencies (pitches) used in singing have a much larger range (about 60 to 1500 hertz or more) than in speaking (about 100 to 400 hertz).

Professional singers may train their voices for additional control and special effects, such as singing very low frequencies (**bass**) or very high frequencies (**falsetto**).

It appears that consonants are very similar in singing and speaking. But vowels are more controlled in singing, especially in terms of frequencies, rhythm and duration. In normal speaking, you don't pay particular attention to frequencies, rhythm and duration: your speaking style (pitch, loudness, speed, rhythm, regional or social accent) is a habit that has grown almost imperceptibly over time to optimize communication with others.

On the other hand, singing should respect the frequencies required by the composer, or at least the frequencies of other singers and instruments playing with you. Clearly, rhythm must also be respected in singing, which influences the duration of individual sounds as well. All this requires more control over the voice, suitable training and practice.

One interesting singing technique is "overtone singing". Although rarely used, the singer makes the two formants F1 and F2 have the same frequency, thus producing a louder sound; this is achieved by continuously adjusting the tongue. An impressive example is available online, along with an x-ray movie showing the tongue movements (by magnetic resonance imaging).[20]

Why are songs and operas harder to understand than normal speech? You may have wondered about this aspect of singing: the difficulty of understanding the words of a song, particularly in pop

---

[20] See "Sehnsucht nach dem Frühlinge (Mozart)" by Anna-Maria Hefele: https://www.youtube.com/watch?v=YIUvX7hebBA

songs and operas. It appears that the vowels are the main issue here. According to Di Carlo[21], an important cause is illustrated at the right in Figure 6-2. When the fundamental frequency f1 = F0 of the vocal cords rises, the harmonics spread out to higher frequencies (see the green lines in Figure 6-2), so that the lower formants F1 and F2 include fewer of the frequencies created by the vocal cords; it then becomes harder to recognize F1 and F2, so that the vowels (and some consonants) also become harder to distinguish. This can be heard very clearly online[22]: when the soprano sings "La, Lore, Loo, Ler, Lee" at low pitch (about 240 hertz) you can easily distinguish these five sounds, but at high pitch (about 960 hertz) they are indistinguishable. Another factor mentioned by Di Carlo is that it is more difficult for our brains to recognize vowels when the sounds are louder.

6.10 Whistling, Yodeling, Whispering, Humming, and Other Human Sounds

Human **whistling** has fascinating properties: **whistling can be very strong, so it can be heard about ten times farther away than shouting, and it is very "pure", with a single, very dominant frequency. Also notable are the high frequencies of whistling, usually in the range of 500 to 5,000 hertz, which is some ten times higher than normal speech.**

Figure 6-16 shows four spectrograms of typical human whistling with varying pitch. The red curves show the fundamental frequency: it is very sharp and very dominant, making it very pure. A weaker harmonic frequency is seen at double the fundamental frequency: it is also very sharp. With some whistlers, a few sharp higher harmonics may also be visible. You can also see and hear whistling in my Animation 6*7.

ANIMATION 6*7 — See my video WA3 at time 11:43 in its section "**Special human sounds**" under the title "**Special human sounds: whistling**". (See details in the section References and Resources below.)

---

[21] "Effect of multifactorial constraints on intelligibility of opera singing (I)" by Nicole Scotto Di Carlo, *Journal of Singing*, volume 63, page 443, 2007

[22] Listen to the two sound files '"La, Lore, Loo, Ler, Lee" at low pitch' and '"La, Lore, Loo, Ler, Lee" at high pitch' at http://www.phys.unsw.edu.au/jw/soprane.html

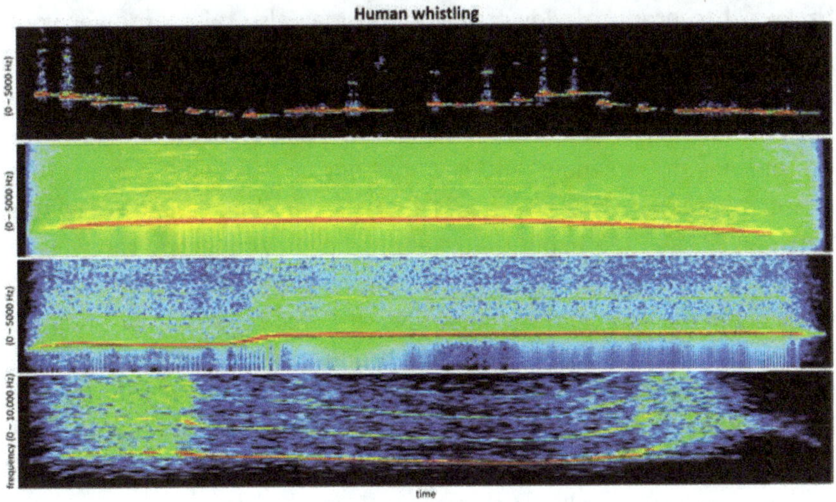

Figure 6-16: Spectrograms for four examples of human whistling. From top to bottom, the time durations are about 4.2, 1.4, 2.4 and 0.7 seconds, respectively. For the middle two graphs, the image brightness is increased to show the second harmonic (yellow line) despite the hissing sound of whistling (the vertical "ribs" at low frequencies are artifacts of the analysis). The bottom sound has less hissing, allowing higher harmonics to be visible (note the doubled frequency range). In the top to lower graphs, the ranges of the fundamental frequency (red line) are about 950 to 2,000 hertz, 800 to 1,500 hertz, 1,000 to 1,570 hertz, and 1,970 to 5,000 hertz, respectively. (*Source*: top recorded with smartphone; second "Male gentle whistle", third "Human long whistle", bottom "Human whistle", all from Mixkit in the public domain; displayed by WaveSurfer.)

There is some "noise" between the harmonics, seen as a green/ blue background: this noise comes from blowing air between the lips, which causes a hissing sound like the letter 'f'. But the main frequency is so powerful that it overwhelms the noise. So, compared to normal speech, the spectrum of whistling indeed looks quite pure.

You can easily detect the mechanism of whistling yourself: everybody should be able to whistle! After you wet your lips, you need to purse your lips to form a small hole (fingers may help). To produce a high pitch, the tip of your tongue should be close to your teeth, forming a small cavity between the roof of your mouth and your tongue. Next, you need to find a good blowing speed, neither too fast nor too slow, to create a pure and loud whistling sound.

Now, try to change the pitch of the whistling sound: notice that you can do this by moving your tongue forward to produce a higher pitch or backward to produce a lower pitch. You may also raise or lower your jaw a little bit, but this is not necessary.

Where do you think the whistling sound is created? Just blowing between the lips would give a hissing noise, like the letters 's' or 'f', or like blowing out a candle, and it would not be very loud or pure: that is not the answer. You may guess that the whistling sound originates between lips and teeth, but that is easily proved wrong: you can whistle while your lips touch your teeth, leaving no space in between. Are the vocal cords the source of whistling? No, the frequencies of whistling go far beyond what the vocal cords can produce. On the other hand, the fact that the tongue's position influences the whistling pitch shows that the whistling sound is created between the teeth and the tongue.

What controls the pitch of whistling? The movement of the tongue in whistling tells us that the mouth serves as a resonator, as with speaking and as with playing a flute. The mouth shape and size define the whistling frequencies. A small enough mouth cavity can indeed produce the highest whistling frequencies. This has been shown by internal imaging of the mouth during whistling.[23] The tongue is seen in the imaging to create a relatively small cavity with two openings: one small opening above the tongue toward the throat in addition to the small opening between the lips. The cavity is large for the low frequencies (from about 500 hertz upward), and small for the high frequencies (up to about 4,000 hertz in the study). However, the flow speed of the air also plays a role: it appears that resonances only occur in a limited range of blowing speeds, which would explain why it is relatively difficult to produce a good whistling sound.

The small flute-like whistles used by sports referees use the same principle, but they have a fixed shape and size, and therefore a fixed frequency (pitch). A whistle with a rotating ball adds another effect: the ball interrupts the air current repeatedly, rapidly changing its loudness, but with a lower frequency than the pitch.

---

[23] "The physiology of oral whistling: a combined radiographic and MRI analysis", by Alba Azola, Jeffrey Palmer, Rachel Mulheren, Riccardo Hofer, Florian Fischmeister, and W. Tecumseh Fitch, *Journal of Applied Physiology*, volume 124, pages 34-39, 2018: https://journals.physiology.org/doi/full/10.1152/japplphysiol.00902.2016.

We saw that all vowels and some consonants rely on certain resonance frequencies of the mouth (F1, F2, F3, *etc.*). And we saw that singing with a higher pitch (higher frequency) makes it difficult to distinguish vowels. Whistling confirms this behavior: whistling creates such high frequencies that it is basically impossible to create different vowels when whistling.

Nevertheless, whistling can be used, and has been used, as a language, forming a **whistled language**. For that purpose, sounds like vowels or consonants are not composed of different mixes of frequencies to tell them apart; instead, the fundamental frequency of whistling is varied in different ways: for example, a particular vowel will have a particular sequence of rises and falls of frequency. This is the same idea as in tonal languages.

Whistled languages are rare. They have typically developed in mountainous regions where it was useful to communicate instantly over long distances, such as between herders or farmers and villagers. For example, in the early part of the 20th century, a whistled language developed in the Turkish village of Kusköy; it apparently started from sounds used to control animals but was then extended to communicate between people, using their normal spoken language as a guide to form whistled sounds.[24] The spectrograms (not included here) show how the whistled word actually follows the tones of the spoken word to some extent (presumably, this makes it easier to memorize and recognize words), but uses much larger frequencies. It is also visible how much loudness is concentrated in the fundamental frequency, compared to the spoken words.

What is yodeling? Yodeling has some similarities with whistling but also many differences.[25] Yodeling probably started with Tibetan monks to communicate over long distances, as it involves high-pitched sounds.

---

[24] See videos by René Guy Busnel (in French, 1967–1968): https://www.canal-u.tv/41227 and https://www.canal-u.tv/46473. The second video includes x-ray movies of the speaker's mouth (by magnetic resonance imaging). A longer movie by Busnel (in French, 1990) includes other examples of whistled languages in humans, and potentially in animals: https://www.canal-u.tv/42019. © CERIMES, Canal-U

[25] Listen to "Oesch's die Dritten — Jodelmedley" at https://www.youtube.com/watch?v=wGexgNKPl0k; or "Melanie Oesch — The Queen Of Yodeling" at https://www.youtube.com/watch?v=BwJNDF2Jfbc

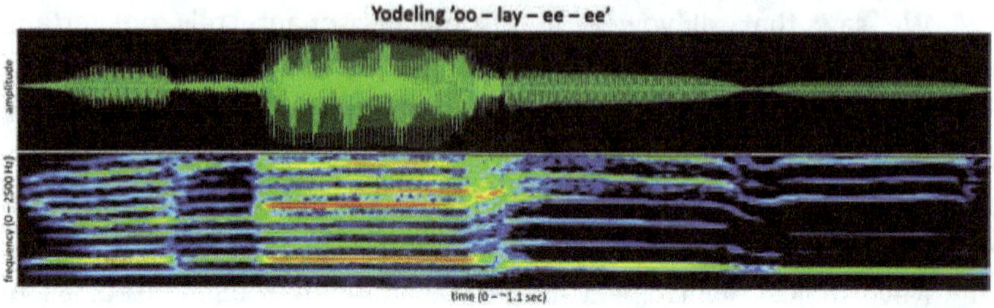

Figure 6-17: Waveform (*top*) and spectrogram (*bottom*) of the four sounds 'oo – lay – ee – ee': the 'ee – ee' is yodeled with high (falsetto) frequencies. The frequency cutoff used in this display is 2,500 hertz. The recording covers about 1.1 seconds. (*Source*: extracted from freeSFX sound file 14980_1460136445.mp3, royalty-free, https://www.freesfx.co.uk/sfx/yodel; analyzed with WaveSurfer.)

Yodeling appeared later in the European Alps, in the 13th century, where it was also used for communication. More recently, it has become a form of singing, especially in the Alps but also in the US.

Whistling mainly produces a single dominant frequency using the lips. By contrast, yodeling uses a few different sounds (such as 'ah – ee – oo – yay – yodel') with multiple frequencies produced by the larynx in the throat and modified by the mouth. The physical principles of yodeling are not very well understood, but they seem to be closer to standard speaking and singing than to whistling. I summarize next an explanation based on extensive research by Timothy Wise.[26]

The two distinctive features of yodeling are as follows: use of high fundamental frequencies, and abrupt switching from one pitch to another, as illustrated in Figure 6-17. The high fundamental frequencies are called "**falsetto**" (or also "**head voices**"): falsetto frequencies are above the normal range of frequencies produced by singers (these are called "**chest voices**"); they can reach around 1,000 hertz. **The falsetto frequencies are thought to involve complex, rapid vibrations of the soft edges of the vocal cords**, rather than vibrations of the vocal cords themselves. **The yodeler switches rapidly between normal frequencies**

---

[26] "Yodel Species: A Typology of Falsetto Effects in Popular Music Vocal Styles", by T. Wise, *Radical Musicology*, volume 2, 2007, available at http://www.radical-musicology.org.uk/2007/Wise.htm ; note that this long article uses much specialized terminology; it also includes nice samples of different styles of yodeling.

and falsetto frequencies, giving the unusual jumps in pitch that are characteristic of yodeling. You can also see and hear yodeling in my Animation 6*8.

ANIMATION 6*8 — See my video WA3 at time 12:04 in its section "**Special human sounds**" under the title "**Special human sounds: yodeling**". (See details in the section References and Resources below.)

Figure 6-17 covers only about 1 second of rapid yodeling: at the center, the fundamental frequency (bottom line) jumps up from about 260 hertz at the left (in the sound 'lay') to about 425 hertz at the right (in the first sound 'ee') within 0.04 second (1/25 of a second); a smaller downward jump to about 345 hertz occurs at right (between the two sounds 'ee ee'). The rapid frequency jumps are clearly exhibited by the changing spacing between the harmonic "ribs". This example illustrates that yodeling does not go as high in pitch as whistling.

What is whispering? Try it now: speak in a very gentle voice, as if speaking only to someone who is close beside you. Your voice should sound like soft blowing, similar to gentle wind. Can you change the pitch of your whispered words or sing while whispering? Probably not. This suggests that your vocal cords are not vibrating; they may open more or less, and you may blow more or less air through your vocal cords, thereby producing stronger or weaker whispering. **But the main effect of whispering is to produce very many frequencies (like the sound of wind). You can still produce different vowels and consonants**, because those are mainly due to formants resulting from the shape of the mouth and are also due to hissing sounds in the mouth and near the teeth and lips.

What is humming? Try saying "hmmm". The "mmm" part is humming: **humming is simply pronouncing the letter "m" for a long time**. You will find that your lips are closed, so your humming comes out through your nose. What happens if you move your tongue? Very little changes! Can you change the pitch of your humming? Yes, you can: how do you do it? It happens in the throat: it is your vocal cords that control the pitch of humming, like that of normal speech.

Can you pronounce other vowels or consonants than 'm' while humming? You can't: why not? Unlike normal speech, in which you change the shape of the mouth, humming uses the nose, which has a fixed shape. You can only change the pitch since that happens in the throat, not the nose. The nose produces formants (resonances), just like the mouth, but you can't change those formants by changing the shape of the nose cavity, so you can basically only produce the 'm' sound and no other sounds. We can produce similar sounds with 'n' and 'ng' instead of 'm', but those are not normally called humming, as they also involve the mouth.

See the sound of a woman humming in Figure 6-18: it has a dominant fundamental frequency (the lowest red line) that varies during her humming a musical tune. We see steady horizontal bands of stronger frequencies: these are the formants of this woman's nose; they never change during the entire recording of about 8 seconds, meaning that the formants are constant.

What are the sounds of laughing, sneezing, coughing, snoring and clicking? These are good sounds for you to explore yourself. As you produce these sounds, pay attention to the shape of your lips, mouth opening and tongue, as well as the use of your nose. One common feature of these sounds is the near-total absence of harmonics in their spectra, so spectrograms show mostly noise and are not very interesting.

There is an amazing variety of ways to laugh, but sneezing, coughing and snoring are much less varied. (To snore, you don't need to sleep or lie on your back: it can happen when you breathe in the right way!)

All people also use clicking sounds even when they are not part of the normal language, such as when kissing. Some languages use clicks:

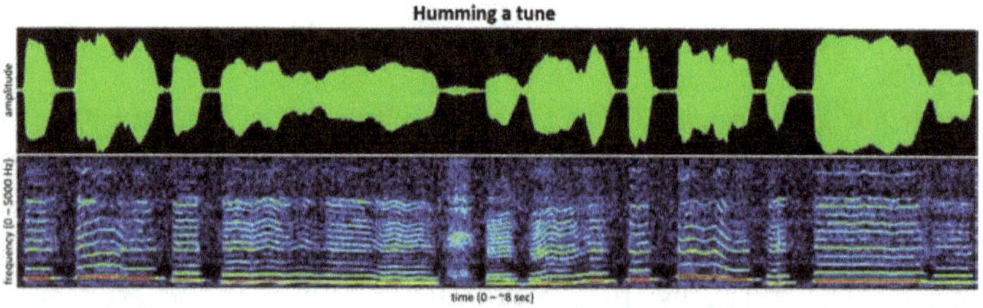

Figure 6-18: Waveform (top) and spectrogram (bottom) of a tune hummed by a woman. The recording covers about 8 seconds. (*Source*: extracted from a sound file by PSsoundproject, in the public domain, https://freesound.org/people/PSsoundproject/sounds/183565/; displayed with WaveSurfer.)

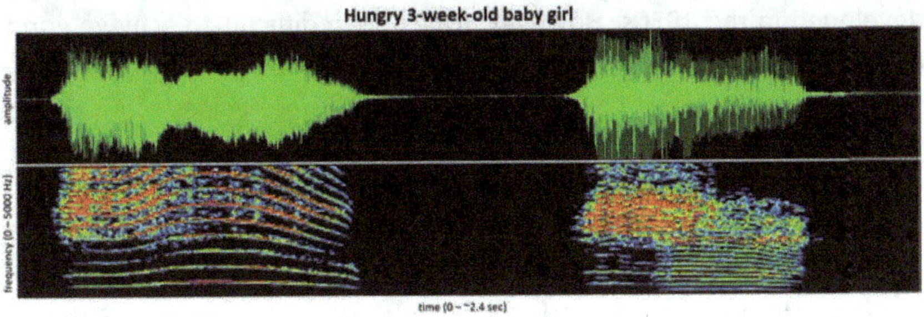

Figure 6-19: Sound of a hungry 3-week-old baby girl. The recording covers about 2 seconds. The lowest frequencies are around 300 hertz at the left and 150 hertz at the right. (*Source*: recorded with a smartphone; displayed with WaveSurfer.)

various clicks are used in some South African languages, such as Zulu and Xhosa. One important click in those languages is called the dental click and resembles the "tut-tut" we say when we disapprove of something. It can be made easily as follows: with a slightly open mouth, stick the front of your tongue against the roof and sides of your mouth so as to block the flow of air; now pull your tongue down until the air rushes in; that produces the click.

What does a baby's sound look like? Figure 6 19 shows the sound of a hungry 3-week-old girl: she clearly varies the pitch. At this age, we cannot call her sound a speech, while we might call it a "universal vowel-like language". You can also see and hear this baby in my Animation 6*9.

ANIMATION 6*9 — See my video WA3 at time 12:30 in its section "**Special human sounds**" under the title "**Special human sounds: baby**". (See details in the section References and Resources below.)

6.11 Speech Synthesis

In Section 5.9, we discussed the electronic **synthesis** of sound in general. Here we focus on the **synthesis of speech**, which we frequently hear in this electronic age.

Voices, including human voices in particular, can be entirely synthesized by electronic means, as described in Section 5.9 with the combination of sine waves. The earliest speech synthesizers were developed in the 1930s. However, it has been difficult to achieve good quality and intelligibility in the resulting voices: many examples of early voice synthesis can be heard on Wikipedia.[27] These voices often sound very mechanical and monotonous.

Natural-sounding voice synthesis was only achieved in the 1990s. An example is the "Speak" option in today's word processors and other software, which can read out typed text. See Figure 6-9 for an illustration of the synthesized sound of "Happy birthday to you", compared with real voices: it is hard to tell from the graphs which voice is synthesized. The actual synthesized sound is quite natural and neutral, but monotonous and clearly distinct from singing or a child's speech (it is also repetitive in that the same word is always pronounced identically): see my Animation 6*5 mentioned in Section 6.6.

The problem with synthesizing voices is the great variety of the spoken language, especially the English language (such as different pronunciations of vowels and many diphthongs, without even thinking of accents and dialects/varieties). First, written text must be converted to instructions for every individual sound, which may need understanding of the text because some words with identical spelling have multiple pronunciations (the English "read" is pronounced differently in "I read now" *versus* "I read yesterday"; "permit" is emphasized differently in "to per<u>mit</u>" *versus* "a <u>per</u>mit"; a number like 1968 may have to be pronounced as "nineteen sixty eight" if it is a year, or as "one thousand nine hundred sixty eight" if it is just counting); also, diphthongs slide smoothly from one sound to another (the English "ay" in "May" is pronounced like 'a'→'i').

A general method of speech synthesis is to collect components of speech (from individual letters to entire sentences) and later assemble their individual sounds to make a full speech. Besides synthesizing these components, for example by electronically producing the formants of vowels, it is also common to record natural voices and extract the individual components from the recordings.

In one approach, the shortest individual sounds (called **phonemes** in linguistics) are combined by stringing them together as needed to form

[27] Listen to the sound clips in "Speech synthesis" at https://en.wikipedia.org/wiki/Speech_synthesis

words and sentences. It is also possible to combine entire stored words, phrases or even sentences. Another approach is to produce a more limited set of **diphones**, which are pairwise sound-to-sound transitions, like 'a'→'i' in "ay" (these transitions capture the changing shape of the mouth from one sound to the next); diphones can be adjusted when needed to produce a wider variety of intonations.

You can also see and hear a synthesized voice in my Animation 6*10.

> ANIMATION 6*10 — See my video WA3 at time 4:11 in its section "**Consonants *vs.* vowels**" under the title "**Consonants *vs.* vowels: a synthesized voice**" and following pages. (See details in the section References and Resources below.)

Singing can also be synthesized electronically. An example of a synthesizer which anyone can use to produce songs is Vocaloid 5.[28] With that synthesizer, you can specify for each sound in a song its pitch and duration, as well as which syllable or letter it is; you may also select various male and female voices, and apply many sound effects.

In Figure 6-20, I illustrate that very complex sounds can be made relatively easily by a computer. You can also see and hear these examples of synthesized whistling in my Animation 6*11. At the top of Figure 6-20 (graph a), I synthesized the simultaneous sweeping sounds of three imaginary whistlers: the spectrogram starts at low frequencies at the left, and then each whistler sweeps up and/or down in frequency for 5 seconds: this is one simple form of frequency modulation. We can imagine this sound being recorded from three separate whistlers and then added together, following the principle of superposition, if we could find whistlers able to produce such pure and well-controlled whistling sounds all the way from 0 to over 10,000 hertz. However, it is much easier to program the combined pure sounds into a computer.[29] We will

[28] See "The Singing Synthesizer That Can Make Realistic Computer Vocals" by Warp Academy at https://www.youtube.com/watch?v=QeFXImLfwPg

[29] The formula for the "sweeping whistlers" is composed of three parabolas for the three frequencies: amplitude $= \sin(2\pi t(440 + 1300t - 200t^2)) + \sin(2\pi t(0 + 4000t - 900t^2)) + \sin(2\pi t(880 - 200t + 100t^2))$, where t is the time from 0 to 5 seconds. The initial frequencies at time $t = 0$ are 440, 0 and 880 hertz, respectively. Negative frequencies are unrealistic and switched to positive. Produced with software Praat.

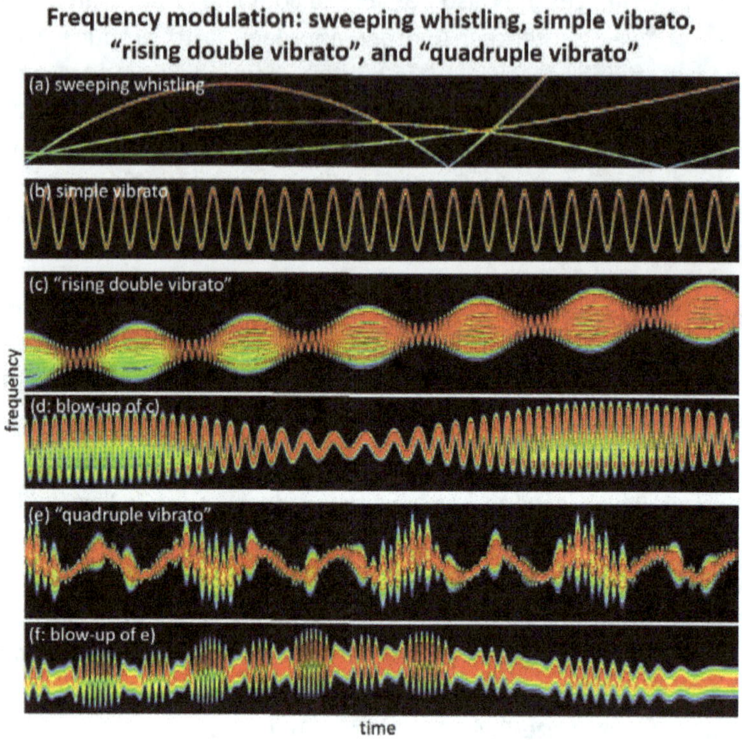

Figure 6-20: Spectrograms for four different synthetic sounds based on frequency modulation. a: three sweeping whistlers. b: simple vibrato. c and d: "rising double vibrato"; d blows up a part of graph c. e and f: "quadruple vibrato"; f blows up the center of e. Each graph covers about 5 seconds, except for the two blow-ups, which cover about 1 second. (*Source*: designed with Praat; displayed with WaveSurfer.)

later see that animals can use modulation in sweeping the frequency of sounds, for instance, birds in Figure 6-23 and dolphins in Figure 6-24.

ANIMATION 6*11 — See my video WA3 at time 12:46 in its section "**Special human sounds**" under the title '**Synthetic sounds: "Sweeping whistling", "simple vibrato", "rising double vibrato", "quadruple vibrato"**'. (See details in the section References and Resources below.)

We have previously encountered the **vibrato** effect that is often used in singing: in Figures 6-4 and 6-9, and in Section 6.9. Let's recall that the vibrato effect is a repeated up-and-down change in frequency. You have certainly heard the same effect coming from the sirens of some emergency vehicles and alarm systems: they sound like a slow "ya-ya-ya-ya", with a repeat frequency that is much smaller than in human vibrato singing.

Interestingly, the vibrato effect is related to FM radio and television transmission: vibrato is indeed an example of **frequency modulation**, as described in Section 5.9. With normal vibrato, the frequency changes repeatedly in the same way at a constant vibrato frequency. With the "double vibrato" or "quadruple vibrato" shown in Figure 6-20, to be explained below, the vibrato frequency itself changes over time.

Figure 6-20b shows a simple vibrato: the frequency oscillates 6 times per second between 2,800 and 5,200 hertz, with an average frequency of 4,000 hertz. (Be aware that this sine-wave-like curve is not the sound wave itself, but the varying frequency; a sound with constant frequency, such as whistling a single tone, would give a simple straight horizontal line in this graph of frequency *versus* time.) This particular vibrato has a very large difference between the highest and lowest frequencies: it would be possible but probably very difficult to produce it perfectly with a human whistler; some dolphins and whales might be able to produce it, given their ability to produce high-pitch sounds.[30]

Now I complicate this simple vibrato so its frequency is modulated in time: see Figure 6-20c and its blow-up Figure 6-20d. First, I make the average frequency increase with time: this makes the sound generally rise in pitch from left to right. Second, I make the repetition rate of the vibrato (which is normally constant) also vary in time: that makes the frequency oscillate faster in the "thicker" parts of the spectrogram (where we see denser vertical "ribs") and oscillate slower in "thinner" parts of the spectrogram (where we see less dense vertical "ribs"); the blow-up in Figure 6-20d shows this clearly. I call this a "rising double vibrato", where the vibrato itself rises over time while the vibrato's frequency varies periodically. This particular rising double vibrato sounds

---

[30] The formula for the "simple vibrato" is: amplitude = sin (2 π t × 4000 + 200 sin (2 π t × 6)), where t is the time from 0 to 5 seconds. The part "sin (2 π t × 4000" gives the average "carrier" frequency 4000 hertz; the part "+ 200 sin (2 π t × 6)" changes that carrier frequency repeatedly up and down with a repetition rate of 6 hertz.

to me like a hoarse frog that turns into a shrill bird. It is defined entirely by a single mathematical formula.[31]

Figure 6-20e and its blow-up Figure 6-20f add two more layers of frequency modulation and adjust the various numbers: I therefore call it "quadruple vibrato". Now the oscillations vary in a more complicated way; the blow-up in Figure 6-20f shows that clearly. This sounds to me like the occasional croaking of healthy frogs (shown by the "thicker" bursts in Figure 6-20e), interspersed with varied chirping of happy birds (shown by "thinner" bursts in Figure 6-20e).[32]

The examples in Figure 6-20 only use frequency modulation, similar to the example shown at the bottom in Figure 5-25. Let's remember that animals (and humans) can also use amplitude modulation: both can in fact be used at the same time, giving even more possibilities.

As shown by these examples, it is possible to produce acceptable animal sounds in a computer. However, human speech is more difficult to simulate, in part because we are highly sensitive to subtle variations, such as intonations, moods and accents.

Furthermore, it is very easy to combine different sounds in a computer: all you need is a "+" sign, thanks to the superposition principle! In addition, we can cut up such sounds into small pieces, like cutting up a sentence into words or letters, and reassemble the pieces into other sound sequences. And we can easily add some noise for a more realistic effect. More importantly, we can also push frequencies and the modulations of sounds beyond the normal capabilities of human voices, thereby creating new sounds.

---

[31] The formula for the "rising double vibrato" is: amplitude = sin (2 π t (1000 + 200 t) + 10 sin (2 π t × 50 + 20 sin (2 π t × 1.4))), where t is the time from 0 to 5 seconds. The part "sin (2 π t (1000" gives an initial frequency of 1000 hertz at time t = 0. The part "+ 200 t" makes that frequency rise in proportion to time. The part "+ 10 sin (2 π t × 50" gives a normal vibrato effect with a frequency of 50 hertz. The part "+ 20 sin (2 π t × 1.4)" modulates the normal vibrato at a frequency of 1.4 hertz.

[32] The formula for the "quadruple vibrato" is: amplitude = sin (2 π t × 2000 + 200 sin(2 π t × 1.8) + 3 sin(2 π t × 2.4 + 200 t + 20 sin(2 π t × 2.8 + 10 sin(2 π t × 0.75)))).

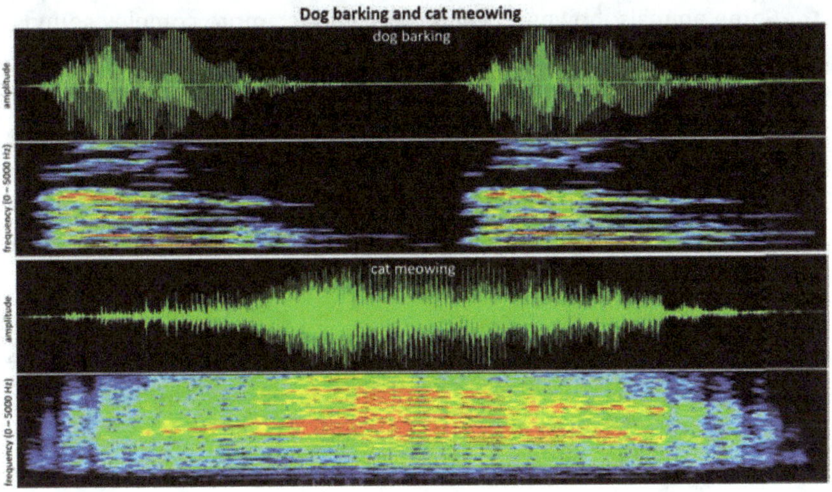

Figure 6-21: Waveforms and spectrograms for a dog barking twice (top) and a cat meowing (bottom), both for about 0.6 seconds. (*Sources*: sound files "Dog barking twice" and "Domestic cat hungry meow" at Mixkit in the public domain; all analyzed by WaveSurfer.)

6.12 Animal Sounds

Can animals speak? Think about this question for a moment. You can break up the question into three parts: Can animals make various sounds? Can animals communicate messages with each other or with us? And can animals exchange thoughts through sounds?

Clearly, most animals make sounds, including most types of **fish** and other underwater animals (their sounds stay under water, so we rarely hear them, even when they are not ultrasounds). Many animals make rather simple or repetitive sounds: think of a **dog's** repetitive barking or a **cat's** slow meowing, shown in Figure 6-21.

You can also see and hear several of these animals in my Animation 6*12.

ANIMATION 6*12 — See my video WA3 at time 13:33 in its section "**Sounds of animals**" under the title "**Sounds of animals: cricket**" and following pages. (See details in the section References and Resources below.)

Some animals, however, can produce much more complex sounds: these include dogs (see Figure 6-22), certain types of **parrots** and other **birds** (see Figure 6-23), as well as **dolphins** and **whales**[33] (see Figures 6-24 and 6-25). Parrots that can mimic human speech are able to make sounds that are as complex as human sounds. A number of animal species have mouths or other organs that can produce sounds which are similar to or as complex as human sounds, for example, some **monkeys.**[34]

We can see in the spectrograms that dogs and cats produce relatively "human" sounds, compared to birds, dolphins and whales. The sample from a dog in Figure 6-22 in particular shows formants (bands of louder intensity), and it actually sounds like high-pitched human "talking". However, it is interesting that the formants of this one dog do not change during this and other recordings, which indicates that he does not modify the shape of his mouth: so he does not exploit the human ability to produce different vowels. Also, we see little evidence of consonants in this dog's "speech" apart from brief "explosions" at the beginning of "words", which already start with many harmonics like vowels, especially no hissing sounds like 's' or 'f'. This dog nonetheless creates a great variety of sounds by smoothly changing both the fundamental frequency and the loudness. We can clearly see the many gradual frequency variations by the ups and downs of the fundamental frequency (the lowest line in the spectrogram), which ranges from a maximum of about 900 hertz to a minimum of about 190 hertz.

Animals which produce complex sounds should also be able to have different languages, dialects or accents within the same species.

---

[33] See video "Haunting song of humpback whales" by Hindustan Times: https://www.youtube.com/watch?v=W5Trznre92c. Also see video "Sea Mammals and their Sounds" by Learning Gone Wild: https://www.youtube.com/watch?v=pz3UytMUrfE

[34] See a "List of animal sounds" at Wikipedia, which includes many sounds that you can listen to: https://en.wikipedia.org/wiki/List_of_animal_sounds. You can listen to other animal sounds at: https://mixkit.co/free-sound-effects/animals/. Also see the website "Discovery of Sound in the Sea": https://dosits.org/, including its Gallery > Audio Gallery > Fishes; listen, for example to the Atlantic Croaker and the Northern Seahorse. (Double click on the various sound recordings under the spectrograms.) At the same website you can listen to Marine Mammals > Baleen Whales, such as the Humpback Whale and the Bowhead Whale.

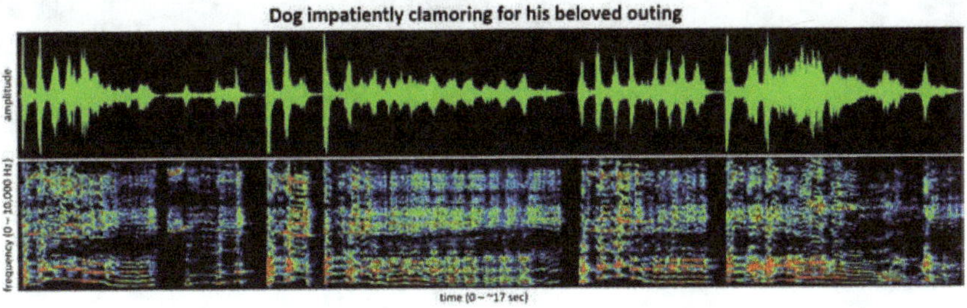

Figure 6-22: Waveform and spectrogram for an impatient mid-size dog asking to go outside, covering about 17 seconds. Frequencies go up to 10,000 hertz. (*Source*: recorded with a smartphone, displayed with WaveSurfer.)

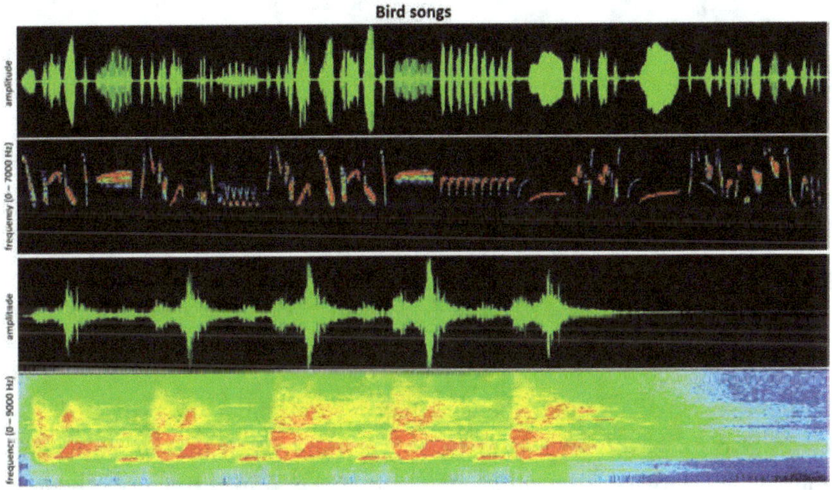

Figure 6-23: Waveforms and spectrograms for birds singing (for about 8 seconds at the top and about 2 seconds in a different recording at the bottom). Frequencies go up to 7,000 hertz at the top and 9,000 hertz at the bottom. The bottom recording seems to show reverberation (which stretches the sound out over time), maybe due to multiple reflections from walls or other structures. (*Sources*: sound files "Little birds singing in the trees" and "Melodic songbird chirp in the wild" at Mixkit in the public domain; all analyzed by WaveSurfer.)

For example, there is evidence[35] that "Orcas 'speak' with a number of different 'dialects' and 'languages'. These differences can be small, like British *versus* American English, or huge, like Japanese *versus* English.

---

[35] See: https://oceanadventures.co.za/can-whales-dolphins-communicate/

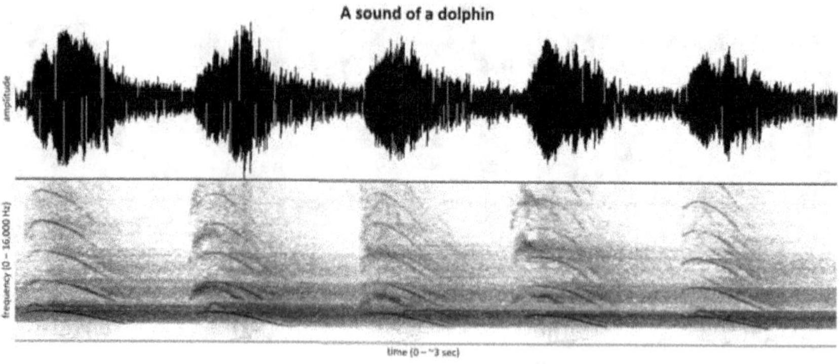

Figure 6-24: Waveform (top) and spectrogram (bottom) of one among many dolphin's sounds: a whistling sound covering about 3 seconds. Note the top displayed frequency of 16,000 hertz: the dolphin also makes sounds above this artificial cutoff. The fundamental frequency arches up to a maximum of about 3,000 hertz, then drops smoothly to about 1,800 hertz, below which this dolphin makes little sound; the harmonics are very sharp. The frequency arches are repeated in some cases, slightly delayed in time, perhaps as reflections from a wall; such delayed reflections may be used by a dolphin for echolocation to measure the distance to a target, similar to radar. In addition, the sound seems stretched out in time, suggesting reverberation due to many reflections from various objects, perhaps rocks or an uneven bottom. (*Source*: extracted from freeSFX sound file 18684_1517422885.mp3, royalty-free, https://www.freesfx.co.uk/sfx/dolphin; displayed with Praat.)

Bottlenose dolphins and sperm whales from different parts of the world also have different dialects. This is one reason that when having animals in captivity from different parts of the world, they may have a hard time communicating". In addition, "it is believed that all cetaceans are able to speak the same language and then have their own unique communication", meaning that dolphins and whales can "speak" among each other, but have their own different "dolphin-only" and "whale-only" languages.

Is making sound sufficient to "speak" or at least "communicate"? Animals often use sound to **communicate** among themselves, for example to find mates or to warn others of approaching danger; they also use sound to scare away other animals. But their "messages" are mostly composed of short or very repetitive signals and consist of only a few words such as "admire me" or "warning: hide". In fact, messages don't need signals of much complexity: think of the old **Morse code** of

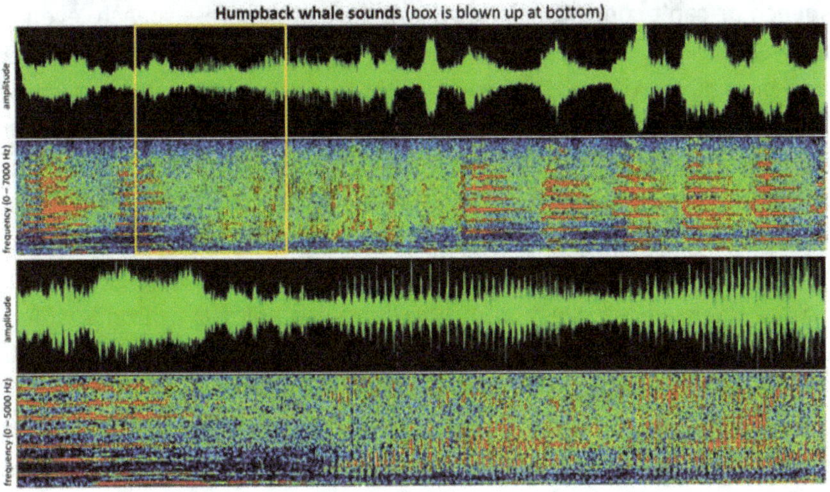

Figure 6-25: Waveforms and spectrograms for a humpback whale. The orange box is blown up at the bottom to show the transition from high-pitched screaming, seen as horizontal ribs, to rapid clattering, seen as vertical ribs (the clattering resembles our teeth clattering in cold weather). The top graph covers about 37 seconds, while the bottom blow-up covers about 7 seconds. (*Source*: sound file by Spyrogumas, in the public domain, https://en.wikipedia.org/wiki/File:Humpbackwhale2.ogg; analyzed by WaveSurfer.)

human telegraphy, in which just two simple beeps (a short and a long beep) suffice to compose a message of any desired complexity (" - - - — — — - - - " = "S O S" = "Save Our Souls" = "Help!"). The **binary code** of 0's and 1's now used in computers does this as well for just about everything, from text and sound to data and movies.

Some animal messages are long and complex, as with some birds and whales. In principle, they could contain messages as complex as those in human speech. There is currently a debate about the question whether such long and varied animal messages are comparable to human "speech" and can even be called "language". Some researchers claim that animals do speak.[36] Others accept that animals can make varied sounds, but maintain that their brains are not able to create

[36] See Eva Meijer in: "Of course animals speak. The thing is we don't listen", https://www.theguardian.com/science/2019/nov/13/of-course-animals-speak-eva-meijer-on-how-to-communicate-with-our-fellow-beasts

speech or can't control their vocal structures to express speech, even if they can understand some elements of speech; other explanations are also possible.[37] After all, the voice is not the only way to communicate: human sign language suffices, so why haven't animals developed sign languages beyond simple signals? Since animals are increasingly found capable of human actions, it is quite possible that some day we will find a way to communicate with and have conversations with some animals. Whether it will be possible to exchange more abstract thoughts depends on the unanswered question of whether such animals can understand and create or at least express such thoughts.

How do animals make sounds? We should remember that we humans not only use our mouth to make sounds: we can also make sounds with other parts of our body, from clapping hands to stomping the ground with our feet, and indirectly with musical instruments. Animals also have several ways to make sounds. In addition, many animals make sounds that fall outside the frequency range of human hearing, especially ultrasound (like some whales, dolphins, bats, crickets, *etc.*).

Closest to humans in sound production are other **mammals** with similar throat, mouth and nose structures. As we have seen, the size and shape of these structures strongly influence the frequency and variety of sounds that are produced. For example, small structures cause high-pitched sounds (as in mice, cats and small monkeys), while large structures produce low-pitched sounds (as in **lions**; see Figure 6-26).

Many birds and **amphibians** (such as **frogs**) have sound-producing structures that are similar to those of mammals. Birds use a **syrinx**. Unlike the larynx, which is located at the top of the main airway, the

---

[37] 'In a 2016 study, a team of biologists from several universities concluded that macaques possess vocal tracts physically capable of speech, "but lack a speech-ready brain to control it"', quoted from Wikipedia: https://en.wikipedia.org/wiki/Animal_language. See also "Monkey vocal tracts are speech-ready", W. T. Fitch, B. de Boer, N. Mathur, A. A. Ghazanfar, *Science Advances*, volume 2, issue 12, December 2016: https://www.science.org/doi/10.1126/sciadv.1600723. And see the commentary on the previous link in "Why can't monkeys talk? Their anatomy is 'speech-ready' but their brains aren't wired for it: neuroscientist": https://nationalpost.com/news/canada/why-cant-monkeys-talk-their-anatomy-is-speech-ready-but-their-brains-arent-wired-for-it-research-suggests

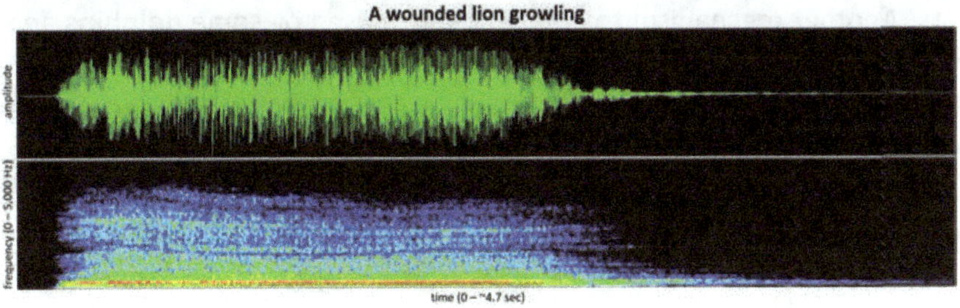

Figure 6-26: Waveform and spectrogram of a wounded lion growling for about 5 seconds. Most of the intense frequencies are around 100 to 300 hertz; the weakness of higher frequencies gives the deep-throated character of the growling sound. (*Source*: sound file "Wounded lion growling" from Mixkit in the public domain; analyzed by WaveSurfer.)

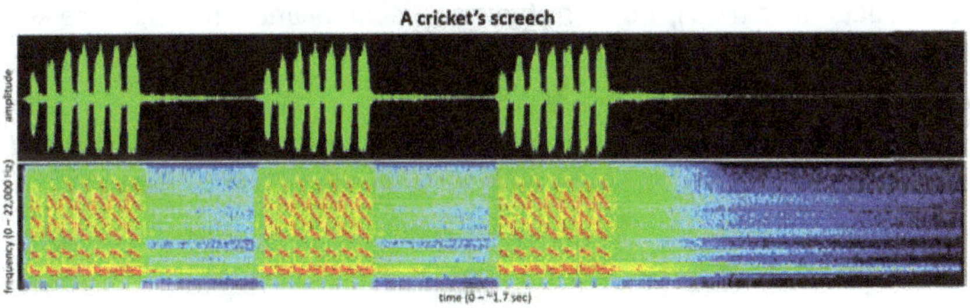

Figure 6-27: Waveform and spectrogram of a cricket screeching for about 1.7 seconds. Note the very high frequencies, shown here up to 22,000 hertz; the lowest of the most intense frequencies, shown in red, are around 2,500 hertz. The clattering itself repeats at the relatively slow rate of about 35 hertz within each burst of 7 repeats. (*Source*: sound file "Single cricket screech" from Mixkit in the public domain; analyzed by WaveSurfer.)

syrinx is found at its lower end, where it branches into the two lungs. One advantage of this location is that the syrinx can produce two different sounds at the same time by activating the two air channels separately.

Some animals scrape hard body parts against each other, including **crickets** (see Figure 6-27), **grasshoppers** and **cicadas**. One body part has a rough corrugated structure, like a rasp or grate; another body part is scraped against it, similar to a stick dragged against a fence, making a strong repetitive clattering sound that can have extremely high frequencies.

Another mechanism to create sound is used by some **dolphins** and **whales**: they have air bags ("air sacs") that blow air through a small hole (usually the blowhole that they use to breathe): air turbulence can give a high-pitched sound, similar to our whistling, or a clattering or clicking sound (see Figures 6-24 and 6-25). Interestingly, toothed whales, including dolphins, have a so-called "melon" in their forehead that serves as a lens to focus sound forward. As we will see in Chapter 9, sound can travel very well through the water surrounding the animal.

This focused sound can be aimed at other animals for communication, or can be used for **echolocation**, namely to "see" the location of other animals or targets or obstructions, as a kind of radar system using sound: this can tell direction and distance to targets. Humans also have this option (which bats use most extensively to hunt for insects in mid-air), but we don't use echolocation much, since we have good vision: try sitting with closed eyes or in darkness in rooms of different sizes in various locations, and try to "hear" the size of that room and the distance of its walls or furniture by listening to echoes of your voice or claps.

Many animals use muscles to cause a drum-like membrane to vibrate, including cicadas (indeed, cicadas have several ways of producing sounds).

It is also very common for animals to hit an external surface to make sounds, just like our stomping on the ground. Elephants, beavers and woodpeckers are good examples.

6.13 What have We Learned in this Chapter?

Our mouth is a rich musical instrument, thanks to the infinitely-variable shapes of the vocal cords, tongue, jaw opening, and lips, besides controlled access to the nose cavity. Most animals have less control over their sound-producing organs, but some (for example, parrots) can produce human-like sounds.

In humans, the vocal cords are the primary source of vibrations, especially for vowels ('a', 'e', 'i', 'o', 'u'): these vibrations set the pitch of our speaking and singing. The mouth cavity serves as an amplifier and filter for certain frequency ranges. For vowels, frequency ranges

called formants are amplified and thereby "form" and distinguish the different vowels.

Consonants can be explosive like 'k', 't' and 'p', or softer like 'g', 'd' and 'b'. Consonants can also be hissing like 's' and 'f', or rolling like 'r', or liquid like 'l', or nasal like 'n' and 'm', *etc.*

Tonal languages, such as Chinese, have a variable pitch within a letter or syllable: the different tones give the letter or syllable different meanings.

Accents are distinguished by different pronunciations of certain words, for example by pronouncing certain vowels differently. Dialects (also called varieties) differ in vocabulary and grammar, as well as pronunciation. Languages have such large differences that they are mutually unintelligible.

Singing often uses higher and lower frequencies than normal speech, in a more controlled, rhythmic and melodious manner. Whistling and yodeling are special examples of the use of high frequencies and rapid frequency changes, respectively.

Voices and songs are commonly synthesized for use in many situations, from simple instructions in transportation to singing. Synthesis can modify recorded sounds or start from electronically produced sounds.

Spectrograms are very useful to illustrate and understand the various sounds made by humans and animals.

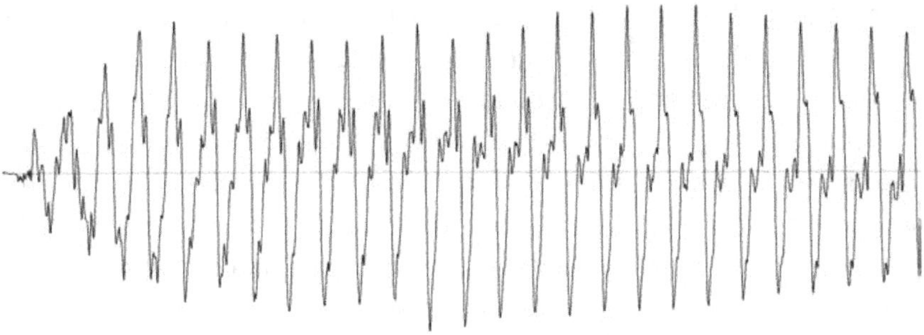

7

Summary of Part A

Part A of this book, in Chapters 1–7, covered waves "from Musical Instruments and Sounds to Human and Animal Voices". We first briefly introduced a wide variety of waves, such as waves on strings and in air, waves in and on water, earthquake waves, as well as electromagnetic, gravitational and quantum waves. While these waves can have very different mechanisms, it is found that the behavior of waves is universal: waves typically have smooth shapes that can repeat in space and in time; waves can travel far through substances or space, with characteristic wavelengths, speeds and frequencies; also, waves can cross each other by superposition and cause constructive or destructive interferences.

An important aspect of waves is resonance and the closely related concepts of standing waves and harmonics: they form the basis for much music, singing, speaking, lasers, atomic structure, *etc*. In particular, we have discussed how the resonances correspond to the familiar notes in music.

After exploring these properties for waves on strings, we turned to waves in air, as well as on sticks, plates, drums and flags. Several of these types of waves are central to music and voices. String instruments — such as pianos, violins and guitars — rely on resonant harmonics on their strings. Similarly, wind instruments — such as trumpets, flutes and saxophones — depend on resonances of the air in their tubes.

Our mouth is a very versatile and capable source of many kinds of sound. This is due in large part to the great variety of shapes that the mouth can take, which greatly influences resonances of waves within the mouth and nose. We have discussed how humans make and use vowels ('a', 'e', 'i', 'o', 'u') and consonants ('k', 't', 'p', 's', 'f', 'r', 'l', 'n', *etc.*), and how accents, dialects and languages differ. We also considered singing, whistling, humming, yodeling, *etc.*, as well as synthetic speech.

Some animals can also create a wide range of sounds, from simple to complex: we analyzed and discussed a series of such sounds.

Waves in air can spread out in all directions, unlike waves that are confined to 1D strings. Moreover, waves in air can turn around corners, in a process called diffraction: this ability makes waves distinctly different from all objects, such as balls.

WAVES – PART B:

From Earthquakes and Tsunamis to Light and the Quantum World

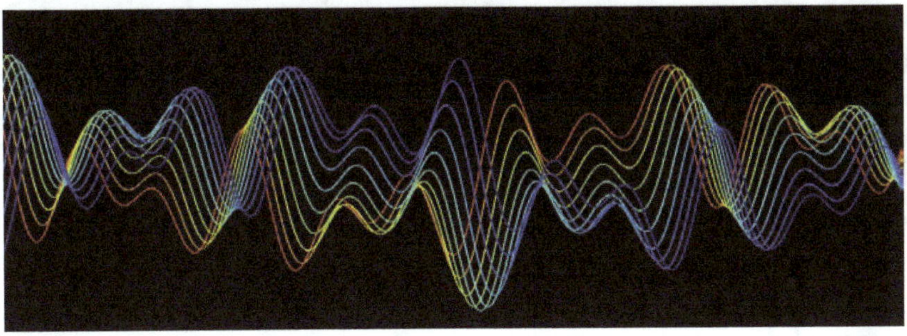

8

Preview of Part B

Part B of this book continues exploring diverse types of important waves with interesting behaviors, from earthquakes and tsunamis to light and the quantum world, as well as beyond to brain waves and human waves. We first extend sound waves from traveling through air (which in Part A was the basis of music and voices) to traveling through liquids and solids: this includes underwater sounds (such as ultrasound) as well as earthquakes, all discussed in Chapter 9.

Next, we look more closely at reflection, transmission and refraction of waves of all types, in Chapter 10: this will be especially relevant when we discuss waves on the surface of water in Chapter 11, including beach waves, rogue waves, tsunamis and boat wakes.

Chapter 12 addresses the wide subject of light and other electromagnetic waves (EM waves). We cover various optical elements such as eyes and lenses, as well as mirrors, prisms, retroreflectors, optical fibers, mirages, rainbows, holography, *etc*. We also consider radio waves, microwaves, infrared light, ultraviolet light, x-rays and gamma rays.

In Chapter 13, we will discuss the fascinating and surprising quantum mechanical waves (QM waves), which are the basis of quantum mechanics. Quantum waves are of great importance in high-tech society, in the form of electronics and advanced biology, but are also essential to understand all atoms and molecules, as well as the colors of substances.

We next consider the very recent and exciting observation of gravitational waves (Chapter 14). We describe that these waves are extremely weak and therefore very hard to detect on Earth, even when caused by gigantic collisions between black holes or neutron stars.

In Chapter 15, we will address a variety of other types of waves, including brain waves, waves in human queues, waves in road traffic, and many more. We will conclude with a more general Section 15.5 that considers the question: "Is doing science like doing sports, or playing video games, or reading science fiction?" This question may interest those readers who wonder whether doing science is very different from doing more common activities: the answer is that doing science is in fact very much like familiar daily activities.

You may start reading this Part B by looking for a topic that strikes your fancy in the following list of frequent questions concerning waves (feel free to jump directly to the indicated chapters and sections). A similar list of interesting topics discussed in Part A can be found in Chapter 1, Section 1.1.

- How different is sound that travels through liquids and solids, compared to sound in air? Such waves include, for example, sound waves in water and earthquake waves in the Earth's crust. These are discussed in Chapter 9.
- Why can "shear waves" travel inside solids, such as rocks, but not in gases or liquids? Shear waves can also be generated by earthquakes. See Section 9.3.
- What happens when waves reach the boundary between different substances in which the waves travel at different speeds? This leads to partial reflection and partial transmission at such boundaries: see Sections 10.1 through 10.5.
- How can a wave cross a wall? What happens to its travel direction? See Section 10.6.
- There is a principle in physics which states that the path followed by a wave between two given points is the quickest

path possible: does this mean that waves can "feel their way ahead" so as to choose the quickest path? We discuss this controversial statement in Section 10.8.

- What makes waves that travel along the surface of water so fascinating? We point out in Chapter 11 that such waves on water are both visible and strong, so that they exhibit a range of striking behaviors, including beach waves, tsunamis, tides and boat wakes.
- How do waves on water differ from other waves, including sound waves inside water? See Section 11.2.
- Why do waves turn toward the coast as they approach a beach? See Section 11.4.
- Why do high waves break into bubbles and drops (whitecaps)? Why do beach waves plunge and crash? See Section 11.5.
- Are tides a kind of wave? How do tides work? Why are there two high tides and two low tides every day? See Section 11.6.
- How are tidal waves (also called tidal bores) related to the tides? See Section 11.7.
- Do you know that "two large ships sink every week on average"? Can you guess why? The main reason is so-called rogue waves; they are also called freak waves, monster waves and killer waves, for obvious reasons: see Section 11.8.
- Tsunami waves also kill: the Boxing Day Tsunami on 26 December 2004 killed at least 230,000 people in 14 countries. What causes them and why are they so murderous? See Section 11.9.
- Why do the wakes of boats (the waves created by a moving boat) have such special shapes? See Section 11.10.
- How is light related to other electromagnetic waves, including radio waves, microwaves, infrared light, ultraviolet light, x-rays and gamma rays? See Section 12.1.
- How do mirrors work? Are left and right really interchanged in a mirror image? What is a one-way mirror? Section 12.2 discusses these issues.
- How do optical prisms and lenses work? See Section 12.3.
- What are mirages? Why does the Sun look flattened at sunset and sunrise? See Section 12.5.
- How do retroreflectors and optical fibers work? See Section 12.6.
- What gives rainbows their colors? How are they related to the solar spectrum? See Section 12.7.

- What are those novel metamaterials with their amazing properties? See Section 12.8.
- Can light turn around corners (by diffraction), like sound but unlike balls? See Section 12.9.
- Is light a wave or a particle? Probably both, as we discuss in Section 12.10.
- How are holograms made and how do they work? See Section 12.11.
- What are quantum waves, which are the basis of electrical insulators and conductors, electronic devices and computers, chemicals and biological substances, together with their colors? Such waves have unique properties which we introduce in Sections 13.1 and 13.2 by analogy with the more familiar light waves.
- How do quantum mechanical waves explain the structures of atoms, molecules and solids? See Sections 13.5, 13.6 and 13.7.
- Are the colors of substances (like green leaves and red tomatoes) explained by quantum physics? See Section 13.8.
- How does quantum physics explain radioactivity? See Section 13.9.
- Why have gravitational waves only been observed recently for the first time? What are they? Why are they amazingly weak, even though we feel gravity very much every day? What do they tell us about black holes and neutron stars? See Chapter 14.
- What is string theory? It is not the theory of waves on musical strings, but a possible "theory of everything", as we discuss in Section 15.1.
- Are neural waves and brain waves similar to other waves? See Section 15.2.
- What other waves exist? Consider human waves, waves in traffic and many other types of waves in Section 15.3.
- Finally, this book will attempt to answer the question: "Is doing science like doing sports, or playing video games, or reading science fiction?" This discussion, in Section 15.5, may interest those readers who wonder whether doing science is very different from doing more common activities: the short answer is that doing science is in many ways a quite familiar activity!

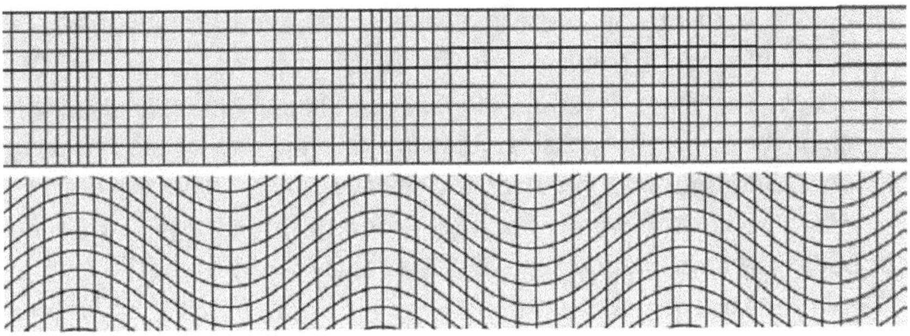

9

Sound Waves Inside Water and Rocks: Earthquakes

Sound is ubiquitous in daily life. It is most familiar as waves that travel through air. In this chapter, we follow sound waves that instead travel through liquids and solids, especially through water and rocks. We will compare such sound to sound in air and discuss how similar and/or different they are. For example, sound ***waves in liquids*** *and* ***solids*** *travel much faster than in air: we will relate this feature to the atomic and molecular structures of liquids and solids. A special type of wave, called a shear wave, can travel only through solids. Earthquakes cause energetic sound waves that can travel deep through the Earth and far across continents: most of these waves have low frequencies, so they may seem silent to many of us, while being very destructive.*

When we swim under the surface of water, or when we dunk our ears into a bath or pail, we can hear sounds, even though they are usually muffled, as if coming from far away: underwater, we can still hear a door slamming shut near our bath or fingers tapping on our bath or pail, or the engines of passing boats in a lake or river, or the knocking of stones under our feet in a river. These sounds must have travelled to our ears through the water, because very little sound travels through our body (the human body is an excellent absorber of sound!).

We are also familiar with sound going through walls, as well as through closed windows and doors: just think of someone knocking on your door, or screaming outside your window, or jumping on the floor above you. These sounds travel to your ears partly through the air, but must also move through the solid substance of a wall, door, window or floor (nevertheless, a large part of the sound is reflected by the surface of those solid substances).

There are in fact many similarities between the motion of sound through liquids or solids and its motion through air: many of the behaviors of sound in air which we have discussed in Chapter 3 can be found in sound traveling through liquids and solids.

But there are also clear differences. We shall address these similarities and differences in Section 9.1.

A very interesting and common situation is a wave on the surface of water, such as on the surface of the sea, or on a lake, or on water in a glass: it is sufficiently different and interesting to deserve its own Chapter 11, which will also deal with devastating "rogue waves" and tsunamis.

9.1 Similarities and Differences to Waves in Air (Gases)

The main similarity between sound waves in **air, liquids and solids** is that these three types of substances **can carry pressure waves** of the kind we saw in Chapter 3 for sound in air. Pressure waves travel equally well through liquids and solids, so sound can travel through water as well as through walls. These are sometimes called **Primary waves** or **P-waves**; the term P-waves is used especially for waves created by **earthquakes** in the soil and rocks. All these waves are **longitudinal**

waves, as illustrated in Figure 3-1 for waves in air: the local back-and-forth oscillation of the air, liquid or solid moves in the same direction as the wave travels.

The principles of wave motion that we have discussed in air remain equally valid for sound in liquids and in solids: we will again have wave amplitudes, wavelengths, wave frequencies, wave interference, wave reflection, *etc.* Moreover, the characteristic smooth oscillating sine-wave shape will also be seen with waves in liquids and solids.

One notable difference is the **speed of sound**. This is seen in Table 9-1, which lists the speed of sound in several different substances, including gases, liquids and solids. The speed of sound in a substance is basically the speed of sound waves. We see in Table 9-1 a broad range of speeds, varying almost a hundredfold from the gas called sulfur hexafluoride to the solid diamond.

Let's now do some detective work to try to figure out where that wide variety of speeds of sound comes from.

If we look at the speed of sound in **water**, Table 9-1 gives three quite different speeds: one speed in **steam** (the gaseous form of water), about 500 meters per second; a 3 times larger speed in **liquid water**, about 1,500 meters per second; and an almost 8 times larger speed in **ice** (solid water), about 3,800 meters per second (we ignore the slower S-waves for now, as they don't exist in gases or liquids).

More generally, we can observe that: **Sound waves travel much faster in most liquids than in air, and sound waves travel much faster in most solids than in liquids.**

However, there are special cases, such as **helium** which behaves differently, having a very low density and being a "noble gas". To make helium liquid, it has to be cooled to extremely low temperatures, namely below about –269°C ~ –452°F; this is quite close to the **absolute zero temperature** (about –273°C ~ –460°F), which is the lower limit of possible temperatures. The reason for needing such low temperatures is that the "glue" between helium atoms is very weak even in the liquid form (this weak "glue" is the reason for calling helium a noble gas, as it does not easily bond to other substances; the "glue" is of the Van der Waals type; see Section 13.6). Combined with the light mass of helium, the weak "glue" favors more complex quantum effects, which result in "abnormal" sound speeds.

Table 9-1: Speed of sound in various substances in typical conditions at the Earth's surface. The speeds are given in meters per second (often abbreviated as m/s); to obtain speeds in kilometers per hour (km/h), multiply the indicated values by 3.6 (or 4 if you don't need precision); for speeds in miles per hour (mph), multiply the indicated values by 5.79 (or 6). P-waves are pressure waves, while S-waves are shear waves: see Sections 9.2 and 9.3. (The values in this table are approximate and depend on many factors such as temperature, pressure, purity, and the frequency of sound; they are sourced from various lists.)

Sound speed (in meters per second)	**in gases**	**in liquids**	**in solids (P-waves/S-waves)**
sulfur hexafluoride	135		
ether	210	1,000	
oxygen (O_2)	325	1,000	
air	343		
nitrogen	350	850	
helium	1,000	200	400
rubber, plastic			100–2,500
flesh			1,500
lead			2,000
water	500 (steam)	1,500 (liquid)	3,800/1,800 (ice)
hardwood			4,000
bone			4,000
glass			4,000–5,600
copper			5,000
iron			5,000–6,000
rock (silica, granite)			5,000–6,000 /3,000–4,000
aluminum			6,000
diamond			12,000

Another observation from Table 9-1 is that the **density of substances** may be important for the speed of sound (density is mass per volume, which we can think of as weight per given volume). It appears at first sight that the speed of sound is lower in low-density substances, such as gases, than in higher-density substances, such as liquids and especially solids.

However, we will see that the reality is a bit different: another aspect of substances is actually more important, namely their hardness, which is related to their softness, compressibility and elasticity, all of which are also related to the "glue" between molecules and between atoms.

Indeed, Table 9-1 also suggests that sound in hard solids, like some metals, rocks and diamond — travels much faster than sound in soft solids, like flesh, rubber and lead. Section 9.2 will explain such differences in more detail.

So far, we have discussed **pressure waves** (**P-waves**) in liquids and solids, similar to pressure waves in gases such as air. However, solids allow an additional different type of sound waves which cannot exist in gases and liquids: so-called **shear waves**, sometimes abbreviated to **S-waves** (especially for waves created by earthquakes in soil and rocks). Section 9.3 explains how this kind of wave depends on the ability of solids to straighten out after being deformed, which is a special form of elasticity.

9.2 Pressure Waves in Liquids and Solids

How could you measure the speed of sound? Could you do it for air? One way is to listen for an echo of your voice reflected from a distant wall: if you measure the distance to the wall and measure how much time it takes before you hear the echo, you can calculate the speed of the sound; that speed would be twice the distance divided by the time. However, this is difficult to do with other gases than air because we need large distances and therefore large amounts of gas!

We can do the same thing under water to estimate the speed of sound in liquid water, but here again, it is difficult to repeat this with other liquids. In any case, sound generally travels so fast that it is difficult for us to get a precise idea of its speed without instruments. We therefore have to rely on measurements performed in scientific laboratories: the results are then listed in publications, from which our Table 9-1 comes.

What influences the speed of sound? We mentioned in Section 9.1 that the speed of sound is considerably higher in liquids than in gases, and also higher in solids than in liquids. In short, **for the speed**

of sound: gas < liquid < solid (but there are exceptions); here we use the symbol "<" which means "smaller than" (similarly, the symbol ">" means "larger than"). This behavior is related partly to the **density** of substances, namely the mass (relative weight) in the same volume, and partly to the **hardness** or **compressibility** of substances, which we will discuss further below.

Consider first the **density** of substances. An example for gases is helium *versus* air *versus* sulfur hexafluoride, as we saw in Table 9-1: helium is lighter than air, which is lighter than sulfur hexafluoride. This results in sound traveling around three times faster in helium than in air, and around three times faster in air than in sulfur hexafluoride (thereby causing people's voices to dramatically change pitch when breathing those different gases; see Section 6.4). The same happens in light *versus* dense liquids, such in water *versus* ether.

When we talk about density of substances, **we have to ask what substances are made of and why substances have different densities: this leads us to look at tiny molecules and atoms.** For understanding sound and its speed, we also must ask how fast those molecules and atoms move.

If you wish to know more about the high **speeds and tiny sizes of molecules and atoms** in various substances, please read Box 9-1. That will give a better idea of the structure and properties of many substances.

BOX 9-1 — HOW FAST DO MOLECULES AND ATOMS MOVE, AND HOW SMALL ARE THEY? Water is composed of molecules, each containing two hydrogen atoms and one oxygen atom, written as H_2O. Air is a mixture of several gases, mainly nitrogen molecules, each containing two nitrogen atoms and written as N_2, and oxygen molecules, each containing two oxygen atoms and written as O_2. Many substances, especially solids like metals and rocks, are composed of individual atoms that do not form molecules; examples are copper and iron, as well as silica found in rocks and glass.

With increasing temperature, the molecules and atoms of all substances move faster: temperature is in fact a measure of the motion energy of the molecules and atoms.

The molecules in air fly around at great speed and often bounce off each other like billiard balls. Their average speed in air (at normal temperature and atmospheric pressure on the surface of the Earth) is around 450 m/s (meters per second), about 30% faster than the speed of sound in air (which is about 343 m/s, see Table 9-1). Some molecules fly several times faster than that average, and some fly much slower, but it is the average speed that matters for sound. However, the molecules can fly in all directions, so only a fraction of this speed contributes to sound going in one particular direction, hence the 30% difference between the speed of the molecules and the speed of sound.

In gaseous steam and air under normal conditions, molecules are widely separated, so they occupy only 0.1% of space. This means that the average distance between neighboring molecules is about 10 times their size: viewed this way, steam and air are fairly dense, as we feel when we wave our hands through the air! The diameter of these molecules is about 0.000,000,000,3 meter = 0.000,000,3 micrometer = 0.000,3 (three ten-thousandths) millimeter = 0.3 nanometer or around 1/200,000 of the diameter of hair; the atoms involved (N, O and H) are about half this size. **This separation allows the air molecules to fly, on average, around 300 times their size between collisions. It is this motion of molecules that moves a sound wave along**: a compression speeds up molecules in one direction; this increased speed then propagates the compression to other molecules by means of collisions, in a domino-like effect.

Let's now compare the molecular size with typical wavelengths of sound in air. As we saw (in Section 3.7), the wavelength of audible sound varies from about 17 meters to about 17 millimeters as the sound's frequency rises from 20 hertz to 20,000 hertz, the normal human audible range. Thus, the wavelength of audible sound ranges from about 50,000 to about 50,000,000 times the molecular size. This means that when we look at a wave, we can't see the tiny molecules, and when we look at molecules, we can't see the gigantic waves.

Another interesting question is how much the density (and therefore the pressure) varies between a wave crest and a wave trough. We answered this already (in Section 3.4): even at the painful level, sound only represents a pressure variation of about 1 part in 5,000, relative to the normal air pressure on Earth; this is only 2% of 1%, too small to draw to scale in a picture.

Now consider liquids and solids, which we can make by cooling gases. A liquid and a solid contain the same molecules and atoms as the corresponding gas: by cooling gaseous air to −194°C (−317°F) or lower, we get liquid air; cooling below −215°C = −355°F freezes air to a solid. By cooling gaseous water steam to 100°C = 212°F or lower, we get liquid water; cooling below 0°C = 32°F freezes water to ice.

(Continued)

(*Continued*)

Cooling slows down molecules and atoms. That gives them a chance to stick together without bouncing off each other, as happens when steam forms drops of water and rain: this is called **condensation**.

The result of condensation is a liquid or solid, in which the molecules (and atoms in metals, *etc.*) are tightly packed, so they occupy almost 100% of space and can't fly around anymore. The diameter of molecules and atoms mentioned above is now also the distance between neighboring molecules and atoms. Now we have a dense substance, liquid or solid, in which the molecules still move at fairly high speeds. However, they can only rattle against their immediate neighbors: they bounce back and forth between their immediate neighbors at speeds that are somewhat slower than in a gas (because of the lower temperature); this rattling motion also transmits the sound through the liquid or solid.

As we already noticed, density is not the whole story for understanding the speed of sound.

Another trend seen in Table 9-1 is that the speed of sound depends on how hard a substance is: **in general, the speed of sound is higher in harder substances**. Hard is simply the opposite of soft, compressible, elastic or "squeezable" (the official scientific term for this kind of hardness is **bulk modulus**, even though this name seems less informative!). As everyone knows, diamond is very hard, while flesh and rubber can be quite soft, compressible, elastic and squeezable; even the dense metal lead is relatively soft.

The data of Table 9-1 thus show that **a more compressible (soft, elastic or squeezable) substance slows down sound waves**.

What are hardness, softness, compressibility and elasticity? We can better understand these closely related concepts by looking at how the smallest parts of substances are linked together. The smallest parts of substances involved in sounds are the **atoms** or **molecules**; molecules consist of atoms. For example, each molecule of water contains two **hydrogen** atoms and one **oxygen** atom, written as H_2O. Each oxygen molecule in the air contains two oxygen atoms: O_2. Each **nitrogen** molecule in the air also contains two nitrogen atoms: N_2. Some substances rarely form molecules, such as metals (lead, iron, *etc.*): they consist of atoms only; such atoms can stick together in clusters that are too large to call molecules.

The atoms in each molecule are strongly linked to each other: scientists use the word **bond** to describe this link between atoms (Section 13.6 describes the bond between atoms in more detail). Molecules will not break apart under normal circumstances: water molecules normally do not break up into oxygen and hydrogen. We can therefore treat the molecules as tiny "hard balls", very similar to billiard balls bumping against each other. Individual atoms in metals, *etc.*, also can be viewed as individual "hard balls" that bump against each other.

Hardness describes how difficult it is to squeeze or compress two molecules or atoms closer together, which is what happens when a sound wave compresses the substance. Softness, compressibility and elasticity are simply opposite to hardness. Air, steam and other gases are much more compressible than liquid water: try to squeeze down to a smaller volume a balloon filled with air and a balloon filled with water. The balloon with water can hardly be compressed; basically, only its shape can be changed but not its volume. Also, a rubber ball is clearly much more compressible than an iron ball.

But, is liquid water more compressible than ice (frozen water)? If yes, then sound should indeed travel faster in ice than in liquid water. To answer this question, we must consider the geometric structure of liquids *versus* solids, which is an interesting topic in its own right.

Let's assume that the molecules and atoms of a substance are hard balls. **An important fact about sound is that molecules and atoms generally like to stick together to form a liquid or solid. However, if they move too fast, they will not stick together but bounce off each other like billiard balls, forming a gas.** There is a "glue" between molecules and between atoms, but this glue is less strong than the glue which bonds together the atoms in a molecule. So we have tiny but "sticky" hard balls. You may imagine making billiard balls sticky and watching them collide at high speeds (they will then bounce off each other) or at low speeds (they will then stick together).

It is also very important to realize that this glue is weak in some substances and strong in others. A weak glue will favor hard balls bouncing away: then we have a gas, such as air or steam. A strong glue will favor hard balls sticking tightly to each other: then we have a rigid solid, such as a piece of iron. An intermediate glue will allow hard balls to hold onto each other but also to slip around each other: this gives a liquid that flows, such as liquid water.

Figure 9-1 illustrates this molecular structure for **water** as it cools from gas to liquid to solid. In a gas (*at the left*), the hard balls fly at high speeds and collide occasionally. In a liquid (*at the center*), the hard balls stick together but can easily flow around each other, at slightly slower speeds. In a solid (*at the right*), the hard balls are stuck rigidly together with a regular alignment, but they still rattle against each other at fairly high speeds, causing some disorder. In the liquid and solid, the hydrogen atoms are attracted to the oxygen atoms of other nearby water molecules: this forms the bonds between molecules.

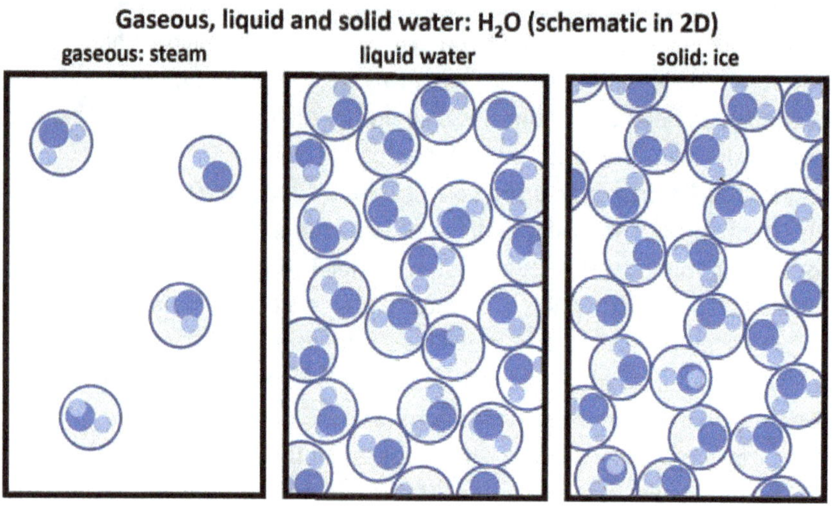

Figure 9-1: Change of water structure from gas (steam, *at the left*) to liquid (*at the center*) to solid (ice, *at the right*). Each molecule of water consists of one oxygen atom (dark disks) and two hydrogen atoms (light disks, some hidden behind oxygen atoms); the molecules form "hard balls" (large dark circles). These artificial 2D models are schematic for clarity; see Figure 9-2 for a true 3D model of ice. The wavelength of audible sound would be thousands to millions of times larger than these images, so we can't show a sound wave on this molecular scale. Caution: The "steam" that we see when cooking or in clouds is not gaseous. Such steam is composed mainly of small droplets of condensed water, so it is a mix of air and water droplets; by contrast, humid air contains water in the gaseous state without droplets. An easy way to distinguish the two forms of steam is that gaseous water is invisible, while water droplets are visible, if only as a cloud.

In the following, we discuss in more detail the changes that occur in **water** as it transforms from steam (gaseous water) to liquid water and finally to ice (solid water).

Let's start with high temperatures (above 100°C = 212°F). The water molecules now move too fast to stick, so they try to fly away, thus forming a **gas** (**steam**); a gas expands if not contained in a tank.

If we cool down the water to freezing temperatures (below 0°C = 32°F), the molecules move slowly enough to strongly and permanently stick together: this makes a rigid **solid**, for example in the form of ice crystals in snowflakes or on cold windows or in a freezer.

However, at intermediate temperatures, between 100 and 0°C (between 212 and 32°F), the molecules not only stick together but also can slide past each other, so they can flow, forming **liquid water**.

We now see that the liquid is intermediate between the gas and the solid: in particular, liquid water is less compressible than steam,[1] and solid ice is even less compressible than liquid water.

In summary, we see that the speed of sound in different substances varies greatly, and that these variations are closely connected with the composition and structure of those substances, particularly whether the substance is a gas, liquid or solid, and whether the substance is hard or soft.

Are you aware that liquid water and ice are very unique compared to other substances? You know, of course, that ice floats on water, which tells us that ice is less dense than water: that is already very unusual, since almost all substances are heavier (denser) when they are solid than when they are liquid. Furthermore, sound travels almost three times faster in ice than in liquid water, although their densities are rather similar: only about 8% different.

Indeed, water is a very special and fascinating material, with unique structures for both the liquid and the solid, even though water is extremely common in our lives in its three forms: steam, liquid and ice. If you are curious, Box 9-2 describes some intriguing and surprising aspects of liquid water and ice. It also shows limits in our current understanding of water: amazingly, we don't know that much about water!

---

[1] Liquid water is compressed by only 1.8% in volume at a depth of 4 kilometers, a typical ocean depth.

BOX 9-2 — THE AMAZING STRUCTURES OF ICE AND LIQUID WATER. Water, in its solid form (ice) and in its liquid form, behaves in some unique ways, different from most other substances. (A somewhat technical, but very complete overview of water properties is available free online in a review article.[2])

Where do the beautiful 6-pointed star shapes of snowflakes come from? Amazingly, those shapes come from the shape and sticking properties of the tiny water molecule, H_2O. The shape of that molecule is a simple V, as shown in Figures 9-1 and 9-2b. Looking down onto this surface (see Figure 9-1 at right and Figure 9-2c), we see an arrangement of molecules similar to a honeycomb. A honeycomb arrangement has a hexagonal structure that can give rise to 6-pointed star shapes when ice grows to large sizes: like snowflakes, it has 6-fold rotational symmetry, meaning that it can be rotated repeatedly by 360°/6 = 60° without changing its structure.

You may have noticed that the H_2O molecules in Figures 9-1 and 9-2 do not look like snowflakes and do not have 6-fold rotational symmetry! It is remarkable that H_2O molecules, without such a shape and symmetry, nevertheless can assemble themselves into structures that do have 6-fold symmetry on the larger scale, leading to the regular star shapes of snowflakes.

Where does this behavior come from? It is the special way in which water molecules stick to each other. Water molecules are "polar", meaning that one end (the oxygen atom) has a negative electric charge, while the hydrogen atoms are positively charged. Since positive and negative electric charges attract each other, the H atoms of one water molecule will be attracted toward the O atom of a neighboring molecule: this is what we see in Figures 9-1 and 9-2. This electric attraction makes the molecular arrangement shown there very logical: **each O atom is linked by single H atoms to its neighboring O atoms, so that each O-O pair has one H atom in between: O-H-O.** This link between two molecules is called a **hydrogen bond** and is of great importance also for biological life. For example, hydrogen bonds are responsible for keeping **DNA** molecules from falling apart. Interestingly, hydrogen bonds are not very strong: that is why you can cut an ice cube with a knife, unlike many other solids. For more information about hydrogen bonds, see Section 13.6.

---

[2] "How Water's Properties Are Encoded in Its Molecular Structure and Energies", by Emiliano Brini, Christopher J. Fennell, Marivi Fernandez-Serra, Barbara Hribar-Lee, Miha Lukšič, and Ken A. Dill, freely available in *Chemical Reviews*, volume 117, issue 19, pages 12385-12414, 2017, at https://pubs.acs.org/doi/10.1021/acs.chemrev.7b00259.

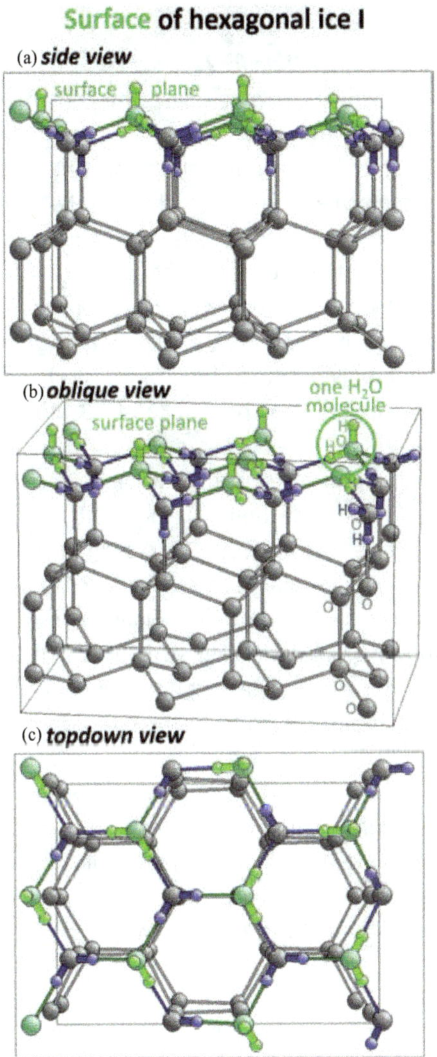

Figure 9-2: The surface (in green) of very cold ice is shown in (a) side view, (b) oblique view, and (c) topdown view. This model of the surface of normal crystalline ice at the molecular scale consists of H_2O molecules, which are shown in green in the outermost layer; the green circle in b shows one H_2O molecule, with its V-shape. The small green H atoms point either straight up out of the surface, or to larger gray O atoms in the layer below; each of these is bonded to two small dark-blue H atoms. In deeper layers, only gray O atoms are shown: for clarity, their H atoms are not drawn; these H atoms would be located on the thin links between O atoms, as in the top layers. The shortest O-O distances are about ¼ of a nanometer. This ice crystal is cut off by the surrounding box for clarity: in a real ice crystal, more water molecules would be present on the sides and below the box. (*Source*: drawn with K.E. Hermann's Balsac software.)

(Continued)

(*Continued*)

Why is liquid water denser than ice? Figure 9-1 also illustrates and explains this unusual fact quite simply. Let's compare the center and right images of Figure 9-1: the solid ice in the right image has large holes in its hexagons. However, the liquid water fills up those holes to some extent because it is able to flow: liquid water molecules can thus slightly slip between their neighbors. This partial filling of the holes also results in a denser liquid and therefore allows less dense ice to float at the surface of water.

Did you know that ice has another special property, namely that it can take on other forms than the commonly observed hexagonal structure shown in Figure 9-2? The commonly observed form of ice is called ice I (with a Roman numeral I). However, **ice can also form at least 18 other crystal structures, labeled ice II to ice XIX.**[3] These other crystal structures only exist under very high pressures, so we never see them in daily life. This large number of distinct crystal structures is quite unusual: most substances can exist in only one or perhaps two different structures. The reason for this richness of crystal structures is the "free space" available between the ice molecules. By squeezing normal ice I, we can collapse it into more compact, denser structures, forming ice II, III, IV, ..., until XIX (and additional structures may still be discovered). There also exist several forms of "amorphous ice", in which the molecules are not aligned but haphazardly arranged: the latest of these was reported in 2023.[4]

Are you aware that a similar compaction happens when ice melts to form a liquid due to heating? The increased vibrations of the water molecules disorient them, so the O-H-O links are less straight and become bent: **those vibrations allow the molecules to come a little bit closer together and make the liquid a bit denser than the solid ice.** This is the reason why ice floats

[3] See "Ice" at https://en.wikipedia.org/wiki/Ice#Phases. A candidate for ice XX was reported in 2022: "Pressure-driven symmetry transitions in dense H_2O ice", Zachary M. Grande, C. Huy Pham, Dean Smith, John H. Boisvert, Chenliang Huang, Jesse S. Smith, Nir Goldman, Jonathan L. Belof, Oliver Tschauner, Jason H. Steffen, and Ashkan Salamat, *Physical Review B*, volume 105, number 104109, 2022, https://doi.org/10.1103/PhysRevB.105.104109.

[4] See description in "Scientists made a new kind of ice that might exist on distant moons", by Jonathan O'Callaghan, in *Nature*, News, 2 Feb 2023, https://doi.org/10.1038/d41586-023-00293-w; the original research is by Alexander Rosu-Finsen, Michael B. Davies, Alfred Amon, Han Wu, Andrea Sella, Angelos Michaelides, and Christoph G. Salzmann, "Medium-density amorphous ice", *Science*, volume 379, number 6631, 2023, https://doi.org/10.1126/science.abq2105.

on liquid water: such behavior is highly unusual because the solid form of substances is normally denser than the liquid form. In fact, at the melting temperature, ice is about 8% less dense than water, so only about 8% of a floating piece of ice sticks out above the surface of water, as do icebergs and ice cubes floating in liquid water. (Further heating of liquid water expands it due to even larger vibrations, but only by 4%, so water remains denser than ice. Sea water is a bit denser than pure water, but not enough to make sea ice sink.)

What is the molecular structure of liquid water? You would think that science knows the answer to this simple question about one of the most common and important substances in our lives! Not so: amazingly, scientists are still hotly debating this question! At one extreme is the answer: in liquid water, the H_2O molecules flow randomly, slipping past each other, similar to their flying around in the gas but now much closer together, as illustrated in the center image of Figure 9-1. At the other extreme is the following model: liquid water consists of tiny pieces of ice (we may call them tiny icebergs) that roll past each other, like dry sand flowing between your fingers; imagine small chunks of aligned molecules like those in the right image of Figure 9-1 appearing in the center image. Both answers are reasonable, as are various other models.

The reason we don't know the molecular structure of liquid water is that we don't have tools to look at water motion on the scale of individual molecules. We do have tools like x-ray diffraction, scanning probe microscopy (see Figure 13-13 in Section 13.9) and various spectroscopies that can "**see**" individual molecules, but only if those molecules don't move.

Why is ice so slippery? Very few solids are as slippery as ice: how come? Amazingly, this question is also still being debated for the same reasons. A common, but incorrect, answer that many people have heard is this: if you stand on ice, your weight melts the surface of the ice (especially on ice skates that have a small contact area with the ice), while **friction** due to sliding helps the melting by adding heat (friction is resistance to sliding that transforms sliding motion into heat). First, it is known that a person's weight, even on skates, is not sufficient to melt ice. Second, this explanation cannot explain why you slip on skis, which distribute your weight over a much larger area of snow.

Before we answer the question of why ice is so slippery, let me relate a fascinating historical disaster which illustrates and gives a strong hint at the currently most likely answer:

Why did Robert F. Scott's expedition to the South Pole in 1910 to 1912 end with his and his team's deaths, while Roald Amundsen succeeded just

(Continued)

(*Continued*)

a month earlier? Many possible causes have been proposed over the years, including Scott's use of motorized vehicles and ponies instead of Amundsen's choice of dogs only, as well as too narrow safety margins in Scott's plan.

More recently, meteorologist Susan Solomon has blamed the weather in her book "The Coldest March: Scott's Fatal Antarctic Expedition".[5] She points out that Scott's expedition, after reaching the South Pole, returned toward the Antarctic coast a month later than Amundsen's team. By then, according to Scott's own measurements, the temperature over Antarctica had dropped by at least 10°C = 18°F, getting even colder as Scott approached the coast.

Solomon notes that, for at least a month, the temperatures experienced by Scott during his return journey were below −34°C = −29°F. We now know that the friction of ice increases markedly below that temperature,[6] so Scott's team progressed at only half the planned speed. The team actually travelled back entirely on foot by dragging sleds, having no more motorized vehicles or ponies: see Figure 9-3.

Notes from Scott's team indeed mentioned that, as the temperature gradually dropped from −26°C ~ −15°F to −40.5°C ~ −41°F, "It has been like pulling over desert sand, not the least glide on the world." "Sandpapery cold snow." "The sledges ceased to glide, and pulling became very heavy." "Impossible friction." As a result, the amount of food that was pre-positioned along the return track was insufficient to feed the team, the last of whom died while stranded in a blizzard about 20 km from the next food supply, and about 300 km from the coast and support team.

This tragic story suggests that **increased friction of ice at very low temperatures was at least partly to blame for the demise of Scott's team**. We can describe how this friction arises as follows:

Consider the surface structure of ice shown in Figure 9-2. The H_2O molecules at the surface, drawn in green, are more loosely linked to other molecules than the deeper molecules: the outermost (green) molecules are only connected to three neighboring molecules, compared to four for the deeper molecules. As a result, the outermost molecules can vibrate more easily, especially as the

[5] Susan Solomon, *"The Coldest March: Scott's Fatal Antarctic Expedition"*, Yale University Press, 2001.

[6] "The Surface of Ice under Equilibrium and Nonequilibrium Conditions", by Yuki Nagata, Tetsuya Hama, Ellen H. G. Backus, Markus Mezger, Daniel Bonn, Mischa Bonn, and Gen Sazaki, freely available in *Accounts of Chemical Research*, volume 52, pages 1006–1015, 2019, https://pubs.acs.org/doi/10.1021/acs.accounts.8b00615.

remaining three links are almost parallel to the surface. This is indeed found[7] to be the situation at very low temperatures around –180°C ~ –292°F. But the surface molecules remain attached to their neighbors. Consequently, and most interestingly, at such very low temperatures, ice is not more slippery than steel, for example; indeed, steel sliding against an ice surface is then not more slippery than steel sliding against steel.

Figure 9-3: Robert F. Scott's team at the South Pole (the fifth team member is behind the camera). (*Source*: https://picryl.com/media/members-of-the-terra-nova-expedition-at-the-south-pole-robert-f-scott-lawrence-1, in the public domain.)

Now imagine raising the temperature: all ice molecules will vibrate more, including the surface molecules, which are more loosely linked; the larger

(*Continued*)

[7] "Molecular Surface Structure of Ice(0001): Dynamical Low-Energy Electron Diffraction, Total-Energy Calculations and Molecular Dynamics Simulations", by N. Materer, U. Starke, A. Barbieri, M.A. Van Hove, G.A. Somorjai, G.-J. Kroes, and C. Minot, *Surface Science*, volume 381, pages 190–210, 1997, http://dx.doi.org/10.1016/S0039-6028(97)00090-3.

(*Continued*)

vibrations of the surface molecules will also increase the vibrations in the next layer of H_2O molecules (shown with dark-blue H atoms in Figure 9-2). At a temperature around –90°C ~ –130°F, the surface molecules will vibrate so vigorously that they break some of their links to their neighbors. Now many of them have only two links rather than three[8]: the surface therefore becomes disordered and begins to resemble a liquid, which we can call "almost liquid". This situation is often called **pre-melting**, meaning melting below the normal melting temperature (here below 0°C = 32°F). Although this film is ultra-thin (only about a molecular diameter in thickness), it starts to act like a banana peel that becomes more slippery as it warms up.

By the time the temperature has reached about –40°C = –40°F, the friction has dropped by a factor of about 10 compared to ice at –180°C ~ –292°F, and also compared to steel against steel (the friction of steel against steel changes less with temperature). This factor 10 means that you can make your foot slip on ice by pushing it with a force that is 10 times weaker than at the lower temperature.

By –16°C ~ 3°F, the next two layers of water molecules also become almost liquid, further decreasing friction by another factor of about 3, and further increasing slipperiness. Raising the temperature even further makes more and more ice layers become almost liquid, until at 0°C = 32°F the entire ice crystal melts rapidly and becomes truly liquid.

With these aspects of ice and liquid water, we have seen that the structure of water varies considerably in interesting and consequential ways. Most other substances have a simpler structural behavior. In particular, the structural behavior is very important in connection with sound and especially the speed of sound: each change in structure can change sound significantly.

9.3 Shear Waves Only Exist in Solids

Look at the old house in Figure 9-4. Its rectangular front wall has been deformed by shearing: the top edge has shifted sideways, so the side

[8] "Some fundamental properties and reactions of ice surfaces at low temperatures", by Seong-Chan Park, Eui-Seong Moon, and Heon Kang, *Physical Chemistry Chemical Physics*, volume 12, page 12000, 2010, http://dx.doi.org/10.1039/c003592k.

Figure 9-4: An old house, deformed by shearing. (*Source*: Wikipedia, by Kotivalo, https://commons.wikimedia.org/wiki/File:Old_wooden_building_about_to_collapse.jpg, under CC BY-SA 4.0.)

edges are leaning over; the resulting shape is called a parallelogram in geometry since opposite sides are parallel to each other.

You can do the same thing with a rectangular sponge: if you push opposite edges in opposite directions, the two other sides will tilt. A sponge, once you release it, will recover its rectangular shape because it is **elastic**. The old house of Figure 9-4 is not elastic. Stronger houses are elastic and do recover their shapes after being pushed sideways, for example, by an earthquake!

Elastic shearing happens in solids, but not in gases or liquids. If you shear a gas or a liquid, it accepts the deformation (for example, it adopts the shape of the container in which you pour it, and if you deform a plastic bag full of air or water, it often does not try to recover its original shape): gases and liquids do not try to recover their original shape after shearing. On the other hand, gases and liquids do try to recover their original density after compression: this causes pressure waves, as we have already seen.

Is shearing similar to bending and twisting? When you think of bending, you think of forcing a straight object to become curved, such as a stick made of wood, plastic or metal. Another example is a string in a musical instrument: such a string is curved into a wave pulse or wave train (however, the mechanism that straightens a musical string is not bending elasticity but tension applied from outside the string). There is no such curve in the old house of Figure 9-4.

Another word that is similar to shearing is twisting, but twisting usually implies rotation, as in twisting a screw. **Shearing is a <u>non-rotating</u> movement of one side of an object**; nonetheless, shearing can cause rotation of other parts of the object, such as the vertical boards of the old house in Figure 9-4.

To clarify these different kinds of deformation, Figure 9-5 compares tension, compression, bending and shearing. Each of these can travel as waves. It may be helpful to view their motion in my Animation 9*1. We have already discussed waves on strings due to tension. We also discussed pressure waves due to compression in gases, liquids and solids. In this section, we will look more closely at these deformations

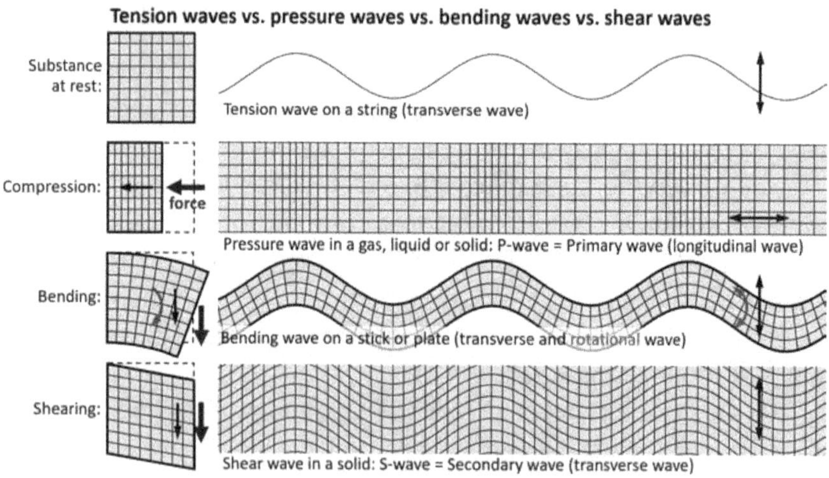

Figure 9-5: Models of four kinds of deformation that can travel as waves: tension, compression, bending and shearing. The waves can travel to the right or to the left. The square at the top left shows a substance at rest, without deformation and without wave: the thin lines outline pieces of the substance that will move and deform in a wave. The thick arrows show forces that cause deformations and waves. The thin arrows give the direction of local motion. The curved arrows show rotation.

and their waves. We will also discuss shear waves in solids. Not shown in Figure 9-5 or discussed in this book is **twisting**, usually called **torsion** (an example of torsion is what happens to your body when you stand firmly on the ground and rotate your shoulders around your spine to the right or to the left: your body then starts to look like a screw).

ANIMATION 9*1 — See my video WB1 at time 9:28 in its section "**Waves in liquids and solids**" under the title "**Waves travel on strings, in gases, in liquids, in solids, *etc.*: animation**". (See details in the section References and Resources below.)

The simplest deformation of a substance is compression (decompression is very similar and alternates with compression in a wave). A force compresses the substance, as shown in Figure 9-5. Here compression happens in one direction: therefore, the square pieces of the substance in Figure 9-5 become rectangular. As we have seen in Chapter 3 for air (and other gases), this causes a **pressure wave** which travels in that direction; Section 9.2 discusses the same type of wave in liquids and solids. The back-and-forth motion of the substance also takes place in the same direction (shown by thin arrows), so it is called a **longitudinal wave**.

The **bending** deformation is more complicated. It is only possible in a solid because gases and liquids do not resist this deformation: they simply flow to any other shape without trying to recover their original shape.

Now consider bending in solids. Try to bend a sponge: it will become curved, as shown at left in Figure 9-5. Its pieces will move in the direction of your force; they will also rotate (see the curved arrows in Figure 9-5). And those pieces will themselves become curved. However, additionally and importantly, there will be regions of compression and regions of decompression: you can see this clearly in a sponge, and in your fingers, elbows and knees, when you bend them. We can thus view bending as a combination of regions of compression and regions of decompression. These can travel together as a wave along a solid stick.

The back-and-forth motion of the substance in bending waves takes place perpendicularly to the travel direction of waves, as shown in Figure 9-5 (thin arrows), so they are called **transverse waves**.

An interesting aspect of the bending wave on a stick is the question: ***What happens when the stick gets thicker?*** You can imagine this case in Figure 9-5, by adding more substance above and below the wavy stick that is drawn. The additional substance will have to be either compressed more or decompressed more. But the substance will resist greater compression and decompression: so, the thicker we make the stick, the more the stick will resist bending. This is why walls and beams that carry weight are made thick in buildings, bridges, *etc.*; supporting cables don't need to be so thick because they become straighter under tension.

Instead of a stick, we could have a thin plate, as in some musical instruments: cymbals, drums, xylophones, bells, *etc.* (see Section 4.2). Here, the musical waves are also bending waves. The bending takes place perpendicularly to the plate, so that the image of bending waves in Figure 9-5 is still valid: that image then looks along the surface of the thin plate.

The shearing deformation only occurs in solids: gases and liquids do not resist shearing, as with bending. Shearing is simpler than bending: in particular, shearing does not involve the increasing compression or decompression seen in thicker sticks. This means that shearing is not limited to relatively thin sticks, walls or beams: **shearing can take place in large blocks of metal, concrete, rock, soil, ice, *etc*.**, just like pressure waves.

The back-and-forth motion of the substance in shear waves takes place perpendicularly to the travel direction of the waves, as shown in Figure 9-5, so they are **transverse waves**.

Both shear waves and pressure waves are common in earthquakes, as we discuss in the next section.

9.4 Earthquakes

Earthquakes are waves inside the Earth, often called **seismic waves**. They are large-scale versions of waves due to everyday collisions between objects in a kitchen or workshop.

What causes earthquakes? Most earthquakes are caused by slippage between plates in the Earth's crust, thereby forming **geological faults** in rocks and soils; other earthquakes may result from volcanic activity or explosions, in particular from **nuclear bombs**; weaker earthquakes can originate in landslides or small explosions that are used to explore the location of oil and gas deposits, for example.

Two main types of waves occur in earthquakes: **pressure waves** (**Primary** or **P-waves**) and **shear waves** (**Secondary** or **S-waves**). S-waves in rocks and soil travel about 40% more slowly than P-waves, as shown in the right column of Table 9-1; an online video nicely illustrates this difference in speed.[9] The difference in speed makes it possible to estimate the distance from the point of detection to the source of the earthquake: the time interval between the arrivals of the P-wave and the S-wave is proportional to the distance traveled by those waves.

The direction from the source can also be estimated from the direction of ground motion at the detector: P-waves oscillate to and from the source, while S-waves oscillate perpendicularly to that direction. By comparing recordings from multiple detectors at different locations, the position, depth and intensity (magnitude) of the earthquake source can be determined with good precision.

How far can earthquakes travel? A common feature of earthquake waves is the enormous distance they can travel before dying out. The most powerful earthquakes and most nuclear explosions are easily detected by instruments on the other side of the Earth.

Figure 9-6 shows "tracks" of P-waves and S-waves inside the Earth due to one earthquake at the top left ("Focus of earthquake"); earthquake waves travel not only near the surface of the Earth but also right across the inside of the Earth.

Thanks to the behavior of earthquake waves, we know that the Earth has an internal structure like "onion peels": it has a mantle outside, an inner core in the center, and an outer core in between. (The mantle is often subdivided into lower and upper mantles, and even finer layers.)

The behavior of earthquake waves shows that these "onion peels" consist of different substances with different wave speeds. For example, the outer core is liquid: we know this because S-waves can't travel through liquids, and indeed don't penetrate this outer core. By contrast,

---

[9] See "P Wave vs. S Wave" by IRIS at https://www.iris.edu/hq/inclass/animation/p_wave_vs_s_wave.

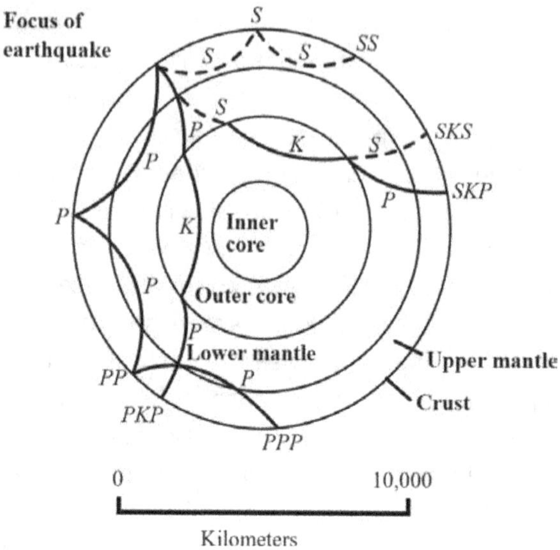

Figure 9-6: Tracks of seismic waves traveling inside the Earth from the source of an earthquake at the top left ("Focus of earthquake"). P and S are labels showing P-waves and S-waves, respectively. P-waves in the outer core are marked with K. S-waves cannot travel through the liquid outer core. Tracks with multiple letters show changes in wave type as the waves travel: for example, the track SKP starts as an S-wave in the mantle, becomes a K-wave in the outer core, and turns into a P-wave on its way out through the mantle. (*Source*: SEWilco at Wikipedia, in the public domain, https://commons.wikimedia.org/wiki/File:Earthquake_wave_paths.svg.)

the inner core is solid, while the mantle is a viscous solid (like hot lava, yoghurt or dense honey that can barely flow). The inner core is about 1,200 kilometers or 760 miles thick (this is the radius measured from its center), while the outer core is about 2,300 kilometers or 1,400 miles thick; they are probably mainly composed of iron and nickel, which cause the magnetic field of the Earth. The mantle (which is about 2,900 kilometers or 1,800 miles thick) is primarily made of compounds called silicates: these are based on silicon and are present in most rocks seen at the surface, as well as in sand.

The surface layer of the Earth is a relatively thin and hard **crust**: it is from about 5 to 80 kilometers thick, which is about 3 to 50 miles (it is too thin to draw in Figure 9-6). The crust contains large and rigid **tectonic plates** that move slowly but surely: their collisions and associated volcanoes are the source of most earthquakes, volcano eruptions and mountains.

In fact, earthquake waves are one of the few ways we can get information about the deep structure of the Earth (the Earth's magnetic field is the other main source of information about the inside of the Earth). It is remarkable that we know so little about the ground deep below us. The main reason is that it is difficult to drill deep holes into the Earth due to high temperatures and pressures there: the record depth reached by drilling is only about 12 kilometers or 7.5 miles, which is only 0.2% of the distance to the center of the Earth (6,371 kilometers or 3,960 miles). Compare that to how much we know about the surface of other planets and moons!

As drawn in Figure 9-6, waves can be transmitted through the boundaries between the "onion peels": at those boundaries, a change of direction takes place, called **refraction** (see Sections 10.3 to 10.6). The curves in the tracks are caused by gradual refraction (see Section 12.5), due to increasing pressure and temperature with depth, which increase the speed of waves.

In reality, seismic waves do not follow narrow tracks or lines as drawn in Figure 9-6: instead, they spread out in all directions, like sound, in the form of spherical waves, similar to Figure 3-1. Consequently, each powerful earthquake can be detected everywhere on Earth. The narrow tracks are nonetheless very useful to graphically show how the directions of those waves change as they progress through the earth: this is called **ray tracing** (see Section 10.7). In particular, once we know the locations of an earthquake source and a detector, we can draw the (curved) line that the wave has followed to reach that detector: this in turn allows us to find the depth and thickness of the "onion peels" inside the Earth.

Actually, there may be more than one track that a wave can follow from its source to a particular detector: the different tracks would then have different lengths, so that different waves from the same source reach that same detector at different times. This is like echoes that have followed different paths to your ears. This provides more information about the Earth's inner structure.

Earthquake waves will also be reflected and deflected by the rock and soil structure, especially in mountainous regions where many earthquakes originate. Analogies are diamonds, glass crystals, chandeliers and groups of drinking glasses or transparent bottles: if you shine light into them (including directed light from a flashlight or laser), you will see many reflections: each reflection has followed a different track from the source to your eye (and your two eyes double the number of possible tracks!).

The different tracks bring different waves to you: in the case of earthquakes, they can interfere with each other (by superposition) and create quite complicated interference patterns through constructive or destructive interference. Consequently, earthquakes can be violent in one spot but gentle nearby, focusing their power very unevenly. These interferences also make it very difficult to predict where destruction is likeliest, since the locations of constructive and destructive interference depend on the directions of the incoming waves.[10] (With light waves, such interferences are rare but possible: the wavelength is much shorter and requires more precision in the shapes of diamonds, glass crystals, *etc.*)

Earthquakes can often be heard, not only felt, because they contain audible frequencies. The frequencies are relatively low, ranging from 1 to 100 hertz (20 hertz is the normal lower limit of our hearing). The corresponding wavelengths range from about 5 kilometers (3 miles) to 50 meters (150 feet). Ground movement is rarely above 1 meter (3 feet).

While living in California, I learned to distinguish earthquakes and estimate the distance from their sources. For example, some earthquakes produce a short and sharp bang, while others cause a lengthy and low-pitched rumbling sound. The motion was similar to the sound: sudden in the first case, and prolonged in the second case; in addition, the motion was more up-and-down (vertical) in the first case, and more back-and-forth (horizontal) in the second case. Gradually, I realized that the sharp bangs come from nearby earthquakes (probably less than 10 kilometers or 6 miles away, but still rather deeper in the Earth, so they arrived from below), while rumbling sounds come from distant earthquakes

---

[10] We have a paradox in wording here! Constructive interference between earthquake waves causes larger vibration amplitudes, which in turn cause most destruction of buildings, *etc*. Conversely, destructive interference between earthquake waves causes smaller vibration amplitudes, which in turn cause less destruction of buildings, *etc.*!

that shake mostly horizontally. This shows that high-pitch earthquake waves die out faster than low-pitch waves, and that distant earthquakes cause mostly slow (but still violent) oscillations.

The P-waves and S-waves which we have discussed are **body waves**: they travel inside the body of the Earth. There also exist **surface waves** which only travel near the surface of the Earth. One example is the **Rayleigh waves**: they are similar to the surface waves on water (see Chapter 11). They have motion both horizontally back-and-forth along the surface and vertically up-and-down, and include shearing, forming roughly circular waves that rotate in the opposite sense compared to water surface waves (water surface waves rotate "forward" in the direction of the wave motion, while Rayleigh waves rotate "backward" against the direction of the wave motion). Another surface wave is the **Love wave**: it is similar to the Rayleigh wave, but the vertical up-and-down motion is now horizontal side-to-side. **Stoneley waves** exist at the boundary between two substances. Examples are solid-solid or solid-liquid boundaries between the "onion peels" of the Earth.

9.5 What have We Learned in this Chapter?

In Chapter 3, we saw how sound travels as waves through air, giving rise to music (Chapter 5) and voices (Chapter 6). In this Chapter 9, we have seen that sound does not only travel through air: it can also go through liquids (such as water) and solids (such as rocks). The speed of such sound is often many times greater than in air; this speed depends strongly on the atomic and molecular structure of the liquids and solids, and especially on the hardness or softness of the substance.

Sound in liquids and solids can be pressure waves, as in air: a wave of compression and decompression carries the sound. In solids, shear waves are also possible, in which the solid substance is distorted but returns to its original shape by elasticity. Earthquakes produce both pressure waves and shear waves, which can travel large distances through the Earth; they even allow us to learn about the structure of the Earth's interior.

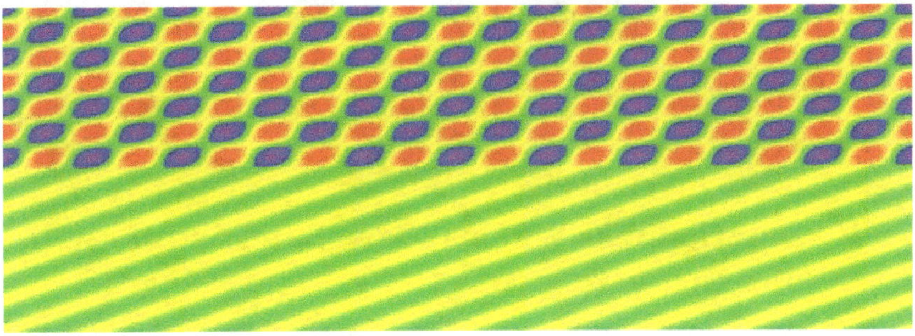

10

Reflection, Transmission and Refraction of Waves in 1D, 2D and 3D

Waves can be reflected, transmitted and refracted ("bent") when they reach the boundary between different substances. Such universal behavior of waves is particularly important in waves on the surface of water (further discussed in Chapter 11*), as well as light waves (see* Chapter 12*). Common examples are string waves of a musical instrument when they bounce from the end of a string; sound waves in air or liquids or solids when they enter another substance like a wall or a warmer or colder layer of air; surface waves on water approaching a beach; and light waves when they enter and exit water or glass.*

This chapter will describe how waves in general behave at such boundaries between substances, starting with simple strings in 1D. In 2D and 3D (for example, with beach waves and light), waves change their direction of travel (which is called refraction). We will see cases where total reflection (and thus zero transmission) is possible: this is used in optical fibers. Using two or more boundaries, we can also make lenses that focus light.

10.1 What Happens When a Wave Enters a Different Substance?

This situation is common with many kinds of waves, such as sound, water surface waves, light and quantum waves. Here are some familiar examples:

- When a sound wave in air hits a door, window or wall, the wave is mostly reflected into the air, but also partly transmitted into the harder material. The sound can also emerge again into the air behind the obstacle. We will discuss in this chapter how such **transmission** happens.
- Light can enter glass, water and other transparent substances in addition to being reflected. We know that light beams can be bent into other directions as they enter such materials, an effect that is called **refraction**. We see this frequently when straight objects like sticks, straws or pens seem bent when partly submerged in water, as in Figure 10-1. This bending also forms the basis for many optical devices, especially corrective eyeglasses, as we will mainly describe in Chapter 12, as well as camera lenses, microscopes and telescopes.
- In addition, different colors of light can be bent into different directions by the same prism or raindrop, giving a color

Figure 10-1: A straight pen is partially submerged in a cup of water, seen from above.

spectrum, such as the spectrum of sunlight and the rainbow. This situation will also be further discussed in Chapter 12.

- Light can be trapped inside a transparent substance by total internal reflection: an example is the optical fiber, in which light can travel long distances and carry information such as telephone calls and internet data without escaping the fiber (see Section 12.6). Similarly, sound can be trapped and travel far inside a long corridor. This Chapter 10 will address those cases.
- You may have heard of mirages and fata morganas: these are amusing visual "reflections" of landscapes due to the bending of light by hot or cold layers of air, as if huge horizontal mirrors were hanging in the air. We will explain them as "gradual refraction" in Chapter 12.
- You may have noticed that the Sun looks flattened into an ellipse when rising or setting close to the horizon; less well known is that the Sun appears to set below the horizon a few minutes later than it should (and even later toward the poles), while rising "too early" in the morning; these effects are closely related to mirages and will be discussed together (see Chapter 12).
- You must have noticed that waves on water usually hit a beach almost head-on, even when the waves traveled in another direction offshore (normally in the direction of the wind): these waves have turned to face the beach. We will look at this intriguing behavior in Chapter 11.

Less familiar is the situation of a wave traveling along a string which at one point suddenly becomes thinner (lighter) or thicker (heavier). For example, imagine a guitar string, one half of which is thicker than the other half: what happens to a wave that hits the point where the string thickens?

We will first consider this one-dimensional case of waves on a string because it is simpler to understand and because it will conveniently build up to the cases of 2D and 3D waves in air, liquids, solids, *etc.*

10.2 Reflection and Transmission in 1D (String)

In Section 2.7, we discussed the reflection of waves from the fixed or free end of a string. In particular, we showed how the wave that is reflected from the end of a string is superposed on the incoming wave: by interference, this creates a standing wave with nodes of no motion and antinodes of maximum motion.

We now replace the end of the string with a continuing string with greater thickness (and thus greater weight or mass), shown in red at the right in Figure 10-2. When the incoming wave (blue) hits the thicker string, only a part of the wave is reflected (green) while the rest is transmitted (red): now the end of the string is a "leaky mirror". The transmitted wave continues in the same direction but is weakened; it has a shorter wavelength because it travels more slowly along the denser part of the string while oscillating at the same frequency (that frequency is given by the original source of the wave and cannot change). The reflected wave is also weaker since the amplitude of the incoming wave is shared between the reflected and transmitted waves.

The reflected wave interferes with the incoming wave, as shown in brown in Figure 10-2. However, the incoming and reflected waves have unequal amplitudes, so they don't form exactly a standing wave like those in Figure 1-5. Indeed, the combined "incoming + reflected waves" (brown) in Figure 10-2 will look more like the incoming wave (blue) than a standing wave when the reflection is weak: now there are no clear nodes or antinodes.

We can vary the thickness (weight or mass) of the continued part of the string: this will change the speed and therefore the wavelength of the transmitted wave; it will also change the reflectivity at the boundary point of the continuing string, and therefore also the transmissivity. The reflectivity is simply the weakening of the reflected wave: thus, on a string, a reflectivity of 20% implies a reflected amplitude that is 20% that of the incoming wave. Transmissivity is the remainder: 100% – 20% = 80% transmissivity gives a transmitted amplitude that is 80% that of the incoming wave. (Reflectivity and transmissivity may be defined somewhat differently for other kinds of waves, for example in terms of energy instead of amplitude.) Figure 10-3 shows the effect of such changes in a condensed form: part (d) condenses all of Figure 10-2, while other parts (b, c, e, f) are the same with different continuing strings. It may be helpful to view this behavior in my Animation 10*1.

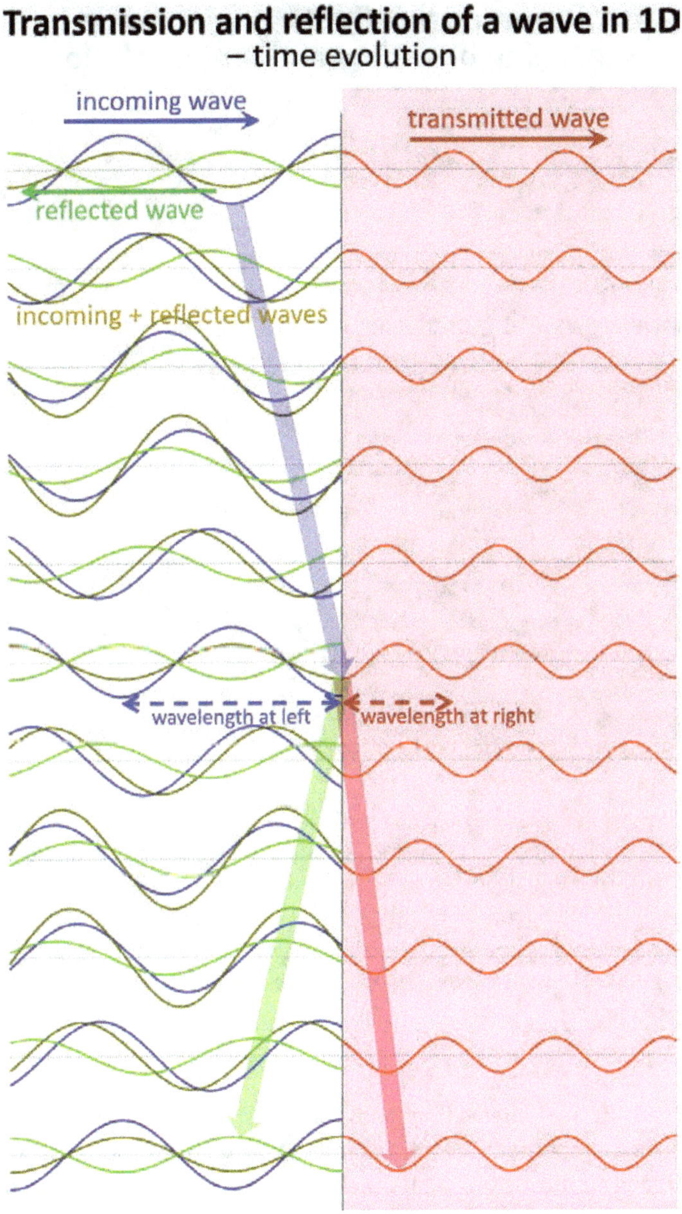

Figure 10-2: Simulation of an incoming wave (blue) on a string arriving from the left and hitting a point where the string becomes thicker, shown in red. At that point, a partial reflection takes place, causing a weaker left-traveling reflected wave (green): it is superposed on the incoming wave to form the brown wave. The rest of the incoming wave continues to the right (red), but with a reduced wavelength (see the dashed double-headed arrows). The snapshots show the evolution of the waves over time from top to bottom, with large arrows illustrating the motion of the separate waves.

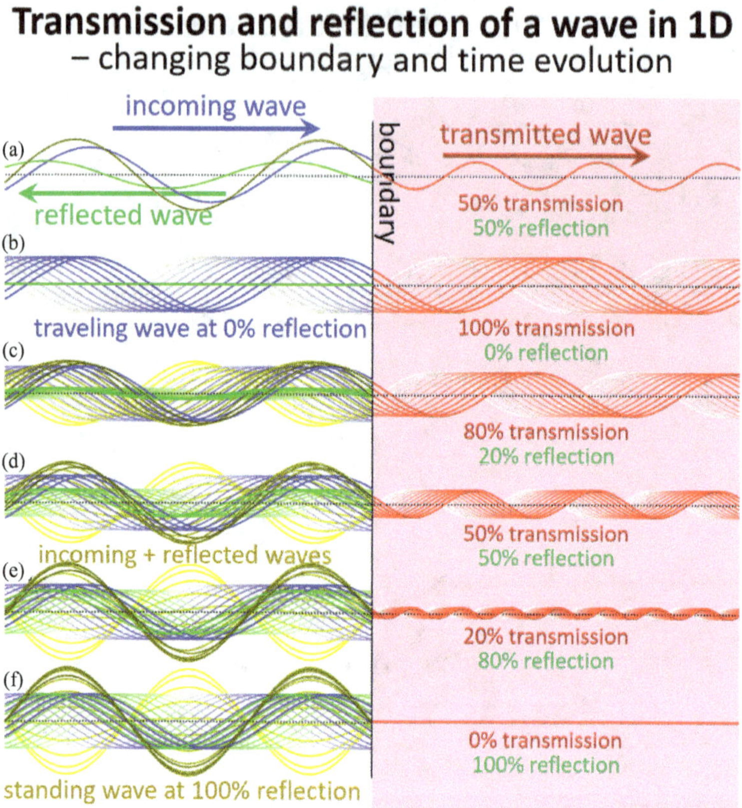

Figure 10-3: As Figure 10-2 for various continuing strings (in red at the right). Here, snapshots are drawn with fading colors to indicate the evolution of the waves with time. In these examples, as the transmissivity drops from graph b to graph f, the wavelength is also shortened. Graph a explains the colors as in Figure 10-2, using one snapshot of graph d; the combined "incoming + reflected waves" are shown in brown-to-yellow colors.

ANIMATION 10*1 — Reflection and transmission of a wave on a string with changing properties: See my video WB1 at time 2:23 in its section "**1D reflection and transmission**" under the title "**A 1D wave can be reflected and transmitted**". (See details in the section References and Resources below.)

If the continuing string is identical to the initial string (as shown in Figure 10-3b), there is no reflection (0%) and total transmission (100%), while the wavelength is unchanged: we have a simple traveling wave continuing undisturbed across a non-existent boundary. In Figure 10-3c, the continuing string is changed so that there is 20% reflection and 80% transmission, with a shortened wavelength. Figures 10-3d, 10-3e and 10-3f continue this trend with 50%, 80% and 100% reflections. The last case has 0% transmission; it is the same as in Figure 1-5, with a perfect standing wave at left (seen in yellow/brown) and no transmitted wave at right: now the continuing string has become so heavy and rigid that it fixes the end of the string, causing total reflection. So, from Figures 10-3b to 10-3f, we see a gradual transition from a simple traveling wave to a perfect standing wave.

10.3 Extension to 2D and 3D

We can focus our attention on 2D cases since 3D cases are essentially the same. Also, 2D cases are easier to picture, both graphically and mentally. An online video is helpful to visualize how waves are reflected and refracted.[1]

We already discussed in Section 3.8 how sound waves in air are reflected by a mirror or wall: see Figure 3-10 and following figures. See also my Animation 10*2.

ANIMATION 10*2 — See my video WB1 at time 4:59 in its section "**2D & 3D reflection and refraction**" under the title "**Total reflection of a 2D or 3D wave by a wall or mirror: motion of pattern**". (See details in the section References and Resources below.)

As above in 1D, we now consider a "leaky" mirror or wall in 2D, such that the incoming wave is partially transmitted and only partially reflected. (We will mainly mention walls from here on; these walls could equally well be mirrors, windows or other hard substances.)

[1] See video "Experiments with water waves" by TIB - Leibniz Information Centre for Science and Technology University Library, at https://av.tib.eu/media/30440. This video shows waves on the surface of water: their reflection and refraction are very similar for other kinds of waves, including sound waves and light waves (see Chapter 11).

Figure 10-4 shows the effect of a "leaky" wall for comparison with the "non-leaky" wall (or mirror) in Figure 3-12. We immediately recognize the checkerboard or egg crate pattern due to the interference of incoming and reflected waves. It may be helpful to view this behavior in my Animation 10*3.

ANIMATION 10*3 — See my video WB1 at time 7:03 in its section "**2D & 3D reflection and refraction**" under the title "**A perspective view comparing partial refraction *vs.* total reflection**". (See details in the section References and Resources below.)

However, inside the wall (toward the bottom right in both parts of Figure 10-4), we see a transmitted wave that travels in a different direction from the incoming wave: it is deflected; it is also weaker. **This change in travel direction is called refraction. The cause of the bending or refraction is the change in wavelength and speed of the wave inside the wall.** Refraction is the reason for many effects, which we will see later: splitting of light into colors, rainbows, mirages, *etc.*

As with the 1D case of Section 10.2, a change in wavelength and wave speed also causes a change in reflectivity and transmissivity: the transmitted wave is weaker than the incoming wave, and so is the reflected wave. The incoming wave shares its amplitude among both the reflected and transmitted waves.

The bottom image in Figure 10-4 is a very interesting and useful special case: it causes **total internal reflection**. In this situation, which we will explain further below, the incoming wave cannot be transmitted through the boundary. Then only weak tails of the wave crests stick into the wall, but they stay close to the wall boundary: these are dying or decaying waves, called **evanescent waves** in physics. This effect is used, for example, in optical fibers to keep the signal from leaking out (see Section 12.6).

Further illustrations of reflection and refraction are given in Figures 10-5 and 10-6. These include the cases shown in Figure 10-4, but now looking "straight down" on maps of the waves. The incoming wave comes from the upper left, as indicated by the white arrows. Wavelengths in

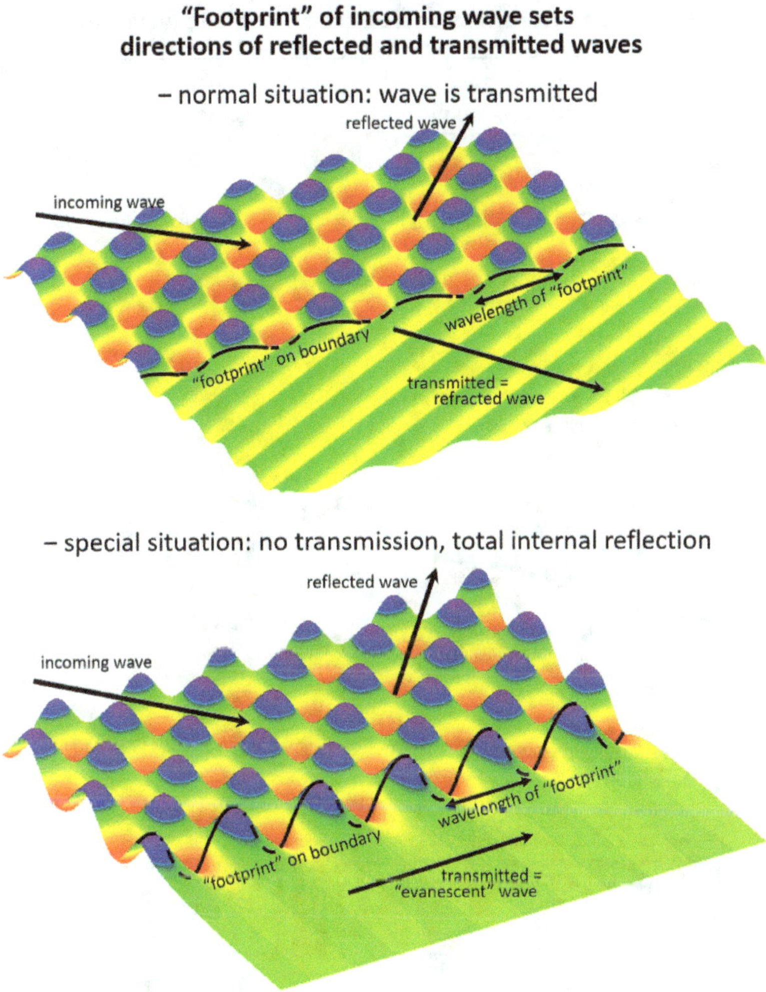

Figure 10-4: Simulation of an incoming wave that hits a "leaky" wall, causing a reflected wave and a transmitted wave. *Top*: typical situation, where the transmitted wave fits the footprint of the incident wave by selecting a particular direction of travel. *Bottom*: special situation, where the wavelength in the wall cannot fit the wavelength of the footprint, resulting in a wave that travels parallel to the wall but decays away from the wall. (For comparison, the bottom image of Figure 3-12 shows the same situation with a "non-leaky" wall.) The incident and reflected waves have the same wavelength, being in the same substance (such as air). In the top image, the transmitted wave has a shorter wavelength (see black arrows) due to a slower speed in the wall; the reverse happens in the bottom image.

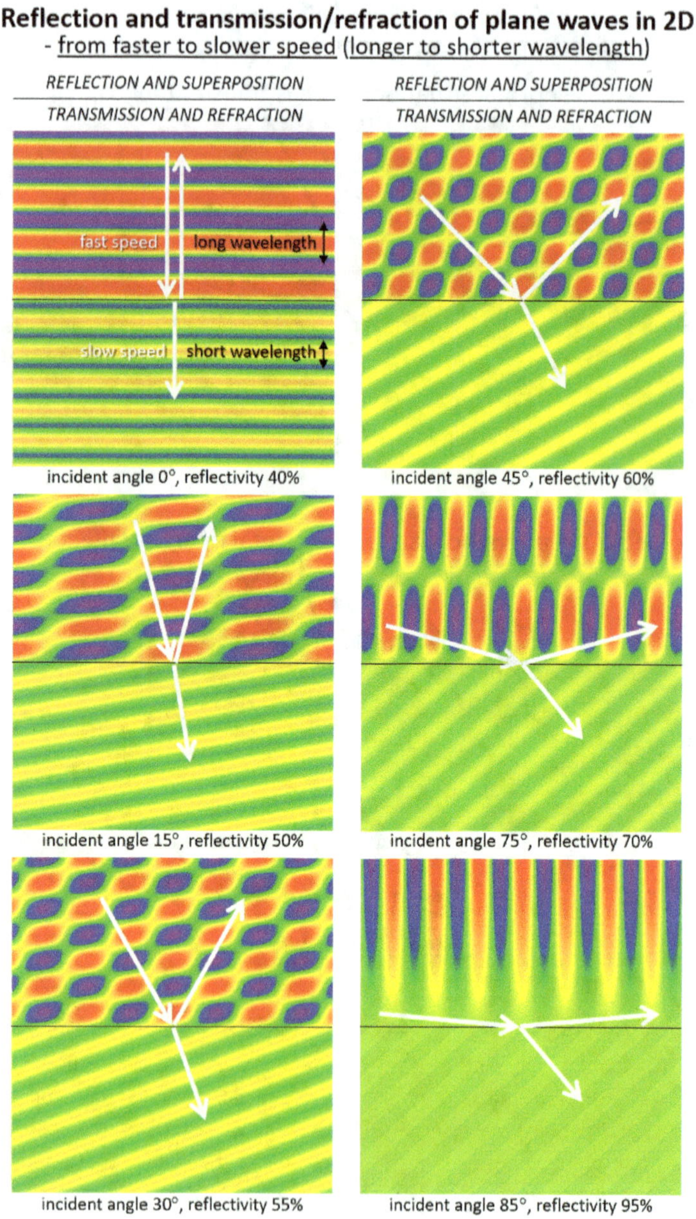

Figure 10-5: As Figure 10-4, but looking "straight down" on several examples of incoming waves hitting a "leaky wall" (shown as a horizontal black line in these views). The incident and reflected waves have the same wavelength across all six frames: only the direction of incidence changes. The transmitted wave has a shorter wavelength (see black arrows), the same across all six frames, due to a slower speed (see white arrows). The reflectivity used here rises with the incident angle (measured from the vertical direction in these views), as is normal with most kinds of waves in nature.

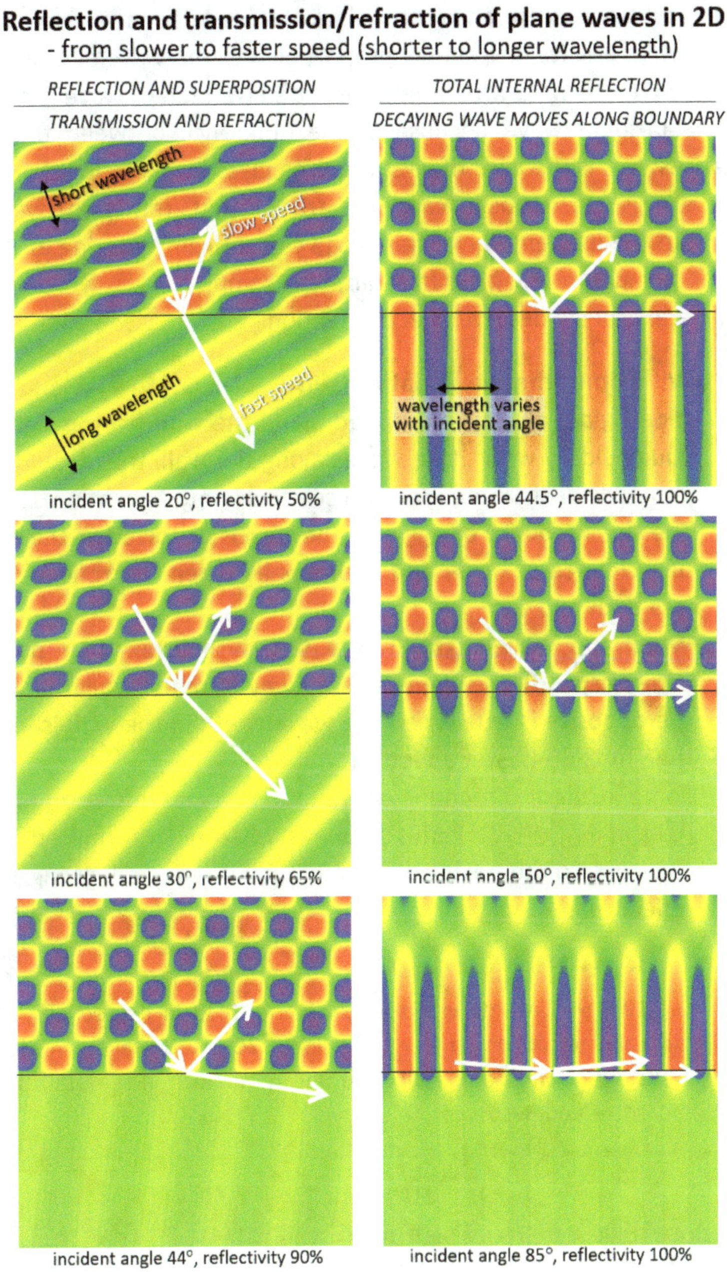

Figure 10-6: As Figure 10-5, but with a longer wavelength (faster speed) in the wall material. In the three left images, refraction takes place, but the bending turns in the opposite direction compared to Figure 10-5 (with a shorter wavelength and slower speed). In the three right images, transmission is impossible, causing dying "evanescent" waves into the wall.

the two substances are shown as black double-headed arrows; wave speeds (white arrows) are simply proportional to the wavelengths.

Figure 10-5 is valid when the transmitted (refracted) wave is slower than the incident wave because of the difference between the two substances: this is the case of light entering glass from air, for example. Figure 10-6 is the opposite case: the transmitted (refracted) wave is faster than the incident wave: this is the case of light entering air from glass, or sound entering glass from air, for example.

Several aspects deserve special mention:

- **With time, the wave patterns in** Figures 10-4, 10-5 and 10-6 **slide rigidly at constant speed along the wall (to the right)**: if you could fly with these waves along the wall, they would look immobile as if frozen. You can see this in my Animation 10*4.
- **The checkerboard or egg crate pattern is distorted as the angle of incidence changes and additionally when the reflectivity is less than 100%.**
- **When the wavelength in the wall is shorter than outside (slower speed), the wave is bent away from the wall boundary; the opposite is true when the wavelength in the wall is longer than outside (faster speed)** (Figure 10-6).
- **Some incident directions cause total internal reflection** (at the right in Figure 10-6): this happens when no transmitted wave can exist, so the full incident amplitude is reflected, giving 100% reflectivity.

ANIMATION 10*4 — See my video WB1 at time 5:30 in its section "**2D & 3D reflection and refraction**" under the title "**A 2D or 3D wave can also be partially transmitted and bent: reflection and refraction**". (See details in the section References and Resources below.)

10.4 Predicting the Travel Direction of Refracted (Transmitted) and Reflected Waves

In Section 3.8, we used the idea of a "footprint" to explain the direction taken by a reflected wave. The result was that the angle of reflection is identical to the angle of incidence (both angles being measured from the perpendicular to the wall). The reasoning was that the footprint of the incoming wave on the wall serves as the source for the reflected wave: therefore, the reflected wave must match that footprint; this requirement sets the direction of travel of the reflected wave.

We can predict the direction of travel of transmitted waves in the same way, by using the footprint of the incoming wave on the wall as the source of the transmitted wave. Again, it is just a matter of rotating the direction of the departing wave until it correctly fits the footprint. If you are interested in how that works, please read Box 10-1. This will in particular explain why we get those somewhat mysterious dying "evanescent" waves that cause **total internal reflection**!

Do you know how to shoot through an optical boundary? You probably know that some **fish** can shoot down low-flying insects by spitting drops of water at them. The waterdrops go nearly straight out of the water, but the fish's view of insects is bent by refraction. This situation corresponds to Figure 10-5, with the fish below the boundary and the insect above the boundary. The fish sees the insect higher up than it really is, so it must shoot the drop below the image of the insect (this is counterintuitive for us since gravity will pull the drop downward, but a fish has less experience with gravity than we do!). The reverse situation is when you shoot a bullet from the air at a fish in the water. You see the fish higher than it really is, so you should aim the bullet lower than the fish's image. Birds plunging to catch fish face the same challenge: they should aim below the image of the fish. (We will also discuss looking into and out of water in Sections 12.3 and 12.4).

How do you or the animals "calculate" the correct direction to shoot? You don't: you do it by trying, missing and improving the aim again and again, as do birds. Likewise, the fish aims its shooting by trial and error. Remember that the angle of refraction depends on the incident direction, so the correction for the aiming direction varies with the light's incident angle. Additionally, gravity and friction will further bend the direction of the drop (in the air for a more distant insect) and of the bullet (in the water for a deeper fish). This is too complex for us and for fish to work out, except by experience.

BOX 10-1 — PREDICTING THE DIRECTION OF TRAVEL OF TRANSMITTED WAVES. Let us first consider more carefully the case of the reflected wave that we described in Section 3.8, Figure 3-12. In this case, we have total (100%) reflection and zero (0%) transmission.

Figure 10-7 simplifies Figure 3-12 by only plotting the wavefronts as viewed parallel to the boundary. The incoming wave is shown in blue, and the reflected wave is shown in green. They have the same wavelength since they travel in the same substance (for example, air). The black line is the footprint of the incoming wave; black dots are on the incoming wavefronts, separated by the footprint's wavelength.

The incoming wave's footprint has two primary properties:

1) its wavelength along the wall (the footprint's wavelength is not the same as the incoming wave's wavelength but is its projection onto the wall, which projection is usually larger); and
2) its direction of motion along the wall (to the right in both figures).

These two properties are sufficient for us to predict the directions of both the reflected and transmitted waves, as follows:

At the right in Figure 10-7 are shown pairs of reflected wavefronts turned to different directions, as trials to fit the footprint. Let's start with the full-blue pair, which is parallel to the incoming wavefronts. One of its two wavefronts must pass through a dot in the footprint marked "pivot". The second blue wavefront then passes through the next dot in the footprint, marked "target". We thus have a good fit for the footprint. However, any wave travels perpendicular to its wavefronts. So, if we let it move to the bottom right (see the rightmost blue arrow), this trial wave moves into the wall and therefore is not a reflected wave; and, if we let it move in the opposite direction, it moves against the footprint's motion (to the left instead of to the right), contrary to our requirement. Thus, this trial wave cannot be the reflected wave.

Now, let's turn the trial wave a bit to the right (clockwise), shown as long-dashed blue wavefronts, while keeping one wavefront at the pivot. Notice how the other wavefront now misses the target (it now passes through the rightmost dashed oval): this does not fit the footprint, which is not acceptable.

Next, we turn the trial wave to the left (anticlockwise): see the blue-green dashed and dotted green trial waves pivot around the pivot dot. Only when we reach the direction of the solid-green lines do we hit the target dot again. With a wave direction to the upper right, we also fit the direction of motion of the footprint: we now have a proper reflected wave, shown as solid-green wavefronts at left. Since the solid-blue and solid-green waves have the same wavelength, the fit to the pivot and target forces them to have the same angle relative to the wall: thus, the angle of reflection is equal to the angle of incidence, as we already saw in Section 3.8.

How about the refracted (transmitted) wave? We can now apply exactly the same reasoning to predict the direction of the refracted wave. This situation is shown in Figure 10-8, which is exactly the same as Figure 10-7 but with refracted waves (red) included in the wall substance behind the "now-leaky" wall boundary. We do not need to specify how "leaky" the wall boundary is: we are only concerned here with the direction of the waves, rather than their amplitudes.

Importantly, we have the same footprint as before because this footprint is given by the direction and wavelength of the incoming wave, which have not changed.

We do need to know the wavelength in the wall substance (more precisely, we need to know the ratio of the wavelengths in the two substances, which is also the ratio of the speeds of waves in the two substances[2]). In Figure 10-8, the refracted wavelength is shorter than the wavelength of the incoming wave (a longer wavelength is treated separately below).

We can now draw a pair of wavefronts (red) in the wall substance, separated by the shorter wavelength. We can try to match these wavefronts to the pivot and target dots of the footprint by rotating the transmitted wave until they fit; remember that this footprint is essentially the source of the refracted (transmitted) wave. This is done at the bottom right in Figure 10-8: only the solid-red wavefronts match the pivot and target points, we have thus found the direction of the refracted wave, at least graphically (this can also be done numerically and exactly with simple trigonometry if the ratio of the wavelengths is given: see Section 10.5).

Related examples are shown in Figure 10-5. There, the angle of incidence is changed from perpendicular to the wall (at the top left) to nearly parallel to the wall (at the bottom right), thereby forcing a change in the angle of reflection and the angle of refraction. In that figure, the wavelengths are kept the same in all six images.

(Continued)

---

[2] As we will see in Section 10.5, the ratio of wave speeds between two substances is called "relative refractive index" or "relative index of refraction". If one substance is vacuum, one speaks of "absolute refractive index" or simply "refractive index". Air is very similar to vacuum for light waves. However, sound does not travel in vacuum, so that there is no "absolute refractive index" for sound.

(*Continued*)

Reflection and transmission/refraction of plane waves in 2D

A. Total reflection (no transmission) from wall/mirror

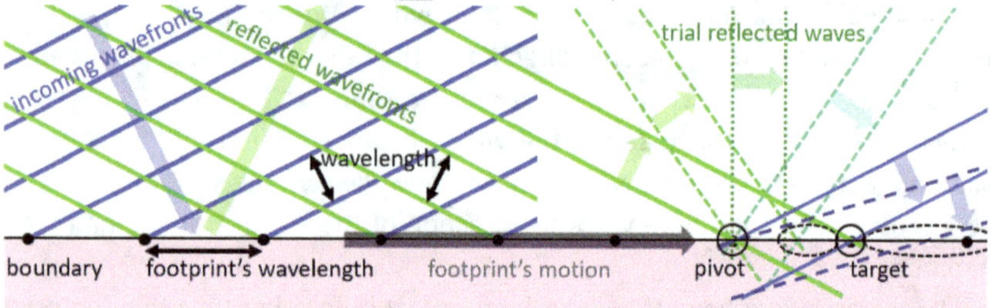

Figure 10-7: An incoming wave (shown as blue wavefronts) arrives from the top left, approaching a totally reflecting wall (red background). The incoming wave makes a footprint (horizontal black line) on the wall; dots indicate the wavefronts on the footprint, which dots move to the right with the incoming wave. Trial reflected wavefronts are shown at right, with different directions. Two successive wavefronts should pass through two successive dots of the footprint, such as the "pivot" and "target" dots, for a proper fit. Only the light-green wave fits the footprint's wavelength and direction of motion.

Reflection and transmission/refraction of plane waves in 2D

B. Reflection and transmission/refraction at wall/mirror

– from longer to shorter wavelength

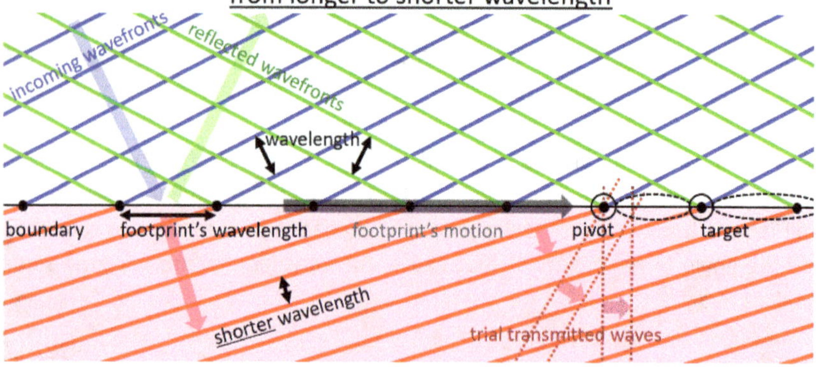

Figure 10-8: As Figure 10-7, but adding refracted (transmitted) waves (red) in the wall substance. The wavelength and direction of the incoming wave are the same as in Figure 10-7. The wavelength of the transmitted wave is shorter than that of the incoming wave (see black double-headed arrows).

In Figure 10-9, we see the case where the wavelength in the wall substance is longer than in the incoming wave. We can repeat the same footprint-matching procedure once more for this case.

At first sight, this appears to be a simple repetition. However, we discover one special difference in behavior. Compare the direction of bending of the refracted wave from the incoming wave: in Figures 10-5 and 10-8, the refraction bends the incoming wave more perpendicularly to the wall (clockwise in this example), while the opposite is true in Figures 10-6 and 10-9 (anticlockwise): more parallel to the wall.

Related examples are shown at the left in Figure 10-6, where again only the angle of incidence is changed. Notice at the bottom left in Figure 10-6 how the refracted wave moves almost parallel to the wall. An increase by half a degree in the angle of incidence (from 44° to 44.5°) leads to the upper right case in Figure 10-6: a dramatic change! We will see what is going on next.

Also surprising is what happens if we make the wavelength in the wall substance even longer: at some point we will get the situation of Figure 10-10. Now there is no way to fit the footprint: no direction of the refracted wave fits both the pivot and target dots. This means that there is no way for the given incoming wave to cause a refracted wave, so there is no transmission at all. One result is that the amplitude of the incoming wave goes entirely to the (green) reflected wave: we get 100% reflectivity and therefore call it **total internal reflection**. This is exactly what happened in Figure 10-6 in the right-hand images.

We can still ask: ***In this case, is there any wave at all inside the wall?*** In fact, there is a special kind of wave right behind the wall boundary. It quickly dies out away from the boundary: it decays exponentially (like a bank account with a constant negative interest rate). It is therefore called an **evanescent wave** in physics; we can also call it a dying wave. Three examples are given at right in Figure 10-6 and another at bottom in Figure 10-4: depending on the incident angle, the decay can be slower (top right image in Figure 10-6) or faster (bottom right image in Figure 10-6). A slow decay exists only in a very small angle range around the so-called **critical angle**.

In addition, the evanescent wave has a wavelength parallel to the boundary: it is simply the wavelength of the footprint rather than the natural wavelength of the wall substance. This is clearly visible in Figures 10-4, 10-6 and 10-10.

(Continued)

(*Continued*)

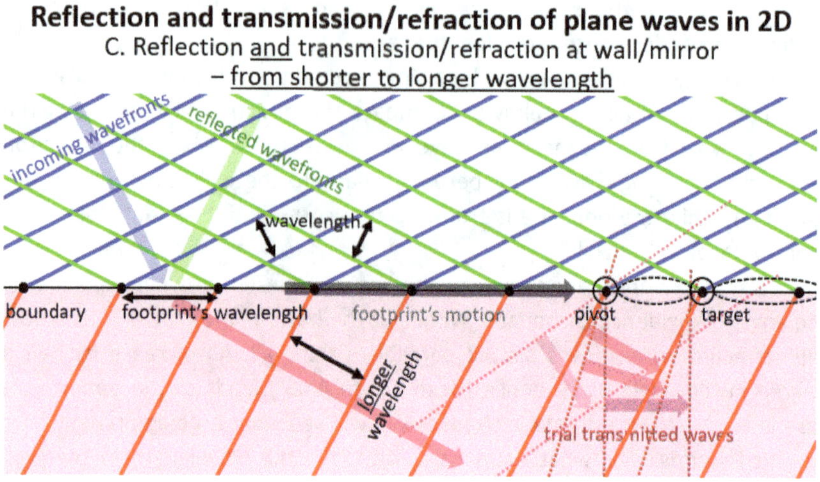

Figure 10-9: As Figure 10-8, but with a longer wavelength in the wall substance than for the incoming wave.

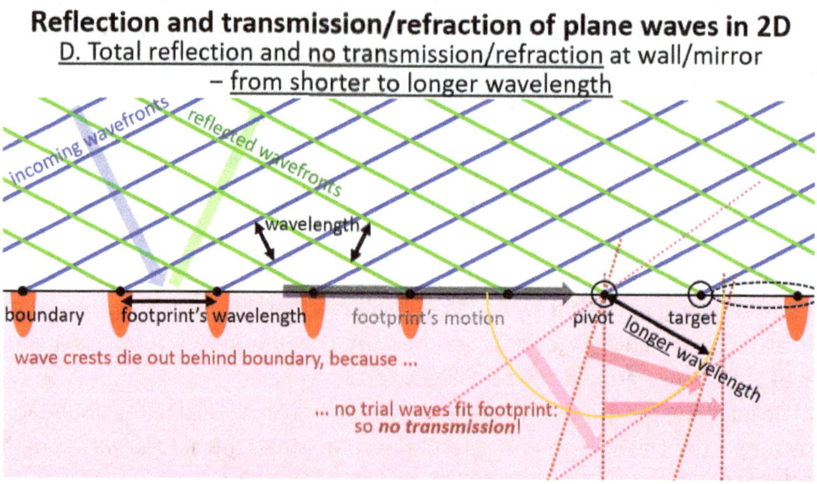

Figure 10-10: As Figure 10-9, but with an even longer wavelength in the wall substance. If the wavelength in the wall substance becomes too long, no fit to the footprint can be found, and no transmission is possible. An "evanescent wave" then exists, marked here by red wave crests attached to the dots (some are omitted for clarity): it dies out away from the boundary while traveling along the boundary.

10.5 Snell's Law of Refraction

We can summarize the discussion of Section 10.4 in Figure 10-11. It will lead us directly to a famous law of physics: **Snell's law** (named after the Dutch astronomer and mathematician **Willebrord Snellius**). Figure 10-11 also displays a simple graphical method to construct the directions of the reflected and transmitted (refracted) waves at the boundary between two substances.

All the steps shown in Figure 10-11 can be condensed into the single formula known as **Snell's law: this law connects the directions of the incoming and refracted waves**. Snell's law can be written in slightly different ways, using the sine function "sin",[3] for instance:

(incoming wavelength) × sin (refraction angle)
= (refracted wavelength) × sin (incident angle)

Equivalently:

sin (refraction angle) / sin (incident angle)
= (refracted wavelength) / (incoming wavelength)

The wavelengths are proportional to the wave speeds in the two different substances, so we can also write:

sin (refraction angle) / sin (incident angle)
= (refracted wave speed) / (incoming wave speed)
≡ (relative refractive index or relative index of refraction)

The last line introduces the **relative refractive index**, which is defined as the ratio between the wave speeds in the two substances.

Snell's law tells us the angle of refraction (which gives the direction of the transmitted wave) if we know the two wavelengths or wave speeds in the two substances (or at least their ratio) and the incident angle. It is one of the simplest powerful laws in physics.

Snell's law results from comparing two triangles in Figure 10-11: the triangles PTU and PTV have a common and equal side PT (which is

---

[3] You will not need to calculate the sine function in this book: it is only shown here as an example of how mathematics allows accurate predictions.

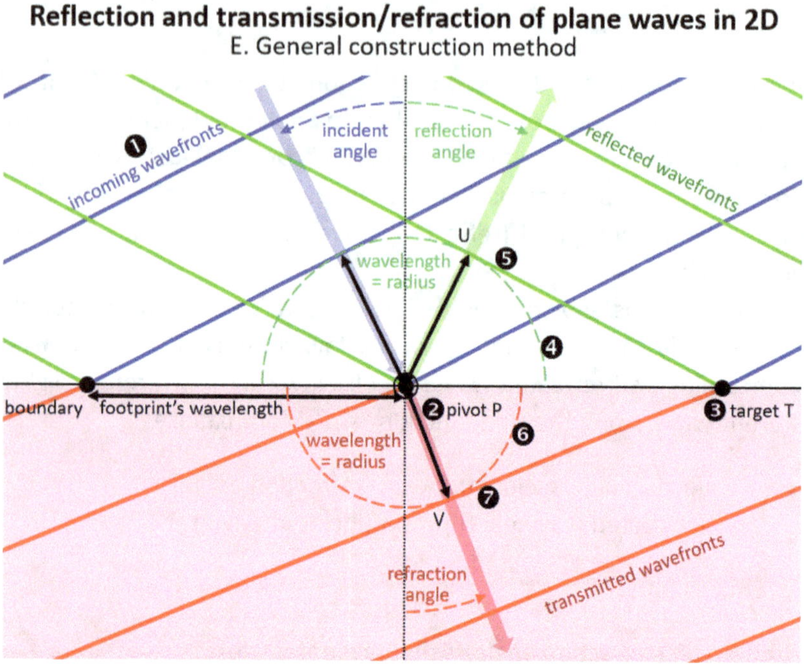

Figure 10-11: Construction method of reflected and refracted wave directions at a boundary. The numbered steps are:

❶ Draw the (blue) incoming wavefronts (they may represent the crests of the wave), spaced by the wavelength of the incoming wave.

❷ Place the pivot point P where an incoming wavefront crosses the boundary between the two substances.

❸ Place the target point T where the next incoming wavefront crosses the footprint.

❹ Draw a (green) half circle centered on the pivot with a radius equal to the wavelength of the incoming wave (the reflected wave will have the same wavelength as the incoming wave).

❺ Draw a (green) tangent from the target point T to the (green) half circle: this is a wavefront of the reflected wave. It has the same inclination relative to the boundary as does the incoming wave, so reflection angle = incident angle.

❻ Draw a (red) half circle in the second substance centered on the pivot, with a radius equal to the wavelength in the second (red) substance; in this example, that wavelength is smaller, so the radius is smaller.

❼ Draw a (red) tangent from the target point T to the (red) half circle: this is a wavefront of the transmitted (refracted) wave. Other (red) refracted wavefronts are separated from each other by the wavelength in the (red) second substance. The refracted wave travels perpendicularly to these wavefronts. This step will fail if the (red) wavelength in the second substance is larger than the footprint's wavelength: then there will be no refracted wave but an evanescent wave, as discussed in Section 10.4.

the footprint's wavelength), while their internal angles at point T are equal to the reflection angle and the refraction angle, respectively. We see therefore that Snell's law results directly from matching the waves' footprints on the boundary.

A word of caution is useful here: it is sometimes not obvious which of the two substances involved in refraction and in Snell's law has the greater wave speed. We saw in Table 9-1 that the speed of sound generally increases from gases to liquids to solids and from soft to stiff substances. However, for light (and other electromagnetic waves), we will see in Chapter 12 that the wave speed decreases in liquids and solids compared to air and vacuum: this is contrary to the case of sound waves and results from the very different mechanisms involved in sound and electromagnetism. What is important and impressive, though, is that a simple relation like Snell's law remains valid for all kinds of waves!

How large is the relative refractive index? For sound, we can look at Table 9-1. Typically, for liquids compared to gases, the ratio of their sound speeds is around 3, so their relative refractive index is also around 3. Comparing solids to gases gives larger ratios, typically from 5 to 40. Comparing solids to liquids gives ratios around 1 to 10. A ratio of 1 gives no refraction.

For light, the relative refractive index is usually referred to vacuum (one of the two substances is simply vacuum): then the word "relative" is dropped and we speak of "refractive index" or "index of refraction". This makes life easier because we don't need to list all pairwise combinations of substances. (Referring to vacuum is not possible with sound, because there is no sound in vacuum; one could refer to air, but "air" then has to be defined precisely.) Vacuum itself thus has a refractive index of exactly 1 for light. In most other substances, light is slower than in vacuum, giving a refractive index larger than 1.

Air under normal conditions has a refractive index of 1.000293 for light: this means that the speed of light is very slightly slower in air than in vacuum, by about 0.03%; consequently, refraction of light is weak between vacuum and air; nonetheless, it can be observed with the naked eye, as we will discuss in Section 12.5! Water has a refractive index of 1.333, causing strong refraction; water ice has a slightly smaller refractive index of 1.31, so it looks almost like water. Nevertheless, you may just be able to see an ice cube floating within liquid water: in Figure 10-12,

Figure 10-12: Ice cubes in water. (*Source*: Cleiton Andrade Andrade, in the public domain, https://commons.wikimedia.org/wiki/File:Ice_cubes_on_water.jpg.)

notice some dark or shiny edges of the ice cubes (however, what you mostly see are the imperfections in ice, such as many tiny bubbles that act like dark clouds). Of course, we can see ice cubes emerging above the water, since ice and air are much more different than ice and water; more precisely: the refractive index between ice and air (~1.31/1 ~ 1.31) is much larger than between water and ice (~1.333/1.31 ~ 1.015).

Typical glass has a refractive index around 1.5: therefore, a glass object is a bit more visible when submerged in water than is ice; see Figure 10-13. Typical transparent plastics also have a refractive index around 1.5.

Figure 10-13: Find the glass teacup submerged in the bowl of water.

A substance with a high refractive index is diamond: 2.417, which means that light in diamond travels at about 40% of its speed in vacuum or air (1 / 2.417 ~ 0.414 ~ 41%). This causes very strong refraction, as well as very strong reflections, at air/diamond boundaries. This in turn makes diamond crystals very flashy and beautiful: **the physical attraction of diamond is due to its slow speed of light! Without its slow speed of light, diamond would look more like normal glass: it would be far less coveted and far less valuable.**

10.6 Double Boundaries

A wave may travel through a wall, door or window and re-emerge on the other side in the same substance (typically air). This wave therefore crosses two boundaries: one in front, the other in back. If, at one boundary, the wave goes from a substance with faster speed to one with slower speed, as in Figure 10-5, then, at the other boundary, the wave goes from a substance with slower speed to one with faster speed, as in Figure 10-6. (It does not matter which boundary comes first and which comes second: the behavior which we will describe below remains much the same.)

Figure 10-14 illustrates such a double boundary, forming a red wall. Let's assume that the substances on both sides of the wall are

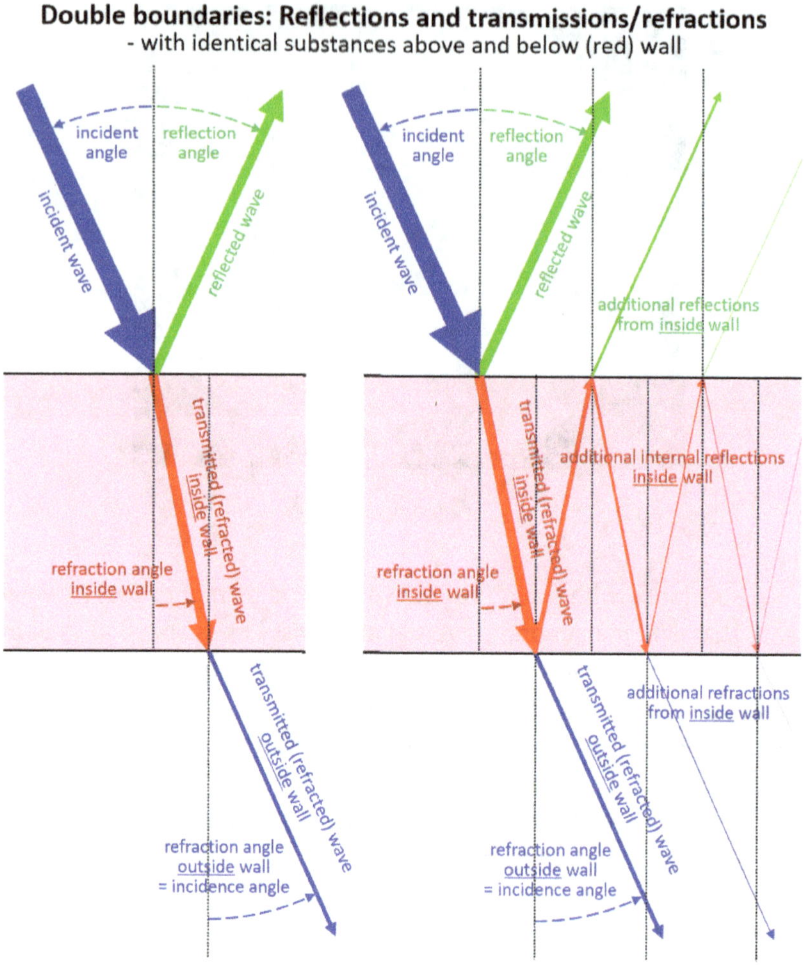

Figure 10-14: Reflection by and transmission/refraction through two opposite boundaries of a wall (red). The substance outside the boundaries is the same on both sides of the wall, for example air outside a door or window. *At left*: no reflection is shown between the boundaries. *At right*: multiple internal reflections are shown between the boundaries. The thickness of the arrows indicates the amplitudes of the different waves: they decrease at each reflection and refraction by 50% in this illustration.

the same, such as air on both sides of a door or glass window. The left part of the figure combines a case from Figure 10-5 at the top with one from Figure 10-6 at the bottom. Figure 10-14 simplifies the illustration by omitting the waves and only showing the wave directions as colored arrows: these arrows are sufficient to understand what is happening. As additional information, the arrow thickness now indicates the wave amplitude, which becomes smaller at each reflection and refraction. It may be helpful to view this behavior in my Animation 10*5.

ANIMATION 10*5 — See my video WB1 at time 7:42 in its section "**Double boundaries**" under the title "**Reflection and refraction by double boundaries: wall — 1**". (See details in the section References and Resources below.)

In Figure 10-14, we see an incident wave (blue) entering from the top left. At the first boundary, it is partly reflected to the top right (green) and partly transmitted (refracted) into the (red) wall. When the red refracted wave hits the second boundary, it again is partly transmitted and emerges behind the wall. Snell's law tells us that this second refraction (emerging from the wall) gives the same angles as the first refraction (entering the wall), because we made the external substance to be the same, so that the external wave speeds are also the same; we only need to interchange the incident and refraction angles. **As a result, the blue wave emerging behind the wall has exactly the same direction as the incoming blue wave.** This is clearly visible in the photograph of Figure 10-15.

We thus see in Figures 10-14 and 10-15 that the incoming wave continues in the same direction after crossing the wall. But it suffers a sideways "jog" or "zigzag" since the red wave in the wall has a direction that is different from the incident direction.

However, if the substance behind the wall is different from that in front of the wall, for example if the wall separates air from water, then the refracted wave would not continue in the same direction! You see this when you look into a glass full of water or a tank full of fish in water: the glass wall grossly distorts the landscape behind it, as seen in Figure 10-16.

Since glass and water give nearly the same speed of light (see Table 12-1), there is little refraction of light between glass and water (similar to the case of ice and water). This is why it is hard to see a glass that is submerged in water: the glass then has little effect on light; see Figure 10-13.

Figure 10-15: Refraction of a ray of light (entering from the left) by a plastic block with parallel sides. (*Source*: ajizai, in the public domain, https://commons.wikimedia.org/wiki/File:Refraction_photo.png.)

Figure 10-16: A straight pen leans in a glass of water, as seen through the glass.

Can an evanescent wave (or "dying" wave) cross a double boundary? This is an intriguing question: if a wave cannot travel inside a wall, can it nevertheless cross through the wall and emerge on the other side? This question seems contradictory! If the wave can't travel in the wall, it should not be able to cross the wall!

However, the key to the answer is the "evanescent" character of that wave inside the wall: it dies out quickly as it penetrates the wall, but not immediately; it has a "tail" that perhaps could reach the other side of the wall and cause a wave to emerge behind the wall by refraction at the second boundary.

Indeed, that is the case, illustrated with two views in Figure 10-17: a wave approaches the wall from the top left and refracts as an evanescent wave at the first boundary (as at the top right in Figure 10-6). But now we have a second boundary: if it is close enough to the first boundary, typically within a few wavelengths of the first boundary, then the tail of the evanescent wave can reach that second boundary since that wave is not completely "dead". Here, a second refraction takes place, creating another wave transmitted behind the second boundary. This transmitted wave is certainly weaker than the original incoming wave, but otherwise it has exactly the same wavelength and exactly the

Double boundaries: Transmission of evanescent wave through wall

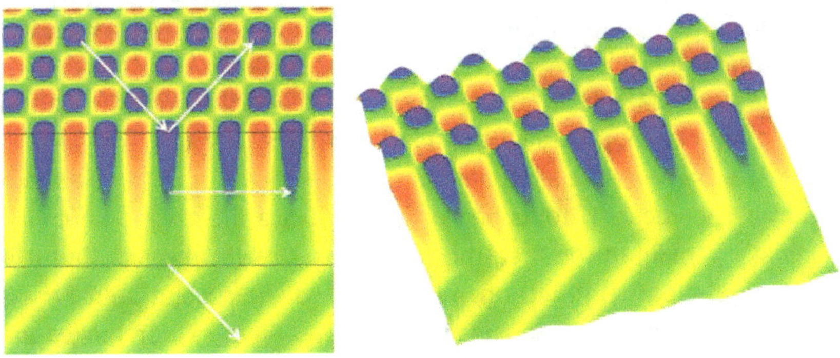

Figure 10-17: "Evanescent wave coupling" by transmission of a "dying" (evanescent) wave through a wall with two boundaries (black lines), shown from two viewing directions. The incoming wave approaches from the top left, as shown by a white arrow. A transmitted wave emerges which travels with the same wavelength and direction to the bottom right.

same direction (toward the bottom right). It may be helpful to view this behavior in my Animation 10*6.

ANIMATION 10*6 — See my video WB1 at time 8:09 in its section "**Double boundaries**" under the title "**Reflection and refraction by double boundaries: wall — 2**". (See details in the section References and Resources below.)

The transmission of an evanescent wave through a wall is called **evanescent wave coupling**. It can be used to transmit part of a wave that travels on one side of the wall to the other side, such as from one optical fiber to another, without disrupting either fiber: all we have to do is place the two fibers very close together; this is done in fiber-optic splitters, for example, as well as in prism couplers. The idea also allows powering devices wirelessly, for example to power smartphones, shavers, *etc.* without direct electrical contact. We will also encounter this idea with quantum waves, where it is usually called **tunneling** or **quantum tunneling**: electrons, because of their quantum wave character, can cross walls that they should not be able to cross as particles (see Section 13.9).

What happens with non-parallel boundaries? This is a very useful situation where a wall actually bends a wave using the same external substance on both sides (such as air on both sides of a glass wall). This effect is central to prisms and lenses, including corrective eyeglasses, microscopes and telescopes.

The principle is illustrated in Figure 10-18, which represents a prism defined by two non-parallel boundaries. At the left in Figure 10-18, we start with two parallel boundaries. We let a wave hit the wall (the prism) perpendicularly to its first boundary. Since the second boundary is initially parallel to the first, the wave goes straight through (but it is weakened by partial reflections). However, **as the second boundary is tilted** (middle and right sketches in Figure 10-18), refraction **makes the transmitted wave rotate** as well.

We can ask: ***Does the rotation of the outgoing wave go in the same direction as the tilt of the second boundary?*** The answer depends on the substance. If the substance of the wall has a slower wave speed than the substance outside, as in Figure 10-6, we get the full-blue

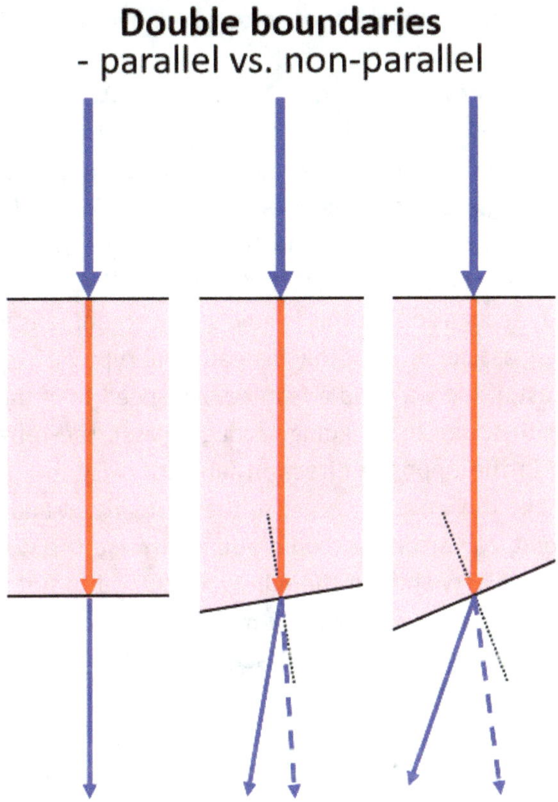

Figure 10-18: Refraction by a wall whose two boundaries may be non-parallel. *From left to right*: the second boundary is increasingly tilted. The substance outside the boundaries is the same on both sides of the wall (as with air on both sides of glass). If the wall substance has a slower wave speed than outside the wall, we get the full-blue transmitted (refracted) waves seen below the wall. But in the opposite case (faster speed in the wall than outside), we get the dashed-blue transmitted (refracted) waves. The dotted black lines are perpendicular to the second boundary. (Only transmission is shown, omitting reflections.)

transmitted (refracted) waves seen below the wall in Figure 10-18: the refracted wave rotates opposite the tilting of the second boundary; this is the case for light passing through a glass wall in air.

But if the wall substance has a faster wave speed than the substance outside, as in Figure 10-5, we get the dashed-blue transmitted (refracted) waves in Figure 10-18: the refracted wave rotates in the opposite direction; this would happen to a sound wave passing through a glass wall surrounded by air.

Can you see how Figure 10-18 leads to lenses? Try to imagine where the refracted waves will go. The full-blue refracted waves seem to point to a single place somewhere below the figure: these waves may converge in a single point. Imagine adding more pieces of wall to the right, with larger tilts of the second boundary: by careful positioning, the pieces of wall could send refracted waves to the same convergent point. You may recognize that this is now a convergent lens, especially if you make the glass curve smoothly instead of abruptly!

By contrast, if the wall had a faster wave speed than the substance outside (as with sound in air going through glass), the refracted wave would rotate in the opposite direction (anticlockwise in Figure 10-18). This would make the waves diverge, so they spread out and don't meet at a single point. Nevertheless, ***could you rearrange the wall pieces to give convergence?*** Try shifting them in Figure 10-18 to put them in the opposite sequence; alternatively, you may tilt the second boundary the other way (clockwise), keeping the sequence of Figure 10-18. Either way, you can make a divergent lens.

Based on this knowledge, we will further discuss prisms and lenses in Section 12.3, specifically in the case of light and other electromagnetic waves.

Can multiple reflections take place inside a wall? Let's look again at Figure 10-14. Its left part actually neglects one effect, which is shown in the right part of that figure. If you follow the incident wave after it refracts into the wall (red wave), it hits the second boundary: there we followed its further refraction out of the wall (blue refracted wave outside the wall). However, the red refracted wave can also be reflected by the second boundary back into the wall: this internal reflection is shown at right in Figure 10-14 as a thinner red arrow pointing to the upper right.

If we now follow this internal reflection, it will hit the first boundary: here it can again both reflect and refract. This refracted wave leaves the wall in the same direction as the earlier green reflection, to the upper right, and with the same direction, wavelength and speed: it simply adds onto the first reflection by superposition (interference). This can cause beautiful colors in oil films, soap bubbles and butterfly wings: see Section 12.2.

On the other hand, the (red) downward-reflected wave goes again down into the wall until it hits the second boundary once more. There, again, reflection and refraction will happen, creating even more waves. This continues again and again, without end, adding ever more reflections and refractions.

However, at each reflection and refraction, the amplitudes of the new waves decrease. As a result, these new waves die out gradually. While they contribute to the total reflection and refraction by the wall, these additional reflections and refractions do not significantly change the behavior of this system.[4]

10.7 Ray Tracing

Figures 10-14 and 10-18 are simplified compared to earlier figures: we have dropped the wavefronts and show only the wave directions. The waves are now simple lines or arrows. This is called **ray tracing** and is very common in optics (the case of light waves): for example, you will often see simple lines representing light moving through lenses and reflecting from mirrors, as in Figures 10-14 and 10-18.

***What gives us the right to replace a wave with a simple line?* The main reason is that we know how a wave behaves at walls, mirrors, *etc.*: we know exactly into which directions waves will be bent. Another reason is that, when we deal with light and everyday walls, mirrors, *etc.*, the wavelength is very small compared to the distances traveled by the light: we then don't need to follow the individual ups and downs of the wave.**

However, we still need to keep the wavelength in mind (or more precisely the frequency of the wave, which is directly related to the wavelength in a given substance). This is because the properties of the substances in walls, mirrors, *etc.*, often depend on the frequency of

[4] The continued reflections and refractions due to repeated reflections within the wall cause an infinity of additional waves that emerge on both sides of the wall. It is challenging to imagine what the summation of all those waves looks like, especially from our graphical construction in Figure 10-14. Fortunately and elegantly, mathematics provides a quite simple solution to these infinite sums: the result is the left part of Figure 10-14, but with somewhat different amplitudes for the "single" reflected and refracted waves. Therefore, the left part of Figure 10-14 still represents the correct situation, if we add one red wave returning from the second to the first boundary.

the wave. But these properties are summarized in the relative refractive index mentioned earlier: thus, if we know the relative refractive index, we can apply Snell's law for refraction, while reflection simply gives equal angles of incidence and reflection. With that information, we can predict the path which the light will follow.

We must also remember that a wave has a width: it is at least as wide as the wavelength is long, and usually much wider. A wave will feel the edges of a narrow hole, even if a simple line can thread its way through it.

With sound, ray tracing can also be helpful: for example, earthquake waves are usually represented with simple rays, as in Figure 9-6. On a smaller scale, such as inside a house, ray tracing is less useful because the wavelength of sound becomes comparable to the size of doors, *etc.* Then wave effects such as diffraction play a larger role: we have seen in Section 3.9 how diffraction can make waves turn around corners without the help of mirrors or lenses. Other examples of light will be given in Section 12.9.

10.8 The Controversial Principle of Fermat

In 1662, the French mathematician **Pierre de Fermat** proposed a controversial principle to explain Snell's law (which had been proposed in 1621 by the Dutch astronomer **Willebrord Snellius**). **Fermat's principle** states that the path followed by a ray between two given points is the quickest path. This principle applies to light as well as other types of waves, such as sound.

We can see Fermat's principle in action in Figures 10-5 and 10-6 (and other figures). Fermat's principle says that Snell's law gives the directions that minimize the time taken for light to go from any point in the first substance to any point in the second substance.

For example, in Figures 10-5 and 10-6, consider the starting point of the white arrow representing the incoming wave and the end point of the white arrow representing the refracted wave. Fermat's principle says that the incoming arrow and the refracted arrow give the fastest path for light between those two points, but only if the arrows' directions (namely the incident angle and the refraction angle) satisfy Snell's law, given the refractive indices of the two substances.

Fermat's principle is indeed correct: it is easy to prove it mathematically for the simple refraction by a single boundary shown in Figures 10-5 and 10-6. It is also correct for double boundaries (Section 10.6) and gradual refraction (to be explained in Section 12.5); in fact, it is correct for all conceivable situations, both mathematically and experimentally.

What then is controversial about Fermat's principle? The controversy becomes clear if we rephrase the principle in this way: light takes the quickest path between any two points. This wording is strictly correct, but it also suggests that light makes a conscious choice to only follow the path that takes the least time.

Try to imagine yourself being that light and making such a conscious choice, or imagine that you want to shine a beam of light from a flashlight onto a fish under water: what do you need to do? First, you must "know" your destination point B; second, you must "look ahead" to where that point B is located relative to your present position A; third, you must analyze the substances between A and B, including where the boundaries are between them and which are the wave speeds in them; and fourth, with all that information, you should somehow calculate which is the fastest path from A to B.

These tasks are already a serious challenge for our human brains: how would a simple light wave (or sound wave, *etc.*) work this out? Does such a wave "try out" many different paths to find the quickest one, before "choosing" one, as you would do by swinging a flashlight until you illuminate your target? Or is there an invisible designer who constantly solves this problem for all the light traveling everywhere in the universe?

Actually, a scientifically satisfying solution to this challenge is "trying out many different paths simultaneously". Imagine starting a wave at point A and letting it go in all directions simultaneously: a spherical wave will do just that (as in Figure 3-1). Each portion of this wave will travel in its own direction according to the laws of physics, including reflecting off mirrors, bending at boundaries by Snell's law, *etc*. However, no portion of the wave "knows" where it will end up. A portion of this wave may reach the destination B: that portion solves the challenge in a purely passive, "unthinking" way; and it is "unconscious" of the fact that it minimized the travel time. In this interpretation, the light has no "goal" or "intent" or "plan": it just travels moment by moment to wherever external influences bring it.

Other **laws of physics** also follow such general principles. One of these is the principle of **minimization of energy**. For instance, a ball on a surface will roll down to find the lowest point it can reach: this minimizes its **potential energy**. Does the ball "explore" or "calculate" the best path to find the lowest point? No! And a ball can't even spread out like a wave to explore all directions simultaneously. What would you do if you were that ball? The ball will passively be pulled in the direction of the steepest slope at its current location and end up at the bottom of the nearest valley: it has no "goal" or "intent" or "plan". Also, the ball may not be able to reach a more distant valley that is even deeper because it cannot "explore" the wider landscape at a distance.

10.9 What have We Learned in this Chapter?

We have learned about the behavior of waves that reach and cross a boundary between substances. For instance, if a string changes thickness (and therefore weight and mass) at some point along the string, a wave will be partly reflected and partly transmitted at that point.

In 2D and 3D, a wave will also be partly reflected and partly transmitted. Transmission will change the wave's direction, which is called refraction: the wave will turn into a new direction inside the different substance. In some cases, total internal reflection is possible, which is used, for example, to keep light inside optical fibers.

The reason for these effects is a change in the wave speed (and therefore wavelength) between the different substances; the speed change is quantified by the refractive index, which is used in Snell's law to predict how large the change of direction is. Diamond refracts light strongly (due to the unusually low speed of light in diamond): this makes diamond flashy!

The change of wave direction can be used to make prisms and lenses (see Section 12.3). In many situations, including prisms and lenses, we can ignore the wave character and draw simple straight lines to mark where the waves go: this is called ray tracing.

11

Surface Waves on Liquids: Waves on Water

Waves on water are the most visible of all kinds of waves. They, as well as waves on other liquids, exhibit most vividly the behavior of waves on strings, sounds in air, liquids and solids, and electromagnetic waves (including light).

However, waves on liquids have very different mechanisms: gravity and surface tension. Also, waves on water often have relatively large amplitudes, so their behavior becomes more complex. Waves with short wavelengths are dominated by surface tension. Waves with longer wavelengths are dominated by gravity; these include such diverse cases as waves at beaches, tides, tidal waves, rogue waves, tsunamis and ships' wakes.

A couple of well-illustrated educational videos discussing waves on water and focusing on ocean waves are available online.[1] *Caution: waves on the surface of water (or other liquids) that are due to gravity, often called gravity waves, should not be confused with gravitational waves that operate on cosmic scales (see Chapter 14).*

11.1 What Shapes do Surface Waves have?

Waves on the surface of water are very visible: their wavy behavior is the most visible of all waves in nature! By contrast, we can't see individual oscillations of sound waves or light waves or quantum waves or gravitational waves, and we can barely see waves on strings. We see waves on water in many places: in our coffee cups and soup bowls, in kitchen sinks and bathtubs, in ponds and swimming pools, in streams and rivers, and on seas and oceans. Waves on water range from tiny ripples up to the worldwide tides, and from gentle swells to mighty beach waves; they also include dangerous rogue waves and devastating tsunamis.

Such waves are not limited to water, but can exist on any liquid, from liquid nitrogen (used as a coolant in many applications) and oil, to molten lava from volcanoes.[2] Since we see water so much more than

[1] See video "Waves and Wave Dynamics" by Sven Holbik at https://www.youtube.com/watch?v=bbmbbFyfXVg, and video "Inside The Navy's Indoor Ocean" by Veritasium at https://www.youtube.com/watch?v=pir_muTzYM8.

[2] For example, see video "Inside Iceland's Volcanic Lava Lake — Mesmerizing Mix of Flowing Lava & Crater by Drone" of July 2021 by Ian in London: https://www.youtube.com/watch?v=Q5TwFGKVNQk. For lava waves, see in particular the times 1:28-2:00, 3:15-4:25 and 8:00-8:50. To give a length scale, the volcano crater and lava lake are estimated to be about 100 meters (300 feet) wide. Another example is the video "Iceland Volcano, drone above Lava pool July 26" of 2021 by SaVids: https://www.youtube.com/watch?v=XCBQu2f8u14. The waves are highlighted by the varying width of bright gaps between dark floating "icebergs". (This lava, although extremely hot around 1,000 degrees Celsius or 2,000 degrees Fahrenheit, is actually near its freezing point, so its surface easily forms solid or semi-solid lava "icebergs" in contact with the cool air; they melt again when they sink into the liquid lava.)

other liquids, we will in this book focus on waves on water. Waves on the surface of other liquids behave very much like waves on water.

Waves on water (and liquids in general) have much in common with the other types of waves that we discuss in this book, including waves on strings, sound waves, light waves, quantum waves and gravitational waves. All these waves can exhibit smooth sine-like shapes that move along at a steady pace; these waves can interfere, reflect, refract and diffract; they have a wavelength and a frequency; they also carry energy and information. **We can therefore apply knowledge about other waves to the case of waves on water.**

Nevertheless, there are significant differences between waves on liquid surfaces and other waves. The most important difference is the mechanism of such waves, as we will soon explore. Moreover, only rarely do waves on the surface of water have a pure sine-like shape: they often have other shapes, many of which are familiar at the beach and on lakes on windy days. Figures 11-1 and 11-2 show a variety of common and special waves on water. **The main reason for this variety of shapes is that waves on water are often strong**, with large amplitudes (wave heights) comparable to their wavelength (crest-to-crest distance).[3]

11.2 How do Surface Waves on Water Work?

Let's first remember how other waves work:

- waves on strings and flags are due to tension and elastic stretching.
- waves on plates and bells are due to elastic bending.
- sound waves are due to compression and decompression of a gas, liquid or solid (and also shear in solids).

[3] The technical term for such behavior is "non-linear". This means, for example, that adding one wave to another does not simply give the sum of their two shapes (except for very weak waves): that would be "linear". Thus, two crossing waves on water can produce a breaking wave that collapses or crashes, as in Figure 11-1e; that usually causes "whitecaps" in the form of droplets and bubbles. It also means that doubling the wave height does not simply double its smooth shape: rather, water waves frequently exhibit sharp crests, also seen in Figure 11-1d-e.

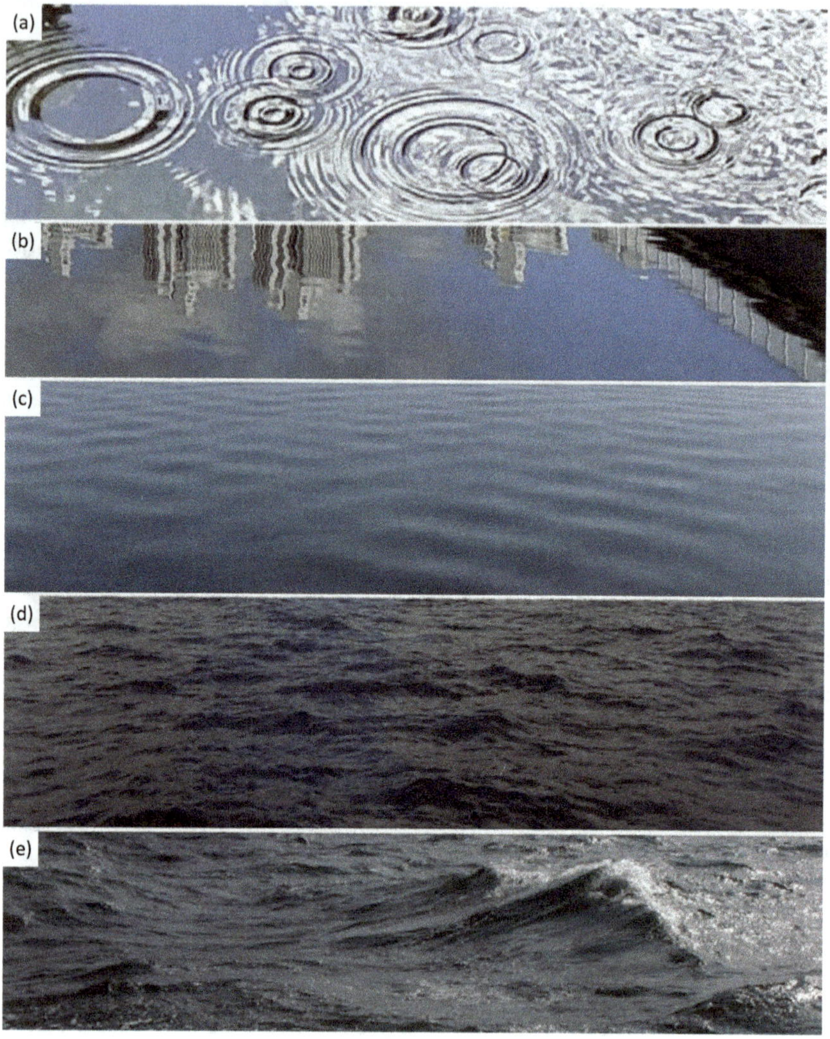

Figure 11-1: Photographs of common shapes of waves on water. a: Circular waves due to drops of water; note how they cross without changing each other. b: Very gentle sine-like waves reflecting buildings, clouds and a bridge. c: Gentle sine-like waves crisscrossing each other and forming interference patterns. d: Rough peaked waves due to moderate wind. e: Breaking wave. The wavelengths (peak-to-peak distances) range from about 1 centimeter at the periphery of the circular waves to a few meters in the breaking wave.

Figure 11-2: Photographs of special waves on water. a: Circular waves due to a falling drop of water. b: A boat's wake. c: A plunging wave breaking at a beach. d: Solitary waves entering a shallow flat-bottomed bay. (*Sources*: a: By homar, for free use, at https://www.needpix.com/photo/805131/drip-close-drop-of-water-macro-water-waves-circles-liquid-wave-drops-of-water. b: By Katarina Miloševic under Unsplash License at https://unsplash.com/photos/MAGrV5JICA8. c: By NOAA, in the public domain, at https://commons.wikimedia.org/wiki/File:Large_breaking_wave.jpg. d: By the Author.)

- light (and other EM) waves are due to interacting electric and magnetic fields (see Section 12.1).

To get a direct feeling for the mechanism of a wave on the surface of a liquid, imagine the following experiments that you can easily do yourself:

- loosely hold a light plastic spoon flat over quiet water in a sink or dish, such that the round bottom of the spoon stays right above the water.
- let the round bottom of the spoon drop into the water, while loosely holding its arm.

- if the spoon is light enough, it will float like a little boat, while bobbing up and down in the water's surface.
- notice the waves radiating away around the spoon.
- the bobbing of the spoon will gradually stop as the energy of its motion radiates away with the waves.

You can also do this experiment with any other floating object, like an ice cube, piece of wood, *etc.*: the object will also bob up and down on the surface of the water. You can also use water drops, as seen in Figures 11-1a and 11-2a or in an online video,[4] or a heavier object like a rock[5] that will sink.

What has happened? Before reading my explanation in the next paragraph, try to imagine the forces that act moment by moment as the spoon (or another object) falls into the water: in your mind, try to build a slow-motion movie of this process.

As it falls, the spoon or other object picks up speed (and thus energy) due to gravity and its weight. That speed pushes the spoon into the water, which pushes water aside. Water cannot easily be compressed, so it must escape sideways and out of the water surface: it is therefore squeezed out at an angle above the surface. This water then falls back into the water surface a bit farther away, acting much like the spoon, and thereby repeats the same process: water gets squeezed out and then falls back down farther and farther away; this forms a wave that travels outward. Some water also pushes the spoon out of the water: the spoon then rises above the water surface and falls back in, also repeating the process.

There are clear similarities between this water wave behavior and the waves on strings (see Section 2.2, in particular Figure 2-6) and in air (see Figure 3-1): an initial deviation from rest causes a disturbance that travels away as a wave.

---

[4] See video "Slow Motion Water Droplet Falling Breaks Surface Tension and Makes Ripples" by stepvideolabs (surface tension is discussed later in this section; the complex shape of the droplet's impact complicates the wave shape in this video): https://www.youtube.com/watch?v=RLn1ErhxOPo.

[5] See video "Slow Motion: throwing a Rock In Water, big ripples" by Handsans (notice the small circular waves created by the splashing waterdrops): https://www.youtube.com/watch?v=RRPP73QM_4k.

With waves on water, **gravity** plays a double role: it pushes down any water that rises above the flat water surface while it also counteracts dips in the water; a dip will be filled by nearby water that tries to go deeper due to gravity (this works through pressure within the water resulting from its own weight).

Another force also acts in waves on liquids: **surface tension.**[6] Indeed, the surface of a liquid acts like a stretched elastic skin, similar to a balloon's skin: that is why drops of water are round like balloons, why water from a faucet breaks up into drops, and why water on a flat surface bunches together into compact puddles. Another way to think of surface tension is that it minimizes the surface area: round shapes have a smaller surface area than pointed shapes. Thus, any bump or dip on the surface of water will be flattened or at least rounded to reduce its surface area: this action also creates a wave.

How do gravity and surface tension coexist in surface waves? **Surface tension wins when the surface is strongly curved, which happens for very short wavelengths, below about 2 centimeters** (surface tension rapidly diminishes at larger distances, hence we don't see larger water drops). Surface tension favors a spherical shape of the water surface. **For wavelengths above about 2 centimeters, gravity wins and tries to maintain a flat horizontal surface.**[7] Around 2 centimeters, both gravity and surface tension act about equally.

How does water move in a surface wave? If you float on water in waves that are longer than your body, you notice that you move both up and down as well as forward and backward; this motion is repeated as each wave crest passes your location. Similarly, if you watch a leaf or twig or duck resting on water with waves, you will see the same motion: the floating object moves up over crests and down through troughs, but it also moves forward over crests and backward through troughs.

---

[6] Surface tension is due to attraction between the atoms and molecules within a liquid. Near the surface of a liquid, the attraction is unbalanced and therefore stronger and one-sided: it pulls the surface molecules into the liquid, like a tight skin. Surface tension also creates capillarity, which is the rounding of a liquid's surface near a solid boundary, as in a glass of water or in a thin tube (for instance in a mercury-in-glass thermometer).

[7] On the scale of the Earth, of course, the water surface due to gravity forms the huge spherical oceans around the center of the Earth.

You will also notice this motion when standing rigidly in the water near the beach while waves pass by you. You see the water rise, then move toward the beach, then drop, and finally move away from the beach before repeating this cycle, which is roughly circular. In this case, you can also feel how the water flows deeper below the surface (or you may observe the motion of fish, plants and dust floating below the surface): you will discover that the water below the surface moves in the same cycle but with a smaller amplitude than at the surface, namely more gently. This amplitude becomes smaller with depth and is almost too small to notice at a depth below about half a wavelength (the wavelength is the crest-to-crest distance).

The roughly circular motion of the surface of water resembles the rotation of a vertical wheel. Objects floating on that water also follow this near-circular motion. The same is true below the surface. This means that the water itself (and floating objects) stays close to the same spot, circling around that spot.[8] Such circular motion is absent in most other kinds of waves (on strings, in sound, in EM waves, *etc.*), but also exists in Rayleigh waves on solid surfaces, as in earthquakes (see Section 9.4).[9]

Circular motion near the water surface is illustrated in Figure 11-3, which shows the depth profiles of waves. In this simple model, each piece of the water follows a circle: it goes round and round that circle, in such a way that the overall crest-and-trough profile travels rigidly to the right. It is very helpful to see this motion in my Animation 11*1.

---

[8] Actually, in real waves the circular path itself moves forward very slowly, resulting in a slow net flow of water in the same direction as the wave travels (it is as if a spinning wheel were slowly sliding forward over slippery ice). There is then also a slow backward flow deeper in the water to compensate for the forward flow near the surface (the forward flow raises the water level ahead of the wave and lowers it behind the wave, which forces some water to flow "downhill" to the back of the wave). These forward and backward flows are very noticeable at beaches, especially with large waves.

[9] Rayleigh waves on the surface of solids (like Earth's crust in earthquakes) have a roughly circular motion that rotates in the opposite sense compared to water surface waves (water surface waves rotate "forward" in the direction of the wave motion, while Rayleigh waves rotate "backward" against the direction of the wave motion).

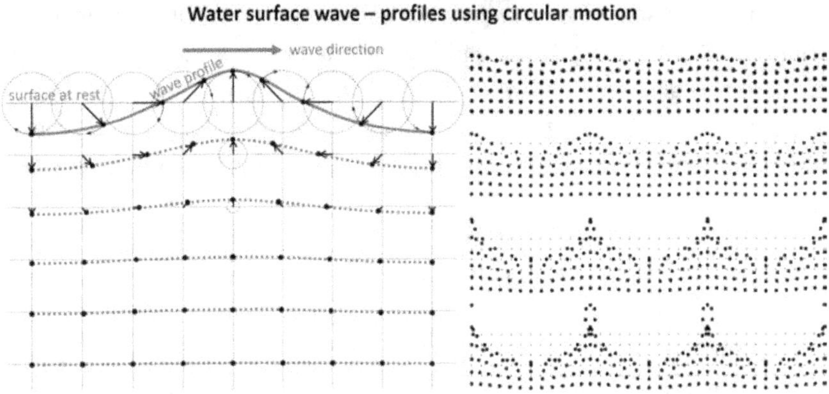

Figure 11-3: Snapshots of wave profiles at and below the surface of water; the waves travel rigidly to the right. The square grid shows the water at rest. *At the left*: one wavelength is shown, from trough to trough. Straight arrows show displacements of the water assuming a clockwise circular motion in the wave (indicated by the curved arrows), resulting in the curved crest-and-trough profile of the wave. Note the sharper crest and flatter trough. Deeper below the surface, the wave profile becomes gradually weaker (dotted gray lines). *At the right*: similar wave profiles showing three wavelengths, from crest to crest. From top to bottom, the circle radius increases, resulting in larger wave amplitudes (the second profile is nearly the same as shown at the left). In the bottom profile, the wave crest explodes.

ANIMATION 11*1 — See my video WB1 at time 11:51 in its section "**Water surface waves**" under the title "**How do waves on the surface of water compare to other waves?**". (See details in the section References and Resources below.)

At right in Figure 11-3, the radius of the circle is gradually increased from top to bottom: this increases the amplitude (crest-to-trough height) of the wave, without changing the wavelength. The top wave, with the smallest amplitude, looks almost sine-like; however, you may notice that the crests are a little bit sharper than the troughs. This becomes clearer in the second wave profile with doubled amplitude (which is close to the profile shown at the left): the crests are much sharper than the troughs, which have become broader and flatter.

Now look at the third profile from the top in Figure 11 3: due to a large circle radius, it has very sharp crests, meaning that the crests are no longer round but have an abrupt corner. Real waves cannot support such spiky crests: real waves with this amplitude "break", forming "whitecaps", including drops and bubbles, as in Figure 11-1e. As a rule of thumb, real waves will break if their height from crest to trough is larger than about 1/7 of their wavelength (for comparison, our model with sharp crests has a ratio of about 1/3, which is too high and must break).

Next, in the bottom right profile, the wave amplitude is even larger. This makes the wave "explode": water is sprayed in all directions. At a beach, such a wave plunges forward, as seen in Figure 11-2c-d.

So far, we have assumed the model of circular motion for waves on the surface of water. This model is correct for very small wave amplitudes. But it rapidly becomes inaccurate for larger amplitudes. We can see one clear sign of this inaccuracy in Figure 11-3. We know that water is essentially incompressible.[10] This implies that, at all times, the volume of the crest should equal the volume of the trough, if measured from the flat "surface at rest". Looking at Figure 11-3, we see that this is certainly not true in this model: the crest is clearly narrower than the trough, so the crest contains less water than is missing in the trough: some water "disappears" in this model!

Fortunately, the circular-motion model correctly predicts the types of behavior of many water surface waves, including: the sharpening of crests with increasing wave amplitude; the "explosion" of waves with large amplitudes; and the reduced motion deeper below the surface.

Moreover, the circular-motion model can easily and realistically be modified for shallow water, such as near a beach. Then, the circles are simply replaced by ellipses (ovals) that are flattened horizontally. This makes the water move less up and down, while it can still move nearly freely forward and backward. At the bottom, the water only moves forward and backward; if there is friction with a rough bottom, the water will move less; it will also gradually lose energy and therefore amplitude.

Figure 11-3 suggests how shallow the water has to be before the bottom is noticed by the wave: the water shown in Figure 11-3 has a depth of about half a wavelength, but we can see that the water at that

[10] Actually, water can be slightly compressed: that is why it can carry sound waves underwater. But this small compressibility is too small to affect surface waves, so we can ignore it here.

depth is only disturbed by waves with very large amplitudes. Hence, we can confirm that **normal waves only feel the bottom if it is less than about half a wavelength deep**. Waves in shallower water also become taller[11]; this results from the broadened troughs, which force more water into the sharpened crests. When such taller waves reach the fatal 1/7 ratio of height to wavelength, they also break into "whitecaps".

Remember that in waves on strings, the oscillatory motion is perpendicular to the string: we called this transverse motion. By contrast, in sound waves based on compression and decompression, the oscillatory motion is outward and inward: we called that longitudinal motion. Waves on liquid surfaces have both motions simultaneously: circular motion is composed of transverse motion (up-down) as well as longitudinal motion (forward-backward).

What starts waves on water? Any disturbance of water will cause a wave. This includes the motion of a spoon in a teacup, of your hand in a washbasin, of your body in a swimming pool, and of wind and boats on rivers, lakes, seas and oceans. Also, earthquakes and underwater landslides can cause waves on water, including tsunamis (see Section 11.9), while the regular motion of the Moon and the Sun causes the tides and tidal waves (see Sections 11.6 and 11.7).

We have described how a spoon starts a wave, by suddenly pushing water aside: most waves start that way, by simple mechanical action. Let's look more closely at how wind causes waves on water, since the mechanism is less obvious in this situation. At the left in Figure 11-3, consider wind blowing to the right across flat water at rest. Constant wind over quiet water cannot suddenly start a wave. Some irregularity is needed: it could be a small object floating on the water, something touching or falling in the water, a fish disturbing the water surface, or a sudden change in the wind itself.

As soon as a little bump of water rises above the flat surface, the wind pushes it along. Imagine the crest at left in Figure 11-3 to be that bump; its exact shape doesn't matter. The wind hits that bump, causing overpressure on its left, which pushes it to the right; also, the wind causes underpressure behind that bump, which pulls it in the same

[11] The wave height in shallow water increases inversely with the fourth root of the depth; for example, if the depth decreases by a factor 16, the wave height is doubled, because $2^4 = 16$.

direction. Thereby, the wind adds energy to the bump of water, making it faster and larger, while making additional water move. This process amplifies the motion and results in short waves that grow over time, both in height and in length. At the same time, other waves are randomly created elsewhere on the water surface. The initial small waves are called **ripples**. As they grow and fill the whole surface, they are called **seas**, especially on large extents of water; such seas are usually chaotic, as new waves are constantly generated by the wind at random spots.

When the waves travel about as fast as the wind, they no longer gain strength since the overpressure and underpressure then become small. Now we have a steadier situation with waves of almost constant size and speed, but they remain chaotic as new waves are created and their directions vary.

However, if the wind is strong enough, the waves will grow too high and start breaking, forming whitecaps of bubbles and flying drops. These patches take away wave energy until an equilibrium is established between the energy input from the wind and the energy output to the whitecaps. The stronger the wind, the more whitecaps are created: in the worst storms, the sea surface looks mostly white.

Waves that do not break can travel remarkably far, even all the way across oceans. Therefore, wind-generated waves will sooner or later leave the area of strong winds and continue traveling through regions with weak winds, propelled by their own energy. Two things happen now: faster waves will outpace slower waves, while waves moving in different directions will also separate from each other. For example, far from a storm, we will find relatively pure waves, not the chaotic mix seen inside a storm: we first see fast waves by themselves, then slow waves by themselves, and they all travel away from where the storm was. Such waves are called **swells**: they look more "peaceful" because they are not chaotic.

Waves cannot continue forever: otherwise, coffee cups, swimming pools, and oceans would remain terribly agitated at all times. ***How do waves die?*** Waves lose some energy due to internal friction, but such friction is weak; friction weakens waves with short wavelengths more than waves with long wavelengths; hence, we see mainly long waves coming from distant storms. They also lose energy through friction with any container, from a cup to a seabed. The breaking of waves also removes energy from the wave motion, as described above.

However, much of the energy of sea waves is lost near the coast: **waves crash onto rocks, break by interference with their own reflections, feel increased friction with the shallow bottom, and self-destruct as they crash onto beaches.**

What happens when a wave crashes onto rocks? When a water wave hits a straight vertical wall, it is simply reflected back, maintaining its wavelength and wave speed. However, if the wave hits rocks that are smaller than the wavelength, very different things happen: the wave is not only scattered into many directions (in the form of plane waves as if from little mirrors, or as circular waves from point sources), but it also can be scattered as waves with all sorts of smaller wavelengths, as well as splashing around as whitecaps. The splashing destroys the oscillating character of the wave. The scattering with smaller wavelengths creates chaos out of an orderly wave: slower wavelets radiate in all directions, largely cancelling each other by destructive interference. The waves' energy is thereby more rapidly converted to heat.

The scattering into waves with shorter wavelengths tells us an important lesson: waves on water (and liquids) normally do not follow the basic rule of wave superposition, which says that we can simply add up waves when they cross each other. We use that rule with other types of waves (waves on strings, sound waves, light waves, *etc.*) to explain and predict the behavior of those waves (like reflection, refraction and diffraction). Very weak waves on water do follow that rule of superposition: a good example is shown at left in Figure 11-4, where **circular waves** are superposed on top of each other. But "everyday"

Figure 11-4: Two sets of circular waves on water. *At the left*: waves due to falling drops of water. *At the right*: waves due to a falling stone. In both cases, the inner waves have a crest-to-crest distance of about 2 centimeters. (*Source* of the photograph at right: Needpix.com, in the public domain, wave-water-stone-2860984_1280.jpg.)

waves on water do not: water waves are too strong, intense and energetic. Scientists call this a "non-linear" behavior.

Fortunately, even though waves on water are "non-linear", we can still understand and predict in a rough way how such waves behave by using the principle of wave superposition: waves on water behave much like the other types of waves, but they can deviate from that behavior, especially when they become stronger.

11.3 The Speed of Waves on Water

We have mentioned several times the speed of waves on water. We have a tendency to think that waves with large wavelengths go faster than waves with short wavelengths. That is generally true for waves on lakes, seas and oceans.

However, look carefully at the two sets of **circular waves** shown in Figure 11-4. The left ones were caused by small drops of water, the right one by a larger stone. Do you see a difference in the order of the circular waves? The difference is this: at the left, the outer waves have a smaller wavelength than the inner waves; the opposite is true at the right! This means that at the left, the waves with smaller wavelengths travel faster than those with longer wavelengths, while at the right, the opposite is true. How is this possible?

An additional hint may help: the inner waves in both photographs have a wavelength of about 2 centimeters. We can therefore conclude that waves with wavelengths around 2 centimeters are slowest: this means that both shorter and longer waves travel faster. No other types of waves have such behavior!

We mentioned 2 centimeters before: for wavelengths below 2 centimeters, surface tension wins over gravity, while above 2 centimeters, gravity wins; thus, the shorter waves on water are dominated by surface tension, while the longer waves are dominated by gravity (2 centimeters is a bit less than 1 inch). Now an explanation emerges: surface-tension waves and gravity waves have opposite speed behaviors. Specifically: **among the surface-tension waves, the shorter ones go faster than the longer ones, but among the gravity waves, the reverse is true, namely, the longer ones go faster than the shorter ones.**

This surprising behavior is not a small effect: 2-centimeter waves on water travel at about 1 km/h ~ 0.6 mph (about a quarter of our walking speed), while very short waves (sub-centimeter) and very long waves (many meters) travel many times faster. For example, waves with millimeter wavelengths travel at about 4 km/h ~ 2.5 mph (our walking speed), while tsunami waves can travel at close to 1,000 km/h ~ 620 mph (comparable to a cruising jet airplane).

In deep water (deeper than half a wavelength), in the circular-motion model, the speed of gravity waves increases with the square root of the wavelength. This increase is seen in Figure 11-4. For example,[12] a gravity wave with a wavelength of 1 meter ~ 3 feet travels at about 1.25 meters/second = 4.5 km/h ~ 2.8 mph (human walking speed); therefore, a wavelength of 4 meters gives double that wave speed: 2.5 meters/second = 9 km/h ~ 6 mph (bicycle speed). Also in deep water, the speed of gravity waves is simply proportional to the wave's period (the time between the passage of successive crests). Thus, a wave with a period of 1 second travels at 1.56 meters/second = 5.6 km/h ~ 3.5 mph; so, a period of 10 seconds implies a wave speed of 15.6 meters/second = 56 km/h ~ 35 mph.

However, in shallow water (shallower than half a wavelength), the speed of gravity waves no longer depends on their wavelength: instead, the wave speed is proportional to the square root of the depth. For example, in the circular-wave model, with a water depth less than 1 meter ~ 3 feet, all gravity waves with a wavelength larger than 2 meters ~ 7 feet travel at about 3 meters/second = 10.8 km/h ~ 7 mph; if the water depth is 10 meters ~ 33 feet, the wave speed increases to about 10 meters/second = 36 km/h ~ 22 mph, but only for waves with wavelengths above 20 meters ~ 67 feet (waves with shorter wavelengths travel as if in deep water, as described above). Note that the tides and larger tsunami waves are in fact shallow-water waves because their wavelength is larger than twice the ocean depth: for these waves, the oceans are shallow (see Sections 11.6 and 11.9).

---

[12] Some data in these two paragraphs are derived from the *"Coastal Engineering Manual — Part II"* by the US Army Corps of Engineers, Books Express Publishing, 2012. This manual gives a detailed overview of the circular-motion (or "linear") model of water gravity waves, as well as other, more complex models. You can also interactively get wave speeds for gravity waves in any depth of water at http://hyperphysics.phy-astr.gsu.edu/hbase/watwav.html#c3 (multiply speeds in meters/second, m/s, by 3.6 to convert them to km/h , or multiply speeds in m/s by 2.24 to convert them to mph).

Due to different forces, the speed of surface-tension waves is proportional to the inverse of the square root of the wavelength: their speed therefore decreases with wavelength, as we saw in Figure 11-4. In principle, they can go as fast as you desire by choosing a small enough wavelength. In practice, however, this would only happen with waves with tiny, invisible amplitudes. Because of their short wavelengths and small amplitudes, surface-tension waves are much less important in practical life (even in a glass of water) compared to the gravity waves.

11.4 Turning of Waves on Water: Refraction and Diffraction

Let's consider the waves visible in Figure 11-5. A common feature of these waves is that they change their travel direction. In image (a) we see ocean swells coming from the upper right and then turning toward the upper left so as to fall head-on to the beach. In images (b-d) we see waves spreading out into many directions; they enter bays and end up heading directly into the curved beaches. These are examples of refraction and diffraction.

In Chapter 10, we saw how a change in wave speed can change the direction of travel of a wave, in the process called **refraction**. That also happens to water waves near beaches: the depth of the water diminishes, thereby slowing down the waves (and reducing their wavelength). In Section 12.5 (for instance, in Figure 12-30) we will see how light waves gradually turn as they move through the air with varying wave speed: the same **gradual refraction** happens here with water waves.

In Section 3.9 (for sound, see Figure 3-16) and Section 12.9 (for light, see Figure 12-55) we discuss how waves turn around corners and spread out widely after passing through narrow slits, in the process called **diffraction**. This is nicely exhibited by Figure 11-5b-c, where waves from the sea enter bays. Figure 11-5d shows both diffraction and gradual refraction: the waves spread out when entering a wide bay and also turn directly toward the beach.

On water, refraction and diffraction (as well as interference and reflection) of waves are easy to see, unlike for sound waves, light

Figure 11-5: Photographs of refraction and diffraction of waves on water. a: Waves turning near shore by gradual refraction. b: Circular waves entering Craster Harbour, UK. c: Circular waves spreading in the Bay of La Concha in Donostia/San Sebastián, Spain. d: Waves entering a bay of the island of Lanzarote, Spain. (*Sources*: a: Extracted from "Ocean washing ashore" by Mathew Waters, in the public domain, https://commons.wikimedia.org/wiki/File:Ocean_washing_ashore_(Unsplash_IBSNW5R1PPM).jpg. b: Extracted from photograph by Oliver Dixon, under CC BY-SA 2.0, https://www.geograph.org.uk/photo/2129205. c: By the Author. d: Extracted from "Mountain view on coastline", by Robert Bye, under Unsplash License, https://unsplash.com/photos/SCUA_9BsQ0c.)

waves, *etc*. Therefore, these characteristic wave-like effects are often illustrated with water waves in classrooms and in videos on the web.[13]

11.5 Coastal and Beach Waves

Waves approaching coasts display a wide range of behaviors, many of which we have mentioned separately in the preceding sections. Let's bring them together here, especially as coastal waves are among the most spectacular forms of waves in nature and are frequently seen and enjoyed.

We will follow a swell as it approaches from a distant storm. Its waves typically have wavelengths of 30 to 100 meters; they travel at speeds of 25 to 50 km/h, comparable to the speeds of many boats, with periods (repetition times) of 5 to 10 or more seconds, and a trough-to-crest height of a meter or two. The swell has roughly constant wave heights.[14]

When the swell hits a vertical cliff or wall, it can be reflected as if from a mirror: this can produce pretty interference patterns in front of the cliff or wall, similar to those shown in Figures 3-10 to 3-12, Figure 11-1c, and Figures 10-4 to 10-6 (ignoring any transmission). The crests in these interference patterns have double the height of the incoming wave crests: this double height may exceed the 1/7 ratio of height/wavelength, so that these crests can break and produce whitecaps, losing much energy in the process; the remaining weakened waves are reflected back toward the sea.

When the swell reaches an island, it can turn around to the back of the island by diffraction. Every point along the island's coast will feel the waves, although less strongly behind the island.

Likewise, if the swell enters a bay, it can expand sideways in all directions by diffraction, as we saw in Figure 11-5b-d. This weakens the waves as their energy is spread out across the bay.

---

[13] For example, see video "Experiments with water waves" by TIB — Leibniz Information Centre for Science and Technology University Library, at https://av.tib.eu/media/30440.

[14] In reality, it is common for an occasional crest to be higher than the others, for reasons that are not well understood; it is sometimes said that every seventh crest is such a "sneaker wave", but that actually varies a lot.

The swell will start to feel the bottom of the sea when its depth is about half of the wavelength, which depth may be about 5 to 50 meters. This slows down the wave, which in turn reduces the wavelength itself (because the time period between crests remains the same): the crests come closer together and become sharper and taller.

If the swell passes over a shallow bottom, its water motion is more elliptical (cycling along a flattened circle). When the depth remains small and more or less constant, the result is short, sharp crests between very flat and long troughs, as in Figure 11-2d. The successive crests are now essentially independent of each other: the crests do not feel their neighbors. Such individual wave crests are therefore sometimes called **solitary waves**.

The slower speed also turns the waves toward the coast by gradual refraction (see Figure 11-5), similar to light turning in air of varying temperature (see Figure 12-30 in Section 12.5). As these waves approach a beach, where the depth is reduced to zero, they turn directly toward the beach, so that wave fronts arrive almost parallel to the beach: that is why waves can crash over a large width simultaneously; it also offers surfers long waves and tubes to surf along.

The waves will break into whitecaps if they exceed the 1/7 ratio of height/wavelength. If this happens close to a beach with a rapid change of depth, the wave can quickly peak very high before it breaks up: we then get an impressive plunging wave, as in Figure 11-2c; such a plunging wave breaks up on the way down.

The breaking into whitecaps and the crashing of a plunging wave convert the considerable energy of the waves into bubbles, flying drops, sound, shifting sand and vibrations of the ground, as well as splitting rocks and shells into more sand; most of this energy ends up as heat, causing a very slight rise in temperature. Little energy is left to be reflected back to the sea: you rarely see waves returning to the sea from a beach.

However, waves do carry water to the beach, most obviously in plunging waves: this water must return to the sea. Water returns to the sea by two means. First, there is a counterflow along the bottom, below the waves. Second, especially with large storm waves that carry much water to the beach, there are **rip currents**, also improperly called rip tides: these are visible “rivers” that form here and there outward from the beach, through which a strong current of water flows back into the sea; rip currents are typically a few meters wide (measured along the beach) and a few tens of meters long (stretching out from the beach to behind the breaking waves). These rip currents are clearly quite dangerous for swimmers (but there is an easy escape for a swimmer caught in a rip current: swim a few meters parallel to the beach, out of the rip current, and then swim toward the beach by following the waves).

We see that waves can have many different behaviors near a coast, making them spectacular.

11.6 Tides as Waves

The **tides** in the oceans are often described as a giant wave with two crests and two troughs circling the **Earth** westward every 24 hours and 50 minutes. The crests are the high tides or flood tides, while the troughs are the low tides or ebbs.

However, this simple model of the tides predicts some surprising behaviors that we will have to modify in our further discussion. For example, the wavelength of that two-crested wave would be enormous: about half the circumference of the Earth, or close to 20,000 kilometers. Its speed should therefore also be very large: 20,000 kilometers in 12 hours and 25 minutes gives about 1,600 km/h (or 1,000 mph), which is twice the speed of cruising jetliners or about 1.5 times the speed of sound; in reality, waves cannot go that fast in the Earth's oceans, because the typical depth of 4 kilometers limits wave speeds to at most 800 km/h. That is still very fast, faster than tsunamis (see Section 11.9). Fortunately, this wave's height is relatively small: on average, about 1 meter from trough to crest; otherwise, tides would be more devastating than the worst tsunamis, twice each day!

By comparing the tides with the apparent rotations of the **Moon** and **Sun** around the Earth, it is clear that the tides are due to both the Moon and the Sun. A better model is in fact two waves: a larger wave due to the nearby Moon and a smaller wave due to the distant Sun (with about half the amplitude of the Moon's wave). These two waves have slightly different frequencies because the Moon and Sun appear to turn at different rates around the Earth. The two waves are superposed on each other, creating a more complex tide profile that can have additional smaller crests.[15] When the two waves are in phase, their crests and troughs pile up on top of each other, giving larger tides called **spring tides**. But when the Moon's and Sun's tides are in antiphase, the crests of one partly fill the troughs of the other, giving smaller so-called **neap tides**.

The tides are due to the attraction of gravity from both the Moon and the Sun. They attract not only the oceans but also the liquid interior of the Earth, together with the Earth's solid crust: tides are also created in the stiff crust, but they are weaker, typically around 30 centimeters (1 foot) in height, and we don't notice them in daily life.

Since our own bodies are also attracted to the Moon and the Sun, should we behave just like the ocean tides? After all, we are mobile like water, so we should also be pulled up! However, we don't feel this effect on our bodies. This shows that the effect of Moon and Sun is very weak compared to the attraction from the Earth itself; that is also why the ocean tides are only a meter or so in height. Unlike human bodies, water flows extremely easily and therefore responds strongly to the weak attraction of Moon and Sun, but only on the large scale of oceans, as we will explain below.

A classic puzzle about tides is this: why do tides generally have two crests and two troughs on opposite sides of the Earth? At any one moment, the Moon (or the Sun) is on one side of the Earth, so it pulls in that direction only: how then can a crest arise on the other side? Is one crest due to the Moon, and the other to the Sun? No, the two crests are not due to the Moon *versus* the Sun; even if only the Moon caused our tides, there would still be two crests and two troughs.

In simple terms, the double crests are due to two opposing effects that nearly cancel each other out: the attraction of the Earth's water by the Moon, and the inertia of that water; inertia "pulls" the water

[15] There are additional slower complications in the tides, for example because of the elliptical shapes of the orbits of the Moon around the Earth and of the Earth around the Sun. Also, the Moon's and Earth's orbits are inclined as they rotate around the Earth and Sun, respectively. These effects repeat after about 28 days or a year, depending on the case.

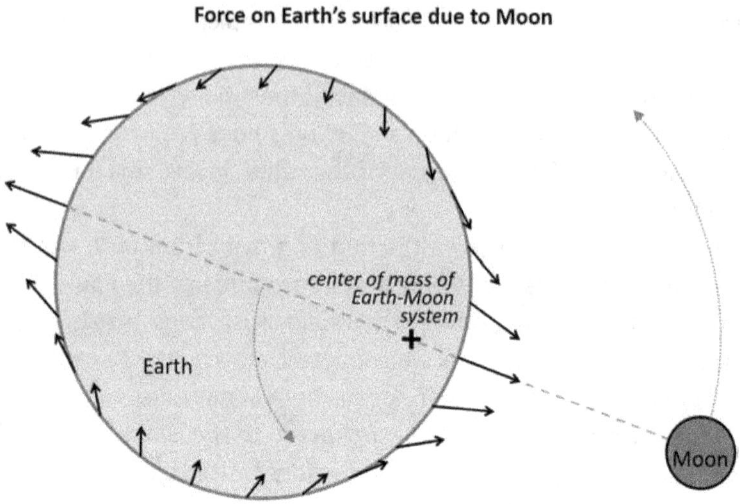

Figure 11-6: The force responsible for tides on the Earth's surface due to the Moon's gravity is shown as arrows, indicating the force's direction and relative size. The curved, dotted arrows show the paths of Earth and Moon as they circle around their common center of mass, marked as a + sign. (Dimensions are not shown to scale: in reality, the Moon is about one quarter the size of the Earth and about 30 Earth diameters away; the center of mass of the Earth-Moon system is located about one quarter of the way down the radius of the Earth's surface, roughly where drawn.)

in the opposite direction.[16] We must realize that both Moon and Earth turn around each other, or more precisely around their common center of mass, which is located below the Earth's surface as shown in Figure 11-6, in about 28 days. As the Earth follows a circular path around that center of mass, water tends to be thrown off by inertia, especially on the far side of the Earth; this is just like a child being thrown off a revolving merry-go-round, or you being "pushed" aside in a turning car, or like

[16] This inertial effect is often called "centrifugal force", although it actually is not a force, but just the tendency of objects to move in a straight line at constant speed when no forces act on them, which is inertia. We feel this inertia on a revolving merry-go-round or in a turning car: our inertia favors going straight, while the merry-go-round and car force us to stay on a circle; the same effect is used in a spinning clothes dryer. The force that pushes or pulls to create a circular motion is called "centripetal force".

water being expelled from clothes in a spinning dryer. It is very helpful to see this motion in my Animation 11*2.

ANIMATION 11*2 — See my video WB1 at time 13:42 in its section "**Water surface waves**" under the title "**Tides: what is their origin?**". (See details in the section References and Resources below.)

The arrows in Figure 11-6 show the net result of those two competing effects. Water nearer the Moon feels a stronger attraction than the effect of inertia, and is therefore pulled toward the Moon, forming a crest. Water opposite the Moon, being farther away from the Moon, feels a weaker attraction from the Moon than the effect of inertia, so it is "pushed" away from the Moon by inertia, forming the second crest. These two crests form the high tides: one crest permanently points toward the Moon, while the other points permanently away from the Moon. The second crest is normally weaker than the first, so the two daily high tides are not equally high.

The water in the two crests is not simply "lifted" up into the air; if it were, our own bodies and any other loose objects would also be lifted into the air by the Moon! Instead, water is pulled in from far distances, where the arrows in Figure 11-6 point into the Earth (forming the low tide or ebb) and along the surface of the Earth: so the high tide is squeezed up by water pulled in from far away, like toothpaste squeezed out of a tube. That is why we don't see tides in coffee cups, bathtubs, lakes or even the Mediterranean Sea or the Caribbean Sea (see the deep blue colors in Figure 11-7, indicating very weak tides): no distant water is readily available to create crests on those relatively small surfaces. It is very helpful to see this motion in my Animation 11*3.

ANIMATION 11*3 — See my video WB1 at time 14:39 in its section "**Water surface waves**" under the title "**Tides: more complex behavior due to continental barriers — 1**". (See details in the section References and Resources below.)

If you are interested, Box 11-1 goes into more details of the mechanism of tides.

BOX 11-1 — THE MECHANISM OF TIDES. When we think of tides, we tend to focus on their daily behavior with normally two crests per day. To better understand tides, we have to remember that we are dealing with the monthly rotation of the Moon around the Earth, and the yearly rotation of the Earth around the Sun, in addition to the daily rotation of the Earth around its own axis.

As shown in Figure 11-6, the cause of the lunar tides is the monthly rotation of the Moon around the Earth, in addition to the gravitational attraction from the Moon. Similarly, the solar tides are due to the yearly rotation of the Earth around the Sun, in addition to the gravitational attraction from the Sun.

These effects form two permanent pairs of tides, one pair pointing toward and from the Moon, the other pair pointing toward and from the Sun. The first permanent pair revolves around the Earth only once each month, while the second pair revolves around the Earth only once each year!

The reason why we see daily tides is that the Earth rotates each day around its axis: the Earth therefore rotates "underneath" the much slower tides. From our human perspective on the Earth's surface, it looks like the tides are rotating around the Earth, while the opposite viewpoint is more accurate.

So far, our discussion has assumed that the water on the Earth's surface can flow freely everywhere, allowing the formation of tides that circle the entire Earth. However, as we will see in Figure 11-7, the continents in practice prevent such a free flow of water and tides. That will favor a different model of the tides, even though their basic mechanism remains the same: gravity and inertia, coupled with the rotations of the Earth and Moon.

As we have seen, the mechanism of the tides is somewhat complex. Unfortunately, you will find various incomplete or inaccurate descriptions of tides in online texts and videos. An interesting discussion can be found in an online video.[17]

How realistic is the above description of the tides? Do the real tides travel smoothly and uniformly like simple sine-like waves around the Earth? Figure 11-7 shows a map of the main Moon-induced tides over a whole cycle of 12 hours and 25 minutes: where is the simple

[17] See video "What Physics Teachers Get Wrong about Tides!" by PBS Space Time, at https://www.pbslearningmedia.org/resource/what-physics-teachers-pbs-space-time/what-physics-teachers-pbs-space-time/.

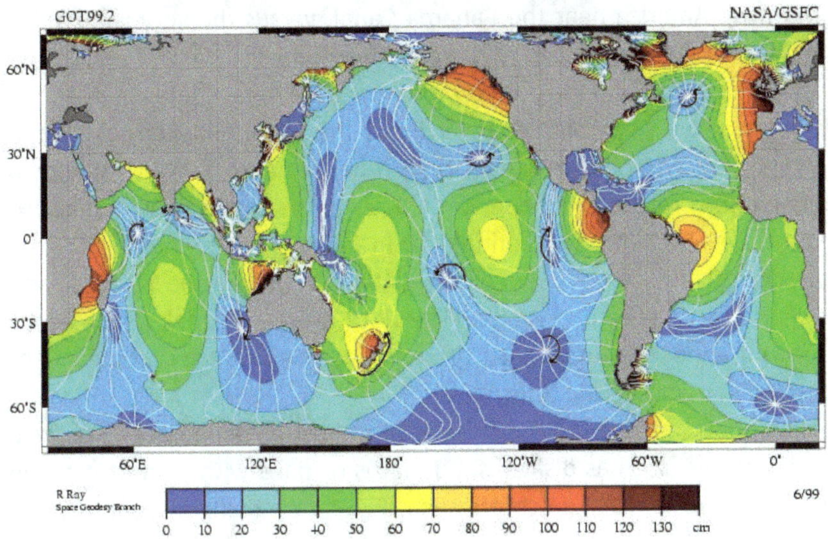

Figure 11-7: Map of tides, including only the main lunar tides (the solar tides would look similar, but weaker, and should be added to the lunar tides). The color indicates the local tide amplitude, using the scale shown below the map. White lines show the time evolution of the tides, with roughly 62-minute intervals (1/12 of the 12h25m period): a white line represents a tidal crest at one moment; 62 minutes later, that crest has moved to the next white line, in the direction given by the black arrows. The white lines converge at points where the tide has zero amplitude: those points have no lunar tides, hence their dark-blue color. (*Source*: R. Ray, NASA, in the public domain, https://commons.wikimedia.org/wiki/File:M2_tidal_constituent.jpg.)

sine-like wave with two crests and two troughs? Reality is clearly much different and far more complex!

The biggest surprise is that there are locations in the middle of the oceans which have no tide at all: on the tide map, those are the points in dark-blue regions where white lines come together. Also, in the middle of the northern Pacific Ocean we see a large patch with almost no tide (less than 10 centimeters of tide). Another surprise is that a tidal wave circles New Zealand permanently: the tide does not move on to Australia, but instead goes round and round New Zealand every 12 hours and 25 minutes; a similar tidal wave circles Madagascar southeast of Africa. Also, there is no obvious connection between the tides in different oceans (notice the very large tides on the Pacific side of Central America near the Panama Canal, *versus* the very weak tides on the Caribbean side, about 70 kilometers away).

So how do the tides go around the world? In fact, looking at Figure 11-7, we see a very complex behavior of the tides. A better description is that **the tides form waves that move around within individual oceans, like waves in independent basins**. That makes sense because tidal waves cannot cross continents: in particular, America blocks the passage of any wave between the Atlantic and Pacific Oceans, and similarly for Eurasia and Africa. (The tides are quite small in the Arctic Ocean and near Antarctica, so these regions do not offer a convenient passage for world-scale tides.)

A good analogy for the tides is therefore a collection of separate round basins, such as dishes, soup plates or pans with water. At home on a small scale, you can create such waves by moving a spoon or your hand through the water in a circular fashion around a basin (or simply by moving the basin itself in circles, in which case the center point may have no tide, like some points in the oceans): this creates a wave that circles the basin, like the tides in Figure 11-7; the complex shapes of the oceans, compared to the simple shapes of round basins, further complicate the shapes of the tides.

The picture of tides is now the following: the Moon and Sun activate waves that circle the oceanic basins with the frequencies of the apparent rotations of the Moon and Sun around the Earth; these waves are strengthened by gravity and inertia in the larger portions of the oceans toward and away from the Moon, and toward and away from the Sun; when Earth, Moon and Sun are aligned, we get stronger spring tides; when the Sun and Moon are seen at right angles to each other from the Earth, we get weaker neap tides.

Can tides resonate in an ocean? This question is important because resonance can increase the amplitude of waves to dangerous levels (see the danger of the breaking up of resonating bridges in Section 1.3). A resonating ocean could easily overflow and destroy human life on land, even on a larger scale than tsunami waves.

Resonances occur when the wavelength fits the size of a container. Thus, the string of a piano or guitar resonates when the wavelength fits the length of that string, forming a standing wave (see our discussion of resonances in Sections 1.3 and 2.8). For ocean waves like tides, the container is the ocean basin: if the wavelength fits that basin's length or width, a resonance can occur. Parts of an ocean may also have suitable shapes and can resonate by themselves.

A famous example of tidal resonance is the large tide in Canada's Bay of Fundy, where the trough-to-crest height of the tides can be as large as 16 meters (about 53 feet). The shape of the Bay of Fundy gives its water a resonance period close to the tide period of 12 hours and 25 minutes; therefore, the tide acts as a periodic force that boosts the natural back-and-forth swinging of the bay's water,[18] similar to repeatedly pushing a child on a swing to increase its swinging. The funnel-like narrowing of the far ends of the Bay of Fundy further increases the height of the tides there, by squeezing the water upward.

How can we extract tidal energy? The amount of energy contained in the tides is enormous, despite their small amplitudes: if we could extract all of that energy, we would get as much power as from about 3,000 large power stations using fossil or nuclear energy (assuming 1 gigawatt of power per station). Perhaps a tenth of that tidal power is technically accessible. The two largest tidal power stations today produce about a quarter of a gigawatt each: the Sihwa Lake Tidal Power Station in South Korea from 2011 and the tidal barrage power station at La Rance in France from 1966.[19] The MeyGen tidal energy project, located just north of Scotland, should deliver about 0.4 gigawatts from 2027, sufficient to power 175,000 homes, using propellers (turbines) in strong tidal ocean currents.[20]

Tidal energy exists in two forms: current and altitude. Tides make water flow from high-tide areas toward low-tide areas: this flow can activate underwater propellers connected to electric generators. Altitude works in much the same way: sea water is captured in reservoirs at high tide and later released through propellers linked to electric generators. This is similar to reservoirs built in mountains for hydro-electric power generation (these have the advantage of a more continuous one-way

---

[18] See "Bay of Fundy Tides: The Highest Tides in the World", including videos, at https://www.bayoffundy.com/about/highest-tides/.

[19] See https://www.britannica.com/science/tidal-power and "Tidal power" at https://en.wikipedia.org/wiki/Tidal_power.

[20] See "MeyGen" at https://en.wikipedia.org/wiki/MeyGen.

stream of water and much larger heights, while they avoid the corrosive sea water).

Areas with higher tides are clearly most favorable to extract power: those are visible in Figure 11-7 in red. However, other factors play important roles as well, such as proximity to dense populations and coastal shapes. Power extraction from tidal waves is relatively efficient: today's turbines can extract about 80% of the power in the water flow (but this fraction must be reduced for the periods between tidal flows, as well as maintenance time). Wind and solar power generation are less efficient and provide a less regular flow of power.

11.7 Tidal Waves: Tidal Bores

Figure 11-8 shows examples of local **tidal waves**, also called **tidal bores**. Such waves are typically seen moving up some rivers near the sea at high tide. They move upstream at speeds of roughly 10 to 40 km/h, like a fast cyclist. Their visible length is quite short: the shortest ones have only one crest (one wave pulse) and pass by in just a few seconds, as in Figure 11-8a; longer ones have a limited series of crests (see Figure 11-8b-c) and pass by within a minute or so. Importantly, their timing is connected with the tides: they occur only with the incoming tide; in fact, they are the incoming tide itself, but compressed into a few seconds or minutes.

In Section 11.6, we discussed tides as being "tidal waves" that span the entire Earth, or at least large parts of the oceans. More often, the term tidal wave is used for the more limited wave that is also called a tidal bore. It can also be described as a **solitary wave**, which is an individual wave pulse or wave packet, rather than a wave train which has a large number of wave crests. In this section, we will focus on tidal bores.

Sometimes, **tsunamis** are incorrectly called "tidal waves" or "tidal bores" because they look alike from the shore. A tidal wave or tidal bore is a consequence of a high tide; therefore, its arrival is predictable and repeated regularly, just like the tides: surfers can easily ride a tidal bore since they can predict its arrival time. By contrast, a tsunami results from an unpredictable earthquake or other sudden deformation of the sea bottom, like an underwater landslide: it is a single event that does

Figure 11-8: Photographs of tidal bores. a: Tidal bore at Silverdale, UK. b: Tidal bore in the Qiantang River, China; notice the many people on the sea wall at bottom left. c: Tidal bore in the Garonne River, France, with surfers and a boat creating smaller wakes. (*Sources:* a: Arnold Price, under CC BY-SA 2.0, at https://www.geograph.org.uk/photo/324581. b: MasaneMiyaPA, under CC BY-SA 4.0, extracted from https://commons.wikimedia.org/wiki/File:20201003钱江潮通过三堡.jpg. c: Reproduced with permission from N. Bonneton, Ph. Bonneton, J.-P. Parisot, A. Sottolichio, and G. Detandt, "Ressaut de marée et Mascaret — exemples de la Garonne et de la Seine / Tidal bore and Mascaret — example of Garonne and Seine Rivers", *Comptes Rendus Geoscience*, volume 344, page 508, 2012. Copyright © 2012 Académie des Sciences, published by Elsevier Masson SAS. All rights reserved.)

not repeat regularly and also has an unpredictable strength. We will discuss tsunamis in Section 11.9.

An important factor in creating tidal bores is the shape of the river mouth: it should be like a funnel that gradually concentrates the incoming tide into a narrowing channel. Most rivers open up to the ocean or sea too abruptly to allow the formation of a tidal bore. Once formed, a tidal bore can travel for tens of kilometers up a river, even around its bends and even through an occasional small lake, as long as the river has a gentle flow. In some cases, a tidal bore can continue upstream for hours because the river current slows it down.

The world's largest tidal bore is found in China in the mouth of the Qiantang River (near the city of Hangzhou): watch it online[21] or search online for a variety of views (mostly in Chinese).[22] This tidal bore is due to the particular funnel shape of the river's estuary, which concentrates a wide offshore tide into an ever-tightening channel. It can reach 9 meters in height and travel at up to 40 km/h.

There is a very large contrast between the enormous scale of ocean-sized tides (thousands of kilometers in size with a period of over 12 hours) and the small scale of tidal bores (meters and seconds). This illustrates our earlier point (in Section 11.2): waves on water often do not follow the superposition principle of waves (they are often "non-linear"). In particular, we see that waves with large wavelengths (tides) can create waves with much shorter wavelengths and much larger frequencies (tidal bores).

11.8 Rogue Waves and the Nazaré Waves

There is a good reason for the sinister name **rogue waves** (they are also called **freak waves**, **monster waves** and **killer waves**): according

[21] See video "World's Largest Tidal Bore Forms in China's Qiantang River" by CCTV Video News Agency, at https://www.youtube.com/watch?v=k6fr6GUSmAA, or video "Incredible Tidal Waves Caught On Camera" by Underworld, at https://www.youtube.com/watch?v=ia9wQFEQ6RM, or the surfing-oriented video "Silver Dragon — worlds most dangerous wave" by Alliance Multimedia, at https://www.youtube.com/watch?v=3dk2IHV2RQo.

[22] Search online for "Qiantang River tidal bore".

to the European Space Agency (ESA),[23] "Severe weather has sunk more than 200 supertankers and container ships exceeding 200 metres in length during the last two decades. Rogue waves are believed to be the major cause in many such cases." "Results from ESA's ERS satellites helped establish the widespread existence of these 'rogue' waves". "In February 1995, the cruise liner Queen Elizabeth II met a 29-metre-high rogue wave during a hurricane in the North Atlantic that Captain Ronald Warwick described as 'a great wall of water... it looked as if we were going into the White Cliffs of Dover.'" And "two large ships sink every week on average": this statistic is truly stunning and rarely reported in the media!

Judging by videos from ships that have survived rogue waves,[24] the sinking of a ship by a rogue wave probably happens in just a few seconds. Rogue waves are unpredictable, and exist only for a few seconds: they are a kind of explosion. Rogue waves are therefore only seen seconds before they hit the ship and they disappear just as fast. They leave little time for the crew to even call for help, and they leave little trace of the accident: ship and crew often simply disappear.

It is reported[25] that "the highest reliably measured ocean waves" had "a wave height of 34 meters (112 feet) peak to trough. The period of the wave was 14.8 seconds, and its wavelength was calculated to be 342 meters." In that instance, "the wave speed is calculated to be 23 m/s [about 83 km/h ~ 52 mph]. [Author Ned] Mayo calculates the power of one meter length of such a wavefront to be 17,000 kilowatts [17 megawatts or 0.017 gigawatts or 1/60 of a large power plant]!" Rogue waves are observed in stormy conditions, in particular near the Agulhas current off the East coast of South Africa, as well as near the Gulf Stream in the North Atlantic. They are also seen in Europe's North Sea around oil rigs.

The simplest explanation for rogue waves would be random constructive interference: whenever different high crests pile up on top

---

[23] See the webpage dated 2004 at https://www.esa.int/Applications/Observing_the_Earth/Ship-sinking_monster_waves_revealed_by_ESA_satellites.

[24] See video "5 Monster Waves Caught On Camera", by Underworld, at https://www.youtube.com/watch?v=nydwk87iEuM.

[25] See the page on "Highest Ocean Waves" at http://hyperphysics.phy-astr.gsu.edu/hbase/watwav.html.

of each other by chance, you could expect them to form a rogue wave. However, calculations show that this mechanism is too rare to lead to the known observations and losses of ships.

ESA's imaging has led to the more likely conclusion that "Rogue waves are often associated with sites where ordinary waves encounter ocean currents and eddies. The strength of the current concentrates the wave energy, forming larger waves — [scientist Susanne] Lehner compares it to an optical lens, concentrating energy in a small area." In this explanation, the larger-scale motion of water (currents and eddies) acts like lenses that steer waves to a focus, causing abnormally large amplitudes there, also by piling up wave crests.

You may know of the giant waves at the Praia do Norte (North Beach) of Nazaré in Portugal, used by surfers.[26] These **Nazaré waves** can reach heights of 26 meters ~ 87 feet. They are also due to a concentration of energy. More precisely,[27] an offshore underwater canyon splits incoming storm waves into a fast wave (moving over the deep canyon) and a slower wave (moving over the shallower bottom outside the canyon); at the end of the canyon, the fast wave turns toward the slow wave by refraction, as if by a lens, and climbs on top of the slow wave, while a previous wave reflected from the cliff ahead climbs on top of the two incoming waves; this superposition can triple the original wave height.

The tallest Nazaré waves are rather narrow and short-lived: each wave spans a few hundred meters and dies within about a minute. This behavior is also typical of rogue waves seen in the open ocean. So, **we can understand such giant waves as the brief superposition of several big storm waves**.

You can easily simulate small rogue waves in your sink or tub by creating waves with your two hands (at the risk of flooding your kitchen or bathroom). A more organized, artificial rogue wave, made by converging

---

[26] See multiple videos at https://nazarewaves.com/en/Home/VideoGallery.

[27] See video "Onda da Nazaré, como se forma" (in Portuguese but graphically understandable, especially from time 2:30 to time 3:00), by Instituto Hidrográfico — Marinha Portuguesa, at https://youtu.be/Yufb2MgcebM.

circular waves, displays the power concentrated in such waves.[28] Their mechanism of superposition is exactly the same as for the natural rogue waves. We saw the same superposition mechanism at work when concentrating sound waves at one point: see Figure 3-9 in Section 3.7.

11.9 Murderous Tsunamis

Tsunamis are tremendously long, fast and often devastating ocean waves that are due to large, unpredictable events such as underwater earthquakes.

The dramatic character of these waves drew world-wide attention with the **Boxing Day Tsunami** on 26 December, 2004, which originated from an earthquake below the Indian Ocean near Sumatra in Indonesia: it killed at least 230,000 people in 14 countries, the highest number of deaths of any tsunami in known history. This was the first tsunami to be extensively recorded, thanks to the advent of cheap cameras and the presence of many vacationers in the region; it thereby also greatly increased our understanding of tsunamis.

The disastrous **Tohoku Tsunami** off Japan on 11 March 2011 killed about 20,000 people, made over 2500 other people disappear, severely damaged a nuclear power station, and displaced over 200,000 people from their homes; its damage cost the equivalent of tens of billions of US dollars. A famous older tsunami occurred near Portugal's capital, Lisbon, in 1755, which killed about a sixth of Lisbon's population and destroyed most of its buildings.

A convenient introductory summary of the evolution of a typical tsunami, from its violent origin to its destructive end, is available online[29]: it offers a good overview of the different phases of a tsunami as it is created, as it travels and as it hits land. Other overviews of tsunamis are also

---

[28] See video "90 ft. Vertical Spike Wave in Slow Mo" by The Slow Mo Guys: https://www.youtube.com/watch?v=iWKFPTgkpXo&t=3s.

[29] See "Life of a Tsunami" by the U.S. Geological Survey, at https://www.usgs.gov/centers/pcmsc/life-tsunami.

available online[30] or in books.[31] Videos understandably focus on the last moments of tsunamis: their devastating effects on and near the shore.[32] Those last moments of tsunamis are spectacular but complex, depending on the situation. We will discuss these phases in more detail further below.

Are tsunamis different from normal waves on water? Tsunamis are basically normal waves on the surface of water, but they are much larger in several respects. The following are characteristic aspects of tsunamis, which we will expand further below:

- Tsunamis are not caused by wind, storms, ships, Moon or Sun. Rather, they result from earthquakes, underwater landslides, volcanic eruptions and other unpredictable large-scale events. Thus, tsunamis are not tidal waves, even though some tsunamis may look like tidal waves when they enter bays and rivers.
- Tsunamis have very large wavelengths, up to about 500 kilometers on deep oceans (such lengths are larger than some countries). These wavelengths are still much shorter than those of the tides, which can have wavelengths over 1,000 kilometers in the open oceans. In shallower oceans and seas, the wavelength of tsunamis drops gradually to about 10 km in water with a depth of 10 meters (10 km is still larger than many cities).
- Since oceans have a typical depth of 4 km, and at most 11 km in deep trenches, tsunamis are shallow-water waves everywhere: they are therefore slowed down considerably by even the deepest ocean bottoms.
- On open oceans, the speed of tsunamis ranges from about 700 km/h for a water depth of 4 kilometers to about 1,000 km/h for a depth of 8 kilometers (this speed is similar to that of cruising jetliners). They slow down near coasts due to shallower water.

[30] See "Tsunami" at https://en.wikipedia.org/wiki/Tsunami and at http://hyperphysics.phy-astr.gsu.edu/hbase/Waves/tsunami.html.

[31] For example: "Tsunami, the Underrated Hazard" by Edward Bryant, Springer, Heidelberg, 2014.

[32] See video "Tsunamis 101" by National Geographic at https://www.youtube.com/watch?v=_oPb_9gOdn4, and video "2004 tsunami" by Fabio Campo at https://www.youtube.com/watch?v=6PrpsB5AuN8, and video "2011 Japan Tsunami — Kesennuma Bay. (Full footage)" by 2011 Japan Tsunami — Kesennuma Bay at https://www.youtube.com/watch?v=JeV8m8OtwRU.

- Tsunamis can easily cross the largest oceans, such as from Alaska to New Zealand or from Chile to Japan. This can take 24 hours for the longest ocean crossings. The spherical shape of the Earth can focus a tsunami down to a small area on the other side of such an ocean.
- The amplitude (wave height) of tsunamis is generally at most a few meters in the open ocean. When tsunamis enter shallower coastal waters, the amplitude can increase to around 20 to 30 meters.
- The period (the repetition time between crests) of a tsunami can range from a few minutes to over an hour.
- The period of a tsunami wave remains the same from its creation to its end, no matter the changes in its speed or wavelength.
- People on ships on the open ocean do not notice the passing of a tsunami, since rising and dropping by a meter or so over many minutes is extremely slow.
- The Japanese name “tsunami” means “harbor wave”: returning fishers would find wave-induced destruction in their harbor even though they had not noticed any unusual waves out at sea.

How does a typical tsunami evolve from its creation to its end? Figure 11-9 sketches the creation of a tsunami by an earthquake on

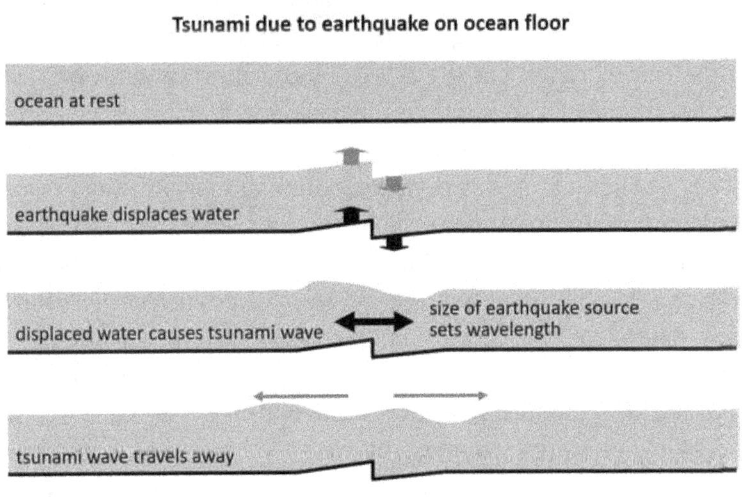

Figure 11-9: Creation of a tsunami by an earthquake on the ocean floor (*from top to bottom*).

the ocean floor. In an earthquake, soil breaks and slips because of the movements of the continental plates (often called tectonic plates): the slipping can be vertical or horizontal or both. If an earthquake raises or lowers the bottom of the ocean, the column of water above it is also rapidly raised or lowered, causing a bump or dip in the ocean surface (both a bump and a dip are possible at the same time, as sketched). This normally happens in a matter of seconds, faster than the water can relax back to its flat surface shape; the change in height of the water surface is usually only a meter or so. An earthquake with only horizontal motion will create a much weaker tsunami because the ocean floor may simply slide along the bottom of the water with little friction.

The width of the earthquake zone shown in Figure 11-9 can be tens or hundreds of kilometers: this is the zone in which the ocean bottom rises or dips. The earthquake thereby causes an initial bump or dip of that size on the ocean surface; it is usually much wider than the depth of the ocean. This horizontal size greatly influences the wavelength of the tsunami. You can easily simulate this size effect in a quiet basin or bathtub or swimming pool: place a fairly flat object (like a spoon or hand or board) horizontally below the water surface, then suddenly lift it up or push it down a bit; this will lift or drop the water surface and create waves with wavelengths similar to the size of that object; for example, a small spoon will create waves with a wavelength of about a centimeter, but if you use your whole hand or a larger wood board, the wavelength will be much larger. (You will probably also generate waves of shorter wavelengths, but these travel more slowly and die off faster.)

An earthquake can continue shaking for a minute or so, often like a zipper, whereby the soil breaks gradually along a line of tens to hundreds of kilometers: this extends the initial tsunami wave over the whole length of the earthquake line. Such a long earthquake is like a long string of multiple wave sources acting with various delay times: the waves from the different sources will be added together, forming a more extended and more complex tsunami. This complexity can increase further if the initial earthquake causes secondary earthquakes or landslides. For simplicity in our discussion, we will assume a short earthquake line: a single source.

We have mentioned the long wavelength of the new tsunami, but what about its wave frequency, which is directly related to its

repetition period? As mentioned above, the period can be very long, from minutes up to about 1 hour: ***how can a tsunami's period be so long?*** This long period is a direct result of the large extent of the earthquake and the initial water rise or dip that it created. To better understand that, Box 11-2 draws an analogy with the more familiar children's swings and the pendulums of clocks.

BOX 11-2 — WHY IS THE PERIOD OF TSUNAMIS SO LONG? Let's consider the initial water profile of a tsunami and ask: how fast does water react to this initial shape?

Imagine an earthquake which lifts water by 1 meter everywhere over an area of 10 by 10 kilometers (this is actually a rather modest earthquake[33]). Assuming an ocean depth of 1 kilometer = 1,000 meters, that is 100 billion cubic meters of water that have been raised by 1 meter; it is equivalent to about 100 billion large bathtubs or 1 billion small swimming pools full of water, with a weight equal to roughly 100 billion small cars, all raised by 1 meter. (To get a feeling for a billion, which is a thousand million or 1,000,000,000, remember that there are about 8 billion people on Earth.)

That water will now fall and sink under the influence of gravity by about 2 meters, while pushing some water outside the initial earthquake zone of 10 by 10 kilometers. (Why 2 meters? Because water falling from a height of 1 meter will first speed up and then penetrate under the normal water surface, where it will next be slowed down and stopped by the surrounding water when it reaches about 1 meter below the normal surface.) The water that is pushed

(Continued)

[33] The Tohoku earthquake of 11 March 2011 broke over an area of about 300 by 150 kilometers, deepening the ocean by 0.4 to over 1 meter, according to the open-access article "2011 Tohoku earthquake and tsunami data available from the National Oceanic and Atmospheric Administration/National Geophysical Data Center", by Paula Dunbar, Heather McCullough, George Mungov, Jesse Varner and Kelly Stroker, in *Geomatics, Natural Hazards and Risk*, volume 2, issue 4, page 305, 2011. This extent multiplies the 100 million cubic meters of my smaller example by at least 5 to reach over 500 billion (half a trillion) cubic meters. Another article published in the same month (December 2011) estimates an area of about 15,000 square kilometers and a lifting (!) of the ocean floor by as much as 5 meters; these numbers also multiply my 100 billion by roughly 5; see subscription-only "Insights from the great 2011 Japan earthquake", by Thorne Lay and Hiroo Kanamori, in *Physics Today*, volume 64, issue 12, page 33, 2011.

(*Continued*)

aside will partly rise above the normal water surface, forming the crest of an outgoing wave.

This motion has some similarity with that of a child's swing or pendulum: the initial earthquake-induced rise of 1 meter is like the initial rise of the swing or pendulum before they are released; the water, swing and pendulum are then pulled down by gravity and start to move sideways; this sideways motion turns next into upward motion further away.

The analogy with a swing or pendulum is very useful because the motion of a swing or pendulum is more familiar to us and easier to experience personally. In particular, it will become clear how the period of a tsunami can become so very long. A crucial aspect of this analogy is that the period does not depend on the mass (or weight) of the object that is swinging: all that matters is its size, so we don't need superheavy swings or pendulums to understand tsunami periods.

An easy experiment to do at home is to use a simple pendulum made of a light string to which you attach any object that you choose, for example, a spoon. Make the spoon swing back and forth as you hold the other end of the string in your hand. Now increase the length of the string: as the string gets longer, you will notice that the swinging period increases as well (specifically, the period grows with the square root of the string's length, so increasing the length by a factor of 4 will double the period).

A string length of about 1 meter will give a period of close to 2 seconds (this is typical of grandfather clocks, in which a 0.994-meter pendulum gives two individual swings of 1 second each). Now think of an earthquake 10 kilometers across: the analog would be a pendulum with string length of the same size, about 10,000 meters: the square root of 10,000 being 100, we get a period of 100×2 seconds ~ 3 minutes. This square-root argument is valid for deep-water waves, but tsunami waves are shallow-water waves: for them, the square root is replaced by direct proportionality, so the period actually grows even faster with wavelength, easily reaching an hour or so in practice.

You can easily verify that the mass (or weight) does not matter in a pendulum. To your string, attach two or more spoons instead of one spoon: you should find the same period regardless of the number of spoons, as long as you keep the string length the same.[34] An ocean of water will not change that!

[34] Three spoons swinging on one string have the same period as three spoons swinging separately on three strings. Therefore, that period will be the same as the period of one spoon swinging on one string.

Can we now understand the high speed of tsunamis? We saw that it takes many minutes to move a wide ridge of a water wave by tens of kilometers. That translates directly to hundreds of kilometers per hour. But we must again remember that the speed of waves on water is limited by the ocean depth, because a tsunami is a shallow-water wave whenever its wavelength is longer than twice the water depth: it cannot exceed about 700 km/h for a water depth of 4 kilometers (which is an average depth for all oceans) or about 1,000 km/h for a depth of 8 kilometers (the deepest points in the oceans are around 11 kilometers).

Now we have tsunami waves with wavelengths of many kilometers (see the third sketch from the top in Figure 11-9), but with amplitudes of only a meter or two. If you were on a boat, you would not notice such a tsunami wave: you would not feel a rise by a meter and then a dip by a meter over a period of many minutes. Also, the wave itself would be invisible: the next crest is beyond the horizon.

Despite its small amplitude of a meter or so, such a tsunami carries enormous amounts of energy: as we saw, this energy is comparable to dropping at least 100 billion small cars by 1 meter each, which is like dropping at least 1 billion small cars by 100 meters each! Such tremendous amounts of energy can cause enormous damage when reaching a coast, even when spread out along entire coastlines.

The simulation in Figure 11-10 shows the spreading of the great 2011 Tohoku Tsunami throughout the Pacific Ocean. The color indicates the maximum wave height (up to 2.4 meters close to Japan) and therefore also the tsunami's relative energy at each point. It is very impressive that the wave heights are well above 10 centimeters along most of the coastline all around the Pacific Ocean. The irregular appearance of the energy flow across the ocean is due mostly to the shape of the ocean bottom, including the presence of many Pacific islands. Since the speed of a tsunami wave depends on the water depth, it can turn (by refraction) into other directions, especially near islands.

Due to the spherical shape of the Earth, a part of the energy would be focused on the point opposite Japan on the globe, namely in the large blue patch in the South Atlantic at bottom right of the map: we indeed see red energy streams aim in that direction (in this map's projection they are curved), but South America prevented that focusing

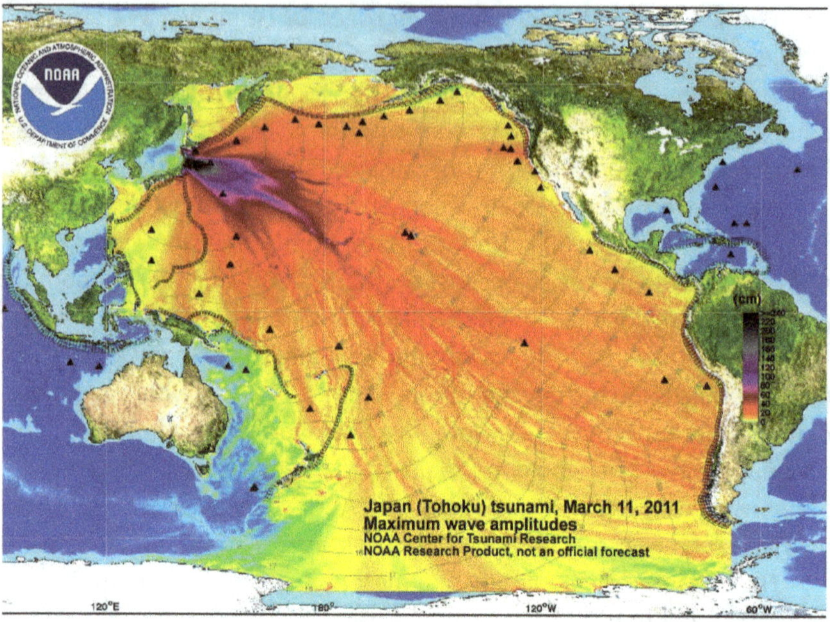

Figure 11-10: Simulation of the Tohoku Tsunami of 11 March 2011 showing maximum wave heights (and therefore also energy) across the Pacific Ocean. The light gray curves mark the arrival times of the tsunami after the earthquake near Japan, up to 21 hours at the Chilean coast, where the wave height was still around 20 centimeters on average. The triangles show tide gauges where wave heights were measured, which helped model this tsunami. A narrated animation of this tsunami is available at https://www.youtube.com/watch?v=Lo5uH1UJF4A. (*Source*: NOAA, in the public domain, http://nctr.pmel.noaa.gov/honshu20110311/Energy_plot20110311_no_tg_lables_cropped_ok.jpg.)

from happening: Chile's long coastline received much of that energy in the form of 20-centimeter waves.

We can now shift our attention to what happens when a tsunami approaches a coast.

Does an approaching tsunami start with a dip or a rise? It is often said that tsunamis arriving at a coast start with the retreat of the water, like a relatively fast low tide that can be deeper than normal low tides. That is generally true; however, it depends to some extent on the shape of the initial earthquake. We must distinguish between tsunamis due to a rising *versus* a dipping ocean bottom. At the left in Figure 11-9, the ocean bottom has risen: this causes a rising wave going to the left;

at the right in Figure 11-9, the bottom has dipped: this causes a dipping wave going to the right. Later, when the tsunami wave reaches shore, it will normally first dip if it came from a dipping earthquake zone; and it will normally first rise if it came from a rising earthquake zone (but it is observed that it may first dip, for reasons that are not yet understood). The same earthquake can cause both a rising tsunami and a dipping tsunami, traveling away in opposite directions, as drawn in Figure 11-9.

This explains why many tsunamis, when seen from shore, often start by dipping: the seawater ebbs away like a fast low tide, often to unusual depths, inviting curious observers to follow it, only to be surprised (and often killed) by a fast rise moments later. Other tsunamis start with a rise, like a rapid high tide, which allows only minutes to escape.

Fortunately, the monitoring of earthquakes gives more warning time because it can take many minutes to many hours for a tsunami to reach a shore after being created by an earthquake.

What happens to a tsunami near a coast? As a tsunami approaches a coast, the water depth decreases more or less gradually. This slows down the tsunami and increases its amplitude (while keeping its period constant). A tsunami slows down to about 160 km/h (100 mph) in a water depth of 200 meters (670 feet), 80 km/h (50 mph) in a depth of 20 meters (67 feet), and 35 km/h (22 mph) in a depth of 10 meters (33 feet); the speed drops to zero only after flooding the coast.

At the same time, the height of the tsunami increases: its height can reach around 20 to 30 meters (67 to 100 feet) in shallow water; that is the height of a building with 5 to 10 stories! People on shore can start to see such a tsunami approaching on the horizon, especially if they know a tsunami is coming, and particularly if the tsunami wave already breaks offshore into a high wall of whitecaps (this breaking happens mainly for the largest tsunamis in very shallow bays with a gentle slope up to a beach).

The wavelength of a tsunami is still large near a coast, such as 10 kilometers (6 miles) at a depth of 10 meters (33 feet). This means that the approaching wavefront seen by a person on the shore is followed by a nearly flat and level water table of several meters in height (around 10 feet) that extends for several kilometers/miles toward the horizon. And since the period of this wave is many minutes (from one crest to the next), it can take many minutes before this elevated water table begins to dip again. During those many minutes, the wavefront continues to go forward up and over a beach, or over docks in a harbor far into the streets of a city; the tsunami wave will flow over any walls or dunes lower than its height; its momentum can also make it overflow higher obstacles. For minutes, it will flood any low-lying land onshore up to its own height, which can be 20 to 30 meters (67 to 100 feet) for the strongest tsunamis.

The weight of water rushing inland is such that smaller houses are easily destroyed (wooden houses often float with the flood); cars float on the flood; boats also can float far inland; see Figure 11-11. Many people are also swept away or trapped if they could not flee earlier.

Only minutes later will the tsunami wave gradually stop rising and slowly start flowing back toward the ocean, as its first crest is followed by a trough. If there are protective seawalls, as is common in tsunami-prone Japan, for example, water can be trapped behind the seawall in

Figure 11-11: An aerial view of the Tsunami stricken Meulaboh, Sumatra, Indonesia (IDN), during Operation Unified Assistance. (*Source*: by Phan Jordon R. Beesley, The U.S. National Archives, in the public domain, https://nara.getarchive.net/media/an-aerial-view-of-the-tsunami-stricken-meulaboh-sumatra-indonesia-idn-during-5434bf.)

city neighborhoods, keeping the water level there high for a longer time, while the water in the sea has already receded to a much lower level.

Several minutes after the trough has reached the coast, the process may start again, as the second crest floods the coast once more. Usually, only the first few crests are high enough to cause much flooding.

Each flow back toward the ocean can bring along large quantities of damaged structures and debris (some of which can even float across the ocean). Many drowned bodies are also swept out to sea, never to be found again.

This kind of slow-motion catastrophe is spell-binding: we can experience it through many videos that are available online, especially for the 2011 Tohoku Tsunami[35]; they show how the giant waves approach, people flee, the water level rises to upper floors of houses, cars and boats float along streets, houses are dragged away and fall apart, and debris moves everywhere; then things quieten down before the water level slowly drops, revealing cars on roofs and boats on streets, while sucking masses of debris out to sea. The 20-minute period of the Tohoku Tsunami means that one such cycle of water rising and dipping takes a full 20 minutes; rarely do we see in videos a second such cycle, as residents flee to safety after witnessing and surviving the first cycle.

A different scenario is seen near rocky coasts. The simplest case is that of steep cliffs with deep water: a tsunami is reflected by such a coastline. This gives rise to extreme interference between incoming and outgoing wave crests, which can explode into towering whitecaps. Reflected tsunami waves going out to sea are typically turned back toward the coast due to gradual refraction in deepening water (see Section 12.5 regarding gradual refraction): they can continue along the

[35] Just search online for videos about the 2011 Tohoku Tsunami. The approach to shore of the tsunami is well shown in "Japan Tsunami 3-11-2011" by Earthquake Engineering Research Institute at https://www.youtube.com/watch?v=3618dZoiaPE. A harbor view with a useful timestamp to time the tsunami, including a weaker second wave, is in "2011 Japan Tsunami — Tonicho Town, Kamaishi. (Full Footage)" by 2011 Japan Tsunami Archives at https://www.youtube.com/watch?v=k6lYBYhkHk4. A dramatic street scene is in "Tsunami in Japan [HD] 3.11 first person FULL raw footage" by Kanpai! at https://www.youtube.com/watch?v=GpuLlIrUYsI. Various situations are compiled in "Japan Tsunami Compilation | Ocean Overtops Wall | Tsunami In Japan" by Ho Quoc Trung at https://www.youtube.com/watch?v=P6QaMoUr2w4.

coast for some time in that manner, creating more destruction along the way.

Another very dangerous aspect of tsunamis is called the **run-up**. We are familiar with this effect as we walk along a beach and try to stay as close to the water as possible without getting our feet wet. We then frequently have to outrun waves that shoot up the beach unexpectedly far. The same "run-up" effect happens with tsunamis, but of course on a much larger scale: it can be tens of meters in height, usually on a hill or rocks: the Tohoku Tsunami caused a maximum measured run-up of 39 meters (130 feet) in one location, the largest ever in Japan.[36] The run-up water will come back down to the sea and may create another wave going outward.

11.10 Bow Waves, Stern Waves and Wakes: Boats and Rocks

Boats produce familiar waves with a very characteristic V-shape; a river flowing past rocks and other obstructions does so as well: see Figure 11-12. These are all **wakes**, because they "wake up" a water surface. We have already seen boat wakes in Figures 1-1a, 11-2b and 11-8c (in front of the tidal wave). Waves in rivers are similar but usually more chaotic because there are many rocks, causing many wakes that crisscross each other.

Wakes of boats consist of different parts: the **bow wave** created by the front of the boat; the **stern wave** due to the back of the boat; and the **wake wave** trailing behind the boat, often white with bubbles. The overall wave pattern is simpler at larger distances: we will focus on that part.

A common aspect of wakes on water is that they are stable, even though they move relative to the water. Wakes are steady and permanent waves if you look at them from the perspective of the source of the waves, whether it is boats or rocks. For example, if you

[36] The largest recorded run-up in the world is not due to a "normal" tsunami, but an on-shore rockslide triggered by an earthquake. In 1958 in Alaska's Lituya Bay (off the Pacific Ocean) the rockslide fell into the bay and pushed water up the opposing mountain to an altitude of 524 meters (1720 feet). See https://en.wikipedia.org/wiki/1958_Lituya_Bay_earthquake_and_megatsunami.

Figure 11-12: *Top*: Wake of a speedboat. *Bottom*: Fixed waves in a stream (flowing from right to left). The stream is about 50 centimeters wide. (*Source of top photo*: extracted from "Boat sailing the Lyse fjord in Norway. Picture taken from the Preikestolen.", by Edmont, under CC BY-SA 3.0, https://commons.wikimedia.org/wiki/File:Fjordn_surface_wave_boat.jpg. *Bottom photo* by the Author.)

are on a boat and look back at its wake, that wake does not change over time (unless the boat changes direction or speed), while the water flows backward from your perspective. It's as if the wake were rigidly dragged along by the boat. However, an immobile swimmer seeing a boat passing by will certainly feel its wake passing by as well.

Similarly, if you look at wakes due to rocks in a river, they also don't change and don't move (unless the rocks move), while the water flows downstream. A single rock sticking out of the flowing water will produce a permanent V-shaped wake downstream that appears to be anchored to that rock. For a leaf or a kayaker floating downstream on a river, the wake due to a rock is a wave that seems to move upstream, even though those waves in fact remain in place relative to the rocks. A waterfall is an extreme example: it is fixed as water falls through it.

The V-shape of wakes reminds us of sonic booms (see Section 3.6): with sound in air, if the speed of the source of the sound is larger than the speed of sound, we get a strong V-shaped wavefront (often called a Mach cone) that produces a sonic boom (see at bottom right in Figure 3-3); such source speeds are called supersonic. With boat wakes, the same happens most of the time: usually the speed of a boat is greater than the speed of waves on the water surface, so boats are "supersonic" relative to the water waves and produce V-shaped waves.

However, we have to be careful here, and this will be important in the following: the speed of waves on the surface of water depends very much on their wavelength, unlike that of sound in the air. So we must ask: what counts as "supersonic" for boat wakes on water? Let's discuss this now.

We can argue that a boat will be "supersonic" relative to all waves that move slower than the boat. Since a moving boat produces waves of all wavelengths and thus all speeds, it should cause many V-shaped "sonic booms" with all the slower waves that it produces: this would be a fan of superposed V-shapes rather than the single V-shape seen in supersonic flight. Indeed, this is observed, especially at slower boat speeds and lower river speeds around rocks: see the left two simulations in Figure 11-13 (we will discuss higher speeds further below).

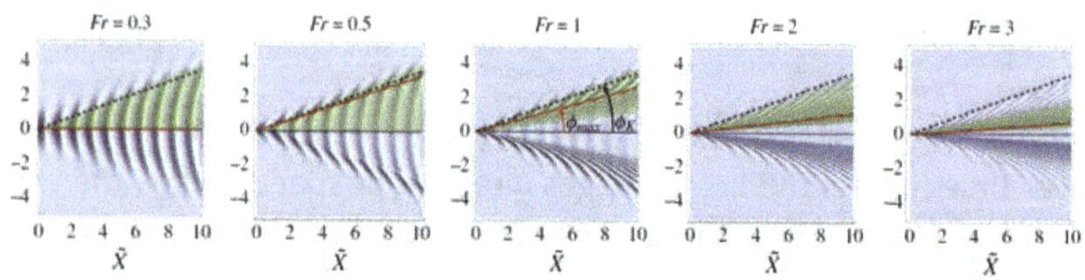

Figure 11-13: *From left to right:* Plots of a boat wake with increasing boat speed (which is proportional to the "Froude number" Fr, such that speeds above Fr = 0.5 are "fast" speeds). In the top half of each graph, the dashed line represents the "Kelvin angle" of the V-shaped wake; the green area is more intense; the red line shows where the wave has the maximum wave height. (*Source*: "Kelvin wake pattern at large Froude numbers", Alexandre Darmon, Michael Benzaquen and Elie Raphaël, *Journal of Fluid Mechanics*, volume 738, page R3, 2014, https://doi.org/10.1017/jfm.2013.607, reproduced with permission, with freely accessible preprint at https://arxiv.org/abs/1309.6751v2.)

To better understand the boat wakes shown at left in Figure 11-13, you may watch a very clear geometrical explanation online.[37] I summarize its main points in Box 11-3.

BOX 11-3 — WHAT IS THE BASIC STRUCTURE OF BOAT WAKES? Consider the boat's wake modeled in Figure 11-14, built up of many straight wave crests shown as very thin lines. First, the boat drags behind it a wave that goes at exactly the speed of the boat itself and in the same direction: its crests are shown as a series of vertical dashed lines. To that speed corresponds a particular wavelength; if the shape of the boat's hull favors that wavelength, it will be amplified. Let's call this the straight-ahead wave.

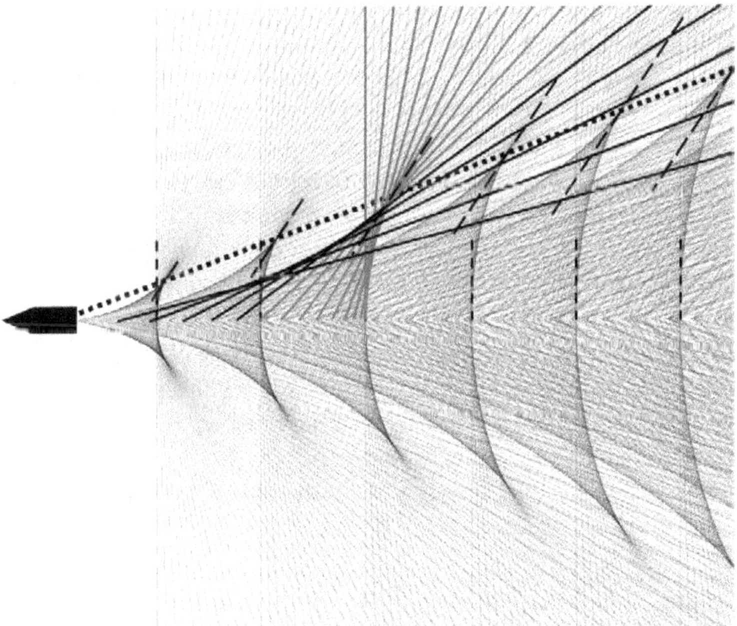

Figure 11-14: Simple simulation of a boat's wake. The entire wake moves rigidly to the left at the same speed as the boat. The very thin lines are wave crests of simple individual straight waves: where they are dense, the wake has high amplitude. Thicker lines mark a few crests of the third set of waves behind the boat.

(Continued)

[37] See video "Why Do Boats Make This Shape?" by minutephysics at https://www.youtube.com/watch?v=95sQcSulRFM.

(Continued)

Waves that go faster than the boat's speed spread out in all directions (not shown), including ahead of the boat: their different wavelengths greatly reduce the chance of constructive interferences, so they tend to cancel each other out. Indeed, you rarely see strong waves moving ahead of a boat.

Waves that travel slower than the boat are in the "supersonic" regime of the boat. As mentioned earlier, each wavelength produces a V-shaped wavefront with a different opening angle, since each wavelength has a different wave speed: these wavefronts are the very thin lines in Figure 11-14. Since the waves repeat periodically with their individual wavelengths, we get a succession of V-shaped wavefronts.

Waves that are slightly slower than the straight-ahead wave go slightly to the left and right of the boat's direction; they ride on top of, and just behind, the wavefronts of the straight-ahead wave, creating a curved wavefront. The slower the wave, the more it bends the wavefronts to the sides, as we also see clearly at left in Figure 11-13: the result is a series of circular combined wavefronts. It so happens that the radii of those circular wavefronts are whole multiples of the distance between the boat and the first circular wavefront, since they are all based on the repetitive straight-ahead wave.

As we add more slower waves, at some point, the circular wavefronts stop growing outward: see how the thick gray lines turn into the thick black lines in the third wave; the black lines together create a new outgoing wavefront closer to the boat. (The same happens in the other waves.) The gray and black lines pile up in a small area, often resulting in a stronger peak in the wake due to constructive interference, depending on the boat's hull shape. Outside this area, the resulting superposition of waves turns from constructive to destructive: this is the outside limit of the V-shape (shown as a dotted line). You can see this situation in the video "Why Do Boats Make This Shape?" referenced above.

We can argue that the faster the boat, the narrower should be the fan of V-shaped waves, because the opening angle of the V-shape is given by the relative speeds of boat and wave (just as with the sonic boom in air). However, this narrowing was not observed in reality for a long time! The famous British physicist **Lord Kelvin** in 1887 proposed an explanation that predicted an opening angle of the V-shape of boat wakes of $2 \times 19.47°$, which actually agreed with the observations (and is seen in Figure 11-14). This V-shape angle was accepted as universally correct for all boats and all speeds until recently: starting in the 1990s, it was pointed out that the V-shape angles of fast boats are clearly smaller. If you look at the wakes of fast boats, you will see that they indeed produce sharper V-shapes, with a smaller opening angle than those produced by slower boats (see the sharp V-shape at the top of Figure 11-12).[38]

In 2013, a simple explanation was proposed[39] for the smaller V-shape angles of fast boats: a boat cannot produce waves with a wavelength longer than the boat itself, which is also related to the change to a plane-like behavior of the boat (the boat rises partly out of the water, adopts a different angle and moves more like an airplane on top of the water). This model puts an upper limit on the wavelength produced by a boat and fits the observations very well. At the same time, the V-shape angle now also depends on the speed of the boat. Figure 11-13 illustrates this with simulations: from left to right, the speed of a given boat increases; the red lines mark the sharper V-shape. You can also get roughly the same result in Figure 11-14 by removing the steeper wave fronts, such as those drawn as thick gray lines: now the zone with dense wavefront overlaps no longer has so many waves in it; this removes part of the sides of the wake, leaving a sharper V-shape; also, the inner part of the wake, where the straight-ahead wave was strong, now only has weaker waves.

The simulations of Figure 11-13 actually show a bit more complexity: the wider low-speed V-shape (marked by the black dashed line at an angle of 19.47°) is still visible in the form of weaker waves that spread out farther than the most intense and visible red V-shape. If you look

---

[38] To best see the opening angle of wakes requires looking straight down. One way to do so is to use aerial or satellite images of boats. For example, in Google Earth you may look at a place where you know that there frequently are speedboats; however, you will only see their wakes when the Sun shines from the right angle. One place I recommend is shiny portions of Lake Mohave (part of the Colorado River) on the California-Nevada border in the US: set Google Earth for the date 9/2010 (to get a good Sun angle and many boats) and look between N35° 13′ to N35° 21′ and between N35° 30′ to N35° 34′; there you will see many very sharp V-shaped wakes due to fast speedboats.

[39] "Ship wakes: Kelvin or Mach angle?", by M. Rabaud and F. Moisy, *Physical Review Letters*, volume 110, page 214503, 2013.

carefully at aerial or satellite photos, you may recognize both V-shapes trailing behind speedboats.

We must also realize that the exact shape of the boat can strongly affect the wave profile, especially at low speeds: a particular hull shape may favor certain frequencies and thus certain wavelengths over others. This is one important reason why the simulated models shown here do not match the observed wakes of all boats, such as those in Figures 1-1a, 11-2b, 11-8c, and 11-12.

An interesting class of boats uses **hydrofoils**. These are essentially underwater wings: instead of creating lift by gliding inside air, they create lift by gliding inside the water. Thereby, these hydrofoils can lift the entire hull of the boat above water, greatly reducing friction and drag and thus considerably increasing speed. Hydrofoils hardly disturb the water surface: they are built as thin sticks that are dragged through the water surface; the foils under the water surface also create very little wavy motion. Therefore, they create almost no waves. The small waves that they do create form very narrow V-shapes, due to the high speed of the boat and the small wavelength and thus small speed of these waves.

11.11 What have We Learned in this Chapter?

Waves on the surface of water (and other liquids) are very visible and diverse: they therefore are highly educational and relevant to the understanding of other types of waves. At the same time, their mechanism is relatively more complex than for other types of waves: both gravity and surface tension play a role. Their motion is also more complicated: in a simple model, the water follows wheel-like circular paths that become smaller with depth, and flatter ellipses (ovals) in shallower water. Furthermore, the speed of waves on water changes dramatically from waves in a drinking glass to the tides in the oceans, also causing intricate boat wakes. Particularly varied is the behavior of water waves that approach a coast, and especially a beach. Special waves of interest are the tides, tidal waves, rogue waves, tsunamis and boat wakes.

12

Light and Electromagnetic Radiation

In this chapter, we focus on light waves and other electromagnetic waves (EM waves such as x-rays and radio waves). The physical principles of EM waves are quite different from those of string waves, sound waves, and waves on water. Nevertheless, EM waves behave in many ways very much like those waves (as well as other waves, which we will discuss in later chapters). For example, we have learned earlier about wave interference, reflection, refraction and diffraction: we will be able to fruitfully apply these concepts to EM waves, and to light in particular.

We can thus discuss a multitude of uses and optical effects of light, such as: mirrors, prisms, lenses, mirages, retroreflectors, optical fibers, rainbows, metamaterials, and holograms. An intriguing topic will be whether light (and other EM waves) can be viewed only as a pure wave or also as a particle: this subject will open the door to quantum physics, which we will deal with in Chapter 13.

12.1 What are Light and EM Waves?

Even though light is abundant in our daily lives, scientists have really understood light only since the early 20th century, about a hundred years ago. ***Why did it take so long to understand light?*** In a nutshell: light is more complex than it appears at first sight! In particular, light waves are more complex than waves on strings or sound waves. Nevertheless, in this book, we will be able to avoid those complications so that we can use the wave concepts of earlier chapters, showing again how general those concepts are. In Box 12-1 (further below), I try to give an intuitive feeling for light, with a simple description of the complexities involved; you may safely skip that box!

How complex is light? **Today, we know that light is a combination of electricity and magnetism, which are connected to each other by relativity, forming electromagnetic waves** (commonly abbreviated as **EM waves**).

It is also known that light always travels at the same speed in vacuum, no matter whether the light source (such as a headlight on a car) moves fast toward you, or fast away from you, or stays with you. By contrast, a ball thrown at you from a car approaching you will fly much faster toward you than a ball thrown back to you from a car driving away. Thus, the speed of light does not depend on its "launch speed". This is one of the remarkable lessons of the theory of relativity introduced by the German physicist **Albert Einstein**.

We also know that x-rays and radio waves are very similar to light. In fact, they are light with different frequencies: so, they are invisible light. Furthermore, light travels not only in substances, especially transparent ones like glass and water, but more remarkably also in vacuum, such as from the Sun or other stars to Earth.

Light is actually even more complex than all that. For instance, as we will discuss in Section 12.10, it is now well established that light can behave either as a wave or as a particle (like a small ball), depending on the circumstances: light can interfere harmlessly like a wave, but it also can deposit energy into your skin like a ball hitting you.

Fortunately, **in many situations of interest, we can represent light by a simple single wave**, just like the 1D string waves of Chapter 2 and the 3D sound waves of Chapter 3. In many cases, we can even use the very simple model of ray tracing from Section 10.7.

Therefore, we will be able to use the ideas of the preceding chapters to describe the behavior of light. We do not need to understand the more surprising aspects of light to describe many of its important properties that we see every day. This allows us to discuss mirrors, lenses, lasers, optical fibers, rainbows, mirages, holograms, *etc.*: those are the main subjects of this chapter.

Let us first explore some important basic properties of light, and of EM waves more generally.

What is the speed of light? In vacuum, such as between planets, moons, stars and galaxies, there is a unique **speed of light**: by any human standard, it is extremely large! **The speed of light (and of all EM waves) in vacuum is exactly equal to 299,792,458 meters per second. This is about 300,000 kilometers per second, or 1,080,000,000 kilometers per hour, or 186,000 miles per second, or 671,000,000 miles per hour.**

The speed of light in vacuum is called a **universal constant** by physicists, not only because it is constant everywhere in vacuum and forever (as far as we know), but also because it is very fundamental to many physical processes, including electromagnetism and EM waves. For example, this value is used to define the length of the meter: the meter is the distance covered in vacuum by light in 1/299,792,458 of a second. Although this definition of the meter may seem impractical in daily life, it has the virtue of being more reliable and precise than any other method to define the meter (earlier, the meter was defined as the length of a metal bar kept safely in Paris, France, which was more convenient but less reliable and less precise).

An important property of the speed of light in vacuum is that it is an upper limit on the speed of any object or signal: this is another fundamental property of the theory of relativity. There are situations where waves do travel faster than the speed of light (for example, x-rays in glass), but such waves cannot carry information at that higher speed.

On the other hand, light and other EM waves travel less fast in most substances. For example, in the Earth's atmosphere, light travels slightly slower than in vacuum, by about 0.03 to 0.05%; in most liquids and solids, light travels much slower than in vacuum, by about 1.3 to 7 times slower, as shown in Table 12-1. As in Section 10.5, we can define a **refractive index** as the ratio of the wave speeds in two substances. With light, the convenient substance to compare with is vacuum. Thus, we may define for light (and EM waves in general) entering a substance:

(refractive index or index of refraction)
≡ (wave speed in vacuum) / (wave speed in the other substance)

Since the speed in vacuum is usually greater than inside substances, the refractive index will be generally larger than 1. Table 12-1 gives examples for visible light.

Table 12-1: **Refractive index** of various substances for visible light. The metals (marked with *) absorb light very strongly: this results in a relatively low refractive index. (The values in this table are approximate; they are sourced from various lists; the speed of EM waves varies with wavelength or frequency, so this table only lists typical values for visible light.)

Substance	refractive index for visible light
silver*	0.15*
gold*	0.27*
copper*	0.46*
vacuum	1
steam	1.000261
air at 0/20 degrees Celsius	1.000293/1.000273
aluminum*	1.0972*
water ice	1.31
liquid water at 0 degrees Celsius	1.333
lens in human eye	1.4
sugar (50% solution in water)	1.42
plastics	1.5–1.6
glass	1.5–1.7
amber	1.55
sapphire	1.77
diamond	2.42
silicon	3.8–6.6 (depending on wavelength/frequency)

A special category of substances is the metals. Their refractive index is often smaller than 1 (metals are marked in Table 12-1 with*): this would imply a speed larger than the speed of light in vacuum. However, EM waves in metals also have very high absorption, meaning that they easily lose their energy to electrons and other processes in the metal. The result is that EM waves only survive very close to the outer surface of metals and are dominated by the behavior of EM waves outside the metal. As a result, such waves in metals cannot exceed the speed of light in vacuum.

It is difficult for us to imagine the speed of light: it exceeds everything we experience in daily life. We do see light coming from lamps, lightning, the Moon, the Sun and stars, but this does not give a feeling for its enormous speed. Table 12-2 may help by showing how fast light travels across familiar distances.

Notice how every distance shorter than the size of the Earth is covered by light in less than the blink of an eye (less than about 1/30 of a second). Radio and satellite signals need more time because of time

Table 12-2: Time needed for light (and all other EM waves) to travel certain distances in vacuum or air. (1 light-year is the distance traveled in vacuum by light in 1 year, which is about 10 trillion km = 10,000 billion km = 10 million million km = 10,000,000,000,000 km = 10^{13} km. See the section References and Resources for more units of length and time, and their abbreviations.)

Distance traveled	Time needed by light
across a computer (~ 30 cm)	~ 1 billionth of a second
across a house (~ 10 m)	~ 30 billionths of a second
across a town (~ 10 km)	~ 30 millionths of a second
across diameter of Earth (~ 10,000 km)	~ 1/30 of a second
from Moon to Earth (~ 400,000 km)	~ 4/3 of a second
from Sun to Earth (~ 150,000,000 km)	~ 500 seconds ~ 8 minutes
from nearest star, Proxima Centauri, to Earth (~ 42.4 million million km ~ 4.24 light-years)	~ 4.24 years
across our galaxy, the Milky Way (~ 100,000 light-years ~ 1 billion billion km)	~ 100,000 years
from the most distant stars observed (~ 13 billion light-years ~ 130,000 billion billion km)	~ 13 billion years

delays in the electronics. The Moon is barely over one second away, at least for light: the moonlight which you see has left the Moon just over a second ago. But light needs 8 minutes to come to us from the Sun: we see the Sun as it was 8 minutes ago. Other stars are light-years away, meaning that light needs years to reach us from those stars (1 light-year is the distance traveled in vacuum by light in one year). The most distant stars detected by our telescopes are over 13 billion light-years away, a staggering distance; and we see these stars as they were over 13 billion years ago: we don't know what they look like now, or even 10 billion years ago! Radio waves (which also are EM waves) take just as long!

Another important aspect of light and EM waves is their frequency. Partly because of the extremely large speed of light, their frequency is also extremely high. For visible light, the frequency is a bit below 1 petahertz, which is 1 million billion hertz. Remember that 1 hertz is one cycle per second, so with visible light we are talking of 1 million billion cycles per second (which is 1 followed by 15 zeroes: 1,000,000,000,000,000 hertz = 1 petahertz = 1 PHz = 1,000 trillion hertz = 1,000 terahertz = 1,000 THz). We will consider the frequencies of light and EM waves later, in Figure 12-4, together with their wavelengths.

Wavelengths of visible light are slightly below 1 micrometer, which is a millionth of a meter, or a thousandth of a millimeter, or a small fraction of a human hair's thickness. Such wavelengths are totally invisible to our eyes: we of course see the light, but we cannot see the individual ups and downs of the waves (an example of wavelengths that we can see is for waves on the surface of water).

An important property of light is color. For this chapter, it is useful to remember the following: white light (such as sunlight) is a mix of many visible waves with different frequencies, including red, green and blue; other colors can be produced by combining those three primary colors, such as to make brown, pink and gray, as well as white. (My book on "*Colors, Light and Optical Illusions*" deals extensively with this fascinating subject.[1])

---

[1] Michel A. Van Hove, "*Everyday Physics — Colors, Light and Optical Illusions*", World Scientific Publishing Co., 2022.

BOX 12-1 — AN INTUITIVE FEELING FOR LIGHT. ***What is the source of light?*** Of course, light can come from lamps, the Sun and other stars, explosions, chemical and biological reactions, *etc.* Here we wish to go a bit deeper and ask how light actually starts, in other words: what is the physical origin of light? There are two main sources of light: moving electric charges and magnets.

We are familiar with electric charges that cause little sparks after we rub a piece of cloth, especially in dry air. These **electric charges** are extremely small in size: they are invisible even with the most powerful microscopes. These charges are the **electrons** that give their name to electronics, electricity, *etc.* They are also present in every atom and molecule that make up all the substances we live with, from air and water to flesh and steel.

Two electrons push each other away by the so-called **electrostatic force**. The name comes from the fact that this force exists already when the two charges are static, which means stationary or immobile. The electrostatic force pushes each electron directly away from the other electron: we can imagine that force as a "field" of arrows around one of the two electrons, such that the second electron will feel a force along the direction of the arrows, as at left in Figure 12-1.

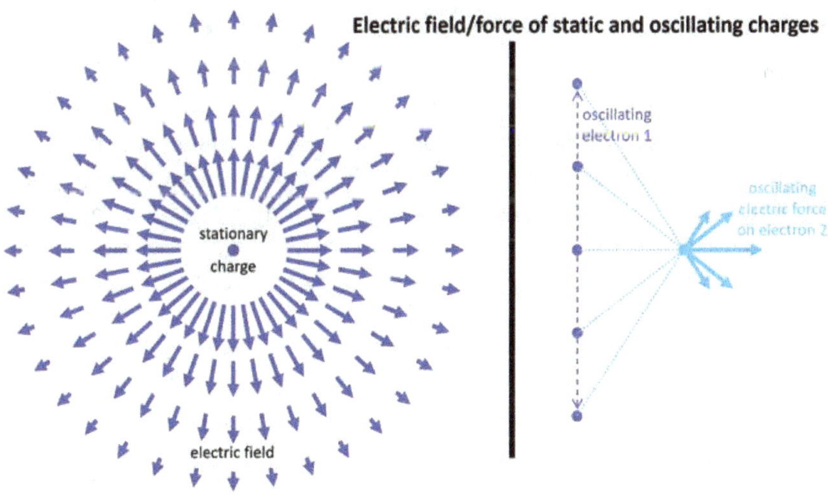

Figure 12-1: *At the left*: Electric field of a stationary charge at center; another charge will be pushed in the direction of the arrows (or in the opposite direction if its charge has the opposite sign, such as negative *versus* positive); the length of the arrows indicates the strength of the force, which becomes smaller with larger distances. *At the right*: electron 1 is oscillating up and down, causing an oscillating electric field and force on stationary electron 2.

(*Continued*)

(*Continued*)

We can imagine this field like a lawn of grass, where the grass blades point away from a central spot; this central spot might have a powerful source of water that pushes all blades of grass to point away from the center. For electric charges, this field is called **electric field**: it shows, at every point, the direction in which another charge will be pushed.

If we now move one electron back and forth, as shown at the right in Figure 12-1, it will push the second electron back and forth as well because the second electron will feel an oscillating force. It is very helpful to see this motion in my Animation 12*1. Using the analogy of a field of grass with a water source, we are moving the source of water repeatedly back and forth: the blades of grass will then swing with the changing direction of the water coming from the source of water. In the language of electric fields: the electric field will oscillate. This oscillation of the electric field is the beginning of an electromagnetic wave. We now need to explain the magnetic part.

ANIMATION 12*1 — See my video WB2 at time 1:59 in its section "**EM waves**" under the title "**An oscillating electric field is created by an oscillating charge**". (See details in the section References and Resources below.)

The second source of light is **magnets**. A typical magnet is the compass: it orients itself approximately toward the North Pole of the Earth, as shown at right in Figure 12-2. We can therefore also talk of a **magnetic field**: at each position on the Earth (and away from the Earth), we can draw an arrow showing in which direction a compass will point if it is placed there; we get a map of compass directions. On this map, it is common to draw lines instead of arrows, as we see at right in Figure 12-2: a compass will orient itself according to the local magnetic field line.

The magnetic field felt by a compass is due to the interior of the Earth. We do not know much about the inner structure of the Earth (see Section 9.4), but it contains liquid metal that forms a giant magnet, which creates the Earth's magnetic field felt by compasses.

Two magnets feel each other, but in a different way than two electric charges: while two electrons push each other apart, two magnets rotate each other and also attract their north poles toward their south poles. Most importantly, for waves, if one magnet oscillates, the other magnet will also oscillate.

We can now ask: ***What is the connection between electric charges and magnets?*** Let's take the example of the **Earth's** magnet: it is due to currents of electric charge flowing in huge circles inside the Earth around its rotation axis. More generally, when an electric charge moves around a circle, it forms a magnet that creates a magnetic field which resembles the magnetic field of the Earth. The magnetic field around a magnet (which is the field felt by another magnet like a compass) is illustrated at right in Figure 12-2. We conclude: **moving electric charges create magnetic fields.**

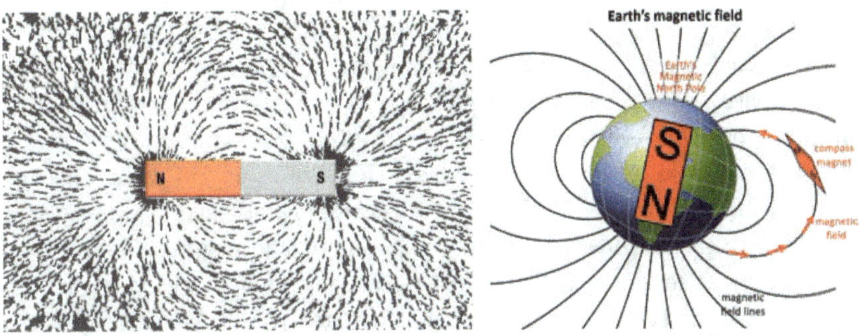

Figure 12-2: *Left*: Iron filings around a magnet. Each filing acts like a small compass: it lines up in the direction of the magnetic field at its location. *Right*: The Earth produces a similar magnetic field on a much larger scale, as if it contained a giant magnet, sketched in red here. Compasses line up with the magnetic field shown here as lines. However, for historical (non-scientific) reasons, the magnet's south pole marked S is actually the Earth's **North Magnetic Pole** (which is located near the Earth's **North Geographic Pole** in the Arctic region).
(*Sources*: *Left*: OpenStax, under CC BY-SA 4.0, https://commons.wikimedia.org/wiki/File:Openstax_college-physics_22.10_ironfilings-coil-magnet.jpg;
Right: adapted from Zureks, in the public domain, https://commons.wikimedia.org/wiki/File:Earth%27s_magnetic_field,_schematic.svg)

Less obvious is that **a magnetic field creates a force on moving electric charges** (such as a current in a metal wire). This electric force can also be viewed as an electric field. Even though this is not so obvious, it is nonetheless extremely important in practice: **electric motors** are based on this magnetically induced force in electric currents. Ask yourself how many electric motors you use every day! Another example is the **aurora** or **polar lights**: charged particles coming from the Sun are deflected by the Earth's magnetic field so that they spiral through the atmosphere and light up air molecules in spectacular displays near the Earth's poles (the lighting up itself is similar to what happens in a neon tube).

Thus, **electric fields can create magnetic fields, while magnetic fields can create electric fields**. In particular, **an oscillating electric field creates**

(*Continued*)

(*Continued*)

an oscillating magnetic field, and *vice versa*. This is directly relevant to an electromagnetic wave: it consists of oscillating electric and magnetic fields that constantly feed back into each other, similar to a couple of people dancing together in lockstep. They are like two inseparable waves, one being electric and the other magnetic, allowing the combined wave to travel forever to infinity, even in vacuum, and without any input of energy after its first creation.

A simple example is a **radio transmitter**: it typically has a conducting metallic wire, along which an electric current oscillates back and forth (this current carries the voice or music signal). The oscillation causes electric and magnetic fields that travel away from the wire as an EM wave. This EM wave can be detected by a distant **radio receiver**, which operates in reverse compared to the transmitter: the traveling electric and magnetic fields cause an oscillating electric current in a wire called an **antenna**; this current then activates a loudspeaker (usually after amplification).

Starting a light wave is more complicated than starting a radio wave because the wavelength of light is very much smaller than that of a radio wave: that would require extremely short wires. Light normally comes from the motion of electric charges inside or between atoms. Moreover, at that small scale, quantum mechanics modifies the motion of electrons; we will look into such effects in Chapter 13.

Sunlight, as well as the light from all other stars, finds its origin in nuclear fusion reactions that constantly take place within the Sun and stars. The nuclear fusion generates immense amounts of heat. Therefore, the interior of stars like the Sun is an extremely hot cauldron of tiny particles (including electrons) that travel fast and also form magnetic fields which become EM waves. The enormous amount of energy carried by these EM waves is sufficient to keep the Earth warm and to supply most of human energy use, including fossil energy from coal, oil and gas (but excluding nuclear and geothermal power used to generate electricity in power plants).

What is the structure of light waves? As we mentioned above and in Box 12-1, light is composed of both electric and magnetic waves that feed into each other: they are therefore called **electromagnetic waves** or **EM waves**. These waves are usually represented as **fields** of arrows or lines which show in which direction charges and magnets would be pushed or oriented (see Figures 12-1 and 12-2). The result looks like the double wave of Figure 12-3, where the electric wave is shown in blue, and the magnetic wave is shown in red.

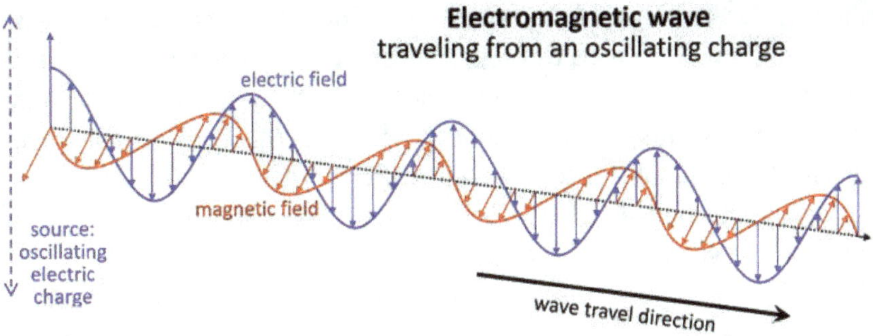

Figure 12-3: Simulation of an electromagnetic wave traveling to the right, created at the left by an oscillating electric charge (such as an electron). In this case, the charge oscillates vertically, so the electric part of the EM wave is also vertical (blue), while the magnetic part of the EM wave is horizontal (red), perpendicular to the electric field.

The origin of such a wave may be an oscillating charge (shown at the left in Figure 12-3): the direction of its oscillation gives the direction of the electric field (vertical in the figure). The magnetic field is perpendicular to the electric field (horizontal in the figure). The two waves are in lockstep with each other: they have exactly the same frequency and wavelength. We see an alternation between the electric and magnetic fields: when one field is maximum, the other is zero, and *vice versa*; this corresponds to the electric and magnetic waves feeding into each other continuously, enabling them to travel together infinitely far. The resulting wave motion is a rigid sliding of the double wave of Figure 12-3 to the right. It is very helpful to see this motion in my Animation 12*2.

> ANIMATION 12*2 — See my video WB2 at time 2:30 in its section "**EM waves**" under the title "**An oscillating charge can create an electromagnetic (EM) wave**". (See details in the section References and Resources below.)

There is an important difference between EM waves, on one hand, and string or sound waves, on the other hand. In string and sound waves, matter is moving back and forth (for example, a guitar string oscillates back and forth, while air molecules move in sound). However, **in an EM wave in vacuum, no matter moves**: the electric and magnetic fields oscillate, but they do not carry matter; this is why EM waves can travel infinitely far through empty space from distant stars. Nevertheless, when an EM wave travels through a substance, including water or glass, it does make electrons and atoms oscillate in that substance, but only a little bit; this causes a loss of energy, so that EM waves do not travel very far through substances, even in water or glass (we will encounter this again with optical fibers in Section 12.6); a substance can also slow down an EM wave, as we discussed earlier, see Table 12-1.

In Figure 12-3, the electric field points vertically, whether up or down: the reason is that the source charge is oscillating vertically. If the source charge had been oscillating horizontally, the electric field would also have been horizontal (and the magnetic field vertical). This preferential direction of oscillation is called light **polarization**. Light can be polarized vertically, horizontally or in any other direction (but always perpendicular to its travel direction). When there are many sources with many orientations, each source produces a different direction of polarization: the combined light from all those sources is then said to be **unpolarized** because no direction is privileged. If the source itself rotates, we can get a rotating polarization, called **circular polarization** (or even **elliptical polarization**). We will discuss the use of light polarization in sunglasses and 3D movies in Section 12.2.

We have mentioned that EM waves can be created by an oscillating electric charge (as illustrated in Figure 12-1). The frequency of that oscillation will then be maintained by the wave: it will be the frequency of that EM wave for as long as it exists. In this way, any desired EM frequency can be generated. We often display such frequencies and wavelengths as an **electromagnetic spectrum** or **EM spectrum**, as in Figure 12-4. In vacuum, each frequency corresponds to a unique wavelength.

An important, but relatively small part of the EM spectrum is the visible spectrum, shown at top in Figure 12-4: it covers the frequencies that the human eye can detect and is also often called the **solar spectrum** (even though the Sun emits EM waves at all frequencies and wavelengths, not only within the visible range). This visible spectrum includes all the **pure colors** which we receive from the Sun: these are waves with a single frequency or wavelength. Among the pure colors, red, green and blue are the most important because our eyes have **cones** that are mainly sensitive to these three **primary colors**. Other

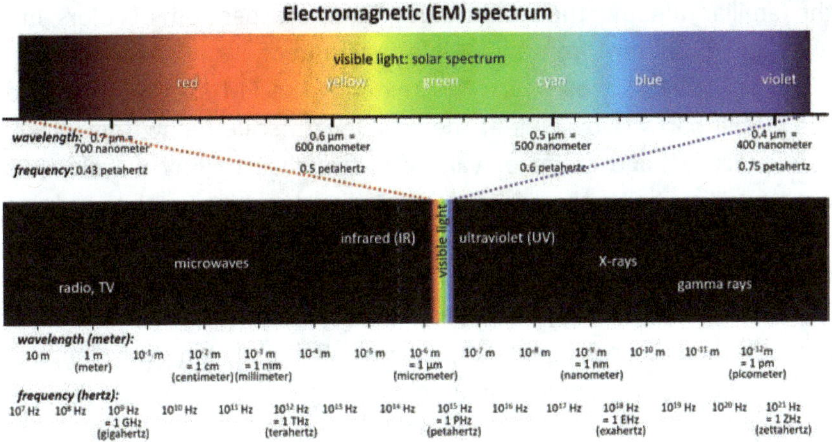

Figure 12-4: Simulation of the electromagnetic spectrum (EM spectrum). *Top*: the visible part of the EM spectrum, often called solar spectrum, showing its main colors. *Bottom*: extended EM spectrum. The extended spectrum continues without end to the left and right; shown here is the part that is most important for humans and technology. There are no firm boundaries between colors or between invisible types of waves. At bottom, the wavelength and frequency scales, including terminology, are only given for reference: they are not needed for understanding this book; powers of 10 are described in the References and Resources section.[2]

colors can be composed by mixing pure colors. For example, white can be obtained as a mix of red, green and blue. My book on "*Everyday Physics — Colors, Light and Optical Illusions*" discusses these and many other interesting aspects of light.[3]

When you see a spectrum due to **sunlight** scattering from a prism or other piece of glass, you see the visible spectrum of Figure 12-4, as we will discuss further in Section 12.3. Section 12.7 below will address

---

[2] The bottom length and frequency scales may be unfamiliar to you: they are called "logarithmic" scales. Each interval is labeled with a number that is 10 times smaller than one of its neighbors; it is like a bank account that each year shrinks to one tenth of its last year's value. Such repeated change by the same factor is called "exponential"; it can shrink or increase. It is difficult to plot numbers that decrease or increase so much: the mathematical function called "logarithm" helps by plotting each interval as equally long. This allows showing a very wide range of values, as in this figure.

[3] Michel A. Van Hove, "*Everyday Physics — Colors, Light and Optical Illusions*", World Scientific Publ. Co., 2022.

the familiar rainbow: the rainbow also displays a spectrum of colors, but it is slightly different from the visible spectrum of Figure 12-4.

The lower part of Figure 12-4 extends the EM spectrum beyond the visible part. If you look at the wavelength scale below this image, you notice a huge range of values: it stretches to tiny wavelengths over a million times shorter than visible light; and it stretches to large wavelengths over a million times longer than visible light. The frequency range is equally wide: see the lowest scale.

The EM spectrum extends even farther than shown in Figure 12-4. In fact, it has no end: to the left, it goes to infinitely long wavelengths, and to the right, it goes to infinitely large frequencies. However, the range shown in Figure 12-4 is of most interest to us as human beings and developers of technology, as discussed in the following.

Starting from the visible range and moving toward the higher frequencies (and shorter wavelengths), we first find **ultraviolet** waves, often abbreviated to **UV** waves. You probably know that many insects are attracted to **black light**, which is ultraviolet light that slightly spills over into the visible blue/violet end of the solar spectrum (it is called black light because most of it is invisible to us): it is used to attract insects to "bug zappers" that electrocute them. Ultraviolet light is also used in **fluorescent lamps**, often called **fluorescent tubes**: they contain a permanent, gentle and quiet lightning stroke that emits constant ultraviolet light, helped by a phosphor coating inside the tube to make it visible, in a color that depends on the gas and coating used. You probably are also familiar with the use of ultraviolet light to combat the forgery of banknotes: proper banknotes use ink that becomes visible only when exposed to invisible ultraviolet light, through conversion of invisible to visible light.

Beyond ultraviolet EM waves, we find **x-rays**, commonly used in medical imaging and health treatments by irradiation, as well as for checking baggage and bodies for security. Figure 12-5 shows the x-ray imaging of a kidney stone.

The higher the frequency, the higher the energy content of a wave: ultraviolet waves and x-rays can be dangerous to the health of human beings. That is why it is recommended to minimize exposure to sunlight and especially to x-rays. Such EM waves are able to damage molecules in the body: in particular, they can cause cancer and genetic defects.

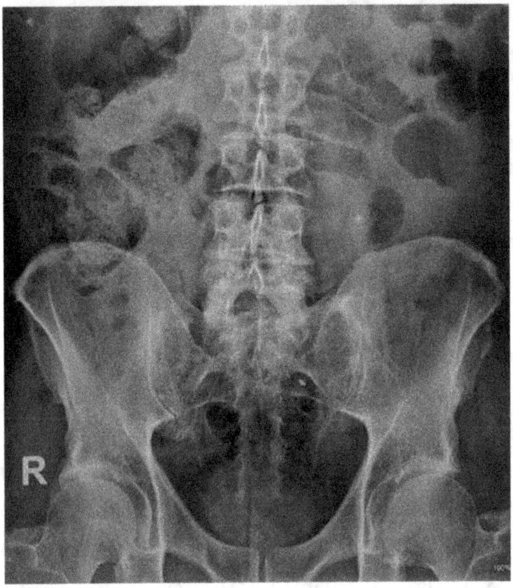

Figure 12-5: An x-ray image shows my kidney stone as a small white dot, slightly above and to the right of the center.

Gamma rays have even higher frequencies than x-rays. Gamma rays therefore also carry more energy, making them even more dangerous and potent. They can even be used to image the contents of trucks and containers, despite their metallic walls.

At lower frequencies than visible light, we first find **infrared** waves, often abbreviated to **IR** waves. When you "feel" **heat** on your skin, you are probably feeling infrared waves: the Sun, incandescent light bulbs, combustion engines, stoves and radiators all emit copious amounts of heat in the form of invisible infrared waves. When you place your hand's palm near your face without touching it, you can feel the heat radiated by your cheek. In fact, infrared waves are sensitive to the temperature of the object that emits them, since higher temperature corresponds to higher energy and therefore higher frequency: this allows **thermal imaging**, as illustrated in Figure 12-6.

Microwaves have lower frequencies than infrared waves. **Microwave ovens** use these waves to heat the water contained in food, and thereby also the rest of the food. Another important application of microwaves is **telecommunications**: we see many telecommunication towers in cities and on hilltops that send data over long distances through the atmosphere in the form of microwaves.

Even lower frequencies are typical of signals sent through the air by cellular/mobile/smart phones, radio and television. These carry very little energy and therefore present a very low risk to health. **Radio** and **TV waves**, like microwaves, can travel long distances through the air, but are more sensitive to weather variations and especially to reflections from buildings, hills, *etc*. Signals from **cellular/mobile/smart phones** travel shorter distances, so we see many of their transmitters and receivers on buildings or near roads.

What are lasers? Lasers are special sources of light waves invented around 1960. The name **laser** is the abbreviation of **light amplification by stimulated emission of radiation**. We have all seen laser beams shining in the sky or being used to scan information from tickets or labels; they are also used in CD, DVD and Blu-ray players. Weak lasers are commonly used as pointers. Others are used in surgery, while powerful lasers can cut thick cloth and metals in manufacturing.

Two characteristics of laser light are: sharpness of the laser beam, and single color, usually red but sometimes green or blue. In Figure 12-7, we see a photograph of a laser beam shining straight into a smartphone camera (this is risky because of the power of lasers: they can damage the camera detectors). This photograph shows another characteristic of laser light: fine "stripes" of bright and dark light, which are interference effects due to its wave nature.

To understand what makes lasers special, let's compare them with normal lamps: see Figure 12-8. A normal lamp emits waves of many colors, in many directions and with random phases. The colors could be those of the Sun spectrum, forming white light. The spreading of this light weakens it as it radiates away. And the random phases cause both constructive and destructive interferences between parallel rays of the same color (meaning the same wavelength): these are called incoherent waves because they are mostly out of sync with each other. It is very helpful to see this motion in my Animation 12*3.

ANIMATION 12*3 — See my video WB2 at time 5:21 in its section "**Lasers**" under the title "**Laser *vs.* normal lamp**". (See details in the section References and Resources below.)

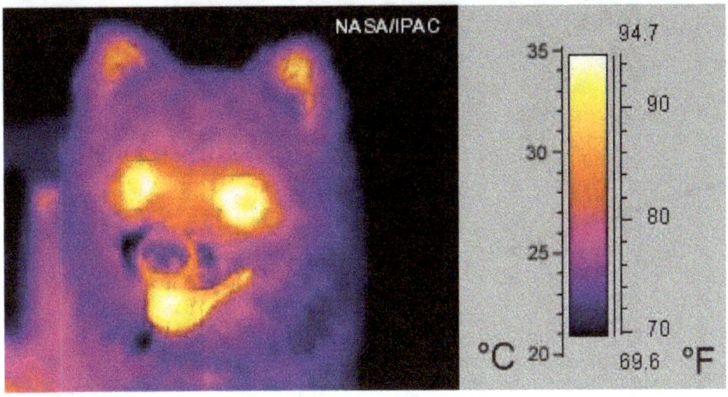

Figure 12-6: An infrared image shows the high temperatures in the mouth, eyes and ears of a dog, while its nose and fur remain cool; the temperature scale at right explains the coloring: white and yellow are warm; purple and blue are cool (these artificial colors are not related to the visible EM spectrum since infrared light is invisible; instead, the infrared frequencies are artificially shifted into the visible range of frequencies). (*Source*: NASA, in the public domain.)

Figure 12-7: Low-power laser (emitting less than 5 milliwatts) shining straight into a smartphone camera (this is not recommended, nor is shining laser light into eyes!). The laser light is purely red (with a wavelength of about 650 nanometers; see the wavelength scale in Figure 12-4), but the camera records it as white in the center, showing overexposure of red, green and blue detectors. The pretty reflections probably occur inside the camera; their fine ribbed structures are due to wave interferences.

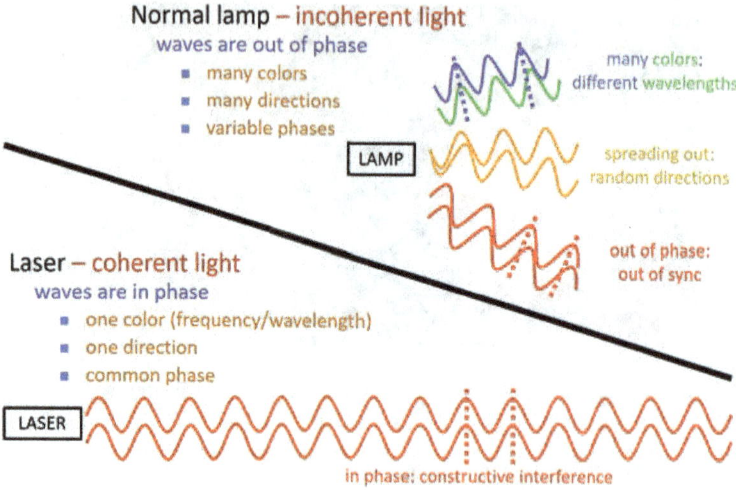

Figure 12-8: *At the top*: Light from a normal lamp. *At the bottom*: Light from a laser.

A laser, by contrast, **emits waves of a single color, in a single direction, and with equal phases, giving more concentrated and powerful light than normal lamps**. The single direction reduces the weakening through spreading of the light as it travels. Common cheap lasers can project a light "spot" of 1 meter across at a distance of about 800 meters, or a spot of about 500 kilometers on the Moon, some 384,000 kilometers away (at the Moon's distance, the intensity would be far too weak to see!). High-power laser light can be reflected from the Moon to the Earth: it is used to measure the distance to the Moon with millimeter precision, by observing the time it takes for a laser pulse to make the round trip, *via* a retroreflector (see later Section 12.6 and Figure 12-32) placed on the Moon's surface.

The single color and equal phases of laser light mean that the laser's waves can interfere constructively and thereby greatly increase the intensity (also by concentrating the light at only one frequency). The equal phases are called coherent, as the waves are in sync with each other.

A laser works by so-called stimulated emission in specially selected materials. This is a kind of snowball effect: an initial light wave travels through the material and triggers other light waves of the same frequency and direction that are in sync with the initial wave. By constructive interference, all these waves together rapidly build up intensity.

12.2 Reflection and Refraction: What is a Mirror for Light?

When we think of a **mirror,** we think of glass. However, ***if we look out through a glass window, do we see the view <u>outside</u> or the scene <u>inside</u>?*** We normally see the view <u>outside</u>. Even if it is dark outside and bright inside, we see only a weak image of the scene inside. So, glass is not a good mirror: it does not reflect much light! Typical glass windows in fact reflect only about 5% of light (but more when looking through the glass at an angle).

And yet, mirrors are made of transparent glass! ***How then does a mirror work?*** You probably know the answer: a mirror is made of a pane of glass with a metallic backing, typically silver or aluminum; and it is the metal backing that does the mirroring. You can easily check that by placing a sharp tip (such as your nail) against a mirror and looking at it from the side: you will see a gap between your tip and its image due to the distance between the tip and the reflecting metal. The glass serves to support and protect the metallic backing while ensuring high transparency of the reflected image; also, glass can be made extremely flat. A purely metallic mirror would also work, but it can easily be scratched and become dull; worse, it may rust (as you often see around the edges of old mirrors).

Sound waves would be reflected by the glass itself: any hard substance will reflect sound. But light is different: it goes through transparent substances (glass, some plastics, water, *etc.*) while being reflected by metallic substances (silver, gold, aluminum, *etc.*). Light is in fact best reflected by electrical conductors: metals. For example, silver can reflect close to 99% of visible light; less expensive aluminum can reflect close to 90%; gold reflects red better than green or blue, making images that look yellowish (but gold has the advantage that it does not rust).

Now you may argue that you see light reflected from clouds, steam, wood, paper, rocks, bricks, cloth and most other substances. However, these substances are rarely flat like a mirror: their surface is mostly rough. They scatter light in many directions so that no mirror image can be formed: this is called **diffuse scattering**.

For mirroring, the surface has to be flat on a scale larger than the wavelength, which for optical mirrors means on a scale larger than micrometers (millionths of meters); this is much smaller than the diameter of a human hair. We then get so-called **specular reflection**, meaning that the wavefronts of reflected waves coincide over large parts of the mirror's surface, and thereby reinforce each other into a single strong mirror image.

How is light reflected by a mirror? Let's compare this situation with waves on strings reflected by the ends of those strings, and with sound waves bumping against a hard wall: in these types of waves, the wave is an oscillatory motion of matter (namely, the string itself or the molecules of air or materials). By contrast, with EM waves in vacuum, the wave is not a motion of matter but an oscillation of both electric and magnetic fields, as sketched in Figure 12-3.

So we need to know how the separate electric and magnetic waves behave when they enter a substance, such as glass or metal (air is almost identical to vacuum for light, so we can ignore their difference in practice). The answer is essentially the same as we have seen for the other kinds of waves: we get reflected and refracted EM waves that fit the "footprint" of the incoming EM wave (see Sections 3.8, 10.3 and 10.4); this follows Huygens' principle that the footprint serves as the source of the continuing waves.

We illustrate this situation in Figure 12-9. Here we see (in 3D perspective) an EM wave approaching from the upper left. We'll assume it is in vacuum or air and approaches a flat substance like water or glass or metal (shown in light blue).

We use **ray tracing** (black arrows), together with double-headed arrows for the electric (blue) and magnetic (red) fields, to simplify the EM wave of Figure 12-3. We choose a polarization in which the magnetic field (red) is horizontal: parallel to the surface; this is called **parallel polarization**. Since the electric field must be perpendicular to both the magnetic field and the travel direction (black arrow), the electric field (blue) is angled due to the incidence direction: it points down at the blue-dotted footprint of the approaching wave.

Reflection and refraction of an EM wave at a boundary

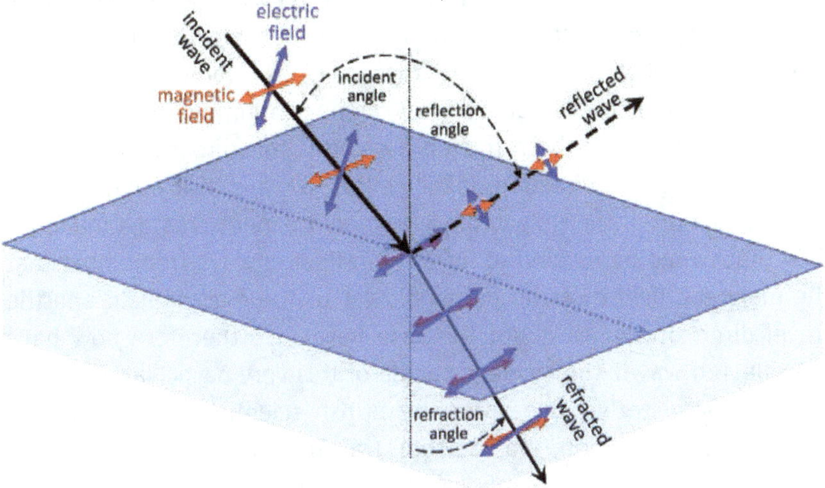

Figure 12-9: Reflection and refraction of an EM wave with parallel polarization at a boundary (the red magnetic field is parallel to the boundary) in 3D perspective.

In this ray-tracing model, the wave hits the boundary in the center of the figure. The footprint argument which we used earlier for sound waves is valid here as well: **we must have a reflected wave with a reflection angle equal to the incident angle; we must also have a refracted wave with a refraction angle given by the wave speeds in the vacuum and in the substance** (we can calculate the refraction angle using Snell's law of Section 10.5).

We now have the geometry of the reflection and refraction of an EM wave: it is just like the geometry of sound waves. Next, we look more closely at the electric and magnetic fields.

Let's follow the incoming crossed electric and magnetic fields as they reach the impact point at the center of Figure 12-9. What happens here physically is that the oscillating electric fields cause oscillations of the electric charges in the substance (water, glass or metal): these charges are mainly electrons of the atoms in the substance. The oscillating electrons (according to Huygens' principle) serve as the source for the reflected and refracted waves. The refracted wave penetrates deeper into the substance in the direction given by Snell's law: the electric and magnetic fields are now perpendicular to this new travel direction, but they are somewhat weaker because part of the wave has been reflected.

More interesting in this scenario is the behavior of the reflected wave. The oscillation direction of the electrons near the boundary is given by the blue double-headed arrow in the center point of Figure 12-9: it is perpendicular to the refracted direction. This direction is more or less pointing in the direction of the reflected wave. If it points exactly in the direction of the reflected wave, then it cannot create a magnetic field that would be needed to form the reflected wave (remember that the magnetic field must be perpendicular to the electric field and the travel direction, which is not the case here): we therefore now have no reflected wave! The incidence angle of this very particular geometry is called **Brewster's angle**, after the British scientist **David Brewster**. In this unique situation, there is total transmission, since there is zero reflection: the refracted wave carries the full energy of the incoming wave.

Let's now deviate from Brewster's special angle of incidence: now the electron oscillation can contribute something to the magnetic field and form a reflected wave. The reflected wave becomes stronger as we deviate further from Brewster's angle.

So far, we have discussed the case of parallel polarization (the magnetic field is parallel to the surface). It is interesting to contrast this with **perpendicular polarization**: we simply exchange the electric and magnetic fields (in Figure 12-9 we can exchange the colors of the blue and red arrows). In perpendicular polarization, the magnetic field is (at least partly) perpendicular to the boundary, while the electric field is parallel to the boundary. With this polarization, the reflected wave behaves quite differently: in particular, there is no Brewster angle giving zero reflection. The reason is that the electric field now causes electrons to oscillate parallel to the surface: in this orientation, these electrons can emit a reflected EM wave for any incident direction. Moreover, the reflected wave is now stronger than with parallel polarization (most obviously near Brewster's angle).

Less obvious from the discussion above is the following very common observation: reflection at "grazing" angles (nearly parallel to the boundary) is close to 100% for a perfectly flat boundary. For example, if you look at a flat lake, you will be able to see into the water near your feet but not farther away: near the horizon, the water acts like a perfect mirror. Similarly, if you look almost parallel through a window, you will see much stronger reflections than if you look perpendicularly through the window.

Another consequence of the reflection behavior of light is the following: **unpolarized light is polarized by reflection**. Take "normal" light, such as **sunlight** or the light from a light bulb: it is unpolarized, meaning that it is composed of many EM waves that have any polarization directions randomly. Let this light reflect from a flat surface, for example, a flat lake or a shiny car or a mirror. From our discussion above, the reflected light now must contain more perpendicular-polarized waves than parallel-polarized waves.

You may be familiar with **polarized sunglasses**: such glasses filter out light with near-perpendicular polarization. The purpose of polarized glasses is to reduce the light coming from the Sun, *via* reflection from horizontal surfaces such as water and shiny cars. These glasses rely on certain substances, especially some plastics like polycarbonates, which absorb light with near-perpendicular polarization more than other light.

How can polarized light be used to view movies in 3D? To see a scene in 3D, our brain relies on receiving from our two eyes two slightly different images of the scene: the two views differ in their viewing angles due to the separation between our eyes. However, when we look at a single 2D picture or 2D movie of a scene, our two eyes send the same picture to the brain, so some 3D information is missing, and the brain cannot easily build the 3D structure of the scene.

To nevertheless see **movies in 3D**, various methods are available to overcome this problem:[4] one method uses light polarization. The idea is to present two views (one for the left eye and the other for the right eye) that are polarized differently: one view is polarized vertically while the other is polarized horizontally. In one approach, these two views can be shown together on the same viewing screen, with alternating lines of the image being polarized vertically *versus* horizontally. The viewer then wears polarized glasses in which one lens blocks horizontal polarization, while the other blocks vertical polarization. Therefore, the left eye sees only the vertically polarized image while the right eye sees only the

[4] See: https://en.wikipedia.org/wiki/Stereoscopy.

horizontally polarized image. The brain then sees both 2D views of the scene and can build the 3D scene as if it was viewing a real scene.

Do mirrors exchange left and right? Looking at yourself in a mirror may be confusing. When you are standing, your right hand appears to become your left hand, and *vice versa*, as shown at left in Figure 12-10. However, your head does not become your feet, and your feet do not become your head. What is the difference between these two cases? Is it because your eyes are aligned horizontally rather than vertically? No, closing one eye makes no difference! There is a difference in the direction of gravity, which points downward: so, does the mirror take the direction of gravity into account when making an image? No, for the following reasons:

If you look more closely at your fingers (upper right in Figure 12-10), you see that they are placed from left to right in the same order in the mirror image as in your real hands. However, if you painted one side of your hands black and the other white (as shown), those colors would be interchanged back to front. This starts to suggest that left and right are not exchanged, while front and back are exchanged!

Consider another situation: if you lie on your side on a bed looking at yourself in a mirror (see at the lower right in Figure 12-10), your feet (which may be pointing to the left) do not seem to become your head

Mirrors: what are left-right, front-back, up-down?

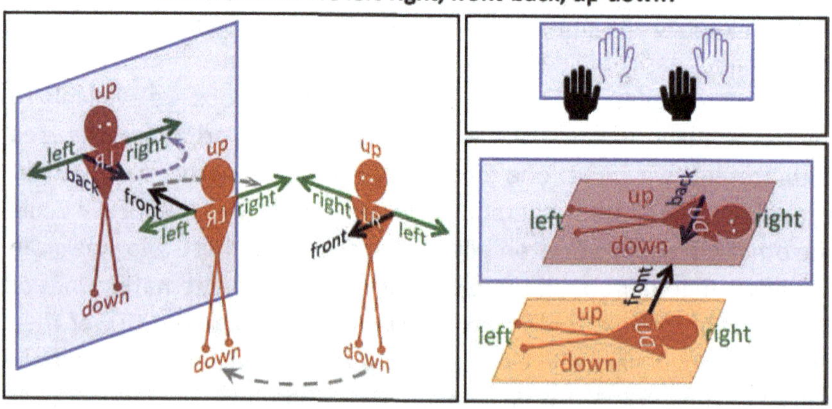

Figure 12-10: Bob, or you, looks at his image in a mirror (blue), while standing (*at left*) or lying on his side (*at bottom right*). On Bob's chest is written "LR" (for left-right) or "DU" (for down-up); these letters are shown in transparency when we look at Bob's back. *At the upper right*: Bob, or you, looks at his hands and their images in the mirror.

(which then is pointing to the right), and *vice versa*: left and right are apparently not exchanged here. However, your hands do seem to be exchanged, even if they are pointing up and down instead of left and right. This contradicts the role of gravity described above. Why would a mirror treat left and right hands differently from head and feet, depending on the direction of gravity?

Also confusing is the text printed on your chest. At the left in Figure 12-10, the letters "LR" are printed in normal fashion on the chest to indicate the left (L) and right (R) ends of the text itself. It becomes reversed when looking at you from behind you (assuming a transparent body), and this reversed text also appears reversed on your image: why is this text reversed and in apparent contradiction with the left and right arms? Indeed, "L" is near the right arm, while "R" is near the left arm! The lying person has "DU" printed on the chest, indicating "down" and "up": it also appears reversed, but not in a contradictory way; why?

All this is indeed very confusing and needs closer scrutiny. As we will see, the answer is actually very simple: a mirror exchanges forward and backward (and not left and right or up and down). In Figure 12-10, forward and backward are shown as black arrows (labeled "front" and "back"): these directions point directly at the mirror. As the figure makes clear, the mirror makes the "forward" direction become the "backward" direction: If you point your left arm at the mirror, then the image of that arm will point back at you from behind the mirror.

We need to clear up one major source of confusion here: the meaning of "left" and "right". In Figure 12-10, there are three points of view: 1) the point of view of the person (or you), whom we call Bob; 2) the point of view of Bob's image (looking from behind the mirror); and 3) the point of view of the artist drawing the figure. Each of these points of view can give different answers to the question, "What is left and what is right?" For example:

- Bob's image looks at Bob from behind the mirror and thereby creates the illusion that left and right are reversed! From the point of view of Bob's image, Bob's right arm is on the left.
- For Bob lying on the bed, as seen by the artist, "left" and "right" point to the feet and head, respectively, and are not related to Bob's left and right arms or to the left and right arms of Bob's image.

- Now imagine that Bob stands in front of the mirror. He sees his head imaged on top, and his feet below: no confusion here. Bob sees his left arm imaged toward the left, and his right arm imaged toward the right. Bob can confirm this by turning his left arm toward the mirror (black arrow labeled "front"): from Bob's point of view, there is no exchange of anything between left and right, or between up and down! However, Bob sees the image of his left arm now pointing in the opposite direction to his real arm: it points back at Bob, not away from Bob. That is the only change due to the mirror. And the same reasoning is valid when you lie down: the only change is that the direction to the mirror is reversed into the direction back from the mirror.

Therefore, "left" and "right" can be subjective concepts. We normally imagine "left" and "right" relative to our own body, but that can contradict the "left" and "right" imagined by another person (including by the image of our body). We must be careful and agree on one choice of "left" and "right", or at least recognize that we may use conflicting notions of "left" and "right".

To test your understanding of mirrors, try to solve this little challenge which I experienced in a bank: I was sitting at a counter while another customer was sitting behind me at another counter, facing the opposite direction (we were back-to-back); mirrored in the window before me, I could see her back as she was signing a document, apparently with her left hand. Was she really left-handed, or was that an illusion due to mirroring? A good way to solve this problem is to draw a simple plan diagram of you facing a mirror and her facing away from it.

Another useful way to think about this topic is the following: without looking at yourself in a mirror, try to imagine what you look like to another person who watches you. After you have done this, you can look at a picture of yourself: that will show you what you look like from the perspective of someone else. Did you guess your appearance correctly? The point of this little exercise is to show that we don't see ourselves the same way that others see us: it is in fact quite difficult to put ourselves in the position of someone looking at us!

Nonetheless, the text on Bob's chest still holds a surprise: it is already reversed before Bob looks at himself in the mirror! This is easy to see yourself: if you look down at your chest (without a mirror), what does the text printed there look like? From your head's point of view, that text starts at the right and goes to the left (it is also upside down!). That is the opposite of the same text seen by someone looking at you from the front: to them, the text will start at the left and go to the right, like normal text. This is also visible in Figure 12-10: Bob standing in the middle has the normal text "LR" on his chest, seen from the front with the "L" at the left; when Bob turns to look at his image in the mirror, that "L" appears on the right and "R" to the left, so "LR" is reversed left-to-right. It is not reversed by the mirror, but by Bob turning from facing the artist to facing the mirror! You can confirm this reasoning with the correct text "DU" printed on Bob's chest while lying on the bed.

There is another important observation to be made. At the left in Figure 12-10, consider Bob's left arm pointing at the mirror (black arrow). Now he turns that arm toward his right arm (following the curved gray arrow): this is a clockwise turn (as seen from above). What does the imaged arm do (it's the black arrow labeled "back")? It turns in the opposite direction — counterclockwise! This is not a new result: it is a direct consequence of the change of direction perpendicular to the mirror that we already saw.[5]

More generally, we can thus say: **A mirror reverses the direction perpendicular to that mirror and therefore also reverses the direction of a rotation around an axis parallel to that mirror.**

Other intriguing questions arise: ***What is left and what is right? Are left and right uniquely defined? Does nature favor left over right, or vice versa?*** Imagine the following scenario: you have to explain what is "left" to people who speak your language but have never encountered the concept of "left" or "right"; in particular, they don't have watches (that define clockwise rotation), they don't know that their heart is (usually) on the left side of the body, they don't know of traffic driving on the right or left side of the road, *etc*. Try to find a way of explaining left and right <u>with words</u>, but <u>without a video link and without drawing a figure</u>! If you find a way to do so for human beings on Earth (perhaps by referring to the sense of rotation of the Sun across the sky?), try to do it for intelligent beings on another planet, who may not have

[5] An interesting and more extensive discussion of all this is given in "Do Mirrors Reverse Left and Right?" at https://math.ucr.edu/home/baez/physics/General/Mirrors/mirrors.html.

human bodies or the same Sun (although they should have a translator to understand you).

Actually, physics knows of a rather complicated way to tell the difference between left and right. It involves the so-called **weak interaction**, which causes **radioactivity**, namely the radioactive decay of atoms. Nevertheless, you would find it difficult to uniquely define left and right with everyday objects. For example, your left and right hands, which are mirror copies of each other, operate in exactly the same way. You could also build the mirror copy of a mechanical watch so it would turn "**counterclockwise**": it would show the same time as the "normal" watch, but displayed left-to-right. So how could you distinguish right from left?

At first sight, it seems that the molecular structure of living beings distinguishes left from right. Indeed,[6] all life on Earth (including humans) uses some molecules but not the mirror image of those molecules, such as "**left-handed**" **amino acids** but not "right-handed" amino acids, or "**right-handed**" **sugars** but not "left-handed" sugars.

Here, "left-handed" and "right-handed" have nothing to do with the preference of most humans to use their right hand, but relate to the mirror properties of the shapes of molecules. Instead, "left-handed" and "right-handed" describe the shapes of our left and right hands: they are mirror images of each other, identical except for the reversal of one direction, front-to-back. Some molecules have the same property: one is the mirror image of the other. A famous example is shown in Figure 12-11. These are called "left-handed" *versus* "right-handed" molecules, or, more scientifically, **chiral molecules**. (Our hands are chiral; the word "chiral" actually comes from the Greek word for "hand", χειρ, pronounced "kheir".)

A living being (human or animal or plant) made of "left-handed" amino acids could not eat and digest a living being made of "right-handed" amino acids. However, according to the laws of physics which apply to life and molecules, living beings made of "right-handed" molecules would be equally possible, as long as "left-handed" sugars would be available. Intriguingly and for unknown reasons, such

[6] See "Must the Molecules of Life Always be Left-Handed or Right-Handed?" by Danielle Sedbrook, *Smithsonian Magazine*, July 28, 2016, online at https://www.smithsonianmag.com/space/must-all-molecules-life-be-left-handed-or-right-handed-180959956/.

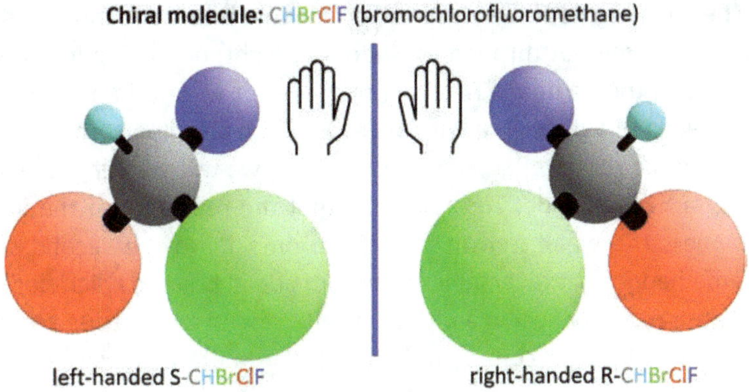

Figure 12-11: Left-handed ("S-CHBrClF", *at the left*) and right-handed ("R-CHBrClF", *at the right*) molecules (called bromochlorofluoromethane), shown with a ball-and-stick model. "S" and "R" mean "sinister" and "rectus", the Latin words for left and right, respectively. Left-handed and right-handed molecules are mirror images of each other, like left and right hands. Left-handed and right-handed molecules cannot be superposed on each other, again like left and right hands. If the green and red atoms were the same, the molecule would be symmetrical and therefore not chiral, which is called achiral. The two molecules could then be superposed on each other after suitable rotation. (*Source*: molecules drawn with K.E. Hermann's Balsac software.)

"alternative" living beings either never developed or, more likely, were vanquished by the "normal" living beings: it is like a tribe exterminating another tribe forever, or a form of evolution.

How do "one-way" mirrors and "one-way" windows work? One-way mirrors are also called "two-way" mirrors; a one-way mirror is really a **two-way window**. These names are confusing for a reason: one-way mirrors and windows in reality involve light passing in two directions through a window.

The ideal one-way window, at least in principle, should allow light to pass through it in one direction, but not in the opposite direction. Actually, such an ideal one-wave window is physically impossible: if light can follow a track through a substance, it can equally well follow the same track in the opposite direction through the same substance. This is true of all the examples of waves crossing boundaries in Chapter 10: in every case, we can reverse the wave direction and follow it through refractions and reflections. The reason is that the reflection and refraction angles, as well as reflectivities and transmittivities, remain the same in both directions.

The trick of a "one-way" mirror or window is in fact a familiar daily situation: looking out through a window at night; or looking in through a window in the daytime. Take the first case: if we are in a brightly lit room looking out through a window at a dark night-time scene, we will mainly see the room reflected in the window; we will see very little light coming in from outside. The reason is that the reflection of the inside overpowers the weak transmission from outside. In the second case, the same happens in reverse: if we are outside in bright daylight looking into a dark room, we will mainly see a reflection of the outside scenery; we will see little light coming out from inside. Both of these cases are almost, but not quite, one-way windows; additionally, the same window will work in opposite ways during day *versus* night.

Figure 12-12 makes this reasoning quantitative. Alice and Bob are in two rooms separated by a window; Alice has a blue light bulb, and Bob has a red one. At top left, using a normal window and lighting of equal brightness, they both see much transmitted light (95% goes through) and little reflected light (5% is reflected), so each sees mainly the other by transmission, and a very weak reflection of themselves.

Bob wishes not to be seen by Alice, so he dims his light 5-fold (middle left in Figure 12-12): now Alice still sees Bob more than herself (in ratio 19:5), while Bob will mostly see Alice and hardly himself (in ratio 95:1); this is not good enough!

So, Bob dims his light further to just 1% of its original brightness (bottom left in Figure 12-12): now Alice sees Bob about 5 times weaker than herself (ratio 0.95:5), while Bob sees Alice about 2,000-fold stronger than himself (ratio 95:0.05). Alice seeing Bob 5 times weaker than herself is good progress, but still not satisfactory. Also, unfortunately, Bob now finds himself in an almost black room, so he can't do much besides looking out through the window. Can we provide enough light for Bob to also watch his computer or read and write, while still keeping himself mostly invisible to Alice?

Let's try increasing the reflectivity of the window to 50% (which reduces the transmission to 50%): this is shown in the right column of Figure 12-12. With equal illumination in both rooms (top right in Figure 12-12), we again get a symmetrical situation: both Alice and Bob see the same thing; now Alice sees both Bob and herself equally (50:50), and so does Bob. Bob still must dim his light (middle right in Figure 12-12): with a 5-fold weaker light, Alice sees Bob 5 times weaker than herself (ratio

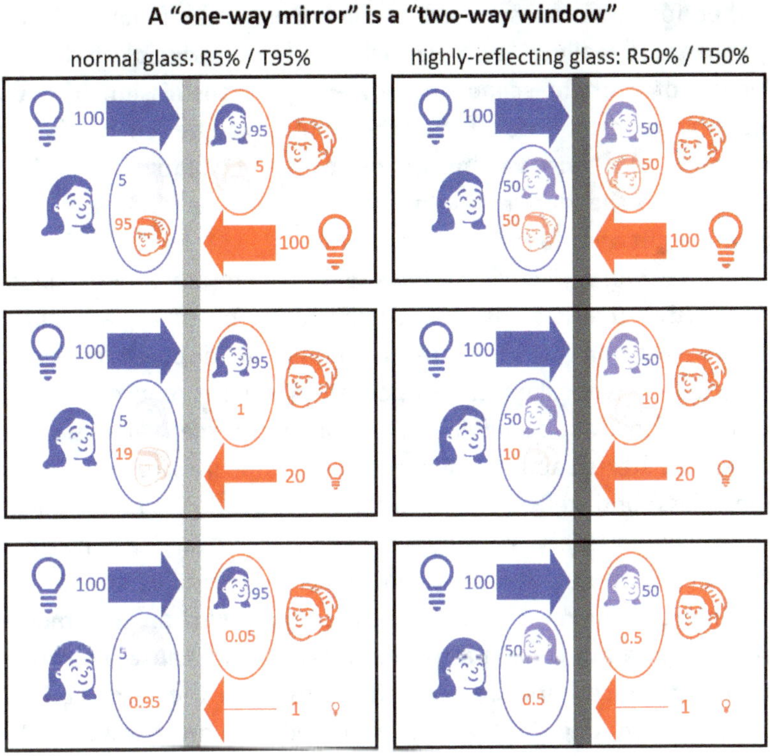

Figure 12-12: Alice and Bob are in two rooms separated by a window (thick gray vertical line). *At the left*, a normal window is used, with 5% reflection and 95% transmission. *At the right*, a window with stronger reflection (50%) and weaker transmission (50%) is used. In the top row, both rooms are equally lit (with light bulbs of intensity 100, shown by the width of the arrows); the light is blue in Alice's room and red in Bob's room. The ovals show what Alice (blue) and Bob (red) see of each other, with numbers giving the brightness of each face. In the second and third rows, Bob has dimmed his light bulb to intensity 20 and 1, respectively (shown as smaller light bulbs).

10:50), while Bob sees Alice 5 times stronger than himself (ratio 50:10). Now Bob has achieved a reasonable result without blacking out his room (see the black-out at the bottom right in Figure 12-12), so he can still do other work.[7] However, Bob has not quite achieved a "one-way" mirror, but almost: he needs to compromise because all windows are "two-way" and symmetrically so. Using tinted glass is also symmetrical: it weakens light equally in both directions.

[7] Opening our pupils allows our eyes to see the same with 4 times less illumination.

Therefore, we can summarize our discussion as follows: **A "one-way mirror" is really a "two-way window". The principle is that the reflection of a bright scene overpowers the transmission of a dark scene. An important aspect is that the window can be symmetrical and still operate as a "one-way mirror" or "two-way window": what is not symmetrical is the illumination on both sides of the window; one side must be bright and the other dark.**

How do anti-reflective coatings work? There are various kinds of **anti-reflective coatings** that reduce the reflection of a glass surface: they are common on computer monitors, TV displays, eyeglasses, camera lenses, binoculars, telescopes, *etc*. They can reduce the typical glass reflection from about 5% to 1% or less. Some coatings rely on an intermediate refractive index: the two boundaries (air-coating and coating-glass) both reflect less, and their combination is more effective than a single boundary. A gradual refractive index varying through the coating is even more effective, although harder to produce.

A more common method uses a coating with an intermediate refractive index but gives it a special thickness that causes destructive wave interference between the reflections from the front and back faces of the coating. This approach, called a **quarter-wave coating,** thus relies on the wave nature of light and is illustrated in Figure 12-13. If the coating's thickness is one quarter of the wavelength of light, the

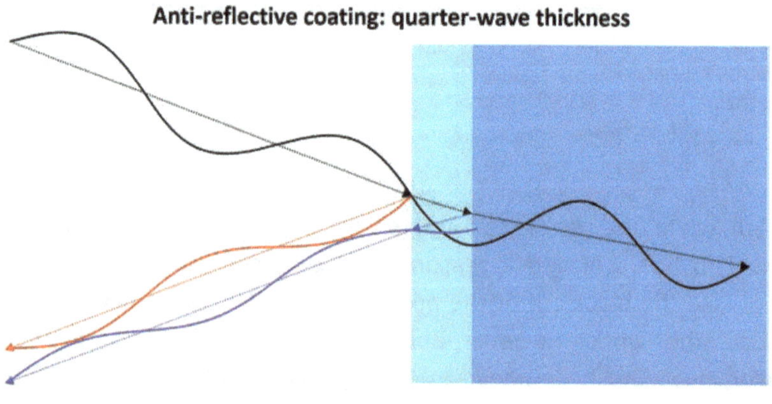

Figure 12-13: A quarter-wave coating (light blue) covers glass (dark blue). The refractive index in the coating is intermediate between those in the air and the glass, giving an intermediate wavelength: the coating's thickness is one quarter of this intermediate wavelength.

round trip from the front face to the back face and back to the front face delays the wave by one half a wavelength: this reflection is then in antiphase compared with the reflection from the front of the coating, which is the condition for destructive interference.

The quarter-wave coating works best for one particular wavelength and one angle of incidence: only then do you get the most destructive interference. When the wavelength or the angle of incidence changes, the condition for destructive interference is not met exactly. Typically, this kind of coating reduces glass reflection from about 5% to 1% or less. This implies a greater transmission into the glass around 99% instead of the usual 95%.

Have you seen colors on thin films of oil? Or in soap bubbles? When we look at a thin film of oil (floating on water, for example), we often see colored rings or bands, as in Figure 12-14. Similarly, soap bubbles with thin soap films exhibit pretty colors that remind us of the solar spectrum. These are called **Newton rings**, or also **thin-film interferences**. Such colors often appear to change as you look from different directions: this effect is called **iridescence** and is also observed with **bird feathers**, with **nacre** in seashells, and with **opals**.

The cause of these effects is similar to the quarter-wave coatings mentioned above. For example, the oil film is the coating and has a thickness that varies smoothly: the destructive reflection due to the

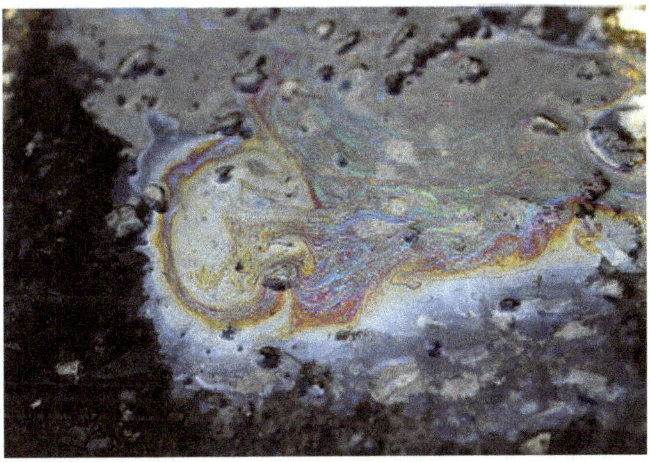

Figure 12-14: Newton rings on a film of oil. (*Source*: Needpix.com, in the public domain, ground-588927_1280.jpg.)

quarter-wave coating then varies with wavelength and therefore with light color. For example, in one spot, the oil film may have a thickness that is a quarter of the wavelength of red light: that will destroy reflected red light, leaving blue and green to dominate. Removing a color in fact leaves the complementary color: thus removing red leaves cyan (a mixture of blue and green); removing blue leaves yellow (a mixture of red and green); and removing green leaves magenta (a mixture of red and blue). As a result, we do not simply get pure colors of the solar spectrum, but mixtures of those pure colors.

This effect is not limited to a film thickness of a quarter of the wavelength, which is extremely thin, less than a thousandth of the thickness of hair: it also acts when we add multiples of the wavelength to the film thickness, because adding a wavelength gives the same interference condition. As a result, color bands can repeat themselves as the film thickness changes.

What causes the beautiful colors of butterflies? Here also, reflection from different boundaries is at play, as with the Newton rings and thin-film interferences. Butterfly wings have layered structures, so there are multiple boundaries one after the other, each of which operates like a quarter-wave coating. The multiplicity strengthens the reflection of light; it also sharpens the colors because the wavelength of the reflected light has to fit several distances between boundaries, not only one such distance.

12.3 Refraction: Optical Prisms, Lenses

In Section 10.6, we saw how double boundaries can bend light (and other EM waves). The bending angle can be changed by varying the angle between the two boundaries: see Figure 10-18. This is the principle of the **prism**. A prism deflects light by an angle that depends on its refractive index, as well as on the angle between its two boundaries, and on the angle of incidence (in Figure 10-18, the angle of incidence was simply set to zero to give perpendicular incidence).

Prisms provide the basic mechanism used in lenses to focus light. Such lenses are found in eyes, eyeglasses, cameras, microscopes, binoculars, telescopes, lighthouses, *etc.* A beautiful example of lenses from a lighthouse is shown in Figure 12-15; in lighthouses, the process

Figure 12-15: Five Fresnel lenses in the Cape Willoughby Lighthouse, Kangaroo Island, South Australia. Each Fresnel lens takes light from a central lamp and focuses it toward a different part of the horizon. There probably are more lenses around this lamp, all of which rotate together around the vertical axis to sweep the horizon. (*Source*: Wikipedia, by Stephen West, in the public domain, https://commons.wikimedia.org/wiki/File:Cape_Willoughby_Lighthouse_Fresnel_lens.jpg.)

is reversed: light from a lamp is focused toward the horizon. Let's see how prisms lead to such lenses!

One logical use of the prism is to concentrate light in one spot, for example, to start a fire for cooking by concentrating sunlight. For that purpose, several prisms must be placed and oriented in such a way that they send different rays of light to converge at the same spot. The center sketch of Figure 12-16 shows one way to do this. Small prisms (blue) are positioned and angled to send parallel bundles of sunlight toward a common point, called the **focus**. Each prism takes a bundle of sunlight and bends it toward the focus. Each bent bundle consists of rays of light that are parallel to each other (because the prism faces are flat): as a result, the light is not concentrated precisely in one focus point, but spread out somewhat, basically by the width of the prisms. We don't get very sharp focusing, but still a considerably larger amount of light close to the ideal focus point.

We can concentrate even more light by adding prisms in circles around the horizontal axis of Figure 12-16, somewhat like the circular prisms of Figure 12-15 in a lighthouse. Lenses in lighthouses are very large and very heavy: they are meters across and can weigh many tons.

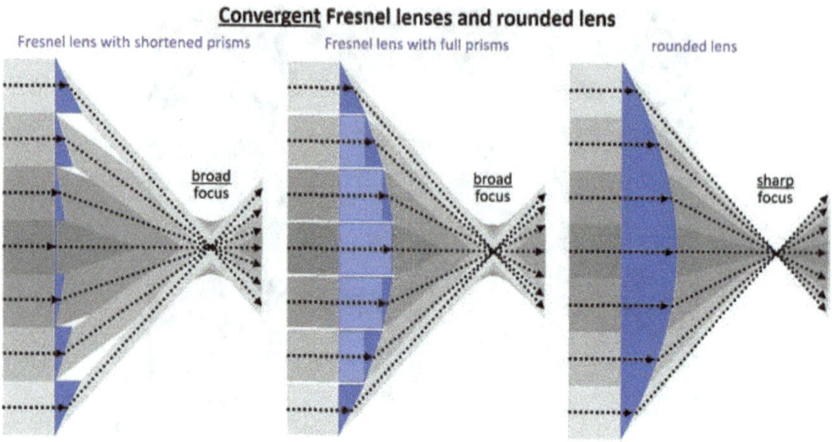

Figure 12-16: Three convergent lenses, from Fresnel lenses to rounded lenses, in air (or vacuum). Glass is shown in blue, in the form of prisms with flat faces at left and center, and in the form of a rounded lens at right. The left lens is obtained from the center lens by removing the glass colored in light blue and shifting the remaining prisms to the left. (The shades of blue only serve to distinguish different parts of the glass; they are not related to the color of the light or of the glass.) Parallel rays of light approach from the left and are focused to the right of each lens. (The gray tones distinguish parallel bundles of light that pass through different prisms.) In lighthouses, spotlights and searchlights, a lamp is placed at the focus: its light rays travel in the opposite direction along the same paths, toward the left through the lens, to form a parallel beam of light aimed at the horizon.

It is worth reducing their weight by using as little glass as possible. We can do this by noticing the glass colored in light blue in Figure 12-16: that part of the glass does not bend light because it has parallel faces left and right; light is only bent at the inclined boundary (colored in dark blue). If we simply remove the light-blue parts in the central model, the light rays will still be concentrated at the same focus area: that is done in the left sketch of Figure 12-16, where the light-blue glass has been removed and the remaining dark-blue glass shifted to the left to make a flatter lens (this shift is not necessary to achieve the same focusing, but the inclination of the prisms has to be adjusted slightly to account for the different distance to the focus area). The result is called a **Fresnel lens**: in 3D, it consists of circular prisms as shown in Figure 12-15. Again, in a lighthouse, the light takes the reverse course: it starts in the focus and is bent by the lens into a beam pointing at the horizon. The lens

Figure 12-17: A flat plastic sheet with concentric prism-like ribs acting as a Fresnel lens. (*Source*: Wikipedia, by Almazi, under CC BY-SA 3.0, https://commons.wikimedia.org/wiki/File:Flat_flexible_plastic_sheet_lens.JPG.)

itself rotates around a vertical axis to sweep its light beam around the horizon as a warning to ships.

You are likely familiar with another form of Fresnel lens, shown in Figure 12-17. Such plastic lenses are often stuck on windows to provide a wider view. Instead of making light rays converge, these make light rays spread out: they diverge. Compared to Figure 12-16, in these diverging lenses, the little prisms are turned around, so they bend the light away from the horizontal axis of the figure instead of toward the axis: see Figure 12-18. As a result, a very wide view outside the window can be shrunk into a narrower view, a bit like a fisheye lens. But the quality of images seen in such lenses is rather low, largely because of the ribs in the surface and the poor optical properties of plastic.

High-quality lenses can also be made by using the same principle of the prism. In Figure 12-16, the center model can be made to focus more sharply by making the prisms smaller and more numerous: with narrower prisms, the light will be concentrated closer to the ideal focus point. We can make the prisms ever narrower, until they form a smooth, rounded surface. This shape is shown at the right in Figure 12-16: it is the familiar rounded lens shape found in many optical devices, including the **eye**, **eyeglasses**, **cameras**, **microscopes** and **telescopes**. In practice,

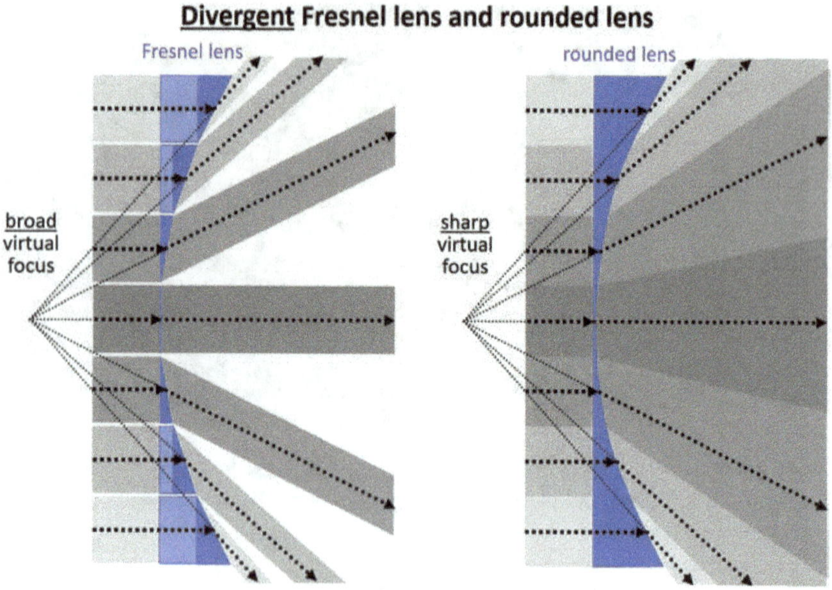

Figure 12-18: Divergent Fresnel lens and rounded lens, in air (or vacuum), drawn like the convergent lenses of Figure 12-16. Here, parallel light rays coming from the left are bent outward, away from the horizontal axis. The diverging rays appear to come from a unique focus behind the lens, called a virtual focus. An eye at the virtual focus can see light coming through the lens from all the directions shown here: this is used in common plastic Fresnel lenses attached to windows. Glass shown in light blue could be removed and the remaining glass pushed left to make a flat lens, as is done with plastic Fresnel lenses. The lens at right joins the prisms in a smooth, rounded shape: this is used in eyeglasses and many other optical devices; the flat left boundary is usually also rounded.

both sides of such lenses are rounded; this avoids that the flat side acts as a disturbing flat mirror.

Similarly, at right in Figure 12-18, a smoothly rounded divergent lens is shown. It is also used in optical devices, in particular corrective eyeglasses. Figure 12-19 illustrates the use of convergent and divergent lenses in eyeglasses to correct farsightedness and nearsightedness, respectively.

What shape is needed for convergent and divergent lenses to achieve perfect focusing? More precisely, the question is: what exact rounded curve should be used to make all parallel rays of light converge in one sharp spot or diverge from one sharp virtual focus?

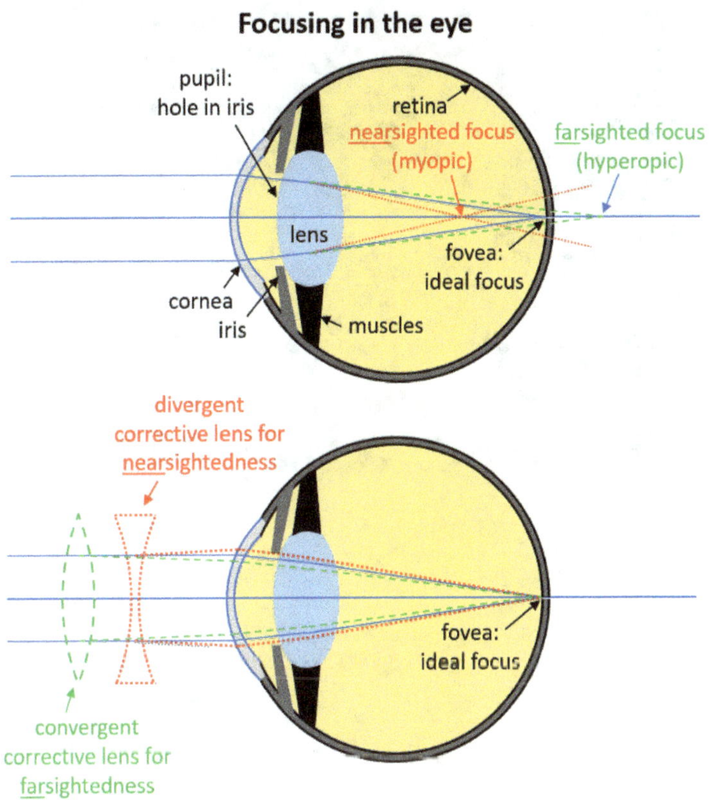

Figure 12-19: *Top*: The human eye receives light coming from far to the left (blue rays). The natural cornea does most of the bending and thus most of the focusing. The flexible lens inside the eye adjusts the focusing to allow viewing nearby objects. If the eye can't focus light on the fovea (where the light receptors of the eye are located), we have farsightedness (green rays) or nearsightedness (red rays). *Bottom*: Eyeglasses can correct farsightedness and nearsightedness by using convergent lenses (green) or divergent lenses (red), respectively.

In Figures 12-16 and 12-18, the shapes of the round lenses look circular in 2D (spherical in 3D). This suggests using a perfect circle (in 2D) or a perfect sphere (in 3D). Let's try this out: Figure 12-20 shows a glass of water with a cylindrical shape focusing sunlight. This glass's shape is very similar to the perfect circle in 2D (we would get a very similar result with a spherical ball of glass).

We can understand the light paths in the photograph by ray tracing, as is done in the lower part of Figure 12-20. We see that rays

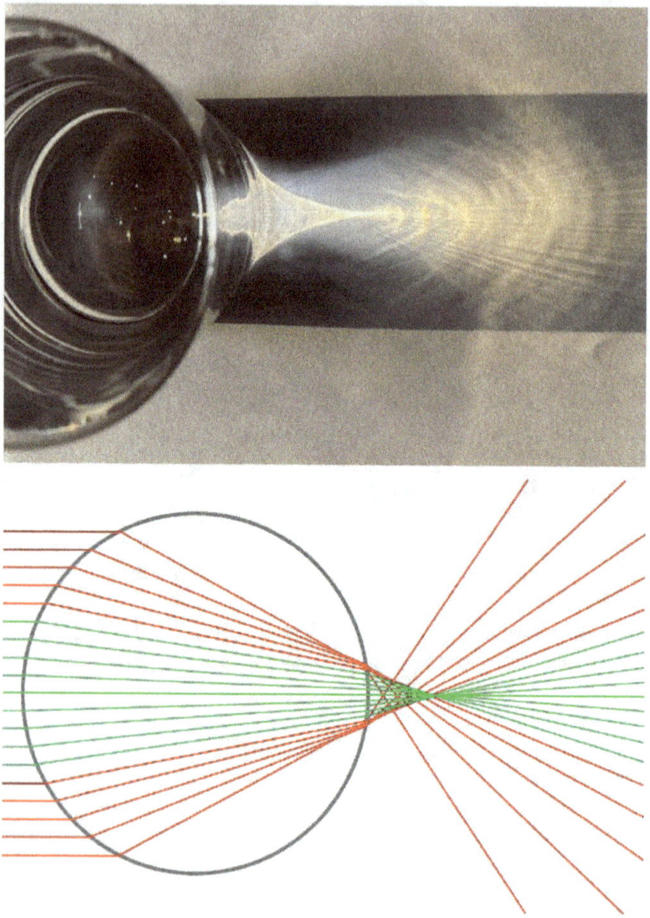

Figure 12-20: *Top*: Photograph of sunlight (coming from left) focused by a glass full of water into its shadow at right; the view is straight down along the side of the glass. *Bottom*: rays of sunlight traced through the glass; green rays focus well, while red rays are more spread out.

close to the horizontal axis (drawn in green) meet at a sharp focus. On the other hand, rays farther from the axis (drawn in red) do not meet at that sharp focus, but increasingly farther away from that focus: this is called **spherical aberration**, which is a deviation from an ideal single focus point due to the circular or spherical shape ("aberration" is a fancy word for "error").

This observation tells us that we can use a circular or spherical lens shape if we avoid rays that are far from the optical axis, so that we only use the green rays in Figure 12-20. In the human eye, the iris blocks the more distant (red) rays (Figure 12-19). But this is still not perfect. The ideally focusing lens shape is clearly not spherical; a non-spherical lens is called aspheric or aspherical.[8] A complication in finding the ideal lens shape is that light of different colors is focused on different points, due to the dependence of the refractive index on frequency (see the discussion of the solar spectrum and rainbow in Section 12.7): this is called **chromatic aberration** (because "chromatic" relates to colors and thus frequencies and wavelengths). **No lens shape can overcome both spherical and chromatic aberration at the same time, so compromises are needed in practice; multiple lenses placed one after the other can also help.**

How does refraction (deflection or bending) of light rays change what we see? Let's take a common, simple case: viewing a **fish in a tank**, as illustrated in Figure 12-21. Light rays do not follow a straight

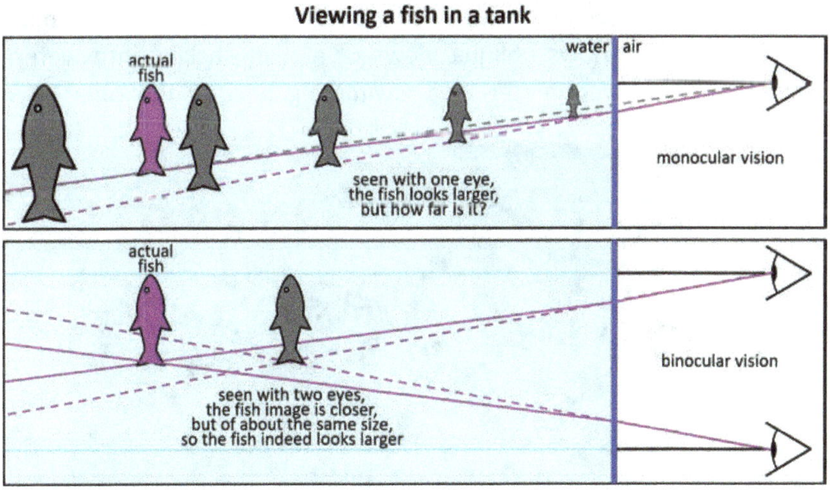

Figure 12-21: The eyes at right look horizontally and perpendicularly into a tank with an actual fish (purple). *At the top*: view with only one eye. *At the bottom*: view with two eyes. A realistic refractive index of 1.333 has been used for water.

---

[8] See "Aspheric lens" at https://en.wikipedia.org/wiki/Aspheric_lens.

line (dashed gray line) from fish to eye but are bent at the boundary between water and air (full purple line). In the top example, using one eye only, the fish's nose is not displaced up or down (because it is on a line perpendicular to the boundary), but the fish's body looks longer: the dashed purple line shows the direction in which the eye sees the tail. However, with a single eye, it is not possible to tell how far the image of the fish is: it could be farther or closer than the actual fish, so the image could in fact be larger or smaller than the actual fish, as drawn in gray.

We need to use our two eyes to tell the distance of the fish's image, as shown at the bottom in Figure 12-21. The apparent directions of the fish's tail (dashed purple lines) cross over at one point, telling us this distance: the fish's image (gray) is clearly closer than the actual fish. We also see that the fish's image has the same size as the actual fish (but, since the image is closer, the eyes see it as larger than the actual fish!). However, this result is correct only with the special perpendicular views of this situation: in general, when looking obliquely at the fish tank, the directions and sizes will vary. **One general result is: the displacement of the fish's image will increase with the refractive index in the tank.**

Figure 12-22 shows oblique views of a ruler sticking into water. Imagine a fish, 1 centimeter long, swimming next to the ruler: you

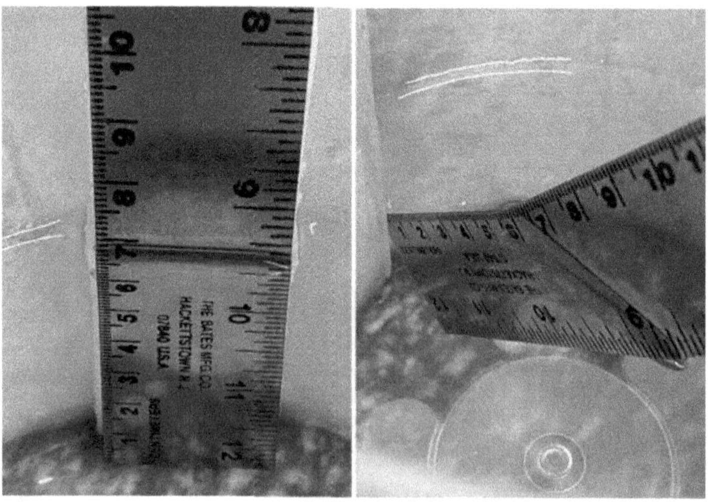

Figure 12-22: Two views of a ruler sticking into water, looking down from above the water.

can get an idea of what it would look like in different positions and orientations. The effect with oblique viewing can be quite different from that with perpendicular viewing, although it all follows the same simple Snell's law! For instance, the ruler sticking straight into the water (at left in Figure 12-22) looks shortened but not narrowed under water: a fish would then look flattened. Also, the inclined ruler (at right in Figure 12-22) looks bent: a fish in the water would here be seen heading in a different direction than it is actually pointing. We will illustrate this further with the view through a diving mask.

If you swim underwater without a diving mask, why do you have blurry vision? You probably tried this: you go for a swim without a diving mask and look under the water surface; you can do the same in a bathtub. Your vision is then very blurred. We can understand this blurring by looking at the top sketch of the eye in Figure 12-19. Imagine that the substance in front of your eyes is water instead of air. The refractive index of water is close to that inside your eyes. This means that the cornea barely bends the incoming light rays: the rays go almost without bending into the lens, which, however, can only slightly correct the direction of those rays and therefore is incapable of focusing the incoming image on your retina. You get extreme farsightedness.

A diving mask with flat glass overcomes this problem. We get the situation of the lower half of Figure 12-23. The air within the mask restores the normal bending of light by the cornea because the rays reach the cornea from air instead of water, so the eye can focus the image as usual; the eye's lens can also still adjust for an object's distance. The rays are now bent at the water-air boundary. The thin, flat glass added to the water-air boundary does not change the rays' refraction while they pass from water into the air space within the mask. (Wearers of corrective glasses can mount them within the diving mask to have normal vision underwater.)

Nonetheless, the diving mask leads to a distortion of the underwater image: see the images of the fish and pole in Figure 12-23. Light rays coming at an angle from the water through the glass into the air of the mask are bent by refraction. The effect is to make rays coming from different directions less parallel to each other, which makes objects appear larger and closer than in reality. Try it out by looking at your hands through a diving mask underwater!

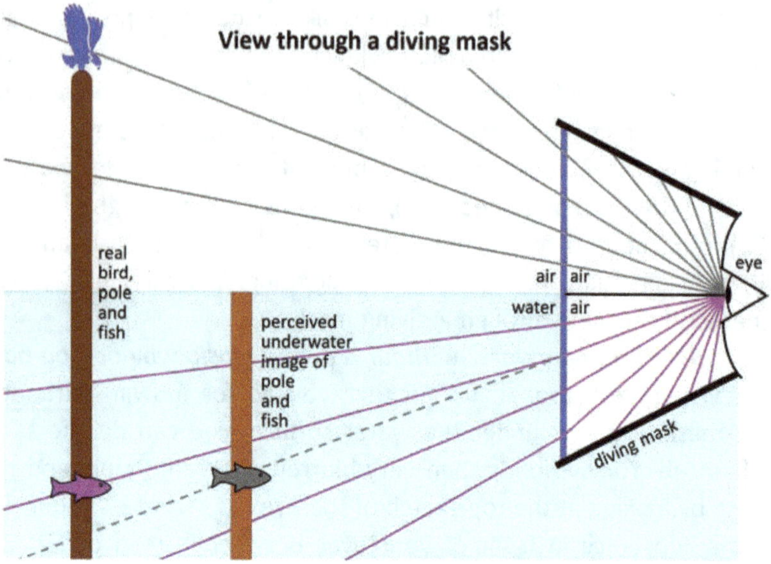

Figure 12-23: An eye (at the right) looks through a diving mask that is half submerged in water. The top half of the view looks through air at a bird on a pole. The bottom half of the view looks through water at a fish near the same pole: due to refraction at the glass (blue) of the mask, the fish and pole in the water appear larger and closer than they really are, similar to Figure 12-21.

How do fish see sharply underwater? To avoid the human's blurry vision underwater, fish rely on a more spherical lens (similar to the shape shown in Figure 12-20) than the flatter lens in human eyes (see Figure 12-19): a spherical lens can bend light more strongly. Lenses of this type can also produce a wider field of view and are therefore used in photography under the name fisheye lens.

12.4 Perfect Mirrors: Total Internal Reflection

The mirrors which we discussed in Section 12.2 do not reflect 100% of light: silver comes close with almost 99%. Nevertheless, we have already seen that certain situations can cause **total internal reflection** of EM waves by a boundary. Examples are shown at the bottom in Figure 10-4 and in the right column of Figure 10-6. These situations are impractical for daily use as mirrors by humans, but they are actually useful for fish, or any other organisms that live in water! More importantly for us, they are extremely useful in optical fibers that guide light along curved paths over long distances (see Section 12.6).

Total internal reflection happens when no wave on the other side of the boundary can match the footprint of the incoming wave because the wave speed in that substance causes a wavelength that is too large to fit the footprint at any refraction angle. As we saw in Sections 10.3 and 10.4, we then have an "evanescent" (or "dying") wave behind the boundary, which travels parallel to the boundary. We can learn more from the following question.

How do a fish and a fisherman see each other? Let's consider the fish and fisherman at the top of Figure 12-24. The purple rays show paths of the light that the fisher sees: in particular, he sees the fish's eye along the black ray. The fish will appear to the fisherman to be higher and closer than it really is.

Next, let's take the fish's point of view, as drawn at the bottom of Figure 12-24. The fish sees the blue rays. In the light-blue cone above the fish, the rays coming from outside the water are bent as they cross the water surface. Among those rays, two are of special interest: the horizontal dashed blue rays travel parallel to the water surface (as at the bottom in Figure 10-4 and in the right column of Figure 10-6). These allow the fish to see the horizon and the fisherman's feet: they are visible to the fish just above the inclined thick dashed blue lines that form the cone above the fish; this cone is limited by the so-called **critical angle**, already mentioned in Section 10.4. This cone of vision above the water is called the **Fresnel window**: the fish can see outside the water only within this window, which forms a circle on the water surface around the fish; interestingly, the fish can see everything above the water, from horizon to horizon, compressed within the Fresnel window.

The dotted blue rays (in the darker blue water) behave differently: they come from below the water surface and are totally reflected by the surface back down toward the fish's eyes. Therefore, outside the Fresnel window, the fish only sees rays coming from below the water surface, including a reflection of what is below the water surface, such as other fish, the bottom of the lake, or a bather's submerged body parts. This

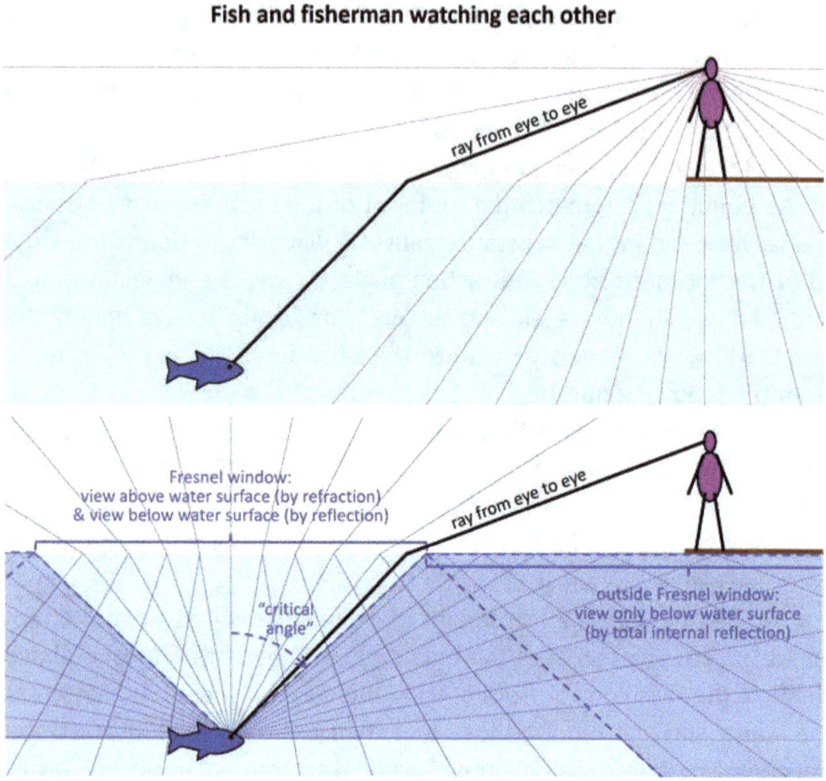

Figure 12-24: A fish and a fisherman watch each other through a perfectly flat water surface. *At the top*: The purple rays show light paths to the fisherman's eyes from below the water surface. These rays are bent by refraction. The black ray is the only ray that goes from the fish's eye to the fisherman's eye. *At the bottom*: The blue rays show light paths to the fish's eyes (more light paths come from below the fish but are not shown). These rays include the same black ray from eye to eye, but in reverse.

can be very helpful for the fish: the water surface gives it a double vision (direct and reflected) that warns it of danger or food! It's as if we humans carried a large mirror that allowed us to see more; but the fish has a clear advantage: the fish also looks behind the mirror through the Fresnel window, whereas a mirror carried in the air by a human would hide the scene behind the mirror.

You can make the same observations when you are submerged in your bath (assuming that you can keep your head submerged long enough to let water waves die out). Looking up, you can see everything that is above the bath water's surface: all of that fits in a cone or circle, the Fresnel window. Looking lower, you will see a perfect reflection of your body and the underwater parts of the bath itself: all of that falls outside the Fresnel window.

Total internal reflection is indeed total (as long as the water surface is flat and clean): it gives a perfect mirror that reflects 100% of the light without loss. Importantly, it is a direct consequence of the wave nature of light.

You can now practice your understanding of total internal reflection with the view seen by a fish swimming in a pool: see Figure 12-25. Imagine what that fish would see as it swam in that pool. For comparison, the right half of the pool has no water, while the fish remains in the same position. Try to explain everything you see in Figure 12-25; afterwards, you may look at Figure 12-26 to find answers.

12.5 Gradual Refraction: Mirages

Have you ever seen reflections like those above a dry road in Figure 12-27, as if the road were flooded? Or bizarre shapes hanging over the horizon as in Figure 12-28, which look like upside-down reflections of the horizon? Or have you noticed how the setting **Sun** appears to be "**flattened**" as in Figure 12-29?

These are examples of the bending of light as it travels through air: they are a gradual form of refraction, as opposed to the abrupt bending at sharp boundaries which we have discussed so far in this chapter and earlier in Chapter 10. The connection is very simple: we get abrupt bending when the boundary is abrupt, while we get gradual bending when the boundary is gradual.

What do we mean by a gradual boundary? Suppose you put a lot of sugar into a glass with water: let the sugar dissolve and settle down. The sugar will concentrate near the bottom but remain spread out toward the top of the water: thus, the sugar density will gradually increase from top to bottom. The result will be a gradual change in the refractive index of the sugary water because the speed of light will gradually decrease toward the bottom: this is a gradual boundary for light.

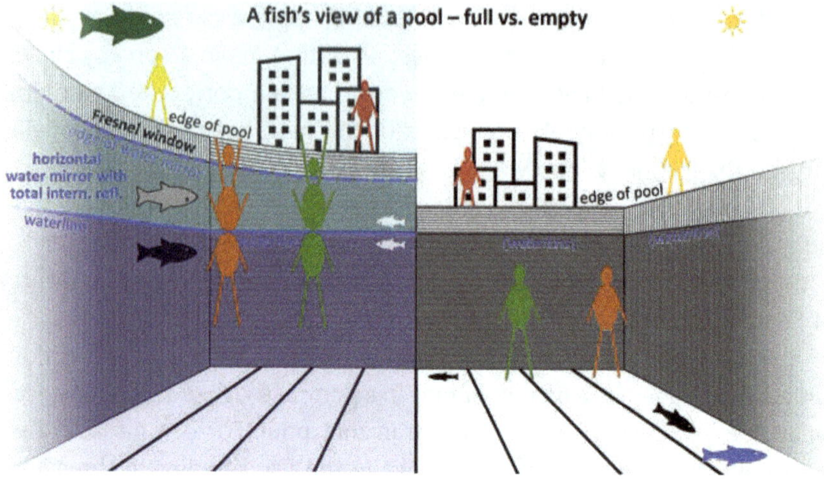

Figure 12-25: Simulated view seen by a fish in a swimming pool (you are looking out through the eye of the fish, which is therefore not shown). In the left half, the pool is full of water (up to the waterline), while the right half shows the same view as if there were no water in the pool (the observing fish remains in the same position on the centerline of the pool). See Figure 12-26 for a 2D sketch of the same situation (with water).

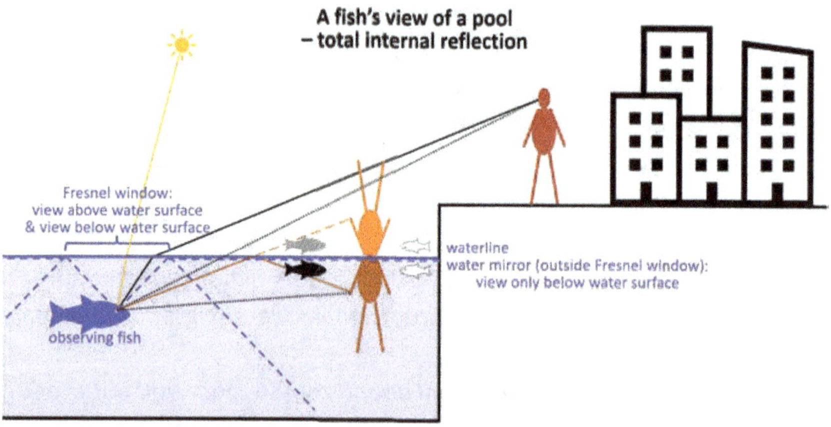

Figure 12-26: A simplified 2D sketch of the situation in Figure 12-25 (with water). This sketch cuts straight through the centerline of the pool. A few notes:

- The observing fish (blue, not drawn in Figure 12-25) sees outside the water in a cone above its eye: this view through the "Fresnel window" compresses everything from the horizon of the water until straight up, such as the red person and the buildings

Caption continued on facing page

Figure 12-26 on facing page

above the pool wall, and the Sun; it also includes the head of the swimmer (not shown here but visible in Figure 12-25).

- The observing fish also sees everything in the water, from straight down to straight up, including the submerged part of the swimmer.
- The observing fish sees a reflection in the water surface, which also shows everything under the water upside-down, including an image of the submerged part of the swimmer, and images of the black and white fish.
- The reflection at the water surface outside the Fresnel window is total (if the water surface is flat), so the upside-down image of the swimmer's body (without the head) looks as sharp and clear as the actual body underwater.

Figure 12-27: Inferior mirage over a hot, dry road. (*Source*: by Yuri Khristich, in the public domain, https://commons.wikimedia.org/wiki/File:Mirage_over_a_hot_road.jpg.)

The **Earth**'s **atmosphere** also forms a gradual boundary for light: both the pressure and temperature of the air gradually change with altitude. So the density of air changes as well, and therefore the speed of light also gradually changes with altitude. The change in speed is small but suffices to cause the effects seen in Figures 12-27, 12-28 and 12-29,[9] because these effects accumulate over large distances. For example, a temperature change from 20°C to 50°C reduces the air's refractive

[9] Many more images of mirages and sunsets are available at "Polar Image" by Pekka Parviainen: http://www.polarimage.fi/, see its pages "Mirages" and "SUN involved".

Figure 12-28: Superior mirage of distant islands due to a colder layer of air below a warmer layer. (*Source*: Pekka Parviainen, http://www.polarimage.fi/, with permission.)

Figure 12-29: Flattened Sun at sunset. (*Source*: Pekka Parviainen, http://www.polarimage.fi/, with permission.)

index from 1.0002718 to 1.0002433,[10] while the change due to pressure variation is even smaller.

Figure 12-30 shows how a gradual change in refractive index causes gradual refraction of light and therefore mirages. The principle is that the portions of wavefronts which travel through denser air travel more slowly: this rotates the wavefront, and therefore the direction of travel.

One way to think of gradual refraction is to replace the gradually changing substance by many narrow slices that have slightly different but constant values of the refractive index, like a "staircase" with small steps: at each of the many sharp boundaries between the slices, the wave turns just a little bit.

Figure 12-30 distinguishes two cases for light in the atmosphere. First, at the left, the air is denser higher up: a ray that aims straight toward the observer's eye is then bent upward into the sky and not seen; instead, rays that go obliquely downward are bent upward, and one of these will hit the observer's eye. Now the observer sees the source of light below its actual height, as if it were lower. We call this an **inferior mirage**.[11]

A special case of inferior mirage is shown at lower left in Figure 12-30: only a thin layer of air, typically below eye level, is less dense, as can happen when the Sun heats up a road, which in turn heats up a thin layer of air above it. This is the situation shown in Figure 12-27: the thin layer of hot air near the ground acts like a mirror, but it is refraction rather than reflection that takes place here. However, we now see a double image: an upside-down "reflection" and an "upside-up" direct view: this shows that light from the same location can reach the eye by two different paths (as shown at the bottom in Figure 12-30). The two paths are analogous to the two views we see in normal mirrors: the original object on our side of the mirror, and the virtual mirrored image that we see behind the mirror.

---

[10] See the interesting page on "Atmospheric Refraction" at http://hyperphysics.phy-astr.gsu.edu/hbase/atmos/mirage.html .

[11] See also "An Introduction to Mirages" by Andrew T. Young at https://aty.sdsu.edu/mirages/mirintro.html.

A different scenario is shown at right in Figure 12-30. Now the air is denser lower down. This curves the light rays downward rather than upward. A ray aimed directly at the observer's eye now curves down into the ground. This time, rays that go obliquely upward will be bent downward, and one of these will hit the observer's eye. Now the observer sees the source of light above its actual height, as if it were higher. We call it a **superior mirage**. This is the situation shown in Figure 12-28: a layer of warmer air above eye level acts like a horizontal mirror stuck to a ceiling (but, again, it is refraction rather than reflection that takes place here). Such a situation occurs especially over cold water or ice: this cools the air near the water or ice, relative to warmer air higher up. As with the inferior mirage, we may see double images, similar to the case of normal mirrors: Figure 12-28 gives a dramatic example.

Many variations of gradual refraction are observed in nature. They differ primarily in the way that the air temperature and density vary with altitude. One spectacular version is the fata morgana: here the air structure is such that vertical walls appear in the refracted image, giving the impression of buildings where none exist.[12]

Gradual refraction is also responsible for the late setting and flattening of the Sun when it is near the horizon, as photographed in Figure 12-29. The explanation is shown in Figure 12-31, where an observer watches a sunset. The Sun has already dropped below the horizon, but the atmosphere bends its light rays around the curvature of the Earth, so that it can still be seen. In addition, light rays from the upper and lower parts of the Sun travel through air with slightly different densities, so that the lower edge of the Sun is more refracted than the upper edge, resulting in the apparent flattening of the Sun. Seeing this flattening of the Sun does not require special conditions: it is frequently and easily seen with binoculars (in good weather and with precautions to avoid looking straight at the Sun when it is still very bright). The late sunset and equivalent early sunrise in the morning are not obvious to

[12] See "Fata Morgana" at https://en.wikipedia.org/wiki/Fata_Morgana

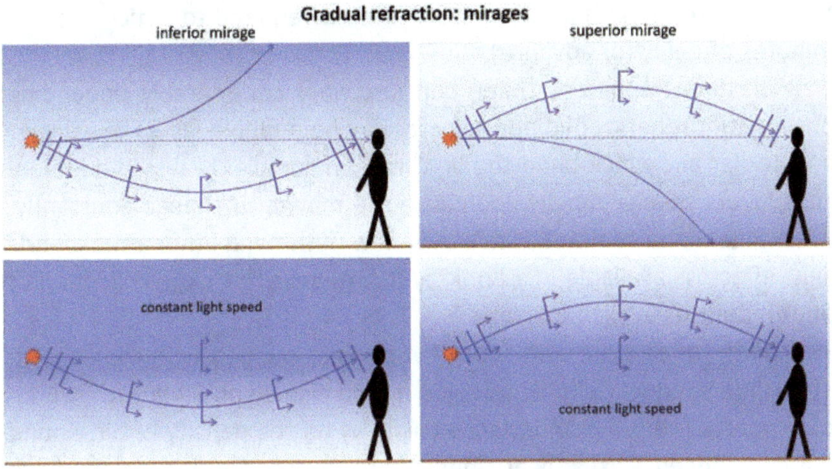

Figure 12-30: Light radiates from a source at the left. It travels through air with varying light speed: darker blue is denser air with a slower speed of light. *At the left*: slower speeds at higher altitude cause an inferior mirage; in the lower sketch, the speed of light only varies below the eye level, remaining constant above the eye level. *At the right*: slower speeds at lower altitude cause a superior mirage; in the lower sketch, the speed of light only varies above the eye level, remaining constant below the eye level. Arrows show directions and speeds of wave motion, which is perpendicular to wavefronts. The effects are much exaggerated in this drawing.

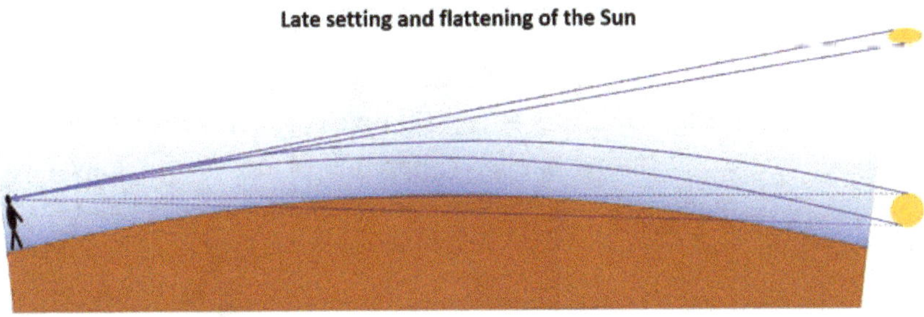

Figure 12-31: Sunlight at sunset (or sunrise) is gradually bent (refracted) as it travels along the curved surface of the Earth, causing the Sun to appear "too high", including above the horizon when the Sun is already below the horizon. Light from the lower edge of the Sun travels through denser air, bending it more than light coming from the upper edge: this makes the Sun appear flattened. The effects are much exaggerated in this drawing. In reality, the Sun appears about 0.5° too high near the horizon. That angle is close to the apparent diameter of the Sun: therefore, at sunset (and at sunrise), the Sun is displaced upward by about its own diameter. The same is true for the moonset and moonrise, since the Moon also has a diameter of about 0.5°. Stars also sink below the horizon by a similar amount when they rise or set. A realistic flattened shape of the Sun at sunset and sunrise is seen in Figure 12-29; the Moon has the same flattened shape at moonset and moonrise (the amount of flattening will depend on atmospheric conditions).

the eye: careful measurement of the time is required to notice the few minutes of delay (or advance).[13]

The Sun's apparent shape can also vary considerably depending on the atmospheric conditions: there may be multiple layers of warmer and cooler air, which bend the light rays in complex ways; in addition, such layers change rapidly and cause the mirage to change constantly. An excellent and comprehensive overview of this and other atmospheric light effects is available in a book by R. Greenler.[14] A large collection of photographs is available online.[15]

Do mirages exist with sound? If mirages happen with light due to atmospheric effects, similar mirages should also be possible with sound: the speed of sound also depends (slightly) on the density of air. Sound mirages indeed exist but are difficult to notice. The reason we have difficulty hearing sound mirages is that our hearing is not directional: we will not notice that a sound source appears to move by a few degrees.

On the other hand, you may have noticed that distant sounds sometimes seem louder than normal. That could be the situation at the right in Figure 12-30: sounds that otherwise would go up and unheard may instead come down to be heard. The reverse may also happen, as shown at the left in Figure 12-30: sounds becoming inaudible at times because they go down. I clearly remember occasionally hearing train horns coming from over 5 kilometers away from my home, although I could normally not hear them. This was likely a mirage effect, perhaps coupled with lensing: air layers at certain times probably bent the sound waves down to my home; similarly, patches of denser air may have acted as converging lenses. Both causes can result in increased intensity of the sound.

---

[13] The amount of delay in sunset depends very much on the latitude of the observer: an observer near the Equator sees the Sun set almost vertically, which shortens the delay time; an observer in a polar region may see the Sun skim along the horizon and with much longer delays; at the North and South Poles the Sun may even sink just below the horizon for days as it circles along the horizon and still remain visible.

[14] Robert Greenler, "*Rainbows, Halos, and Glories*", Cambridge University Press, Cambridge, 1980.

[15] See the website "Polar Image" by Pekka Parviainen: http://www.polarimage.fi/, and its page "SUN involved".

Similar effects can take place with earthquake waves. In Figure 9-6 of Section 9.4, we saw that earthquake waves gradually curve as they travel through the "onion peels" of the inner Earth; this gradual refraction is due to increasing pressure and temperature with depth, which increase the speed of waves and make them curve upward. Near the Earth's surface, the soil conditions also vary and deflect such waves. But here again, we are not well equipped to observe such effects, especially because during earthquakes our thoughts are more on survival and less on observation and measurement.

Why do stars twinkle? A more familiar and closely related effect is the **twinkling** of stars. It is due to rapid local variations in the density of air in the atmosphere. These are mainly thermal variations, for example in the turbulent hot air rising above a fire or hot object, or in the hot exhaust of a combustion engine. We have all seen objects through hot air appear to shake as the air rises, while sharp light sources twinkle just like stars. Such variations can cause lensing of light (as well as deflection of a star's image). Dense parts of the air can focus light toward your eye, as if you were wearing convergent eyeglasses, while less dense parts can spread out the light like a divergent lens: the intensity of light that you see then varies rapidly. You may know that planets normally do not twinkle: the reason is that they are much closer to us than the stars, so the planets appear as slightly larger disks that are less affected by the atmospheric variations.

The twinkling of stars is quite similar to seeing a lamp through waves in a swimming pool: such a lamp, installed underwater, appears to "twinkle" and jiggle around as its light is refracted by the ever-moving waves on the surface of the pool. The surface of water produces much stronger refraction than do thermal variations in the atmosphere, so large underwater lamps twinkle while planets in the sky do not.

Twinkling is a serious problem for telescopes looking at stars in the sky: it reduces the sharpness of star images because images are usually recorded over much longer time intervals than the variations in the atmosphere and are then smeared out by twinkling. One elegant solution, although complex, is called **adaptive optics**: it splits up the main mirror of a telescope into many small mirrors whose orientation can be rapidly adjusted to compensate for the atmospheric variations.

Thereby, the small mirrors together form a large single flexible mirror that is continually adjusted so as to focus all the light arriving from a star at a single sharp point. This approach requires light sensors that record how atmospheric variations bend the light away from its average position; then, a computer directs micromotors to tilt each mirror to bend the light back to where it should go. This can improve the recorded sharpness of a star's image tenfold.

Adaptive optics can also be combined with artificial **laser guide stars**, especially when the star of interest is too weak to be used to adapt the optics of the telescope. These guide stars are artificially created by sending a sharp laser beam into the sky close to the direction of the star of interest and observing its light coming back from the atmosphere: this light will suffer almost the same atmospheric variations and can thus be used to quickly adapt the optics of the telescope.

12.6 Retroreflectors and Optical Fibers

A flat mirror reflects light in one particular direction: to see myself, I have to stand directly in front of the mirror. ***Is it possible to design a mirror that allows me to see myself from other directions?*** Indeed: **retroreflectors** do that. A familiar example is a pair of mirrors in a corner, such as in an elevator: if I look in that corner, I see my image in that corner, no matter where I am standing. A very clever use of retroreflectors is on the back of vehicles: they reflect light coming from other vehicles back toward those vehicles, no matter where they are, and, most importantly, toward the drivers of those vehicles, as a warning.

How do retroreflectors work? The basic idea is to use two or three mirrors that are perpendicular to each other. In the 2D example at top left in Figure 12-32, we see a ray of light (red) reflected once from each of two perpendicular mirrors. With a bit of geometry, it is found that the two reflections change the direction of the incoming ray by exactly 180°, so the ray returns in the reverse direction toward where it came from. The elegant feature is that the ray may come from any direction between those mirrors. This arrangement works in a 2D plane. For example, if you look at yourself in such a 2-mirror corner with vertical

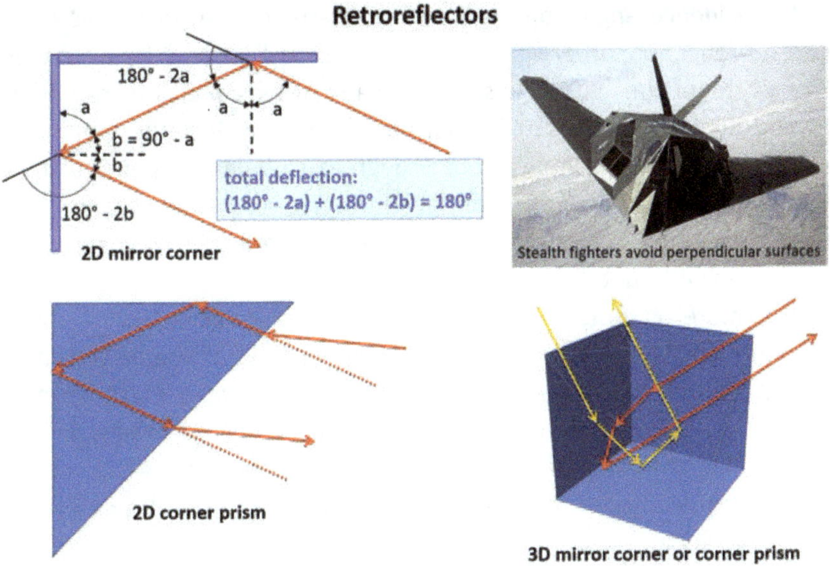

Figure 12-32: Examples of retroreflectors. *Top left*: a pair of perpendicular mirrors forming a 2D mirror corner, with ray tracing that shows how an incoming light ray is turned around with two reflections; "a" is the incident angle on the first mirror, "and "b" is" the incident angle on the second mirror. *Bottom left*: Another mirror corner, but using total internal reflection, forming a corner prism. *Bottom right*: A 3D arrangement of three perpendicular mirrors; here all light rays are reflected three times in 3D, so they end up going in the opposite direction of where they came from; similar retroreflectors have been placed on the Moon's surface to measure the distance from Earth to Moon. *Top right*: perpendicular pairs of surfaces can reflect radar signals directly to the radar source (as shown at the *top left*), exposing the location of an airplane; this can be avoided with surfaces that are not perpendicular to each other, as in this photograph. (*Source of photograph*: U.S. Air Force, in the public domain, https://en.wikipedia.org/wiki/File:F-117_Nighthawk_Front.jpg.)

mirrors (as in an elevator), you will see your eye in the corner at its correct height above the floor.

A similar effect is obtained with a rectangular prism, as shown at bottom left in Figure 12-32. Light entering the prism from the front face (opposite the rectangular corner) is first refracted but is then reflected twice within the prism into the reverse direction. The refraction upon entering and exiting the prism compensates itself. This setup works best if the incidence angles on the two inner faces of the prism give total internal reflection; otherwise, some of the light will leak out from another face of the prism. With a glass prism made of typical glass surrounded by air, this condition is satisfied for both internal reflections with initial angles of incidence within about 5 degrees of being perpendicular to the front face of the prism. More oblique angles of incidence cause loss of light due to leakage; this loss can be much reduced by coating the back faces with metal, as with normal mirrors.

A more complete 3D retroreflection is achieved with three perpendicular mirrors, such as the three inner sides of a cube shown at bottom right in Figure 12-32. Now any ray entering this 3-mirror corner, including if it is angled up or down, will be reflected exactly in the reverse direction. In this case, you will see your own eye's image in the corner of those three mirrors. You may experience this in an elevator that also has a horizontal mirror against its ceiling, in addition to two vertical mirrors on two adjoining sides.

Seeing our image in mirrors reminds us of the discussion in Section 12.2 about left *versus* right, up *versus* down, and front *versus* back. We can ask the same questions about retroreflectors: do we see ourselves "inverted" left-to-right, up-to-down and/or back-to-front? Figure 12-33 illustrates the three cases of a single mirror, a pair of mirrors and a triplet of mirrors: it challenges you to select which image of your face you will see in each case!

We have discussed several cases of total or near-total reflection of waves. For example, the end of a string causes total reflection (Section 2.7); a very hard wall can reflect almost all the sound falling on it (Section 3.8); a normal mirror can reflect nearly 100% of light (Section 12.2). With total internal reflection, it is possible to reflect 100% of light (Section 12.4). We have also seen that gradual refraction can produce 100% reflectivity (Section 12.5).

Using such walls and mirrors, we can channel sound and light through tubes over large distances. We mentioned the case of the speaking tube (voice pipe), still used on ships, in Section 3.10. For light, we are familiar with **optical fibers** used for decoration (see Figure 12-34), or for imaging internal cavities in the body using a device called an **endoscope**, or for telecommunications, including over the internet. Let's look more closely at optical fibers.

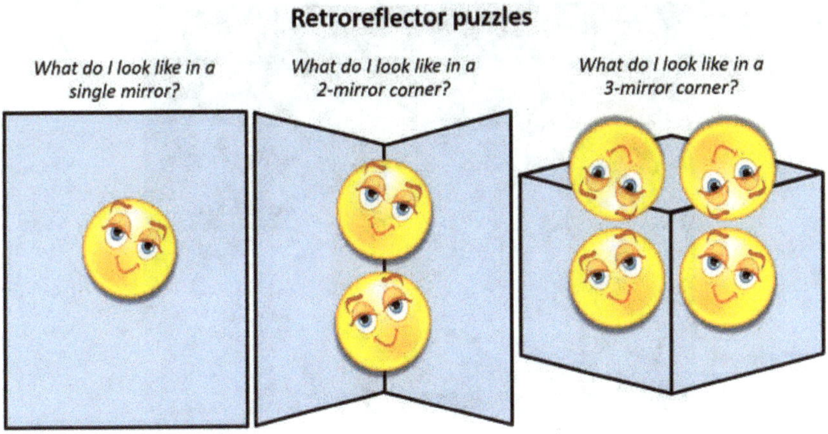

Figure 12-33: If you see your face as the image in the single mirror (at the left), which face image do you expect to see in front of the 2-mirror corner and the 3-mirror corner? Hints are in the footnote.[16]

How do optical fibers work? Figure 10-14 shows, at the right, a wave being channeled between two boundaries by being reflected repeatedly and thus traveling along a zigzag path inside the "red wall". However, at each reflection, the wave loses amplitude as it leaks out by transmission (refraction): the wave inside the wall will then not survive for a long distance, such as from one building to another, or from one continent to another.

To avoid this amplitude loss, we ideally need 100% reflectivity: so we need total internal reflection within the red wall. We have a solution at right in Figure 10-6: there the incident angle is relatively large (more parallel to the boundary). If we now place two such boundaries opposite each other, we can make the wave bounce back and forth with 100% reflectivity at each bounce, so it travels far between those two boundaries, as in a leak-free tunnel. This is the principle of the optical fiber for light.

---

[16] The leftmost face image in Figure 12-33 is the left-to-right reverse face that a friend looking at you would see, or that you will see on your passport photo. The 2-mirror corner will reverse left and right again, because each mirror in effect reverses the image, showing (at the bottom) your face as seen by a friend or in your passport photo. The third mirror on the ceiling reverses top and bottom (but not left and right): you should therefore see the top left image.

Figure 12-34: Bundle of optical fibers with red light emerging from their ends. (*Source*: in the public domain, https://www.rawpixel.com/image/5925561/photo-image-background-light-public-domain.)

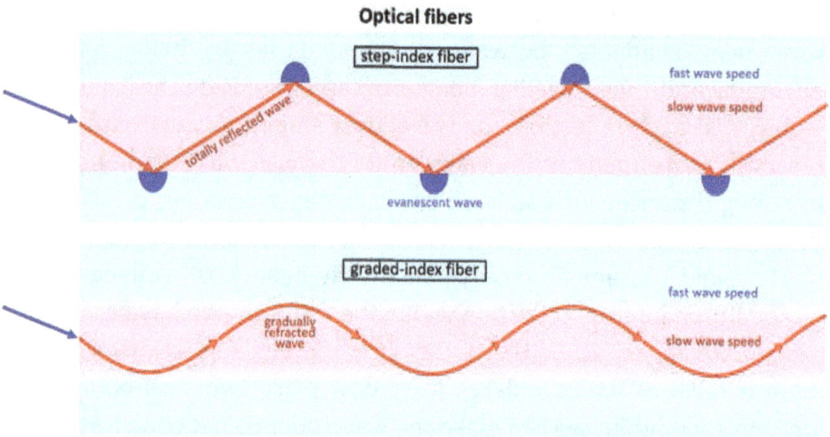

Figure 12-35: Two types of optical fiber: step-index and graded-index. An external wave (blue) enters the left end of the fiber, then stays within the fiber either by repeated total internal reflection (*at top*) or by gradual refraction (*at bottom*). The dark-blue half-circles represent evanescent waves dying out into the fast-wave substance (light blue). Ray tracing is used here; the snaky red curve at the bottom is not the wave itself but only its path, to be compared to the zigzag path at the top. The wavelength of light is much too small to draw on the scale of the optical fiber.

Two models are shown in Figure 12-35. The upper model has a sharp boundary (as in Figure 10-14). Here, the internal (red) substance has a slower wave speed (and shorter wavelength) than the external substance (blue), producing total internal reflection for suitable directions of the light. The wave now repeatedly bounces back inside the fiber with no or little loss of amplitude (imperfections in the material will inevitably still cause some loss). This model uses a step-like sudden change of the refractive index and is therefore called **step-index fiber**.

The lower model in Figure 12-35 uses the principle of gradual refraction that we discussed in Section 12.5, as illustrated in Figure 12-30: it is therefore called **graded-index fiber**. Here, the sharp boundary is replaced by a gradual change in refractive index. Light that travels away from the centerline of this fiber will turn back gradually toward the centerline, similar to the mirages in Figure 12-30.

The successful operation of an optical fiber requires that the light rays always undergo total internal reflection: otherwise, light will leak out and the signal will weaken. This requirement implies that the fiber must have no sharp bends. A sudden bend in the fiber could make a light ray hit the fiber's wall at an angle that allows partial refraction out of the fiber instead of being totally reflected back into the fiber.

12.7 Solar Spectrum and Rainbows: Dispersion

Rainbows are splendid natural events: Figures 12-36 and 12-37 show beautiful examples. They remind us of the solar spectrum of Figure 12-4. Indeed, there is a direct resemblance between rainbows and the solar spectrum, but what is the connection? Before we address this question, note the second rainbow in Figure 12-36: it is weak but visible at the upper left and upper right; it is called the **secondary rainbow**, to distinguish it from the brighter **primary rainbow**. The secondary rainbow is normally present but often not seen: either it is too weak compared to the background light or our attention is fixed on the primary rainbow.

How do rainbows come about? What causes their display of colors? We all know that rainbows are seen with **sunlight** in rain. Under good conditions, they are also seen in water spray from sprinklers, in fog, in clouds, in dew drops on grass, *etc*. Clearly, water drops play an important role. In fact, water drops split up sunlight into its colors, from

Figure 12-36: A double rainbow photographed in Alaska. The inner, more intense rainbow is called the primary rainbow, while the outer, weaker rainbow is called the secondary rainbow; it is seen at the upper left and upper right. Both rainbows can form full circles, when not obstructed by the ground. (*Source*: Eric Rolph, under CC BY-SA 2.5, https://upload.wikimedia.org/wikipedia/commons/5/5c/Double-alaskan-rainbow.jpg.)

Figure 12-37: A rainbow against a dark sky. (*Source*: Needpix.com, in the public domain, rainbow-2760167_1280.jpg.)

red to blue. This splitting of the colors is due to refraction of light in the waterdrops, just like the splitting of light by a prism. The prism being simpler, we start by examining the case of the prism.

Figure 12-38 sketches a glass prism. From far left, five parallel rays with typical colors approach the prism in the same direction: they are partly refracted (bent) into the prism (we ignore here the part that is reflected out of the prism). After crossing the prism, these rays are partly refracted out of the prism, to the right (again, we ignore the part that is reflected). However, after each refraction, the rays are no longer parallel to each other: that is because of their different colors. **The refraction angle of light depends on the light's color, which depends on the light's frequency or wavelength, and ultimately on the wave's speed through the refractive index. This effect is called dispersion because parallel rays are dispersed into different directions due to their different wave speeds.**

If the incoming rays with different colors coincide, they will still be refracted into different directions, as drawn at the right in Figure 12-38. This is the case of sunlight: it contains all the colors shown in Figure 12-4, and each color is bent into a different direction, forming the **solar spectrum** seen at right in Figure 12-38.

You likely have seen such colorful solar spectra at home due to sunlight being refracted and dispersed by glass objects. Those objects that have irregular angles and corners are most likely to form solar spectra, such as drinking glasses, chandeliers and other sparkling pieces of glass.

Where does this color-dependent bending of light come from? The most important factor in refraction is the speed of the wave on both sides of the boundary. With EM waves in substances other than vacuum, this speed depends on the frequency of the wave because waves of different frequencies interact differently with the substance. This dependence on frequency was not mentioned in Table 12-1, which lists only average values for visible light! This dependence is large enough to cause the colors of rainbows and the colorful reflections in prisms.[17]

---

[17] Quantitatively, the speed of light varies by about 1% in water and about 1.3% in typical glasses as light slows down from red to blue. For example, the refractive index of water increases from 1.3318 for red light to 1.3435 for blue/violet light.

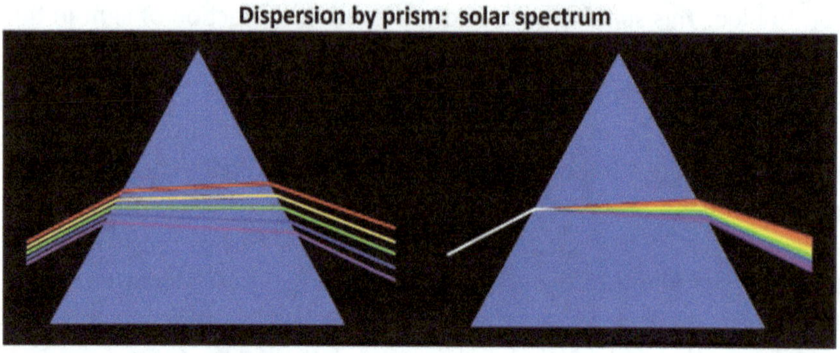

Figure 12-38: A prism disperses light rays of different colors. *At the left*, parallel rays of different colors are refracted by a prism into somewhat different directions. *At the right*, a beam of white light which contains all visible colors is split by the same prism into all its colors.

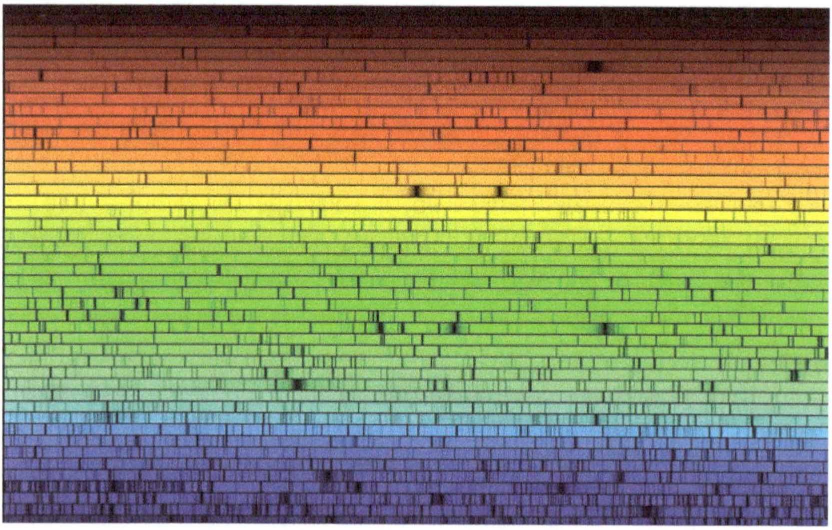

Figure 12-39: Photograph of the solar spectrum, by the US National Aeronautics and Space Administration (NASA). This solar spectrum is much more detailed than the one shown in Figure 12-4. The spectrum, covering the wavelengths 400 to 700 nanometers, is cut up into 50 slices laid one below the other; each slice covers 6 nanometers. (*Source*: This photograph is in the public domain; it is available with better detail/resolution at https://solarsystem.nasa.gov/resources/390/the-solar-spectrum/.)

An important property of prisms with perfectly flat faces is that they separate colors without any overlap: all waves of a given color (frequency) will go in a particular direction that is different from all waves of other colors. We will see below that waterdrops do overlap rays with different colors, thereby mixing colors and making the rainbow visually different from the solar spectrum.

The non-overlapping dispersion of prisms allows us to study in great detail the kind of light coming from various sources. Figure 12-39 shows an impressive example: it shows a highly detailed spectrum of solar light from NASA. We see that the real solar spectrum is much more complicated than the simple, smooth distribution of colors drawn in Figure 12-4. Indeed, the actual measured spectrum of light coming from the Sun also has dark **absorption lines**; these narrow dark lines are due to nuclear reactions in the Sun and to gases or dust in the Earth's atmosphere: they show up at those wavelengths or frequencies where light is absorbed and does not reach us on the surface of the Earth. These absorption lines are not normally visible, because they are too narrow to be detected with normal cameras or our eyes. We will therefore also ignore them in the following.

Here we need to make clear what we mean by colors. The solar spectra of Figures 12-4 and 12-39 contain an infinite number of colors from red through yellow, green, cyan (green-blue), blue and violet. However, they do not contain other colors such as gold, silver, navy blue, salmon pink, turquoise, pink, indigo, white and gray; those colors are combinations of the "solar colors" from red to violet, which are pure colors. For example, gold combines red and weaker green (while equal red and green give yellow); white is an equal mix of red, green and blue; pink combines red and white; turquoise combines green with weaker blue and much weaker red; indigo combines dark blue with a small amount of red; and gray is simply weaker white. (See my book on *"Everyday Physics — Colors, Light and Optical Illusions"*[18] for more details.)

In this section, we focus on the pure colors of the solar spectrum: each of these colors has one specific frequency, and therefore also one specific wavelength, together with one common wave speed in

[18] Michel A. Van Hove, *"Everyday Physics — Colors, Light and Optical Illusions"*, World Scientific Publ. Co., 2022.

vacuum (the speed of light). Consequently, all the pure colors are separated from each other by a prism (due to dispersion). By contrast, a prism would split up a combined color into its component pure colors; for example, a prism splits up white sunlight into the pure colors of the solar spectrum. As we will see, a rainbow recombines different pure colors (coming from the Sun) into combined colors: a rainbow thus does not produce pure colors.

How different are rainbows from the solar spectrum due to a prism? Besides the similar-looking colors, there is a clear structural similarity: a ray of light can enter the front of a raindrop and exit from the back; refraction will then also bend rays of different colors into different directions. However, the round shape of a raindrop also causes other effects that we wish to discuss next.

Figure 12-40 (drawn by C.M.G. Lee) gives an excellent overview of the origin of rainbows. These involve several interesting aspects:

- We must have sunlight, or another distant bright source of light producing nearly parallel rays of light, preferably containing many colors.

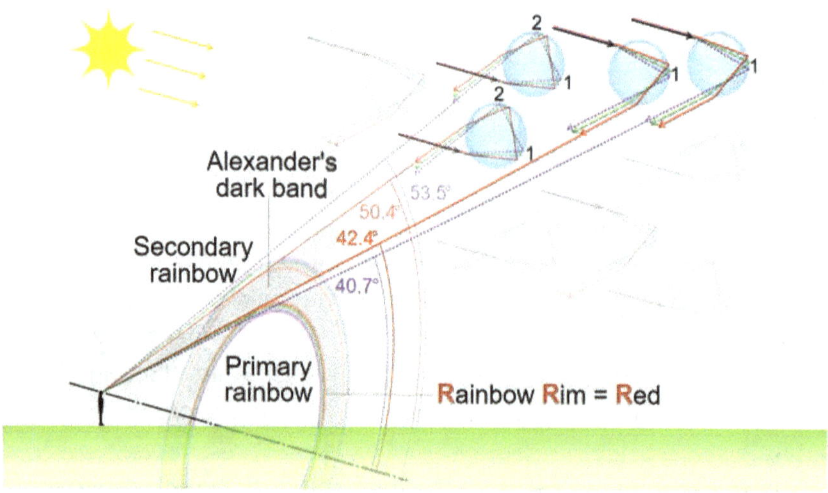

Figure 12-40: Seeing rainbows: only raindrops in the correct directions from the observer produce the colors of the primary and secondary rainbows. Alexander's dark band is a relatively dark region between the two rainbows, due to light having been redirected outside this region. (*Source*: "Rainbow principle" by C.M.G. Lee, under CC BY-SA 4.0, https://commons.wikimedia.org/wiki/File:Rainbow_principle.svg).

- There must be many raindrops (or other drops of water such as from sprinklers or fog or dew, in the right direction compared to the Sun; these drops do not need to be far from the observer). Figure 12-40 shows many raindrops, but only some of them (marked with "1" and "2") are correctly lined up to create a rainbow seen from the observer's position.
- Light rays penetrate the raindrops and are reflected from their back side, emerging from the front side to form the **primary rainbow** (see the two drops labeled "1" at the far right in Figure 12-40).
- A second internal reflection is possible: it creates the **secondary rainbow**, after emerging in a different direction from the primary rainbow (see the two drops labeled with both "1" and "2" in Figure 12-40).
- Dispersion (the varying angle of refraction for different colors) splits the colors into different directions, similar to the solar spectrum seen with a prism (this is marked with red, green and blue rays in Figure 12-40).
- The round shape of the drop mixes the colors (unlike the prism), so the rainbow colors are combined colors, not identical to the solar spectrum (we will discuss this further below).
- An observer will see the primary or secondary rainbow of a particular raindrop only if his/her eye is on the corresponding cone of that specific drop.
- The observer will see an arc of a rainbow only if there are many raindrops in many different directions.
- If the observer moves, the rainbows will appear to move with him/her, as if the rainbows were attached to that person. The rainbows do not grow in size if the observer gets closer to the raindrops.

We explore these separate aspects one by one in Box 12-2. You may also consult very useful webpages for further details.[19–21]

---

[19] See "Rainbow" at https://en.wikipedia.org/wiki/Rainbow

[20] See "Rainbows" at https://atoptics.co.uk/bows.htm

[21] See "The Calculus of Rainbows" by Te-Sheng Lin at https://teshenglin.github.io/post/2020_cal_s4p1/

BOX 12-2 — HOW WE SEE RAINBOWS. Let's first look at a single sun ray approaching a water drop from the Sun, as drawn at the left in Figure 12-41. That ray is partly reflected outside and away from the drop: this outside reflection produces uninteresting white light.

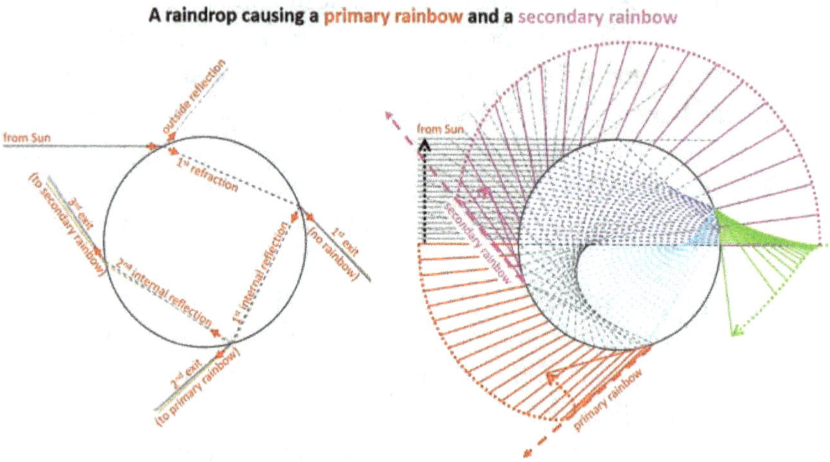

Figure 12-41: Light rays reflected and refracted by one raindrop. *At the left*: One sun ray approaches a raindrop horizontally from the left; the ray is reflected and refracted each time it hits the surface of the drop. At each refraction, the white sunlight is split into multiple colors that spread out, as shown by red, yellow, green, blue and violet lines. (More reflections and refractions are possible but weak, and therefore not shown.) *At the right*: Multiple sun rays (of a single color or frequency now) approach horizontally from the left at different distances from the central axis of the drop; these sun rays are shown as dotted gray lines (the line colors here do not indicate light color). Each incoming ray follows a different path, causing "fans" of outgoing paths: the outside reflection gives the light-gray fan of paths toward the top; the 1st exit gives the green fan; the 2nd exit gives the red fan (which produces the primary rainbow after one internal reflection); and the 3rd exit gives the purple fan (which produces the secondary rainbow after two internal reflections).

The incoming ray is also partly refracted into the drop (marked "1st refraction"): it then crosses the drop. The ray next approaches the back of the drop, where it is partly refracted out ("1st exit"). Rays with different colors are refracted into different directions, just as with a prism: close inspection of the left sketch of Figure 12-41 shows this dispersion as five rays of different colors going to lower right behind the drop. We have already seen this situation in Figure 12-20 (Section 12.3) for sunlight passing through a cylindrical glass:

however, as the photograph in Figure 12-20 shows, no rainbow is seen behind a cylindrical glass, or even behind a spherical glass ball! The reason is that the rays of different colors are not bent into privileged directions, resulting in all pure colors from various incoming rays being mixed in all directions so that only white light is seen.

(Incidentally, it is sometimes said that light has total internal reflection in a raindrop: that is not the case, as the angle of incidence on the internal surface of the drop never reaches the critical angle for total internal reflection, at least for light entering from outside the drop.)

We must look further to find a rainbow! Instead of refracting out through the back of the drop, the ray may be reflected into the drop (marked as "1st internal reflection" at left in Figure 12-41): it can now hit its inner surface a second time. Here again, refraction out of the drop is possible (marked as "2nd exit"): we will see that this causes a rainbow when we take the round shape of the drop into account; it is called the **primary rainbow**. Before we do that, we can follow the ray inside the drop through a second reflection and a third refraction out of the drop (marked as "3rd exit"): this also causes a rainbow, called the **secondary rainbow**.

We could continue like this with further reflections inside the drop and further refractions outward, causing a "tertiary rainbow", a "quaternary rainbow", *etc.*, but these are much weaker and are almost impossible to observe because each outward refraction weakens the light staying within the drop. Note how much weaker the secondary rainbow already is in Figure 12-36. These third and fourth rainbows would actually be seen in very different directions, namely toward the Sun rather than away from the Sun like the first and second rainbows.

We must now look more closely at the effect of the round shape of the raindrop, compared with the flat faces of the prism. The right sketch in Figure 12-41 shows rays coming horizontally from the left (thin, dotted black lines; only rays above the centerline are shown, because those below that line behave identically, but upside down): some of those rays hit the drop close to its centerline, others near its edge. Unlike the prism, the spherical drop imposes a varying angle of incidence at its surface. Now each parallel ray is refracted into a different direction, shown as the dashed dark-blue lines inside the drop. These rays hit the back of the drop in different places and at different angles of incidence, where they can exit (1st exit) and form the green fan of rays (compare this with the equivalent situation in Figure 12-20). The curved, dotted green arrow relates these paths to the series of incoming sun rays marked by the dotted black arrow.

(*Continued*)

(*Continued*)

Internal reflection is also possible, giving the light-blue fan of internal rays. These rays can exit (2nd exit): they form the red fan of rays going to the bottom left. However, note the reversal of exit directions, marked by the sharp turn of the dotted red arrow: this is the main cause of the colorful rainbow that we see in the sky! What is happening is that successive sun rays (marked by the dotted black arrow) sweep the red light rays first to the left (back toward the Sun) and then increasingly to the bottom left; however, at some point, the sweep slows down, stops and reverses, near the red arrowhead. This accumulates many light rays near that reversal direction, giving strong light in that direction: this is the reason why the primary rainbow is bright in that direction. However, this does not yet explain the colors of the rainbow. We'll come to that shortly.

The secondary rainbow is created similarly: at the 3rd exit from the drop, the rays form the purple fan at the top in Figure 12-41. Here again, the fan has a reversal, near the purple arrowhead.

It is helpful to watch in a video how these light rays form fans with reversals: see my Animation 12*4.

ANIMATION 12*4 — See my video WB2 at time 7:45 in its section "**Refraction**" under the title "**Rainbows: formation — 2**". (See details in the section References and Resources below.)

We have found two privileged directions (the reversal directions) where outgoing rays are concentrated and cause the primary and secondary rainbows. We still need to explain the colors of these rainbows. To that end, Figure 12-42 focuses our attention on the two direction reversals that cause the rainbows.

In the central sketch of Figure 12-42, we observe that the sun rays which approach the upper edge of the raindrop are mainly responsible for the special reversal behavior that leads to the primary and secondary rainbows. This is not surprising, since the incidence and refraction angles change rapidly as light approaches the edge of the drop.

At right and left in Figure 12-42, we blow up the fans producing the primary and secondary rainbows. In addition, we now include the dispersion effect: rays of different light colors are refracted by different amounts (specifically, blue/violet light is refracted more strongly than red light; Figure 12-40 gives precise values for the resulting directions of the primary and secondary rainbows).

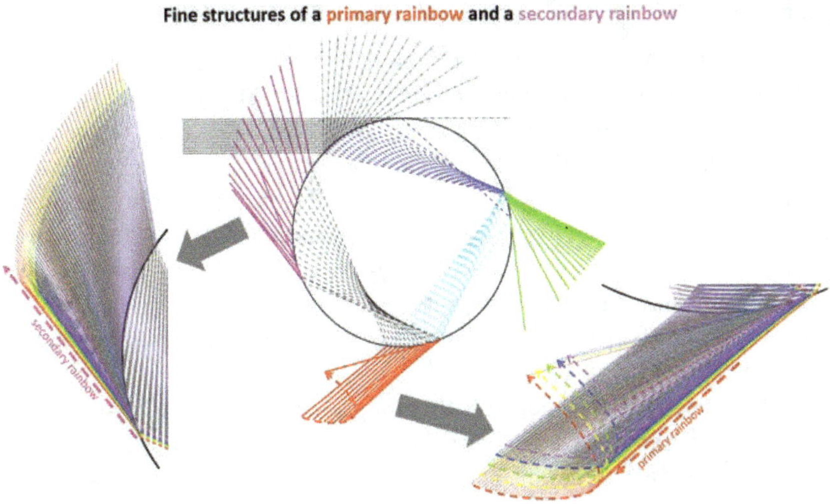

Figure 12-42: Fine color structures of a primary rainbow (*at the bottom right*) and a secondary rainbow (*at the left*). *At the center*: like the left sketch in Figure 12-41 but focusing on sun rays that come close to the upper edge of the drop. *At the bottom right*: blowup of the fan of rays in the primary rainbow, showing the different directions of five different pure light colors; the reversal direction depends on the light color (due to different refractive indices), causing the splitting of colors in the rainbow seen near the bottom right edge. *At the left*: same for the secondary rainbow.

Let's now look more closely at the primary rainbow (bottom right in Figure 12-42). We see what looks like five colored fans that fold back on themselves (in reality, there would also be intermediate fans with other pure colors). At the bottom right edge, the red fan is exposed. To its left, we see a yellow fan, followed by a green fan, a blue fan and a violet fan. These fans overlap each other partially. They look like overlapping semi-transparent colored plastic films: where they overlap, we will see a mix of their colors. For example, the rightmost edge of the red fan does not overlap other fans, so its red color remains pure red; the right edge of the yellow fan overlaps the red fan, giving orange; the right edge of the green fan overlaps both the red fan and the yellow fan, *etc.*

Another way to describe what we see is the following: imagine painting five lines close to each other, in this order: red, yellow, green, blue and violet. Before the paint dries, you brush the paint from red toward blue and beyond. This will mix red paint into all the other colors; mix yellow paint into green, blue and violet; mix green paint into blue and violet; mix blue paint into violet; also mix all paint colors together beyond violet.

(*Continued*)

(*Continued*)

As a result, in rainbows, we don't see the simple solar spectrum of Figure 12-4. Rather, we see a washed-out version where colors are more mixed and less "pure": especially the blue end of the spectrum is often weak, and it can become almost white/gray due to the mixing of all colors. We see this in Figure 12-36.

Another reason for a washout of colors in the rainbow is that raindrops are often not exactly spherical, especially as they drift and get deformed in the wind: this is one reason that real rainbows do not have a sharp red leading edge, but a gentle edge; another reason for this soft edge is that our vision does not cut off sharply at one color, but gently fades away, as we see in the solar spectrum of Figure 12-4.

A further aspect of rainbows is that they are usually seen against a bright sun-lit sky: the colors of the rainbows are then relatively weak. Figure 12-37 has the advantage of a dark, stormy backdrop, giving stronger contrast. Figure 12-43 shows a blow-up of that picture, cutting across the rainbow. Very obvious is the absence of a sharp red edge: the colors are clearly washed out. The colored lines below the photograph tell us the amount of pure red, green and blue colors across this primary rainbow (all humanly visible colors, including combined colors, can be produced by mixing red, green and blue). We clearly see a high red content at the left, then a medium-level green content near the center, and finally a weaker blue content to the right. Also clear is the increasing overlap of colors toward the right in the rainbow: where more green is added to red, we get yellow; and where more blue is added, we get green-blue and then blue-violet. There is also a contrast between the left and right colors outside the rainbow: darker to the left (corresponding to Alexander's dark band) and lighter to the right, which is also very visible in Figure 12-37; these are darker gray and lighter gray, respectively, basically weak whites (due to nearly equal amounts of red, green and blue).[22]

[22] In Figure 12-43, the area below the lowest colored curve is the gray background of the sky because it is common to the three basic red, green and blue components of the light: we perceive equal amounts of red, green and blue as gray, which becomes white when it is intense. Therefore, only the area above the lowest colored curve gives real non-gray color. For example, at the far left, there is a little bit of green and more red above the blue curve: this means that the perceived color is gray + a bit of green + more red = gray + orange; no blue is seen as such because it is part of the gray. Analogously, further right where the red curve is lowest, the perceived color is a mix of green and blue (and gray), with no perception of red. At far right, where the green curve is lowest, we perceive a mix of blue and red (and gray), without green. At the right edge, only gray remains. For details on how colors work, see my book: Michel A. Van Hove, "*Everyday Physics — Colors, Light and Optical Illusions*", World Scientific Publ. Co., 2022.

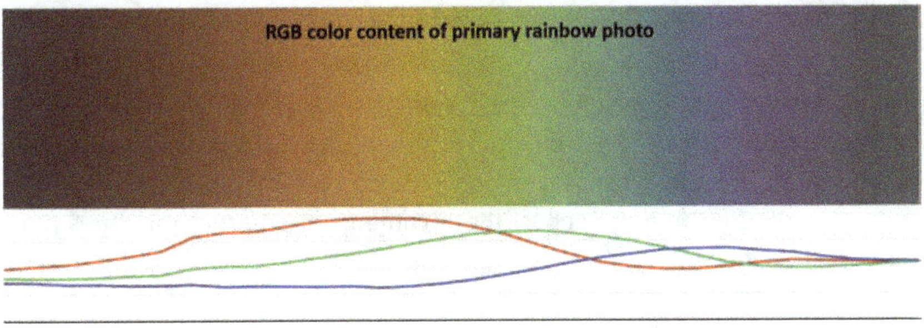

Figure 12-43: Blow-up of a cut across the primary rainbow of Figure 12-37 (*top*), showing (*bottom*) the amount of red, green and blue light (taken from the photo's RGB content, with zero amount at the flat gray line). A special observation is that the violet at right seems to contain some red: this is because violet looks to us a bit like purple and magenta, which colors do contain red; in fact, our eyes' "red" cones are slightly sensitive to blue/violet. (*Source* of image: extracted from Needpix.com, in the public domain, rainbow-2760167_1280.jpg.)

The case of the secondary rainbow is very similar to that of the primary rainbow. One difference is that the order of the red-to-blue/violet colors is reversed in the secondary rainbow, as is visible in Figure 12-36: red is inside, and blue-violet is outside. The reason is that the secondary rainbow exits the drops in a very different direction, so we need to look at Figures 12-41 and 12-42 with sun rays entering the drop below its centerline (it's like mirroring those two figures upside down).

So far, we have only considered a single raindrop. And in Figures 12-41 and 12-42, we have only sketched sun rays approaching the upper half of the drop. Because the drop is a sphere, we can take these figures and rotate them around their horizontal centerline by 360 degrees. This rotation transforms the two rainbows into two full cones opening up toward the Sun, as shown in Figure 12-44 for many raindrops (this figure is oriented like Figure 12-40 for easy comparison). Each drop creates two cones: the narrower cone (shown in red) is the primary rainbow cone, while the broader cone (shown in purple) is the secondary rainbow cone. The cones are inclined to match the sunlight direction. These cones are real: they are cones of increased light intensity radiating away from each raindrop in the sunlight.

Now we can combine all the individual rainbow cones from the different raindrops to build up the rainbows seen by an observer (at the bottom left in Figure 12-44). The important point is this: the observer will see increased light whenever a rainbow cone from any raindrop hits his/her eye. For example, in Figure 12-44, notice the line-up of primary (red) cones along the upper red

(*Continued*)

(*Continued*)

arrow: these come from raindrops that contribute to one point in the primary rainbow seen by the observer. Cones along the large red arc (at the right) contribute to other points around the primary rainbow, forming the arc of the rainbow. Similarly, the secondary rainbow (purple) is built up by the alignment of many secondary (purple) cones; these come from different raindrops than the primary rainbow.

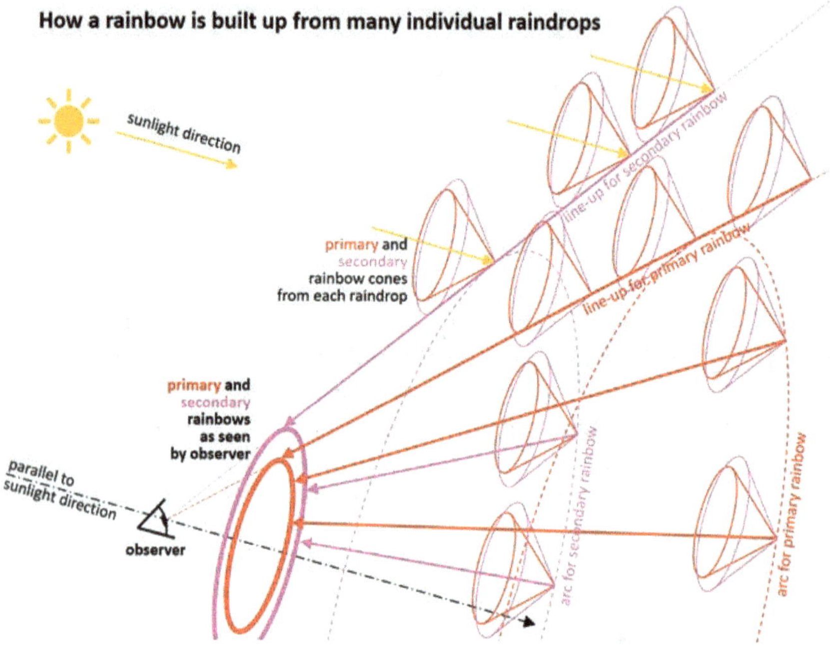

Figure 12-44: Many raindrops combine their individual rainbow cones to form the primary and secondary rainbows seen by an observer. The raindrops are not drawn; they are located at the peaks of the cones. For simplicity, this sketch does not show the dispersion of the color spectrum.

If the observer moves, he/she sees the rainbows move along as well, as if the rainbows were attached to the person. In that case, different raindrops contribute to the rainbow.

In Figure 12-44, we have ignored the color spectrum of the rainbows. In fact, each rainbow cone shown there is a set of many cones with different colors spread out over a small angle. C.M.G. Lee's drawing (Figure 12-40) adds this aspect to show how the observed rainbows obtain their colors.

The darker region between the primary and secondary rainbows, marked as **Alexander's dark band** in Figure 12-40, is the result of the reversals which are illustrated in Figures 12-41 and 12-42. Both the primary and the secondary rainbows refract sunlight away from this region, hence its darkness. The width of Alexander's dark band is about 8°, measured between the leading edges of the rainbows, namely the red arcs. This angle depends on the refractive index: it increases to about 13° between the violet edges of the two rainbows.

We may now ask: ***Could we close Alexander's dark band by changing the refractive index, thereby collapsing the primary and secondary rainbows into a single rainbow?*** The answer is yes! By reducing the refractive index for red light to about 1.3125 (from the 1.3318 used in Figure 12-44), the primary and secondary rainbows are pushed against each other along their red arcs, closing Alexander's dark band: this very distinctive "joint rainbow" would have a single joint central red arc (due to the primary and secondary rainbows touching each other), with yellow, green, blue and violet arcs spreading out on both sides of the red arc.

Which substance has a refractive index close to 1.3125? In Table 12-1, we find that water ice has a refractive index of about 1.31! (Depending on the wavelength, this value ranges from 1.3049 to 1.3147.[23]) So we should see a "joint rainbow" from droplets made of ice rather than liquid water. Have you ever seen or heard of such a "joint rainbow" or "icebow"? Nor have I. When raindrops freeze, making ice crystals and hail, they lose their spherical shape and completely disrupt the paths that produce the classic rainbows. Ice crystals, due to their more angular shapes, in fact produce a variety of other optical effects in the atmosphere, including halos, arcs, spots, pillars, circles and crosses.[24]

One subtle feature in the primary rainbow of Figure 12-36 is weak lines below the rainbow: these so-called **supernumerary rainbows** are wave interference effects due to light following different paths that produce constructive and destructive interferences.[25] Supernumerary rainbows need water drops of uniform size because the wave interference depends closely on that size: that is why they are not always seen, such as in Figure 12-37.

[23] See "Index of Refraction of Ice" by Munif Hussain at https://hypertextbook.com/facts/2005/MunifHussain.shtml.

[24] Robert Greenler, *"Rainbows, Halos, and Glories"*, Cambridge University Press, Cambridge, 1980.

[25] More precisely: In Figure 12-42, for a given color and for a given exit direction near the rainbow, two paths lead to that same exit direction, thanks to the fold in the colored fans. Since the two paths have different lengths, they can interfere constructively or destructively, producing a series of more intense and weaker rainbows. And since each color has a different wavelength, each color produces a different set of such rainbows, weakly seen in Figure 12-36.

We have seen that the beautiful rainbow is actually a very rich physical phenomenon. It involves several important steps, including especially the combination of dispersion (refraction dependent on color/frequency) with the rounded shape of raindrops.

12.8 A Novelty: Metamaterials and Negative Refraction

What can brand-new man-made materials do for us? Here we will follow the story of a recent discovery and its promises for the future, from hiding objects and eliminating sounds to making perfect lenses. This story exhibits how science can progress from wild ideas to practical uses in optics, acoustics and other areas. Even though such applications are not yet used "every day", some of them may show up in stores "any day" in the near future.

In the 1990s, it was realized that we are not limited to the familiar natural materials, such as iron, bricks, wood, glass, water, air and biological substances: we can go beyond nature and manufacture novel materials that have very unusual and useful physical properties.

Traditional materials consist of atoms and molecules, as we discussed in Section 9.2 with the example of water and ice (see Figure 9-1). In Section 12.2, we have explored how electromagnetic waves, including light, behave in such materials. In particular, refraction — the bending of light as it enters or exits a material — depends on how the atoms and molecules change the oscillation of the electric and magnetic fields in an EM wave.

The crucial realization in the 1990s was that **it should be possible to engineer artificial atoms and molecules that change EM waves in other ways than the familiar refraction that we see every day** in our drinking glasses, fish tanks and swimming pools. This idea by British physicist **John B. Pendry** and his colleagues at GEC Marconi Materials Technology in the UK[26] led to the term **metamaterials**; "meta" comes

[26] "Magnetism from Conductors and Enhanced Nonlinear Phenomena", by J. B. Pendry, A. J. Holden, D. J. Robbins, and W. J. Stewart, *IEEE Transactions on Microwave Theory and Techniques*, volume 47, page 2075, 1999.

from the Greek word μετά which means "beyond", implying materials beyond those found in nature.

The artificial atoms and molecules of a metamaterial would still be made of natural atoms and molecules, but these would be arranged in new ways within the material: natural atoms and molecules would be assembled into larger units with new properties. A rough analogy is Lego blocks: each Lego block is composed of normal atoms and molecules, but the larger blocks have special shapes with unique functions, so they become bricks, beams, wheels, propellers, *etc.* In the case of metamaterials, the blocks are usually small compared to the wavelength of the light or sound that they will influence, so those waves still see the metamaterials as smooth and structureless.

It is quite remarkable that the proposed metamaterials obey pre-existing theoretical knowledge, namely the well-established principles of electromagnetism which British physicist **James Clerk Maxwell** proposed in the mid-1800s: the new metamaterials in fact extend the validity of Maxwell's theory well beyond the materials for which his theory was designed. This is an excellent example of the power of physical understanding: a physical theory can predict new and unsuspected physical behavior. Another example of such predictive power is Albert Einstein's theory of relativity: for instance, it predicted the bending of light by large masses like the Sun and stars, which effect was observed later, giving support to the validity of the theory of relativity.

Frequently, such new out-of-the-ordinary predictions are controversial at first: many scientists (and other people) find them so unfamiliar as to doubt, or at least question, the correctness of the new ideas. Only after actual measurements are found to agree with the predictions will scientists feel comfortable with the new developments; this can take years or decades, during which time intense discussions and debates often occur. A good example is the measurement of gravitational waves a century after their prediction by the theory of general relativity (see Chapter 14).

Already in 1968, Russian scientist Victor Veselago had proposed that substances could have a **negative refractive index**: this switches the refraction direction in the transmitted wave, as shown by the black arrow in Figure 12-45, compared to normal substances (we will explain below why this black arrow also seems to point in the wrong direction,

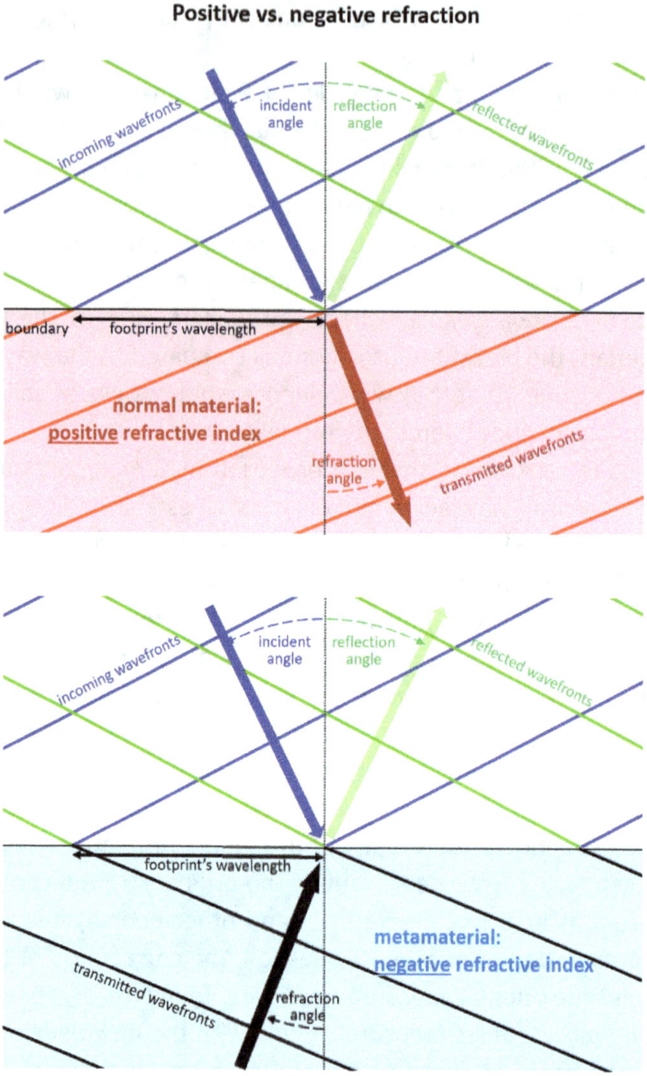

Figure 12-45: *At the top*: normal refraction by a material with a normal positive refractive index (red background). *At the bottom*: negative refraction by a metamaterial with a negative refractive index (blue background). The incident and reflected waves are the same in both cases. Counter-intuitively, in the metamaterial (*bottom*), the refracted wave travels toward the boundary from below, as indicated by the black arrow; nevertheless, energy travels away from the boundary.

up instead of down). However, for lack of natural materials with this property, Veselago's idea was almost forgotten.

Thirty years later, David Smith and his colleagues set out to check the properties of some metamaterials proposed by Pendry. In doing so, they realized the old dream of Veselago and found **negative refraction**.[27]

Let's see how this surprising and counter-intuitive behavior works. Snell's law, given in Section 10.5 and repeated here, predicts the refraction angle of waves as they pass from one substance to another:

sin (refraction angle) / sin (incident angle)
= (refracted wave speed) / (incoming wave speed)
≡ (relative refractive index or relative index of refraction)

The physical properties of the two substances on either side of the boundary are contained in the relative refractive index, which can be measured for each pair of substances. For light, we assume that one of the two substances is vacuum, for which the refractive index is 1: we then drop the word "relative" and speak of **refractive index**.

Known substances found in nature have a positive refractive index: this corresponds to the refracted wavefronts continuing into the quadrant opposite the incoming wave, as they do at the top in Figure 12-45, and at the left in Figures 12-46 and 12-47. More explicitly, if the incoming wave comes from the upper left, the refracted wave goes to the lower right; and if the incoming wave comes from the upper right, the refracted wave goes to the lower left.

According to Snell's law, however, metamaterials with a negative refractive index change the sign of the refraction angle, making it negative: that switches the direction of the refracted wave, as shown at the bottom in Figure 12-45, and at the right in Figures 12-46 and 12-47. It is very helpful to see this motion in my Animation 12*5. As a result, an image like that of a pen resting in a glass of water (Figure 10-16) would

[27] "Composite Medium with Simultaneously Negative Permeability and Permittivity", by D. R. Smith, W. Padilla, D. Vier, S. Nemat-Nasser, and S. Schultz, *Physical Review Letters*, volume 84, page 4184, 2000, https://doi.org/10.1103/PhysRevLett.84.4184. These developments are further described in the free-access article "Reversing Light With Negative Refraction", by J. B. Pendry and D. R. Smith, *Physics Today*, volume 57, page 37, June 2004, https://doi.org/10.1063/1.1784272.

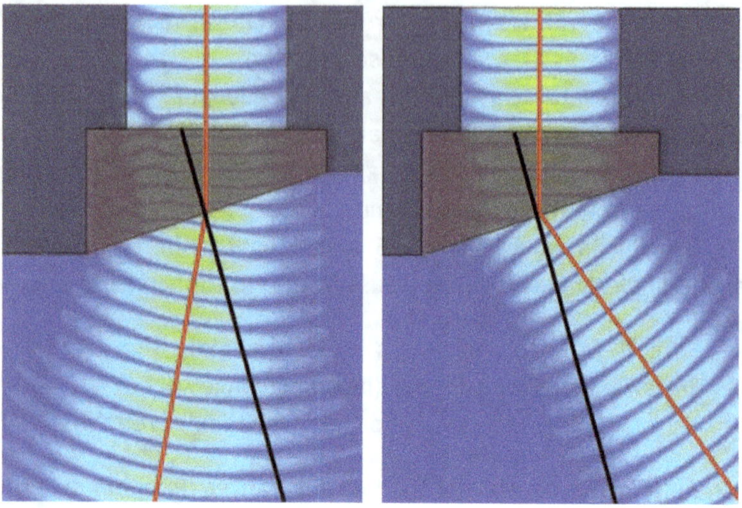

Figure 12-46: Refraction of a light beam of limited width arriving from the top of the two sketches through a wedge-shaped slab. *At the left*: normal refraction by a wedge with a positive refractive index. *At the right*: refraction by a wedge with a negative refractive index. The incoming direction is perpendicular to the first (upper) boundary, where no refraction occurs. The black line is perpendicular to the second (lower) boundary, where refraction occurs in different directions. (*Source*: Reproduced from "Reversing Light With Negative Refraction", J. B. Pendry and D. R. Smith, *Physics Today*, volume 57, page 37, June 2004, https://doi.org/10.1063/1.1784272, with the permission of AIP Publishing.)

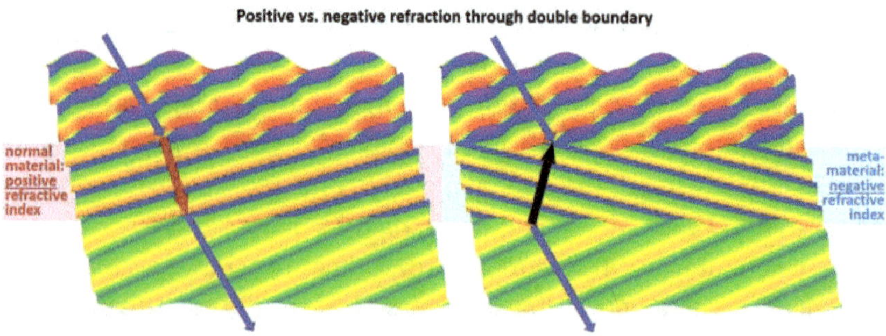

Figure 12-47: As Figure 12-45, but for transmission through a slab with two parallel boundaries (reflections within the slab are ignored), in perspective view. *At the left*: normal slab with a positive refractive index (red background). *At the right*: metamaterial slab with a negative refractive index (blue background). In both cases, energy flows downward through the slab, while in the negative refraction case (*at the right*) the wavefronts within the slab move upward (black arrow). The wave motion can be easily imagined as follows: smoothly and rigidly slide the two images to the right, or smoothly sweep your eyes to the left.

change dramatically if the water were replaced with a substance with negative refraction: it would show the pen tilting the other way under water, from the upper left to lower right in that image.

ANIMATION 12*5 — See my video WB2 at time 9:07 in its section "**Refraction**" under the title "**Metamaterials: negative refraction — 1**". (See details in the section References and Resources below.)

More surprising is that negative refraction also switches the direction of motion of the wavefronts in the refracted wave: they now travel toward the boundary, not away from the boundary as we are used to seeing on water surfaces, for example. This follows from the formula for the refractive index which we have already seen:

$$
\begin{aligned}
&\sin(\text{refraction angle} < 0) / \sin(\text{incident angle} > 0) \\
&= (\text{refracted wave speed} < 0) / (\text{incoming wave speed} > 0) \\
&\equiv (\text{relative refractive index or relative index of refraction}) < 0
\end{aligned}
$$

A negative refractive index implies that the ratio of the refracted and incoming wave speeds is also negative, so that one of the two wave speeds must be negative. Since we define the wave speed in the normal substance to be positive, the wave speed in the substance with a negative refractive index must now be negative. This means that the wavefronts travel in the opposite direction in that substance.

Our intuition conflicts with this behavior: we expect that a wave coming from above the boundary will produce a transmitted wave that continues in the same general direction, away from the boundary! Actually, energy carried by the wave does continue away from the boundary as we would expect; it is only the wavefronts that move in the opposite "abnormal" direction. Information follows the direction of the energy, so it also travels in the proper direction, away from the boundary.

The reason for this apparent contradiction lies in the behavior of the light inside the "Lego blocks" of the metamaterial: waves interfere in a complex manner with themselves inside the more complex structure of each block, and this can result in waves apparently moving backward. You can get a feeling for such complexity in a hall of mirrors at a fair: when you are surrounded by many mirrors with many orientations, everything looks very confusing, and you may well lose your sense of direction. Similarly, sound traveling through a building with many open doors can become very confusing.

An interesting use of negative refraction is the **perfect lens**, or **superlens**. We have seen how lenses work in Section 12.3 (Figure 12-16). Not discussed there is a limitation of such lenses: since they use light waves, the wavelength limits the sharpness of the convergent focus, thereby restricting the image details that can be distinguished. A lens made of a metamaterial with a negative refractive index can overcome this limitation: **a superlens can produce perfectly sharp images as long as the lens itself is perfectly constructed**. Moreover, such a lens can be thin, saving space and weight.

The concept of a superlens has been demonstrated experimentally with visible light: one of the first experiments achieved a refinement of the image by a factor of 6 over a normal lens. As is common in the first stages of technological development, early results are greatly improved later. It should be noted that such superlensing works only at specific frequencies, rather than for all colors, and that losses in the material will also limit the image details.

The superlens is but one example of **metasurfaces**[28]: these are thin metamaterials that can transform light in a variety of ways, such as by rotating the polarization of light, focusing light, generating 3D images, changing the rotation of light, and decomposing light. These properties can improve the performance of information technology, for example.

There are other interesting new materials besides those that produce negative refraction (indeed, many metamaterials do not produce negative refraction, but cause other useful effects). These include **photonic crystals**, which have "Lego-like" structures of size comparable to the wavelength of light,[29] which is close to 1 micrometer (1/1,000 of a millimeter, smaller than the diameter of hair); these structures are repeated regularly like real atoms or molecules in a crystal. See the ice crystal model in Figure 9-1 and imagine Lego-like

[28] "Metasurfaces for quantum technologies", by Kai Wang, Maria Chekhova, and Yuri Kivshar, *Physics Today*, volume 75, issue 8, page 38, August 2022.

[29] Wave packets of light are often called photons.

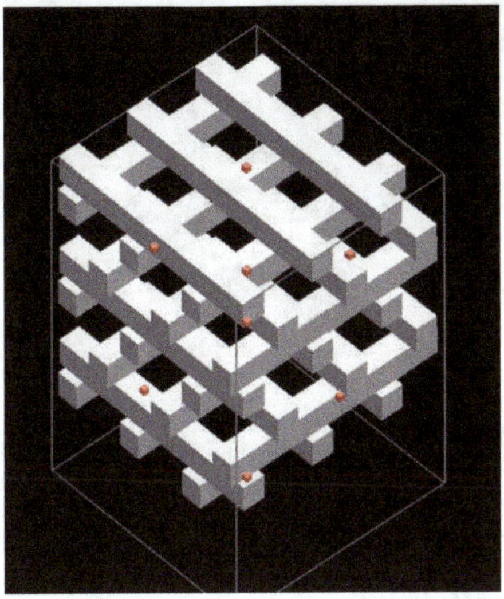

Figure 12-48: Model of a photonic crystal composed of regularly stacked bars of uniform material, spaced by about a wavelength. The bars may be composed of a variety of "normal" materials. (*Source*: Wikimedia, by Vahagn Mkhitaryan (Վահագն Մխիթարյան), file 3D_ֆոտոնային_բյուրեղ).png under CC BY SA 3.0, https://commons.wikimedia.org/wiki/File:3D_%D6%86%D5%B8%D5%BF%D5%B8%D5%B6%D5%A1%D5%B5%D5%AB%D5%B6_%D5%A2%D5%B5%D5%B8%D6%82%D6%80%D5%A5%D5%B2.png.)

blocks a thousand times larger than the water molecules, each block acting like a hall of mirrors.

A simple model of a photonic crystal is shown in Figure 12-48. Photonic crystals are also under intense study and are capable of many new possibilities, including: negative refraction; thin coatings with low or high reflectivity; color-changing paints (such as automobile paints that change color with angle of view); secure printing (with inks that also change color with angle of view); photonic-crystal fibers for refined applications; couplers to feed data into and out of optical fibers; use with solar cells and as sensors.

Metamaterials can be tailored to hide objects in such a way that the cloak itself is invisible: **in invisibility cloaks, metamaterials guide light around the object to be hidden so that the viewer sees no clue to the existence of the cloaked object.** This technique is similar to gradual refraction, discussed in Section 12.5, but the waves traveling along different rays all arrive with the same phase behind the cloak, as if they

Figure 12-49: Light rays (red) enter a metamaterial (blue) that guides them around a cloaked region (black), when viewed from any direction. The cloak and the cloaked region are both invisible.

had not taken a detour; this requires that the waves speed up in the metamaterial to catch up with the waves that go straight past the cloak. Figure 12-49 shows a schematic design for such a cloak. The black region is totally invisible from outside in any direction; also, no light enters it or leaves it.

It is easy to guide light around an object with mirrors and prisms. Such an arrangement, however, will only work when viewed from some particular directions. And it is harder to make those mirrors and prisms invisible themselves. It is even harder to synchronize the waves and thereby erase all hints of the presence of an object with a cloak: this requires a carefully designed metamaterial.

Cloaking is one example of the more general concept of **transformation optics**: in this technique, light rays are steered wherever desired by designing metamaterials that cause bending and speed-up where needed; in addition, the electrical and magnetic fields that form the light rays are also "tailored" to achieve the desired effects.

Extending the notion of negative refractive index, it has been proposed to consider **near-zero-index metamaterials**. A small index implies a long wavelength. An analogy is the tides on the oceans, compared to the much shorter familiar waves due to wind and ships: the tides (see Section 11.6) are waves with ocean-size wavelengths and very high speeds, even larger than tsunamis. A light wave entering a pipe containing near-zero-index material is similar: it will become nearly uniform inside that pipe and suffer no reflection or turbulence due to structural defects of the pipe.

Do other types of waves exhibit negative refraction? **Interestingly, the ideas of negative refraction and metamaterials can also be applied to other types of waves, besides light.** These include, for example, microwaves, sound waves and earthquake waves. One advantage of these waves over light waves is that their wavelengths are much larger, permitting easier fabrication of metamaterials with "Lego-like" blocks of sizes that are still near the corresponding wavelengths. In fact, microwaves were used before light waves to prove the possibility of negative refraction, perfect lenses, *etc.* because the wavelengths of microwaves (centimeters) are thousands of times larger than those of light (micrometers). Metamaterials have also been used to harvest microwave energy, making it possible to wirelessly send and collect energy as microwaves into small devices.

Sound waves have also been extensively used with new acoustic metamaterials. An early proposed example of an acoustic metamaterial consists simply of soft rubber balls in water[30]: the size of the balls is smaller than the sound's wavelength, such as centimeters. At certain sound frequencies, both the density and compressibility of the rubber become in effect negative; this seems counter-intuitive, but corresponds to intense resonances (shaking) that can occur in any substance[31]. Consequently, the refractive index for sound in such a metamaterial becomes negative. As with EM waves, one can then observe negative refraction, which allows acoustic superlenses, for example.

More generally, metamaterials enable novel ways to shape sound; such metamaterials are often called **phononic crystals** or **acoustic metamaterials**. One possibility is to reflect sounds without using hard and heavy walls. Another is to cancel sounds and produce a silent zone in a noisy environment.

---

[30] "Double-negative acoustic metamaterial", by Jensen Li and C. T. Chan, *Physical Review E*, volume 70, article 055602(R), 2004.

[31] More precisely, a negative density means that the substance moves in the opposite direction when pushed, while a negative compressibility means that the substance is expanding under compression, while shaking.

An intriguing suggestion is to use metamaterials to control seismic waves in earthquakes (we discussed earthquakes in Section 9.4). The idea of seismic cloaking is to deflect earthquake waves away from buildings or other structures in order to minimize damage.[32] A suitable metamaterial in this case would be a collection of empty boreholes dug around the structure to be protected.[33] Analogous cloaks could be designed against storm waves and tsunamis on water.

Metamaterials can also be designed to have specific structural properties, such as mechanical strength, independently of their electromagnetic or acoustic behavior. These may resemble the internal lattice structure of natural bones, but on a smaller scale: very small and tight structures tend to be stronger. Another application of structural metamaterials is the **unfeelability cloak**: a flexible lattice-like structure can operate as a perfect cushion that completely hides what is below it.

We see in metamaterials a field of research and development in full expansion: many ideas are proposed and tried out; some of these ideas, after modifications or refinements, become practical for technological applications; someday, such ideas may evolve into devices for common use in daily life.

12.9 Turning around Corners: Diffraction

In Section 3.9, we discussed how sound waves can spread out by turning around corners, a capability that we call **diffraction**. Objects like balls cannot spontaneously turn around corners. Light and all other waves can turn around corners: they thus also exhibit diffraction. An online video is helpful to visualize how waves are diffracted.[34]

---

[32] "Invisibility cloak could hide buildings from quakes", by Colin Barras, *New Scientist, Earth*, 26 June 2009, https://www.newscientist.com/article/dn17378-invisibility-cloak-could-hide-buildings-from-quakes/.

[33] "A Step Towards a Seismic Cloak", by Ping Sheng, *Physics*, volume 7, page 34, 2014, https://physics.aps.org/articles/v7/34.

[34] See video "Experiments with water waves" by TIB — Leibniz Information Centre for Science and Technology University Library, at https://av.tib.eu/media/30440. This video shows waves on the surface of water: their diffraction is very similar for other kinds of waves, including light waves.

Probably the most spectacular display of diffraction by light is the range of colors we see reflected by CDs and DVDs (compact and digital video disks), as photographed in Figure 12-50: these colors are due to diffraction by the many tiny grooves or lines of bumps on the disks, as we will explain further below. Another example of diffraction by light is provided by sheer curtains, mosquito nets and sieves, made of regular nets of threads or wires, and best seen at night, as in Figure 12-51a-c: a strong, distant sharp light shining through them will split up into a regular pattern of light points, which may also exhibit colors like a solar spectrum. You may look for this effect yourself: fine and tight nets are best. In Figure 12-51d-h, there are diffraction patterns due to red laser light shining through 1D and 2D nets.

We have already seen the basic mechanism of diffraction in the closely related situation of two spherical waves interfering with each other: see Figure 3-6; there, two waves from two point sources interfere to produce a characteristic pattern of strong and weak outgoing rays.

Let's consider next a very analogous situation. Suppose now that we have a plane wave hitting an opaque wall that has two small holes in it, as at the left in Figure 12-52: we can view those two holes as two sources of waves going into the dark space behind the wall. These two waves will be spherical in shape, going in all directions behind the wall: the waves have thus turned around the corners of the two small holes, which motion we call diffraction. These two spherical waves now interfere in exactly the same way as the two spherical waves in Figure 3-6: hence the very similar interference patterns of Figures 3-6 and 12-52. These are also the patterns that we see with laser diffraction in Figure 12-51d-e: we can get many light points on either side of the bright laser beam; they correspond to the directions of strong oscillation in Figure 12-52.

The above situation (shown in Figure 12-51d-e and at left in Figure 12-52) is famous in physics: it is called **Young's double-slit experiment**[35] (in practice, one often uses long parallel slits rather than round holes to increase the intensity of the resulting pattern). **This experiment is famous because it distinguishes very clearly and convincingly between**

[35] While this experiment is called after Thomas Young, many science historians believe that Young never did this exact experiment himself, but a related experiment.

Figure 12-50: CDs producing colors by diffraction.

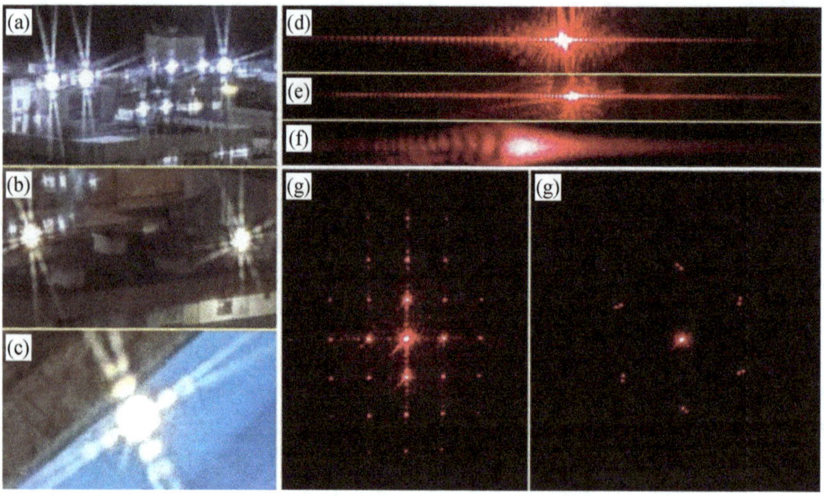

Figure 12-51: Photographs of various diffraction patterns. *At the left* (a-c): A sieve (a tea strainer) splits up sharp distant lights by diffraction. Hints of color are seen in the streaks of images a and b; image c is from a lamp inside a swimming pool. The V-shapes of the streaks are due to the zigzag structure of the thin wires in the sieve. *At the right* (d-h): a laser is used as the source of light to form 1D or 2D diffraction patterns on a white surface in a dark room. *At the top right*: images d-e show diffraction by parallel thin pencil leads, while image f shows diffraction from a knife edge; the knife and its shadow are to the right. *At the bottom right* (g-h): 2D laser diffraction patterns through regular nets of ~35 × 35 tiny transparent dots on a ~2 × 2 mm square of a 35 mm black slide (computer-generated). Image g is for a square net of dots, h for two hexagonal nets of dots, one being slightly shrunk and rotated relative to the other.

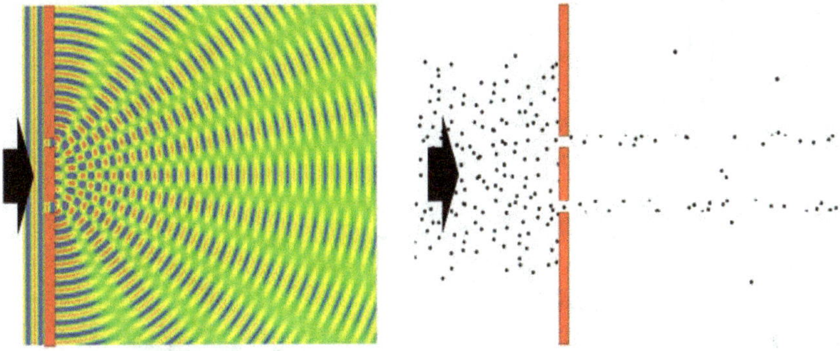

Figure 12-52: Diffraction of a wave (*left*) *versus* passage of balls (*right*) through two identical sets of holes (or slits) in a wall (red). Some balls hit the side of a hole and bounce in any direction.

waves and objects: Figure 12-52 contrasts what you see when sending a wave (at the left) or when throwing objects (at the right) in the same direction through the same pair of small holes (or narrow slits). While a wave fans out as distinct strong *versus* weak streaks, balls primarily go straight through one or the other hole (those balls that hit a side of a hole can bounce in any direction.) In particular, notice that the wave interference creates a strong streak straight ahead midway between the two holes or slits, whereas the balls rarely go there. It is very helpful to see this motion in my Animation 12*6.

> ANIMATION 12*6 — See my video WB2 at time 11:22 in its section "**Diffraction**" under the title "**Bending of light: the double-slit experiment distinguishes waves from particles**". (See details in the section References and Resources below.)

What happens if we change the separation between the two holes? This is illustrated in Figure 12-53 (here the peaks show the directions of strongest oscillation): we see that a larger spacing between holes causes a tightening of the diffracted streaks; indeed, the streaks become narrower and also come closer together. The four curves, from left to right, show what happens when the separation between holes is doubled (from the leftmost curve to the rightmost curve), as marked by gray arrows: the strong wave streaks orient themselves more tightly

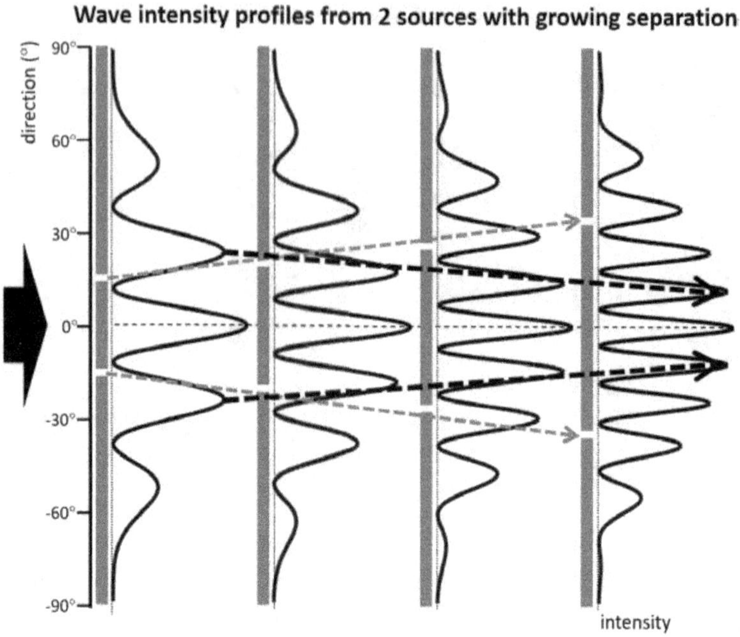

Figure 12-53: Calculated intensity profiles of waves coming from 2 sources/holes/slits (shown as openings in the gray walls), with growing separation between the sources: the separation between the sources at right is double that at left, with the rightmost case being closest to Figure 12-52. The presentation here is different from Figure 12-52: the wavy lines show the wave intensity averaged over time at a large distance from the sources (in the so-called "far field"), so the peaks show maximum oscillations as in the red/blue streaks of Figure 12-52. Also, the vertical axis now gives the direction from the sources: the direction 0° goes straight through (perpendicular to the wall), while the directions –90° and 90° go to the bottom and top parallel to the wall, respectively.

toward the forward direction, as shown by dashed black arrows. To get a closer feeling for this form of interference, you can simulate it on the web.[36]

Importantly, balls do nothing like that: if we increase the separation between the holes, the two beams of balls will remain strictly parallel to each other.

What happens when we have more than two holes/slits? When we look at a distant light source through a net, like a sheer curtain,

[36] See an interactive simulation of double-slit diffraction at "Apps on Physics" by Walter Fendt at https://www.walter-fendt.de/html5/phen/doubleslit_en.htm.

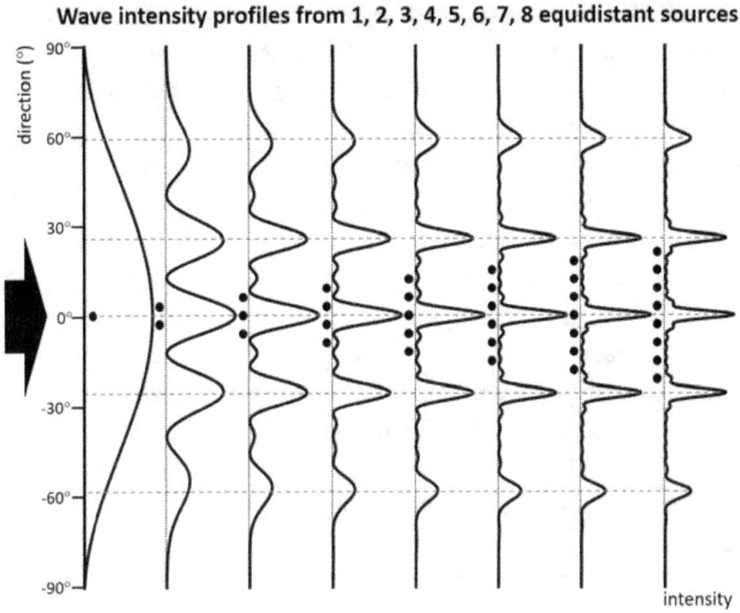

Figure 12-54: Calculated intensity profiles of waves coming from 1, 2, 3, 4, 5, 6, 7 and 8 equally spaced sources/holes/slits (symbolized as black dots) in a wall. The direction 0° goes straight through (perpendicular to the wall), while the directions –90° and 90° go to the bottom and top parallel to the wall, respectively. (In reality, the main peak heights would grow with the number of sources, but this factor has been divided out for better visibility).

mosquito net or sieve, the light passes through many holes or slits that are equally spaced: in Figure 12-54, we explore what happens then. At left, Figure 12-54 displays the special case of a single hole or slit (symbolized as one black dot): it gives a smooth and wide distribution of light; in this model, it forms a very wide peak in the forward direction and falls off gradually to both sides at +90° and –90°, which are up and down parallel to the wall (most balls thrown at this one hole or slit would go straight through, horizontally toward the direction 0°).

The second curve from the left in Figure 12-54 shows the case of 2 sources/holes/slits (symbolized by 2 dots): that is the same situation as shown in Figure 12-52, but with a different spacing between sources, so that there are only 5 privileged directions (strong streaks) in Figure 12-54, compared to 11 in Figure 12-52.

Let's now add more sources: the third curve from the left in Figure 12-54 shows the case of 3 sources. We notice that the 5 peaks of the 2-source case keep almost the same directions, but smaller peaks appear between them. Also, and more importantly, the 5 larger peaks have become sharper (narrower). As we add more sources, up to 8 sources, this trend continues: the main peaks maintain their directions but become sharper, with additional weaker intermediate peaks. Again, this behavior does not happen with balls instead of waves: most balls will go straight through the holes, as shown at the right in Figure 12-52, in multiple parallel streams of constant width. (The case of 7 sources corresponds to that shown at the left in Figure 3-8, which depicts the wave interference close to the 7 sources, in the "near field": we already see there the build-up of strong beams in certain directions marked by arrows; these beams become more dominant farther from the sources, as seen in Figure 12-54.)

If we continued adding many more sources, we would end up with extremely sharp peaks and essentially no intensity between the peaks: we would thus get very sharp light beams. We clearly see such very sharp peaks made with laser light in Figure 12-51g-h. That is what we should see ideally when looking at a sharp light source through a perfect net. In practice, sheer curtains, mosquito nets, sieves, *etc*. are geometrically imperfect, so that sharp beams are washed out somewhat and become wider, as we see especially in Figure 12-51a-c; it is as if we looked through only a few holes, seeing a pattern more like that of 3 or 4 sources in Figure 12-54.

Figure 12-51 shows photographs of real diffraction patterns, as opposed to the idealized situations of Figure 12-54. Real nets, such as sheer curtains, mosquito nets and sieves, usually consist of two perpendicular nets: threads are lined up in two perpendicular directions with equal spacing between them. Then diffraction happens in both directions: we therefore get side peaks in both directions. In fact, the diffractions in both directions can also interfere among themselves: this gives a 2D grid of side peaks; the "combined" peaks tend to be weaker, but are sometimes seen in photographs, for example in Figure 12-51g.

Why don't we see such diffraction effects all the time? After all, we often see light passing by a corner, or going through a window: those obstacles cast shadows, with edges that should also produce diffraction effects. Do we see ripples like those in Figure 12-51? We don't normally see these diffraction effects. The main reason is that these effects are usually overshadowed by other light: they are therefore more visible at night or in very dark places. Another reason is the short wavelength of light, which is smaller than a micrometer (much smaller than the diameter of a hair). A small wavelength causes diffraction effects that are strongest at very small angles. As a result, the most visible effects are only seen very close to bright lights, so our eyes tend to be blinded by those bright lights. A further reason is that there are also other causes for streaks of light like those seen in Figure 12-51, but without ups and downs in intensity. In particular, water (tears) on our eyes frequently causes such streaks, which are not due to diffraction: they are mostly due to reflections and refractions at the edges of eyelids, iris, *etc.*, and can easily overwhelm the diffraction effects.

In Figures 12-50 and 12-51a-c, we see colors: why? The directions of the diffraction peaks that we saw in Figure 12-54 depend not only on the spacing between the sources, but also on the wavelength of the waves. More precisely, it is the ratio of spacing to wavelength that determines the directions of peaks: thus, doubling both the wavelength and spacing will not change the peak directions. However, with a fixed spacing of a net, light of different wavelengths, and therefore of different colors, will go into somewhat different directions, just like a prism that causes refraction and dispersion of colors. The net thereby splits the colors by wavelength; in particular, sunlight will be split into the solar spectrum.

A CD or DVD disk has circular tracks of data packed closely together: these close tracks also form a net that can split light into its colors (Figure 12-50); in this case, light is reflected in addition to being diffracted. The curved shape of the tracks around a disk causes a rich variety of diffraction conditions that make a feast for the eyes, especially when the disk moves. That flashy feast also scares birds, making CDs and DVDs ideal scarecrows.

Nets that are used to diffract light are also called **diffraction gratings**: these are often made by scratching or imprinting a surface with a regular net of lines that act as equally spaced sources of light waves.

How does light diffract from a corner or edge? At the beginning of this section and in Section 3.9, we mentioned that waves can turn

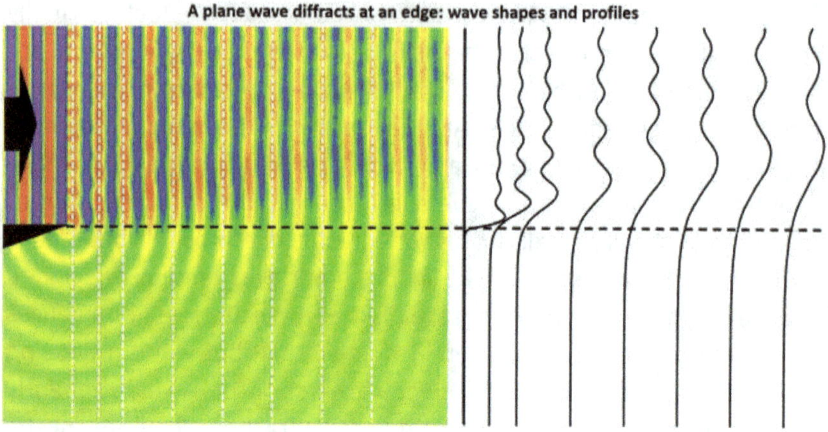

Figure 12-55: *At the left*, a simulated plane wave arriving from the left passes an edge (the tip of the black triangle), similar to Figure 3-16 for a corner: diffraction makes the wave spread out and turn around the edge, like a corner. *At the right*, intensity profiles are drawn along the corresponding white dashed lines at the left, showing how the ripples expand away from the edge into the plane wave at larger distances, roughly proportionally to the distance. (The oscillations of the spherical waves are smoothed out by averaging over time to obtain the intensities at right.)

around a corner. We illustrated this important aspect of diffraction in Figure 3-16, where we let a spherical wave radiate into the space behind the corner. If we take a sharper edge, such as a knife edge instead of a corner, the wave can wrap around this edge entirely, as shown at the lower left in Figure 12-55. A better description, according to Huygens' principle, is to let spherical waves start along the whole wavefront of the plane wave as it arrives at the edge: this result is simulated in Figure 12-55. Now we see ripples in the waves, spreading across the plane wave as it moves further; we also see a weakening of the plane wave near the corner or edge, as it leaks intensity to the spherical wave around the corner or edge.[37]

The ripples in the plane wave are the most characteristic aspect of such diffraction at corners and edges, but they are difficult to see in

[37] An amusing aspect of diffraction at corners or edges is that it does not depend on the wavelength. That is because a corner or edge has no particular size: a given corner or edge looks the same to a wave whether it has a wavelength of a micrometer or of a meter. Therefore, Figure 12-55 is valid for any wavelength.

practice with the naked eye: the ripples are weak, while the relatively strong plane wave overwhelms the view. A picture obtained with a laser is given in Figure 12-51f: we see ripples on the "open" side (to the left) and a gradual decay on the shadow side (to the right).

Knife-edge diffraction, as it is often called, plays a significant role in a rather unexpected situation: observing exoplanets, which are planets circling stars other than our Sun. A problem with observing an exoplanet is that its parent star is far brighter than that planet. It is therefore necessary to block the parent star's light with a shade: this shade causes knife-edge diffraction that perturbs the image of the exoplanet. Astronomers therefore try to compensate for this effect when observing exoplanets, especially as these are extremely far away and look extremely small at those distances.

How about diffraction through wider holes or slits? Knife-edge diffraction plays a more daily role when we think again of shining light through holes or slits in a wall. Such holes and slits must be wide enough to allow enough light to shine through for it to be seen. From that point of view, a hole or slit is really a pair of corners or edges facing each other.

This is clearly demonstrated in Figure 12-56, where a plane wave passes through a slit: both edges of the slit produce diffraction like that of a knife edge. Now, however, these two sources of diffraction overlap each other and therefore interfere with each other: this is very visible near the center of Figure 12-56. Now the size of the slit relative to the wavelength becomes important: a narrow slit, comparable to a few wavelengths, will produce strong wave interferences; by contrast, a wave passing through a slit that is many wavelengths wide will survive much farther behind the slit.

Furthermore, both edges of a slit leak out intensity from the plane wave, which therefore becomes narrower. Figure 12-56 follows this plane wave until 100 wavelengths away, 25 times the slit width: by then, the plane wave looks more like a spherical wave that spreads out rather than continuing as a narrow beam; indeed, a small hole seen from far away looks like a point source emitting a spherical wave. You can simulate the diffraction by a single slit online.[38]

---

[38] See an interactive simulation of single-slit diffraction at "Apps on Physics" by Walter Fendt at https://www.walter-fendt.de/html5/phen/singleslit_en.htm .

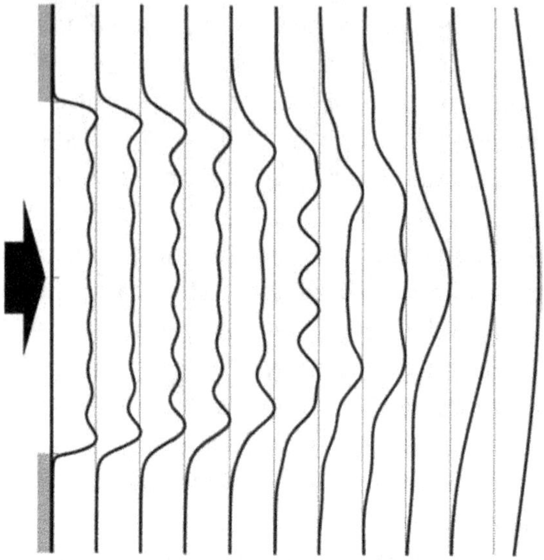

Figure 12-56: A plane wave arriving from the left passes through a slit in a wall (gray). The slit width is 4 wavelengths. From left to right, the 11 curves give the intensities at the following distances from the wall: 0.04, 0.2, 0.4, 0.8, 1.6, 3, 6, 12, 25, 50, and 100 wavelengths.

When we discussed multiple slits, we treated each slit as being a point-like source of light (for example, in Figures 12-52 to 12-54), rather than as a slit with a non-zero width as shown in Figure 12-56. To properly describe the diffraction by multiple slits, we should also take their individual width into account. This leads to a diffraction pattern that is the combination of a single-slit pattern and a multiple-source pattern. The result is seen in Figure 12-51e: there we notice that the string of light points has gentle ups and downs in brightness, unlike the more uniform points of Figure 12-51d. These gentle ups and downs are the single-slit diffraction profile: they change the multiple-source pattern.

We have seen that diffraction of waves has a behavior that is absent from the motion of balls and other objects: diffraction is therefore an important and relatively simple test of whether we are dealing with waves or particles. In Section 12.10, **we will find out that nature still has some surprises for us: waves can also behave as objects, so light can also be made of objects!**

An important scientific use of diffraction is with x-rays: **x-ray diffraction** can tell us the atomic and molecular structure of materials, for example the structure of ice (Figure 9-1) or all sorts of biological molecules such as DNA (see Figure 13-8 in Section 13.6). Such knowledge helps us to understand the physical, chemical and biological properties of all materials, and to develop new materials with desirable properties. For that purpose, one uses x-rays with extremely short wavelengths around 0.1 nanometer, which is about 10,000 times shorter than the wavelengths of visible light: the reason is that such short wavelengths are very close to the distances between atoms, which therefore give strong diffraction effects that tell us the structure of the material.

The basic principle of structure determination by diffraction is that waves bouncing off different atoms interfere in a way that depends on the relative positions of those atoms, just as the diffraction patterns we have seen depend on the distance between sources/holes/slits: an analysis of these interferences can reveal the atoms' relative positions, in particular the distances between them; such an analysis is done in computers. X-ray diffraction is, however, different from imaging with x-rays, which is used medically to see bones, teeth and organs in our bodies: imaging uses absorption of x-rays by matter to produce darker or lighter shadows, as in Figure 12-5; no diffraction is involved.

12.10 Is Light a Wave or a Particle? Wave-Particle Duality

The question "Is light a wave or a particle?" is still difficult to answer satisfactorily today, even though we experience light abundantly every day. In this section, we will therefore explain why this question is so difficult. In fact, you will probably feel much confusion at times! Not to worry because physicists are also still confused; that makes this story all the more interesting!

In Section 12.9, we have seen that diffraction of light is very strong evidence to conclude that light consists of waves: only waves can behave that way. By contrast, reflection of light by a mirror does not prove that light is a wave: balls are also reflected in much the same way from a mirror. Refraction (namely the bending of light when crossing a boundary) is less direct evidence for the wave character of light:

historically, understanding refraction has relied more directly on the density and other properties of substances, and on the speed of light in those substances, rather than on the wave character of light; light was generally treated as simple rays, as we described in Section 10.7, without the need to think about waves.

In the 17th century, it was proposed that light consists of waves, primarily by **Christiaan Huygens**, **Robert Hooke** and **Augustin-Jean Fresnel**. Even then, many scientists still believed that light is made of very small bits of matter, like extremely fine grains of sand: these are often called **particles** in physics. For example, **Isaac Newton**, one of the greatest physicists of all time, believed in light particles, which he called corpuscles; he explained refraction as a change of their speed perpendicular to the boundary (the speed parallel to the boundary would not change).

Today, we use the name **photons** for the particles of light, in analogy with electrons: we will see in Chapter 13 that electrons can also be regarded either as particles or as waves, specifically quantum waves. Therefore, our discussion here about whether light is formed of particles or waves is of much wider importance: all our electronics, and much more, also depend on it!

In 1801, the British polymath **Thomas Young** performed the double-slit experiment (modeled at left in Figure 12-52): this was the first direct evidence from nature that light has the character of a wave. Remember that an experiment (a measurement of nature) is the "truth" that all science has to obey: a model, hypothesis or theory is only valid if it agrees with experiment, which is the objective evidence from nature.

James Clerk Maxwell found that his theory of electromagnetism worked perfectly with electromagnetic waves, thereby strengthening the argument for light waves, as well as other EM waves (infrared, ultraviolet, x-rays, *etc.*). Nonetheless, a long debate about the particle *versus* wave character of light continued. You may enjoy a beautiful and thoughtful online video[39] discussing this debate.

---

[39] See video "I did the double slit experiment at home" by Looking Glass Universe at https://www.youtube.com/watch?v=v_uBaBuarEM, as well as follow-up videos. This video mentions quantum mechanics as well as light diffraction: these two phenomena are indeed very similar, as we will discuss in Chapter 13.

Let's consider one major reason why, in the 19th century and beyond, some scientists still believed that light is made of particles.

Imagine doing Young's double-slit experiment, as shown at the left in Figure 12-52. Let's record the light on a screen behind the two slits: we expect that the streaks of strong *versus* weak light drawn in Figure 12-52 will illuminate the screen as bright and dark bands, as in Figure 12-53.

We start with a very weak light wave. To our great surprise, we will see something like the left bar in Figure 12-57: it is uniformly black with some light points scattered about. We see only a few light points hitting the screen haphazardly; we will not see broad and smooth bands of light, like those of Figure 12-52. In fact, we can dim the light to the level where only one light point at a time hits the screen. Digital cameras, including those in smartphones, can detect such single light points: if you point your camera at a dark scene and let it automatically adjust the exposure, it may show individual dots that look like snow or television noise; these dots are due to single light points.[40]

Returning to the left bar in Figure 12-57, such points of light are also observed with a single slit: for example, if you try to photograph a very dark scene, you will just see random points of light, not a broad wavy spread of light.

We next increase the light intensity, or we just wait longer to record more light on the screen. We will then see more light points on the screen, as in the second bar from the left in Figure 12-57. This continues as we let more light in: see the next bars toward the right in the figure as they fill up with more light points. It is very helpful to see this action in my Animation 12*7.

ANIMATION 12*7 — See my video WB2 at time 15:03 in its section "**Wave or particle?**" under the title "**Is light wave or particle? Evidence for both!**". (See details in the section References and Resources below.)

---

[40] Remarkably, our eyes are also able to detect individual light points when it is very dark! See the freely available article "Direct detection of a single photon by humans", by Jonathan N. Tinsley, Maxim I. Molodtsov, Robert Prevedel, David Wartmann, Jofre Espigulé-Pons, Mattias Lauwers and Alipasha Vaziri, *Nature Communications*, volume 7, article 12172, 2016, https://doi.org/10.1038/ncomms12172.

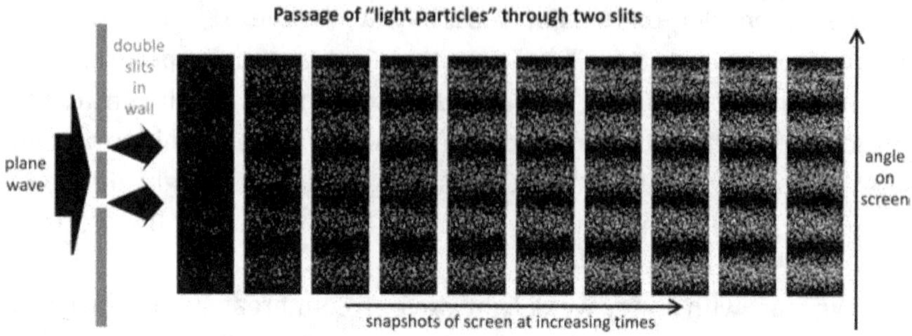

Figure 12-57: The simulated double-slit experiment showing impacts of light arriving through two slits (or holes) on a screen: ten snapshots are shown from left to right, for increasing recording time.

Interestingly, the light points on the screen seem to be more concentrated in some parts of the screen than in others: we see bands with dense points, and bands with very few points. This is already true in the left bar, at low light intensity, even though there are so few light points that it is hard to see bands on the screen. And, most importantly, these broad bands of light points in fact correspond directly to the broad streaks of diffracted light of Figure 12-52. In particular, these bands would shift when changing the separation between the two slits, as expected from diffraction. However, they certainly do not correspond to the two streams of balls shown at right in Figure 12-52.

The sharp light points are similar to raindrops: in a rainstorm, it is essentially impossible to predict where each raindrop will fall, but meteorologists can forecast in which areas rain will fall; these areas of rainfall are much more predictable than the location of each raindrop. In principle, if we had enough precise information about the water, wind and temperature distributions in clouds, we could forecast where each raindrop would fall, but in practice, we don't have such information.

What does Young's double-slit experiment tell us? The patterns of light in Figure 12-57 are actually extraordinary; moreover, they are confusing!

First, what we see are sharp points of light, instead of the broad bands of light of Figure 12-54. If you think of sound waves, for example, do they ever hit you like sharp needles, or do they feel more like a broad cushion of sound? Similarly, do waves on water ever form sharp needle-like peaks, rather than the broad waves that fall on a beach? Furthermore, the sharp points of light are really random: physicists do not know how to predict where they will show up, even though the wave has a well-defined, smooth wave-like shape which we can predict. So, we already have several puzzles here.

Second, the patterns of light exhibit both the character of a wave, as broad diffraction-like bands with more or less light, and the character of particles, as sharp points of light. This seems to tell us that light is both a wave and a collection of particles! More surprises follow...

Third, as you may realize (see Section 3.7), the bands of bright and dim diffracted light are due to constructive and destructive interference. In particular, we see many bands of near-zero light due to destructive interference. Can you imagine a way for particles (balls, grains of sand, *etc.*) to produce such bands of near-zero intensity, especially when only one particle arrives at a time? It would be quite amazing if there were a mechanism for particles to bounce from the edges of the slits in just such a way as to produce these diffraction-like bands of near-zero intensity.

Fourth, to form these bands of light, which must be due to diffraction, the light must have passed through both slits: if the light had passed through only one slit, we would have obtained a different diffraction pattern, namely that of a single slit (such as the very broad leftmost curve without streaks in Figure 12-54 or the very broad rightmost curve in Figure 12-56, neither of which have alternating strong and weak bands of light). On the other hand, if light consisted of particles only, they could not pass through both slits: each light particle would have passed through only one of the slits. All of this appears to be full of contradictions...

Fifth, if we could accurately measure the brightness of each individual light point on the screen, we would discover that each light point has essentially the same brightness! (This assumes that the light has a single frequency, as in laser light, for example.) Indeed, when we increase the intensity of the incoming light, we get more light points on the screen, not brighter light points! These observations suggest that light hits the screen as equal packets of energy: more packets then mean more intense light.

Moreover, if all the light points have the same brightness (intensity, energy), there must be a minimum possible amount of light: light then has a non-zero minimum brightness. Similarly, there is nothing smaller than one person or smaller than a one-cent coin: there is no "half person" or "half-cent coin" (you can have half a cent on paper, but not as a coin). The minimum possible amount of light is called a **quantum** of light, introduced by Albert Einstein in 1905.[41] The word quantum gives rise to the terms quantum waves, quantum mechanics, quantum theory, *etc.*, as we will see in Chapter 13. It is thus impossible to have a fraction of one quantum of light. And that quantum of light is precisely the **photon**. The photon is the smallest possible packet of light: it cannot be divided; and we can't have a fractional number of photons, such as 1.3 photons. The most confusing thing is that "smallest packet of light" sounds very much like "particle"!

After all this evidence from nature, is light finally a wave or a particle? Well, we have evidence for both explanations: **light seems to behave (at least part of the time) as a wave, but light also seems to behave (at least at other times) as a particle**.

For the last century, the scientific consensus has been to accept this "dual" explanation: we call it the wave-particle duality. It is unsatisfactory to have two explanations for the same thing, but we have not yet found a better single explanation! Moreover, the same wave-particle duality has been accepted also for electrons and other particles because they also exhibit the character of a wave, such as diffraction (see Chapter 13). Some physicists dislike the "dual" explanation and prefer to say that photons are neither waves nor particles, but something different in its own right; that is fine, but it is still convenient to explain that photons sometimes behave as waves, and sometimes as particles.

---

[41] As Einstein wrote in 1905 to introduce the quantum of light: "According to the assumption to be contemplated here, when a light ray is spreading from a point, the energy is not distributed continuously over ever-increasing spaces, but consists of a finite [i.e., limited] number of energy quanta that are localized in points in space, move without dividing, and can be absorbed or generated only as a whole." (In "On a heuristic point of view concerning the production and transformation of light", by A. Einstein, *Annalen der Physik*, volume 17, page 132, 1905; translated into English and available online at https://einsteinpapers.press.princeton.edu/vol2-trans/100.)

So how does this dual character of light work? In simple terms:

- When light travels from one place to another, it behaves like a wave. This includes light interfering with other light (with constructive and destructive interference), entering substances (refraction), and turning around corners (diffraction). Light has no mass: it is weightless.
- When light hits a substance, it can give all its energy to that material, just like any particle that has mass (and therefore weight). This includes creating a light point on a screen, or a light point in an eye or a detector in a camera; it also happens in a collision of light with a small particle such as an electron. After hitting a substance, the wave no longer exists: it is totally converted to energy which the substance will use in another way, such as heating itself (vibrating). Nevertheless, the substance can use that energy to emit a new photon of light, immediately or later, which then travels away as a wave, similar to the emission of a radio wave (see Section 12.1): this completes the cycle from a wave back to a wave.

A most remarkable feature of the conversion of a wave to a particle is the following. Somehow, a light wave suddenly "collapses" into a particle when it hits a substance. One amazing aspect of this is that a wave is spread out over a large space (wide enough to pass through multiple slits at the same time), while that wide wave can quickly shrink to a point-like particle (giving a tiny flash of light on a screen or detector). At present, physics has no proper explanation for how this wave collapse happens. It is usually called wave function collapse in scientific publications.[42]

As we will see in Chapter 13, there are also other, more important problems in understanding quantum mechanics.

How large or small is a quantum of light? The quantum of light, which is a **photon**, is very small in terms of energy. The German physicist **Max Planck** in 1900 proposed the following relation between

[42] You will find online and in books various mathematical descriptions for the **collapse of waves**, but the underlying physical process is debated.

the energy and frequency of a light wave: (energy) = (Planck's constant) x (frequency); here Planck's constant is a tiny number, which makes the energy of a photon extremely small, even with the very high frequencies that we see in the EM spectrum in Figure 12-4.

Let's try to get a feeling for how small this energy is. A single photon of red light can cause a single chlorophyll reaction, which converts sunlight into chemical energy that sustains life (chlorophyll gives plants their green color). That one photon converts only one molecule of chlorophyll: it takes billions of such molecules and therefore billions of photons to make just one leaf. Another example is: a typical radio transmitter emits photons, which are about a million times less energetic than the reaction of a single chlorophyll molecule.

On the other hand, a single photon of ultraviolet light is sufficient to change one DNA molecule in our body: this may be sufficient to induce cancer or genetic defects. This is why humans must protect themselves from sunlight.

A crucial aspect of Planck's relation "(energy) = (Planck's constant) × (frequency)" is that the right-hand side is fixed once we know the frequency of the light (for example, the frequency of red light): it does not depend on the intensity of the light. To make intense light, we must therefore shine many photons, as we see in Figure 12-57.

12.11 Holography: Reconstructing the 3D View of an Image

You have likely seen **holograms**: some credit cards and banknotes have an embedded hologram showing a 3D image of a bird, face, *etc*. (the main purpose of these holograms is to make forgery difficult, not to entertain you). Some toys and artwork use holograms to offer intriguing 3D views.

Holograms are special 2D images that appear to be 3D images, in the sense that you can look at them from various directions and see the correct 3D perspective of objects; thus, you could also look sideways at an object. For example, a hologram can let you see not only the frontal view of a head but also the view from the left side or the right side of the head.

Holograms contrast with normal pictures, which only present a view in a single direction, such as a passport photo. When you look at a normal photograph or drawing, you see the scene from one particular direction only. The only 3D information that you get out of a normal picture is what your brain imagines, based on familiar sizes of people, houses, shadows, hidden parts, *etc*. A stereo pair of views (like those seen by your two eyes) further helps your brain interpret the two views in terms of positions and distances between the objects in the scene, but holography is different.

The technique of holography uses the wave character of light to record and display 3D information so that objects can be seen in 3D as if they were in their correct positions. You can watch holograms online,[43] or make your own.[44] The principles of holography are shown in an old online video,[45] but I suggest that you first read my simpler description of the basic idea below.

There exist other techniques for 3D viewing which we do not discuss here, such as shining light into a cloud of smoke or onto a semi-transparent sheet of glass (this actually only produces a 2D image but gives an impression of 3D). Our focus is on the original holographic idea of the Hungarian-British engineer and physicist **Dennis Gabor** from the late 1940s, which uses wave interference to record and display a 3D scene.

The holographic principle is basically very simple: we first "freeze" and record the light coming from a scene. Later we re-activate that "frozen" light and let it continue its path to the eye: we then see the original scene as if it was still there. The trick, compared to standard

---

[43] See video "REAL holograms are finally here! (And they look very cool)" by CNET at https://www.youtube.com/watch?v=7oGtgbsmmg8, video "Amazing Must See Technology 7D hologram Shown in Dubai, Poland and Japan" by Tech World at https://www.youtube.com/watch?v=4N0Ewb_OVsU.

[44] See video "How-To: Holography" by "Make:" at https://www.youtube.com/watch?v=lJVhWwlNovY, and video "Making Real Holograms!" by The Thought Emporium at https://www.youtube.com/watch?v=aTB2ryoWIFU. Both videos describe the use of the commercial LitiHolo Hologram Kit.

[45] See video "Introduction To Holography — 1972" by LitiHolo at https://www.youtube.com/watch?v=-2SdHYrMlkc.

photography, is that we record the interference pattern of the light, not just the light intensity and color. That allows seeing the scene in 3D, unlike standard photography. Let's discuss this step by step in the following.

In Figure 12-58, we compare normal vision with holographic vision for a simple object: a ball. In the top sketch, light illuminates the ball from the left: some light is scattered by the ball as spherical waves in all directions. The eye receives light scattered by the object, thereby seeing the object in a particular direction, which depends on the location of the eye. The scattered light can also interfere with the incoming light, but the wavelength is so short, and the wave frequency is so high, that the eye cannot see the interference pattern of crests and troughs: the eye only sees light intensity and colors. The same happens with a standard camera replacing the eye.

The idea of the holographic method is to record and thereby "freeze" the wave that passes, for example, in the black box at the top in Figure 12-58. That recorded wave is an interference pattern between the incoming wave and the scattered wave: that pattern is called a **hologram**. This works best at a single wavelength, hence lasers are used for this purpose; the incoming light is then a plane wave if the light source is distant.

A very-high-resolution photographic film is needed to record the fine structure of the pattern, which is on the scale of the wavelength of the laser light and therefore smaller than a micrometer (much finer than a hair). The recorded pattern is thus a "frozen" copy of the interference pattern, which copy can be transported and stored, like a normal picture. However, the hologram itself does not resemble the original object at all: the hologram is a confusing network of crests and troughs that do not look like the object.

To see the object, we can place the hologram in the same plane wave with which we originally lit the object: see the bottom sketch in Figure 12-58. For example, we can shine the original laser onto the hologram. This plane wave will now "re-activate" the scattered light, as if it came from the original ball, and recreate the image seen by the eye: this creates a virtual (imaginary) ball in the place of the original ball. The "re-activation" of the light is just Huygens' principle: each point in space acts like the source of the continuing wave; here the hologram is

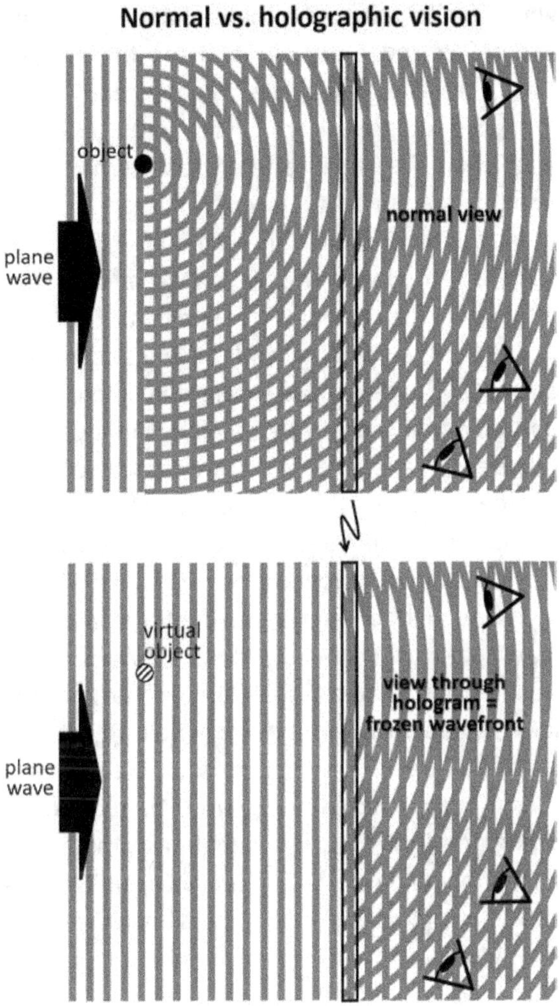

Figure 12-58: View of an object (ball at the left) in normal vision (*top*) and holographic vision (*bottom*). *At the top*: The object is illuminated with a plane wave and scatters light as spherical waves in all directions (only forward directions are shown here). The wave interference in the black box can be recorded with a suitable film. The plane wave is often called a reference wave or reference beam, while the scattered wave is often called an object wave or object beam. *At the bottom*: The recorded wave interference pattern is illuminated with the same plane wave as used before (but now without the object): the pattern now serves as the source and recreates the "old" waves coming from the object, allowing the eye to see the original scene from any direction.

the source, which emits waves like those that came from the original ball. We can therefore move our eye and see the ball from different directions, as if we were looking at the original scene.

As mentioned, holograms are made with laser light: single-wavelength light gives sharper interference patterns than mixed-color light, for example white sunlight. Once a hologram is made with laser light, the best is to look at it with the same laser light because it has the proper wavelength. However, the hologram can also be viewed with mixed-color light, for example with sunlight or a lamp's light: the resulting view will be less sharp and will have less contrast, but it may still be satisfactory if you don't need to see fine details in the image.

Multi-color holograms are also possible, although less detailed. They can be made, for example, with three lasers of different colors (typically red, green and blue). The three lasers shine simultaneously at the object, mixing three interference patterns at three different wavelengths. The resulting hologram can again be viewed in mixed-color light, such as sunlight or lamp light. The three laser colors produce three different virtual images that combine to give realistic colors.

So far, we have only imaged a single simple ball. For a more complex object, the idea remains the same: a complex object can be viewed as a collection of balls, each of which produces a similar interference pattern. The resulting combined hologram is then extremely complex, but the recording and imaging steps remain as simple as for one ball. The incoming light falling on the hologram will again recreate the waves due to the original complex object, so that its image will become visible, and it can be viewed from different directions.

An amusing feature of holograms is that you can cut parts of them away: half a hologram still gives you the same view as a complete hologram (although you lose details). This is because, by Huygens' principle, every part of a hologram is part of the source for the waves behind it and therefore contains information to recreate the complete view.

Above, we have neglected one aspect that confuses the final image: the recording of the "frozen" interference pattern is actually not quite complete. Indeed, since the interference pattern is "frozen" in time, it is not possible to tell in which direction — forward or backward — the original wave was traveling. When we illuminate this frozen pattern with a plane wave, we therefore in fact create two copies of the original object: one where it was originally, and another in the opposite direction (on the other side of the plane of the hologram), as if the original light had come from the opposite side. In other words: you can look through the hologram back-to-front or front-to-back and see the same thing twice: two identical virtual images. Often you know which is the correct image because you are familiar with the scene, but in some situations, you cannot distinguish the two images and therefore cannot tell in which direction the original object was located.[46]

12.12 What have We Learned in this Chapter?

Light, a form of electromagnetic (EM) waves, is quite different from the string waves, sound waves and water waves that we discussed earlier. Even so, EM waves behave in very similar ways. In particular, the earlier concepts of interference, reflection, refraction and diffraction remain valid for EM waves.

Thus, we have applied these ideas to reflection and refraction of EM waves, which are important for understanding how mirrors, windows, prisms and lenses work. An interesting case is what a swimmer (or a fish) sees under the water's surface, especially with total internal reflection. This also explains how optical fibers transmit messages over extremely long distances. Another interesting case is mirages, which are gradual refractions in the atmosphere that appear to create imaginary objects. Rainbows also are due to reflection and refraction: we have discussed how the rainbow's colors differ from the solar spectrum seen with a simple prism. So-called negative refraction is caused by novel man-made materials: we have explored some new applications of these "metamaterials", such as superlenses and invisibility cloaks.

We have also explored how holograms are made and used to produce 3D images that can be viewed from multiple directions.

---

[46] This issue is called "phase problem" by scientists because it is related to the inability to measure the absolute phase of the waves: what is recorded is the intensity of the wave, which neglects part of the information available in the wave itself, needed to tell in which direction the wave was moving.

Diffraction, whereby a wave turns around a corner, is very characteristic of waves. We have discussed how diffraction is used to distinguish waves from objects. However, we have also learned that it is necessary to assume that light sometimes acts like a wave, and sometimes like a particle (called a photon), even if this seems confusing: this is called wave-particle duality. Furthermore, those photons have a fixed energy called a quantum. We will encounter these ideas again in Chapter 13 in connection with the quantum theory of particles.

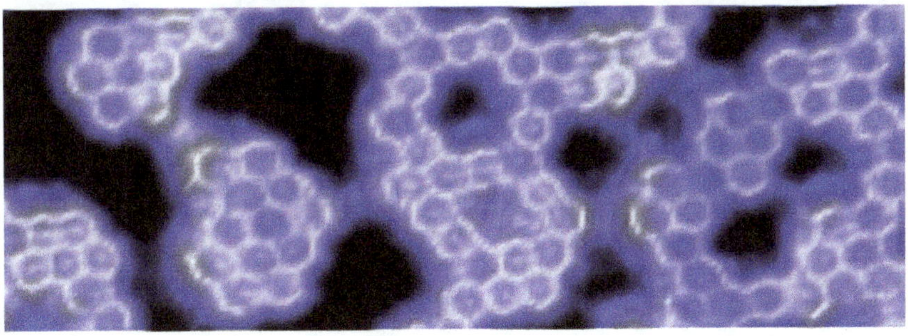

13

Quantum Mechanical Waves: From Atoms to DNA and Electronics

Quantum mechanical waves (QM waves), sometimes called quantum waves or matter waves, are part of the "modern physics" that was developed in the early 20th century. They gave rise to a revolutionary change in our worldview of nature and to whole new areas of science and technology in wide use today, from electronics to genetics.

This chapter aims to explain some of the main mysteries of quantum physics, by close analogy with the more familiar behavior of light waves described in Chapter 12. It makes clear that quantum waves operate mainly on the submicroscopic level of atoms and related particles, but nevertheless are responsible for all the materials and many processes of our daily lives: these include electrical insulators and conductors, electronic devices and computers, chemicals and biological substances, together with their colors.

13.1 What are Quantum Mechanical Waves?

In the early part of the last century, a new type of waves was discovered that led to amazing technological developments and biological understanding: **quantum mechanical (QM) waves**, sometimes called **quantum waves** or **matter waves**. These waves are at the core of **quantum physics**, also called **wave mechanics**, **quantum mechanics** and **quantum theory**, and are also fundamental to **quantum chemistry**. Together with relativity (which we do not discuss in this book), these waves form the basis of the so-called **modern physics**. The earlier so-called **classical physics** includes Newton's laws of mechanics, Maxwell's laws of electromagnetism as well as the laws of thermodynamics.

QM waves (quantum mechanical waves) are now understood to provide the basis of all substances, natural or man-made, from gases and chemicals to semiconductors and biomolecules. These waves have made possible modern electronics and its myriad uses today; they are also at the root of all the atoms and molecules of chemistry and life, including DNA, which transmits genetic information from one generation to the next, and motor molecules that power all life. Electrons are at the heart of the behavior of electronics and molecules: therefore, we will often use electrons as examples.

QM waves are characteristic of objects that are very small compared to the human scale, including atoms and molecules, as well as elementary particles, such as electrons, quarks and others. Humans have no direct experience of their behavior and are therefore frequently surprised and confused by how they act. Our daily experience is with large objects on the human scale. Our intuition therefore breaks down rapidly as we observe the apparently odd behavior of much smaller objects.

Rest assured: physicists are also constantly surprised by this microscopic behavior of matter! For scientists, even after studying these quantized matter waves for decades, they are invisible, strange and mysterious waves! As the famous American physicist **Richard Feynman** wrote[1]: "Even the experts do not understand it the way they would like to, and it is perfectly reasonable that they should not, because all of direct, human experience and of human intuition applies to large objects. We know how large objects will act, but things on a small scale just do not act that way. So we have to learn about them in a sort of abstract or imaginative fashion and not by connection with our direct experience."

If quantum waves are new to you, I suggest that you first read Box 13-1, which gives advice on how to approach this unfamiliar subject.

BOX 13-1 — HOW TO READ THIS CHAPTER. Quantum physics is unfamiliar to most people. I wrote this chapter with this fact of life very much in mind: I thus tried to make the subject of quantum physics as accessible as possible.

As you read this chapter, you should therefore open your mind as if you were visiting a foreign culture with totally different and mysterious customs. Just as physicists do, you may have to simply accept the unfamiliar properties of small objects. Physicists only accept them because the deep insight which these properties give is constantly verified to very high precision by their impact, from natural biology to man-made electronics. You must also appreciate that different scientists describe quantum physics in different ways, as they attempt to make this subject more understandable in familiar terms.

Perhaps an analogy will help: suppose we had no experience of seeing waves in daily life. Instead of seeing waves on water, we would be familiar only with powder; powder can flow like liquids, but doesn't make visible waves (with the possible exception of dunes being blown slowly across a desert). Now imagine learning about waves, without visual experience of wavelengths, wave periods, wave speeds, wave superposition, wave interference, refraction, diffraction, beach waves, boat wakes, tides, tsunamis, *etc*. These new wave concepts would open a vast new range of behavior that would be completely alien to you: not surprisingly, you might feel confused; moreover, you might question whether invisible waves are real! The same happens with the QM waves that we will introduce in this chapter: you may feel confused, and you may be tempted to question whether those strange waves are real! That is

(*Continued*)

[1] Richard P. Feynman, Robert B. Leighton and Matthew Sands, "*The Feynman Lectures on Physics*", Basic Books, 2011, volume III, page 1–1.

(*Continued*)

a completely normal reaction. Hopefully, you will be intrigued and want to know more about this topic. If you are interested in the more human aspects of science, you are invited to read Section 15.5, which compares doing science with doing sports or playing video games or reading science fiction.

One challenge in reading about quantum physics is that several unfamiliar concepts are combined: it is easier to become familiar with one concept at a time. Another analogy can help understand this aspect: in human society, we deal with family-related issues such as family ties and inheritance. To deal with family ties and inheritance, we must first understand family structure: we must first learn the meaning and role of parent, partner, husband, sister, brother, sibling, grandparent, aunt, cousin, niece, half-brother, sister-in-law, ex-husband, *etc.*, and all their mutual relationships (how is your niece related to your grandmother?). Only with that complex family tree firmly in your mind can you properly understand family ties and then how inheritance works.

Much the same happens in some areas of physics, especially quantum physics. Therefore, it is quite understandable if you have some difficulty following the reasoning in this chapter. Don't worry: you may just read the text and watch the illustrations (and animations in my video) to get a flavor of quantum physics and its impact on daily life. Of course, you may also skip this chapter in your first reading.

Sections 13.2 to 13.4 try to familiarize you with the nature and behavior of QM waves, mainly by analogy with EM waves. You may prefer to skip to Section 13.5 if you wish to read about atoms and their structure. You may also read Section 13.6 about how atoms bond together to form molecules. Sections 13.7 to 13.9 cover the practical impacts of QM waves, such as electronics and biomolecules, as well as the origin of the colors of different substances.

13.2 How do EM Waves and QM Waves Operate?

In Chapter 12, we discussed electromagnetic (EM) waves, in particular light waves. We have seen that they can behave not only as waves but also as particles called **photons** (see Section 12.10). If EM waves can behave as particles, a natural question follows: could particles of **matter**, such as electrons, billiard balls, and planets also exhibit this wave-particle duality? In other words, ***could all matter also behave sometimes as particles and sometimes as waves?***

The answer, discovered in the early 20th century, is **yes, matter can also behave as waves**: this is especially evident for very small objects like electrons! By analogy with EM waves, we can thus say that **QM waves describe the behavior of matter, especially at small scales**. We will explain this by starting with the analogy of EM waves and by contrasting with other types of waves which we discussed in this book.

To avoid confusion, **this book uses the term "quantum mechanical waves", or "QM waves" for short, for quantum waves of matter; the term "QM waves" thus excludes EM waves.** We avoid the frequently used term "quantum waves", which often also includes EM waves, since these waves are quantized as well. We also avoid the less common but more convenient term "matter waves" that is also used for QM waves, but can include the classical non-quantum waves such as waves on strings, in air, water, *etc.*, since these also involve the wavy motion of matter.

As we saw in Chapter 12, EM waves consist of oscillating electric and magnetic fields that create each other, since a time-varying electric field creates a magnetic field, and *vice versa*. This behavior thereby perpetuates their motion and allows them to travel infinitely far in vacuum. "Traveling in vacuum" is a surprising idea, since electricity and magnetism are tightly connected with matter (such as with particles carrying electric charges and with magnets): such matter does not exist in vacuum (that is also one reason we speak of "fields" instead of "forces", since there is no matter in vacuum that could feel those forces). But we accept that light travels through the vacuum of outer space from as far as the Sun and distant stars.[2]

Likewise, **QM waves can travel in vacuum: QM waves don't need a substance (matter) to travel in or on.** Indeed, you can throw any material object (such as a space vehicle) into the vacuum of outer space and see it travel forever. However, this behavior contrasts with other types of waves: string waves in musical instruments need strings to exist and travel; sound waves need gases, liquids or solids; waves on water need the surface of water.

[2] Nevertheless, physicists now understand that the vacuum is more than "nothing": vacuum does have physical properties such as energy and electric and magnetic permeability, which explain the speed of light.

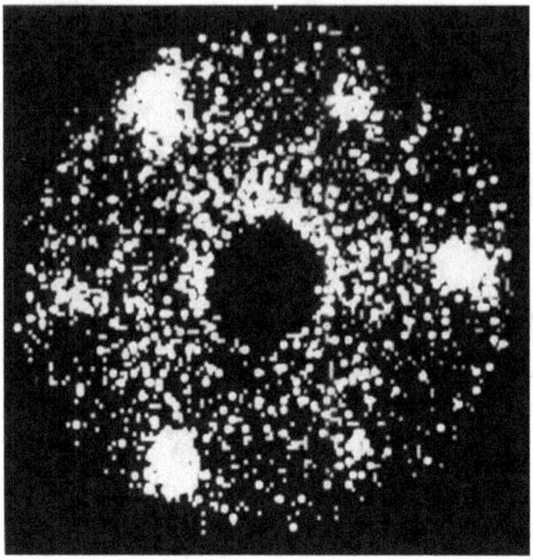

Figure 13-1: An actual diffraction pattern of electrons that were reflected from the surface of a platinum crystal. The individual impact points are due to single electrons, while the overall pattern with six denser regions is formed by wave diffraction of the same electrons in the platinum surface. The detector is a circular disk with a diameter of about 10 centimeters, and with a hole at its center (this hole is needed to direct the electrons toward the crystal): without the hole, there would be another dense region of impacts in the center, somewhat like the diffraction pattern of laser light in Figure 12-51h, which is however much sharper. (*Source*: Courtesy of U. Starke and D. F. Ogletree)

EM waves can also behave as particles, as illustrated in Figure 12-57. We saw there that EM waves hit a detector (like a screen) as small impact points, not as spread-out waves. However, they form denser as well as less dense bands of points, corresponding exactly to the constructive and destructive interference of waves.

The same is true of QM waves: Figure 13-1 shows individual impacts of electrons measured by a circular detector; we see six regions with denser impacts due to wave interference. Figure 13-1 **proves the wave character of electron particles.** I invite you to watch an entertaining video showing this wave-particle duality.[3] We conclude that **QM waves hit a detector as small impact points, not as spread-out waves, while they also exhibit wave interference**.

---

[3] See video 'The observer effect of quantum physics says: "Your thoughts affect reality"' by Art of Spirit — Awaken, at https://www.youtube.com/watch?v=Bq69-MI9TA0.

We also mentioned in Section 12.10 that the EM particles, called photons, are the smallest possible packets of EM waves: photons are indivisible. Similarly, **electrons are also indivisible packets of matter**: when a stream of electrons hits a detector, the smallest impact is that of a single electron, not of a fraction of an electron; the same is true of other elementary particles (quarks, *etc.*).

Atoms and molecules, and indeed most substances in nature, are more complex because they are assembled from multiple elementary particles (electrons, as well as protons and neutrons which are composed of quarks). Nevertheless, atoms and molecules also can act as QM waves, for example by diffracting like electrons, even though they can be divided into multiple smaller particles. By extension, large objects such as billiard balls, humans and planets, while being composed of atoms and molecules, also can act as QM waves; however, we will learn in Section 13.3 why we never notice such behavior with large objects.

A further analogy between EM and QM waves is that ray tracing of light corresponds to the paths of particles like electrons: we can do ray tracing of electrons when their wavelength is too small to notice; that is similar to tracing the path of a car as a line on a map, for example.

Nevertheless, a significant difference exists. On one hand, photons are easily destroyed (absorbed) and created (emitted); here we mean totally destroyed and fully created; that is what happens in a camera or eye, and in a lamp, respectively. On the other hand, electrons and other elementary particles are much more durable: instead of being destroyed or created, an electron normally collides and changes its direction of travel, usually at a different speed. Nevertheless, such particles can also be destroyed or created in very-high-energy collisions that transform elementary particles; for example, an electron and a positron can "annihilate" each other and be completely transformed to EM waves of very high energy.

A very important common feature of EM and QM waves is shown in Figures 12-57 and 13-1: the impact points of photons and electrons are randomly distributed within interference bands or regions. In addition, the impacts are totally unpredictable in location and timing, although their distribution strictly respects the wave interference pattern. We can therefore call the magnitude of this interference pattern of more intense and weaker bands a **probability distribution**: this means that we can't predict where and when each impact will happen, but we can predict how the impacts will be distributed, namely according to the interference pattern. The more impacts take place, the more the distribution will look like a simple interference pattern, as in Figures 12-57 and 13-1. The probability distribution expresses our ignorance of where and when a particular photon or electron will impact and transfer its energy.

The unpredictability of where electrons will impact is a fundamental aspect of quantum physics. Before **modern physics**, we had **classical physics**: in classical physics, everything was completely predictable for all future times, as long as we knew the initial situation precisely. For example, we could predict the exact paths of billiard balls before hitting them; and we could, at least in principle, predict the exact weather a year into the future if we precisely knew the present air temperature, humidity, wind speed, *etc.*, everywhere on Earth, including the position and speed of every molecule.[4] Classical physics includes Newton's laws of mechanics and Maxwell's laws of electromagnetism: they are "classical" laws that are modified in modern physics for extreme situations that we are less familiar with. Thus, the laws of quantum physics give the same behavior as the laws of classical physics for human-scale objects and situations: physicists call this the **correspondence principle**. Similarly, relativistic physics (also part of modern physics) agrees with classical physics in human-scale situations; the relativistic laws differ from classical physics, however, for extremely high speeds, close to the speed of light.

QM waves also have the same property of impacting unpredictably, as we see in Figure 13-1: like EM waves, **QM waves should be viewed as probability distributions**. As with EM waves, this means that we cannot predict where and when a particle described by a QM wave will impact other matter, while the impacts will nevertheless be distributed exactly as given by wave interference. In particular, we can make the

---

[4] In Section 12.10, regarding the wave-particle duality of EM waves, we made the analogy with raindrops: we could predict where each individual raindrop would fall if we knew beforehand everything about each molecule in the atmosphere. But in practice, because of limited knowledge, we can only predict (to some degree) in which areas rain will fall and in what quantities.

same double-slit experiment of Figure 12-57 with particles such as electrons and obtain the same result as with EM waves: unpredictable impact points distributed precisely according to the double-slit wave interference pattern.

The unpredictability of locating a particle and its motion at the same time is more generally called the **uncertainty principle**, proposed by the German physicist **Werner Heisenberg** in 1927. Loosely phrased, **the uncertainty principle tells us that we cannot know all that we wish to know about the position and movement of particles**. It is a very fundamental aspect of quantum physics; it is not due to a lack of measurements, but it is an unavoidable limit on how much we can know. **The uncertainty principle is a major fundamental characteristic of quantum physics.**

If you are not convinced that nature is unpredictable, you are in good company: one of the greatest physicists of all time, **Albert Einstein**, famously wrote: "I, at any rate, am convinced that he [God] does not throw dice."[5] He also expressed this opinion as "God does not play dice with the universe". Einstein never fully accepted this "gambling" aspect of quantum physics, even though he was well aware that quantum physics was and remains extremely successful at describing and predicting properties of matter at the small scales of molecules, atoms, electrons, *etc.*, where classical physics fails. All his life, Einstein strongly favored determinism, namely predictability. To this day, some physicists propose deterministic mechanisms to avoid unpredictability and uncertainty: however, no such mechanism has yet been verified experimentally, so we just have to accept unpredictability and uncertainty as the reality on small scales.

While we cannot predict where an electron will impact a screen, we can predict with full precision the probability of impacts at every point of the screen, at least if we know the starting conditions exactly. So there is full predictability at the level of the probability distribution.

There is another common property of EM and QM waves: we cannot record the extremely rapid oscillations of these waves without

---

[5] Originally in German: "Jedenfalls bin ich überzeugt, daß der nicht würfelt." From "Letter to Max Born", written in 1945, published in *"The Born-Einstein Letters"*, Walker and Company, New York, 1971, Irene Born (translator).

destroying the waves; if we try to measure these waves, they "collapse" and immediately convert to particles (an electron can next convert back to a new wave and travel further). This contrasts with other types of waves: we can photograph the individual oscillations on a string; we can record the individual pressure variations in a sound wave; and we can easily see the individual crests and troughs of waves on water. **Since we cannot record the oscillations of EM and QM waves, they can be thought of as an abstraction or mental image.[6] Even so, these waves are very relevant, since we can use them to predict what happens when many photons and electrons move around, as in electronics and biomolecules.**

For example, when intense EM waves go through a double slit, we can tell with great precision how many photons will reach each part of the screen in Figure 12-57 because they will be distributed according to the interference pattern. The same is true with QM waves: if we throw many identical material objects, like electrons in Figure 13-1, through slits or any other obstacle, we will also be able to predict quite accurately how many will impact each point of a screen, even if we cannot predict where each individual electron will impact.

An interesting difference between EM waves and QM waves is that we can imagine electric and magnetic fields in EM waves interacting with each other to keep the wave going, as in Figure 12-3: the electric field creates the magnetic field, and *vice versa*, making the wave continue on its way. There is no known equivalent mechanism in QM waves. For example, **QM waves are not gravitational, electric or magnetic waves**, even though they may show the result of how two particles (such as two electrons) interact with each other through gravitational, electric or magnetic forces. QM waves only tell us the distribution of the positions of the particles in space and time; they are not a map of forces or interactions between particles.

---

[6] To emphasize the distinction between a physical wave (such as a wave on a string or in air or on water) and a probability distribution (such as a quantum wave), physicists call quantum waves "wave functions".

13.3 Why are QM Waves Only Important for Very Small Objects?

In reading the explanations above, you may have asked yourself: "Why have I never seen all that happening in front of my eyes? No balls or other objects that I have ever seen behave like those QM waves!"[7]

The fact is that in nature, the wave character of matter only shows up on the very short length scales of small molecules, atoms, electrons, and other small particles. This is due to the extremely small wavelength of those particles' waves, which is called the **de Broglie wavelength** (after the French physicist **Louis de Broglie**,[8] who in 1924 proposed these QM waves in analogy with EM waves). This wavelength of QM waves is proportional to the inverse of the object's mass and the inverse of its speed; therefore, the larger the mass and the larger the speed, the smaller the wavelength of the QM wave.[9]

For example, a person of 70 kilograms walking at a normal speed of 5 km/h would have a wavelength of about 10^{-33} meter (which is 0.000,000,000,000,000,000,000,000,000,000,001 meter, with 32 zeroes after the decimal point). This wavelength is unimaginably smaller than atomic dimensions: no detector could measure such small lengths. More importantly, the wavelength is much smaller than the object itself: if

---

[7] Actually, it has recently been found that humans can detect single photons, although you would not be aware of it. See the freely available article "Direct detection of a single photon by humans", by Jonathan N. Tinsley, Maxim I. Molodtsov, Robert Prevedel, David Wartmann, Jofre Espigulé-Pons, Mattias Lauwers and Alipasha Vaziri, *Nature Communications*, volume 7, article 12172, 2016, https://doi.org/10.1038/ncomms12172.

[8] The name "de Broglie" has two pronunciations in French. One pronunciation sounds somewhat like "de Bro-gli" in English, and is often used in English; the more common French pronunciation sounds somewhat like "de Broy" in English.

[9] You may know that "mass" is not identical to "weight", although they are closely related. The mass is the amount of matter in an object, while the weight is the gravitational force that pulls on that mass. Therefore, the weight of an object is directly proportional to its mass. But the weight of an object is different on the Moon or other planets than on Earth, because gravity varies in strength, while the object's mass is the same everywhere.

you and I collide, your and my QM waves should interfere, but our physical sizes (many centimeters) don't allow our QM waves to come close enough to interfere with each other. As a result, we can ignore the wave character of human-scale and larger objects.

On the other hand, a typical electron in a metal has a wavelength of about 10^{-9} meter (or 0.000,000,001 meter = 0.001 micrometer = 1 nanometer): this is comparable to the size of atoms and molecules (between about 10^{-10} and 10^{-8} meter, which is at most 10 times smaller than that wavelength). This scale of wavelength very much changes the way electrons interact with each other and with the atoms; this fact is most important for the structure and properties of atoms and molecules, including biomolecules, and for electronic materials, including all their uses from smartphones to cameras, computers and telecommunications, as we will discuss in Sections 13.5 to 13.7.

13.4 What is the Meaning of the Quantum Mechanical (QM) Wave?

Various attempts have been made to interpret QM (quantum mechanical) waves. As we have seen, we can only speak in terms of probability, which is the likelihood (or "chance") of finding a particle in a certain position at a certain time.[10] In the case of the double-slit experiment, the QM wave can tell us how likely it is that the electron will hit the screen at a particular position; the next electron coming through the same slits will have the same chance of impacting the screen at the same position, but it could hit the screen anywhere else instead, and it most probably will hit elsewhere. Another example is that if two electrons, labeled 1 and 2, collide with each other somewhere in space, the QM wave for that pair

---

[10] We have not mentioned that this wave can additionally tell us the particle's "spinning" state, analogous to the spinning of the Earth around its own polar axis, which makes the particle act like a tiny bar magnet. We have also not mentioned that the QM wave is actually the square root of the probability, together with a phase that causes interference; mathematically, these two quantities are usually combined in a so-called "complex number". In this book we don't need to be aware of any of these details.

of electrons can tell us how likely it is to later find electron 1 at position A and electron 2 at position B.[11]

In other words, **the QM wave tells us the possible outcomes of experiments that try to pinpoint the location of electrons, but again only as probabilities, since the exact outcome is unpredictable. The QM wave itself gives us the interference pattern, not the individual impact points.**

13.5 Atoms: Nuclei and Electrons

We are all familiar with the atom logo shown in Figure 13-2. We will discuss what exactly it tries to represent and why it actually is an incorrect picture of real atoms. As we will see, the planet-like orbits in that logo should be replaced with 3D waves representing the motion of the electrons.

Figure 13-2: The standard logo symbolizing an oversimplified atom.

[11] Note that the probability of finding electron 1 at position A is not independent of the probability of finding electron 2 at position B! Think of two colliding billiard balls, which also obey quantum physics, but for which Newton's laws can be applied with great accuracy: if, at some time after their collision, ball 1 is found at position A, then ball 2 can only have gone to one very specific position B, given by Newton's laws; the probability of any other outcome is negligibly small. Likewise with two electrons: if electron 1 is found at location A, then electron 2 must be found at a specific position B given by Newton's laws. The difference relative to billiard balls is that we don't know beforehand where electron 1 will go, while we do know (with extremely high likelihood) where billiard ball 1 will go. In both cases, the final position of ball 2 is then also known.

First, you are probably aware that the name "atom" is also basically incorrect. Indeed, the word atom comes from a Greek word that means uncuttable or indivisible: "ἄτομος" (pronounced "atomos") is composed of "ἀ", meaning "not" (as in atypical), and "τομος", meaning "cut from" (as in tomography, an imaging technique that cuts up the human body into virtual slices). Until the 19th century, it was believed that all matter (such as rocks, wood, flesh, water and air) is made of particles that are indivisible and could therefore rightfully be called atoms.

However, in 1897, British physicist **Joseph J. Thomson** discovered that much lighter and negatively charged "corpuscles", as he called them, can be extracted from atoms. We now know these as electrons; we also know that all electrons are identical. This discovery suggested that there is some sort of internal structure inside the atom.

Around 1910, physicist **Ernest Rutherford** of New Zealand found that most of the mass of an atom, including its positive charge, is concentrated in a nucleus at its center and that this nucleus is much smaller than the atom. Around 1920, Rutherford further discovered that the positively charged nucleus of the smallest atom, hydrogen, is also found in the nucleus of all other types of atoms (such as carbon, oxygen, silicon, copper, *etc.*): he called this nucleus of hydrogen a **proton**. All protons are identical, and each different type of atom has a different but whole number of protons in its nucleus. It became obvious that **the "atom" is not indivisible**.

Next, in 1932, British physicist **James Chadwick** discovered that most nuclei also contain **neutrons**, particles that are about as heavy as the protons, but have no charge and are thus electrically neutral. Three decades later, in 1964, American physicist **Murray Gell-Mann** and Russian-American physicist **George Zweig** proposed that the proton and neutron are made of three quarks each; this was later confirmed experimentally. Clearly, the atom is very much divisible! However, so far, quarks are considered to be indivisible, and are thus called elementary particles (together with electrons).

Another misnomer is "atomic energy", which is usually meant to be "nuclear energy". The enormous energy due to the explosion of so-called "atom bombs" actually comes from changes in the nuclei of atoms and therefore should be called nuclear energy coming from nuclear bombs. Atomic energy, properly speaking, is due to the chemical interactions between atoms in molecules (see Section 13.6): for example, a chemical reaction that burns fossil fuel (such as gas, oil, or coal) is the result of changes in the bonding between atoms, including bond breaking and bond making. For clarity, this kind of energy is better called chemical energy; it is very much weaker than nuclear energy.

***How do electrons behave in a limited space, such as within atoms?* Atoms contain electrons that surround the nucleus: the electrons are most important in binding one atom to another to form molecules and other substances, from metals to glass and rocks.**

Most electrons are contained within the limited space of an atom. Some electrons can travel to neighboring atoms when the atoms are bonded together. As we have seen, all electrons act like waves, so we have to ask ourselves how these waves behave in the limited space of the atoms.

In this book, we have already seen similar situations. For example, a wave on a string with fixed ends is limited to the length of that string; we know that this wave will bounce back and forth from both ends; most importantly, we know that this can create standing waves (also called resonances), as shown in Figure 1-5. Sound waves in a musical instrument or a mouth also bounce back and forth and can form standing waves with particular frequencies that dominate music and speech.

We learned that such standing waves can only form when the wavelength fits the size of the container (specifically, the wavelength must be a whole multiple of twice the container size so that the wave can repeat itself after each round trip). Waves with other wavelengths and frequencies will die out soon in that container.

The QM wave of an electron will do the same thing: it will bounce back and forth between the ends of a container, in this case an atom (the "walls" of an atom are due to the electrical attraction of the electrons toward the protons). And here also, only standing waves can exist permanently: this means that **electrons in an atom are limited to some particular wavelengths**.

This limitation is extremely important for understanding atoms, as well as molecules and other substances. Here is why: let's go back to Figure 1-5, which shows standing waves with 1, 2, 3, 4, and 5 loops along a bridge, board or string; it is repeated at left in Figure 13-3, with up to 8 loops. Those standing waves could be QM waves as well. We see

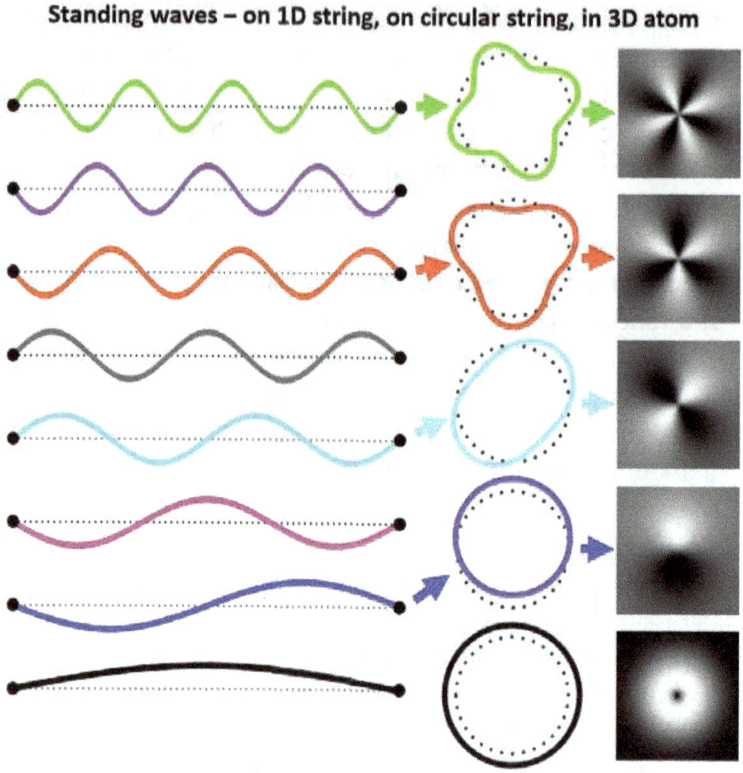

Figure 13-3: *At the left*: Standing waves on a 1D string with fixed ends (black dots). *At the center*: Corresponding standing waves on a circular string, without fixed points. *At the right*: 2D cuts through 3D standing waves in an atom: these are "electron orbitals", centered on the atomic nucleus. Brightness indicates the probability of finding an electron at that location. The two waves at the center bottom and right bottom correspond to zero wavelengths around the circle (they can be viewed as rhythmically and uniformly expanding and contracting balloons, as if breathing, unlike the 1D wave at bottom left, which oscillates up and down with fixed ends).

that the QM waves also can only have distinct states of vibration with a whole number of loops; fractional loops, such as 0.7 or 1.25 or 3.8 loops, are not possible.

This limitation of the wavelength of electrons in an atom to only some values is called **quantization**. It is a form of quantum behavior due to the wave nature of matter: only very precisely defined states of vibration are allowed; no other repetitive vibrations between those states are possible.

The separate quantized states differ in the number of loops that they have within the limited space of the atom. As we have seen, the more loops a standing wave has, the faster it vibrates: it then has a higher frequency, which is most familiar in the case of music and voices. A faster frequency also translates to higher energy. In the case of atoms, we therefore usually speak of **quantized electron states** with different **quantized energy levels**.

What is the shape of electron states inside an atom? On a 1D string, standing waves are simple: they are the sine waves shown at left in Figure 13-3, with fixed ends.

In the case of an atom, we can start by imagining the string to be wrapped as a circle around the nucleus at the center of the atom, drawn as dotted circles in the central column of Figure 13-3. The string vibration can now be imagined to fit along the circular string, in such a way that the wave rejoins itself after one full turn around the circle (you can also imagine a spherical water drop oscillating in this manner as waves travel along its surface). You can recognize the peaks and crests added to the circular string. The wave must now repeat itself without break or kink as it goes around the circle: it should be smooth everywhere because it now has no fixed points where it could be reflected. For this reason also, the 1D standing waves with odd numbers of loops (1, 3, 5, 7) don't exist on the circular string: they don't have a whole number of wavelengths, so they don't fit smoothly on a circle. We get the same connection as before between the wavelength and the string length: **the wavelength must be a whole multiple of the round-trip distance, which is now the length of the circle** (the circumference of the circle).

We have now mapped the standing waves of the 1D string onto similar standing waves of a circular string surrounding an atom's center. Each different number of wavelengths is now a different possible state of an electron. In the quantum language, we have drawn different probability distributions of electrons around the atomic nucleus.

The circular-string model at the center of Figure 13-3 is a simple version of the famous **Bohr model of the atom**, proposed in 1913 by the Danish physicist **Niels Bohr** before a more detailed quantum theory was developed. His model was built upon a miniature version of the solar system with its planets rotating in different orbits around the Sun, which was proposed earlier by Ernest Rutherford. The familiar atom logo of Figure 13-2 tries to portray this planetary model of the atom in a simple way that is easy to understand; however, it is very simplistic. First, in Bohr's model for the "planetary" atom, there is a constraint limiting the possible wavelengths of the electrons in the atoms, making the atom a quantum system; there is nothing like this limitation in the solar system because of its huge size compared to the planets' tiny de Broglie wavelengths.

Second, in reality, an atom does not have circular strings on which electron waves can travel. The electron waves are three-dimensional and look more like sound waves in the air: they spread out in all directions. In particular, electron waves in atoms are more like the sound waves in boxes that we discussed in Section 3.13: the waves fill all the available space rather than staying on lines. **Moreover, the atom has a more complex shape than a rectangular 3D box: this results in electron waves with different shapes, often called atomic orbitals**. Simple examples are shown at right in Figure 13-3 (the word "**orbital**" reminds us of the planetary orbits, but it is meant to describe more generally "wherever an electron can be found in an atom or molecule").

You may notice a resemblance between the standing waves on the 1D and circular strings, on the one hand, and corresponding lobes in 3D in the right column, on the other hand. This is best seen with multiple loops: for example, the top wave, with four crests and four troughs on the 1D string and on the circular string, gives rise to four bright lobes and four dark lobes in the 3D wave at right.

We have shown in Figure 13-3 how the quantization of standing waves causes the quantization of quantum waves for electrons in atoms. To simplify the story, we focused on 1D standing waves (the waves at right in Figure 13-3 still have a 1D oscillatory character, as we follow the waves around the nucleus). However, as with sound in 3D boxes, there can be standing waves in three separate dimensions in atoms. In the case of atoms, as with any spherical object, there can be three separate sets of standing waves: one around the "equator" of the atom (as in Figure 13-3); another along lines joining the two poles; and a third along straight lines through the center of the atom. These three sets of waves can mix into a wide variety of standing waves: Figure 13-4 shows some of these electron states or atomic orbitals for a real atom, **hydrogen**.

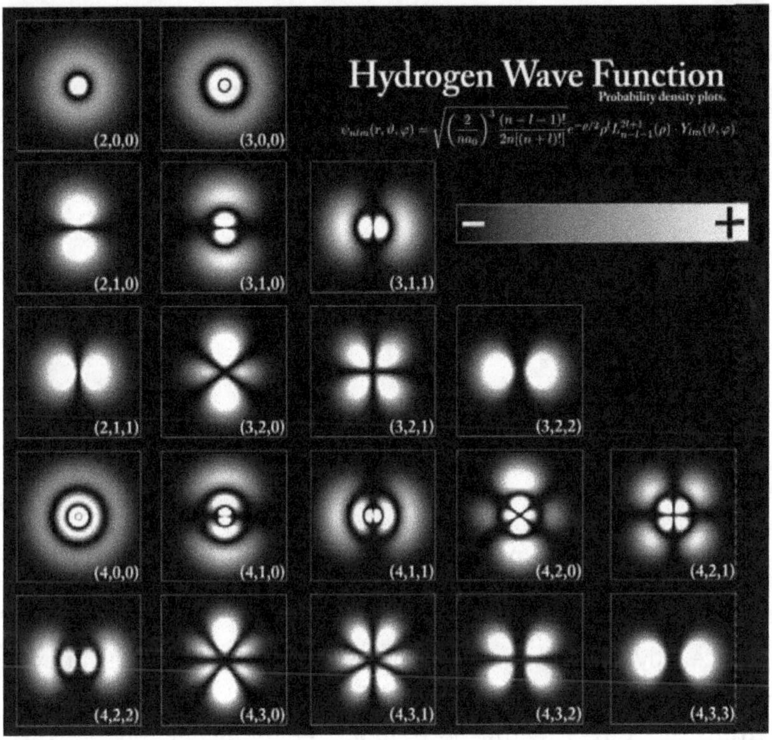

Figure 13-4: Each box shows a 2D cross-section of a 3D electron probability distribution calculated for the hydrogen atom: these are possible electron states or orbitals. Brightness indicates the probability of finding an electron at that location. The numbers (*n*,*l*,*m*) refer to the general mathematical formula at upper right, which gives the corresponding wave function describing each orbital and may be ignored here. In each box, the nucleus is at the center, but it is not shown because it is about 100,000 times smaller than the box size. (*Source*: Gray-scale version of "Hydrogen Density Plots" by PoorLeno, in the public domain, https://commons.wikimedia.org/wiki/File:Hydrogen_Density_Plots.png.)

One helpful way to look at electron orbitals is to imagine them being electron density maps. In that spirit, the brightness in Figure 13-4 indicates directly where the electrons can mostly be found.[12]

---

[12] Remember that the orbitals only give the probability of finding an electron at any particular position; electrons are not smeared out matter with higher or lower densities in different places. Strictly speaking, the probability is the square of the orbital: even after squaring the orbitals, as plotted in Figure 13-4, we still get peaks and valleys in the same places.

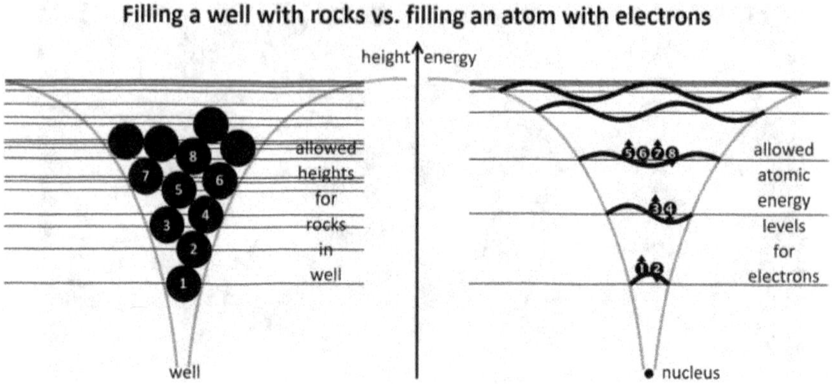

Figure 13-5: *At the left*: Filling a well with rocks (black circles) on the human scale. The gray lines are the walls of the well, mimicking the case of the atom at the right. *At the right*: Filling an atom with electrons (black circles), as a rough sketch. The gray lines are the "walls" of the atom. The allowed energy levels in the atom, marked by horizontal lines, show the electronic structure of the atom. The electron energy in the atom at right corresponds to the height in the well at the left.

Since waves with more and shorter loops vibrate faster, they also have higher frequencies and therefore higher energies. Therefore, **electrons in states with more loops have higher energies than electrons in states with fewer loops**. This is sketched schematically at right in Figure 13-5, which represents electronic states that are allowed for electrons. These energy levels become important when we build an atom one electron at a time, as we will do in the next paragraphs.

How are atoms structured? We will now build an atom by adding one electron at a time to a nucleus. At the left in Figure 13-5 is an analogous situation in classical (non-quantum) physics: filling a well with identical rocks on the human scale. The first rock that is dropped into the well, labeled 1, goes as far down into the well as its size allows: that is the position with the lowest **energy**[13] that the rock can find; it gets there by giving up energy (by producing heat or sound or damage,

[13] This energy is commonly called "potential energy" in physics. It is the energy an object gains by being lifted higher up (for example, against gravity). When you walk upstairs or ride an elevator to a higher floor, you gain potential energy. You can use that potential energy by jumping down, thereby gaining speed and breaking other objects, while possibly killing yourself and maybe others at the same time. Similarly, when a rock drops into a well, it loses potential energy, while gaining speed until it hits something that stops it.

for example). When the second rock 2 is dropped into the well, it will also try to find the deepest position, but rock 1 already occupies that deepest position because of its size: so rock 2 takes the second-best position, above rock 1; it therefore has somewhat higher energy. We continue filling the well in the same manner with rocks 2, 3, *etc.*: they land in gradually higher positions.

Next, we try to do the same thing with electrons added to a nucleus, as sketched at the right in Figure 13-5. The wall of the "well" is now formed by the electrical attraction of the negative electrons toward the positive nucleus. The rock's height now is replaced by the electron's energy. The electrons try to get as close as possible to the nucleus due to electric attraction, which produces a well shape with a deep center. Like the rocks in the well, the electrons will drop as deeply as possible.

Suppose that we want to build an atom around a nucleus that contains a number Z of protons, each of which carries a positive charge. **Hydrogen**, the smallest atom, has just one proton, so $Z = 1$ for hydrogen; the **oxygen** atom has 8 protons, so $Z = 8$ for oxygen. Because each electron carries exactly the opposite charge as each proton, we will need to add Z electrons to neutralize the Z positive charges of the nucleus. Thus, we will need to add one electron to the hydrogen nucleus, or eight electrons to the oxygen nucleus. (If we don't neutralize the electric charge of the nucleus, the atom becomes less stable and therefore chemically reactive: it is then called an ion.)

When we add the first electron to the nucleus, it adopts the state with the minimum number of wavelengths, namely the lowest in the model at the right in Figure 13-5: it is labeled 1. That is because this state has the minimum energy: an electron inserted into an atom will spontaneously fall to the state with the lowest energy, just like a rock that will roll to the bottom of any well you place it in (when an electron falls like this, it releases energy, typically as light or heat).

Importantly, the electron state with the minimum energy also gives the electron a particular wavelength: that wavelength is much larger than the size of the nucleus and therefore prevents the electron from falling into the nucleus. This is why most atoms are permanent: **their quantum character prevents atoms from collapsing**.

For hydrogen, we have now completed our task of creating an electrically neutral atom: it has one positive proton and one negative electron in the lowest-energy orbital; the proton and the electron neutralize each other. However, for any other atom type, such as oxygen, our task is not yet complete.

To make an **oxygen** atom, with 8 protons in its nucleus, we need to add 7 more electrons, so we next want to add a second electron. Since electrons are tiny (as far as is known, they may even have no size at all), we hope that the second electron can drop to the same level as the first. Now, nature plays an unexpected trick on us: it does not allow two identical electrons to adopt the same state! This is called **Pauli's exclusion principle**, after the Austrian physicist **Wolfgang Pauli** who proposed it: this adds one more surprise to the quantum world; we just have to accept it as the way nature works. Pauli's exclusion principle therefore forces the second electron to adopt a different state. This effect is additional to the electrical repulsion between the negative electrons.

However, in one more twist of nature, electrons can have a "spin", roughly similar to the spinning of the Earth around its axis, which makes the electron act like a tiny magnet. In quantum physics, the electron spin can only be "up" or "down" (corresponding to spinning one way *versus* spinning in the opposite direction). Furthermore, the size of the electron spin is also quantized: it can only have a single value "up" or the opposite value "down".[14] In Figure 13-5, we show this spin with an "up" or "down" arrow. An electron with "spin up" is counted as being different from an electron with "spin down", such that these two electrons are allowed to have the same state in an atom. So, in our story of adding electrons to build an atom, we can allow two electrons of opposite spin to have the same state.

As a result, we can place the second electron, labelled 2 at the right in Figure 13-5, in the same lowest-energy state but with the opposite spin. The third electron, labelled 3, is not allowed in the same state as the first two (due to Pauli's exclusion principle): therefore, it must go into the next-higher energy level, which is the second deepest; it may have

[14] Specifically, an electron can only have a spin of +½ ("up") or –½ ("down"). For comparison, photons can only have a spin of 0 or +1 or –1: plane EM waves have spin 0, while other EM waves can have a circular polarization to the "right" or to the "left", corresponding to spin +1 or –1, respectively. The unit of spin is the so-called reduced Planck constant h/(2π), which is a very small quantity that is fundamental to all of quantum physics. Thus, an electron can only have a spin of +h/(4π) or –h/(4π). Similarly, a photon can only have a spin of 0 or + h/(2π) or – h/(2π), while its energy is hf, where f is its frequency.

either "spin up" or "spin down". The fourth electron can fit in the same state as the third, but with the opposite spin to electron 3. Electrons 5 and 6 go into the next-higher energy level, also with opposite spins. For electrons 7 and 8, there is a new opportunity: in the real oxygen atom, the orbital of electrons 5 and 6 can have three different orientations in space (such as from left to right in the figure, or from front to back, or from top to bottom), with equal energies. These three orbitals count as different for Pauli's exclusion principle, so six more electrons are allowed to be at this energy level. That allows electrons 7 and 8 to have the same energies as electrons 5 and 6. (We could add two more electrons at this energy level, but that would exceed the number $Z = 8$ of protons needed in oxygen.)

In general, this game of adding electrons into higher-energy states continues under the peculiar quantum rules that we have seen. This goes on until the Z necessary electrons have been added to the atom, neutralizing it: after adding a total of 8 electrons to the nucleus with 8 protons, we have made a neutral oxygen atom, as shown in Figure 13-5.

One aspect that we have ignored here is that most actual atomic nuclei also contain **neutrons**: these are similar to protons, except that neutrons have no electric charge. Oxygen nuclei, besides 8 protons, also have neutrons. Hydrogen nuclei normally don't have neutrons. All other atomic nuclei have neutrons.[15] The number of neutrons in a nucleus does not affect the number of electrons needed to neutralize the electric charge of the protons; the neutrons also don't affect how atoms bond together to form molecules and other substances, so we can indeed ignore the neutrons here.

[15] The number of neutrons in an atomic nucleus can vary somewhat: that variation gives rise to "isotopes", which are atoms with different numbers of neutrons but the same number of protons in the nucleus. For example, oxygen, with 8 protons, has isotopes with 8, 9 or 10 neutrons in their nuclei. Hydrogen also has isotopes: the three most stable ones have 0, 1 or 2 neutrons in the nucleus. Hydrogen with 0 neutrons is usually simply called hydrogen, or sometimes protium (from the Ancient Greek word πρῶτος, pronounced prôtos and meaning "first"). Hydrogen with 1 neutron is called deuterium, while hydrogen with 2 neutrons is called tritium. Both deuterium and tritium are very important in powering the Sun and in nuclear fusion reactions, such as those that occur in thermonuclear bombs ("hydrogen bombs"), and hopefully in future electric power generation from nuclear fusion. Isotopes of uranium are already used for power generation in power plants that use nuclear fission, as well as in nuclear fission bombs.

13.6 Molecules and Other Assemblies of Atoms: Bonding

How do we assemble atoms into a molecule? Above we have made a hydrogen atom and an oxygen atom. Let's now make a **water molecule** by placing two **hydrogen** atoms next to one **oxygen** atom, forming H_2O. Figure 9-2 shows many molecules of water: one molecule is encircled in part (b) of that figure; another is shown at the bottom of Figure 13-6, as a small red ball and two small gray balls. It has a V-shape, with an O

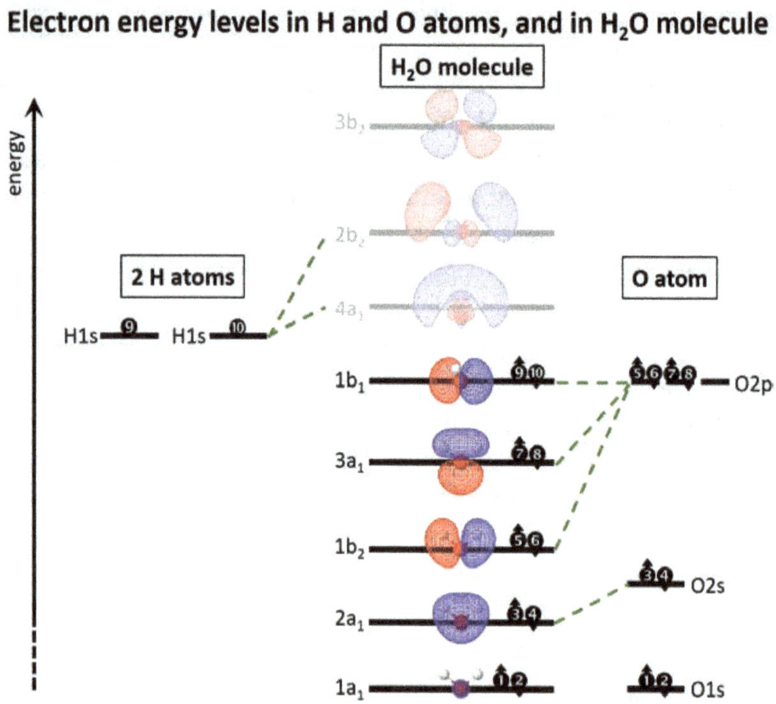

Figure 13-6: Molecular orbitals of electrons in the water molecule H_2O, ranked by energy. The molecular orbitals are illustrated in the center column, with labels $1a_1$, *etc.*: the red and blue "bubbles" correspond to crests and troughs of the wave functions. Non-interacting atomic orbitals are listed at the left and right for 2H and O, respectively, with labels H1s, O1s, O2s and O2p. The O2p orbital can have three equivalent orientations. Inclined lines show which atomic orbitals give rise to the molecular orbitals at the center. Empty orbitals are obscured: they contain no electrons. (*Source of colored molecular orbitals*: "Molecular orbital diagram for water" by ChiralJon, under CC BY-SA 2.0, https://commons.wikimedia.org/wiki/File:Molecular_Orbitals_for_Water.png.)

atom at the point of the V and two H atoms at the ends of the V. So we need to put the three atoms side by side against each other, as H-O-H (here we don't need to worry about the bend in the molecule).

When we place two or more atoms next to each other, their electronic structures interact and modify each other. An analogy is the resonance chamber of a violin: the large wood box of a violin has a narrow segment between two wider segments. The wider segments have standing waves which interact and modify each other through the narrow segment that links them. This results in a richer set of combined standing waves, which are further enriched by the unequal sizes of the two wider segments. The same happens when we place atoms next to each other: we get a richer set of combined electron orbitals, called **molecular orbitals**.

Figure 13-6 shows how this works for the water molecule. Notice toward the left a pair of short lines labeled H1s (H here stands for hydrogen): these two lines correspond to the lowest allowed electron energy level in Figure 13-5, one line for each of the two hydrogens in the molecule (remember that only one electron is present in a hydrogen atom, and that it is in the lowest allowed energy state). This pair of lines represents the two identical hydrogen atoms <u>before</u> they interact with the oxygen atom.

To the right in Figure 13-6, we see several short lines labeled O1s, O2s and O2p (O here stands for oxygen).[16] These lines correspond to the three deepest electron energy levels of the oxygen atom which we saw in Figure 13-5: they again contain 8 electrons. These levels represent the oxygen atom <u>before</u> it interacts with the two hydrogen atoms, each of which contains one electron (labeled 9 and 10 in Figure 13-6).

Now we let the oxygen and hydrogen atoms interact with each other, simply by bringing them closer together: they will form new combined electron states (corresponding to the combined standing

---

[16] The labels 1s, 2s and 2p are used by physicists and chemists to uniquely identify the individual electron states in atoms; we use them here as convenient names. The same holds for the labels $2a_1$, *etc.*, used for molecular orbitals. We see at right in Figure 13-6 that the O1s state in oxygen has a much lower energy than the H1s state in hydrogen: that is because oxygen has 8 protons instead of 1 proton, which electrically pulls the electron state closer to the nucleus, making it smaller and thereby allowing it to drop deeper in energy.

waves of the violin's resonance chamber). The combinations are shown in the central column of Figure 13-6, labeled $1a_1$, $2a_1$, $1b_2$, $3a_1$, *etc.* These combinations are new standing waves that chemists call molecular orbitals.

We have now constructed a molecular structure, except that we still need to place the electrons in the different molecular orbitals. We do this in the same way as with a single atom: we fill electrons into the allowed energy levels one by one from the bottom up, while respecting the quantum rules, until all electrons are in the lowest possible energy levels. In this case, we need to insert 8 + 1 + 1 = 10 electrons to complete the molecule. These 10 electrons fit in the levels up to and including $1b_1$: they are numbered 1 to 10 in Figure 13-6. Higher levels, such as $4a_1$ and $2b_2$, are "empty" of electrons and are therefore obscured in the figure.

The colored diagrams in Figure 13-6 represent the molecular orbitals themselves in a 3D rendition (the large red balls are the O atom, and the smaller gray balls are the H atoms): each "bubble" indicates that electrons are most likely to be found inside that bubble, rather than outside it. For example, the bottom diagram labeled $1a_1$ is a slightly modified version of the 1s state of O: that state has the deepest energy and is tightly concentrated near the O nucleus due to the strong combined attraction of its 8 protons: therefore, the diagram shows only a small bubble around the O nucleus, with almost no contribution from the two H atoms.

It is impossible to draw a 3D probability distribution in a 2D diagram; therefore, such 2D images only provide partial information about where electrons can be found. But we must also remember that there is no way to actually go into an atom or molecule and "see" each electron's location: we can only speak in terms of probabilities of finding electrons here or there. Therefore, we don't need to draw the full details of a 3D map of electronic states.[17]

What causes bonding between atoms in a molecule and other substances? We have seen above how electrons behave when atoms are brought together: they can occupy combined electron states, the

[17] As mentioned earlier, we cannot measure wave functions, including these molecular orbitals. If we want more details of the molecular orbitals, we have to calculate them: these are complex calculations that are done on computers.

molecular orbitals. That redistributes the electrons, and that in turn can lead to attraction between the atoms themselves. This is all due to the wave character of the allowed electronic states and is therefore a quantum mechanical effect. Such attraction can arise in different ways, as follows.

One type of bonding between atoms is called **covalent bonding**. This strong attraction between atoms results from placing electrons primarily between neighboring nuclei. The reason is that opposite charges strongly attract each other when they are close together. For example, in the H_2O molecule, which has the linear structure HOH, it would be electrically attractive to form a line of positive and negative charges like +–+–+. This alternation of positive and negative charges can be achieved if negative electrons place themselves between the atoms, while the H and O nuclei are positive, forming the string +–+–+. Such a line of nearby charges is in fact achieved in H_2O by the orbitals $2a_1$, $1b_2$ and $3a_1$ shown near the bottom of Figure 13-6: these orbitals, when filled with electrons, thus attract the H atoms and the O atom toward each other, forming strong bonds between them.[18]

These three molecular orbitals have an additional interesting property: they allow both an electron from a hydrogen atom and an electron from the oxygen atom to concentrate between the hydrogen and oxygen nuclei, which we could symbolically write as +––+––+, with double negative charges between the positive nuclei. That further increases the negative charge between the atoms, also increasing their mutual attraction. **Covalent bonding thus basically shares electrons among neighboring atoms and attracts those atoms to each other.**

Covalent bonding is also important for many other substances besides water. For example, carbon atoms are bonded together mainly through covalent bonding, and so are silicon atoms. Carbon atoms are fundamental to biological molecules: they form the backbone of many larger molecules important for life; carbon atoms also form most fossil fuels, from coal and methane gas to petroleum (most of which are derived from biological molecules).

---

[18] You could argue that the two H atoms may attract each other by placing an electron directly between them. However, there are not enough electrons available to fill that position in H_2O, resulting in a repulsion between the two somewhat positive H atoms.

Carbon atoms can assemble to form diamond crystals, where they are also bonded by strong covalent bonds, as are silicon atoms in pure silicon crystals. Carbon and silicon atoms have four electrons in higher-energy orbitals that are available for bonding to other atoms: each of these four electrons is shared with another atom, so each atom bonds strongly to four other atoms. This is visible in the crystal structure of silicon, shown at the top left in Figure 13-7 (diamond also has this crystal structure, where each silicon atom is replaced by a carbon atom). Silicon is the basis of most electronics, especially computer chips.

A second type of molecular bonding is called **ionic bonding**. It also contributes to the bonding of water molecules. With ionic bonding, electrons are transferred from one atom to a neighboring atom, leaving some atoms to be charged more positively (they become positive ions) and other atoms to be charged more negatively (these become negative ions): **in ionic bonding, the oppositely charged atoms strongly attract each other**.

For example, in H_2O, there is a slight net transfer of electrons from both H atoms to the O atom: this makes the H atoms somewhat more positive and the O atom somewhat more negative. Again, positive attracts negative, so this net charge transfer also causes the O atom to attract both H atoms to itself. Bonding in water molecules thus has a mixed character, with covalent bonding dominating over ionic bonding.

In many substances, ionic bonding dominates: examples are salts (including table salt, NaCl, shown at the top right in Figure 13-7), metal oxides (including rust such as Fe_3O_4, at the center in Figure 13-7, and many rocks) as well as glasses. In these materials, the individual atoms can be strongly charged, either positively or negatively, so they become ions that attract each other strongly. For example, in sodium chloride, the chlorine atoms are negative ions, while the sodium atoms are positive ions. (You may wonder what keeps the positive and negative ions apart: it is the electrons in the ions that repel each other when they get too close together.)

These substances normally do not form small molecules composed of just a few atoms, but find it energetically more advantageous to aggregate into larger solid pieces. The reason for this aggregation is that each strongly positive ion can then surround itself with several nearby strongly negative ions, and *vice versa*: each negative ion can surround

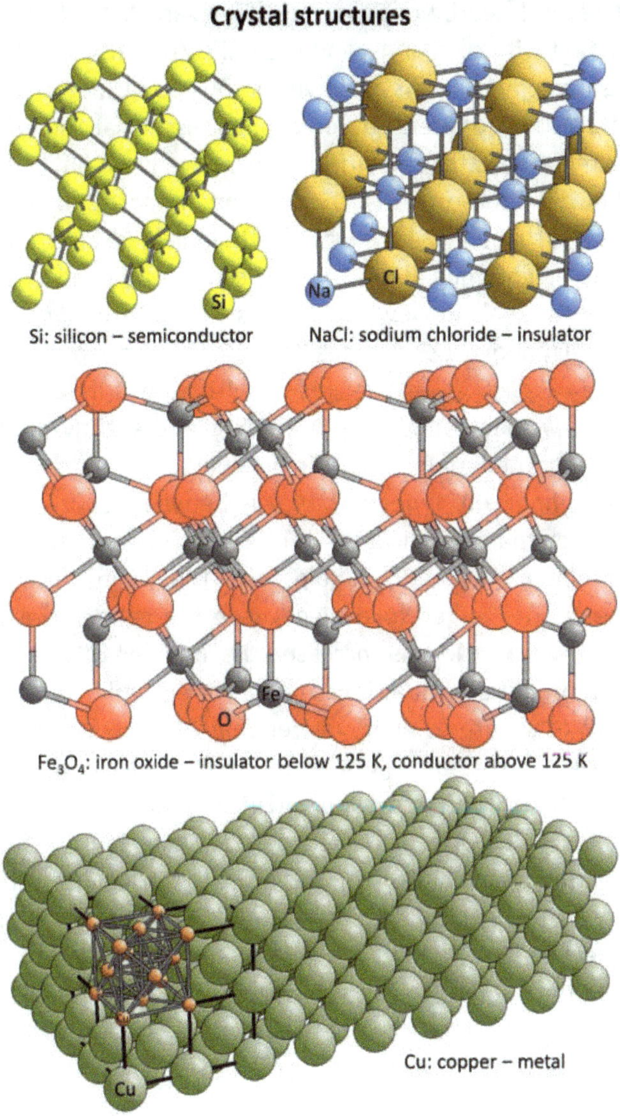

Figure 13-7: Representative crystal structures, with atoms shown as balls. *Top left*: silicon (Si), a semiconductor. *Top right*: sodium chloride (NaCl, table salt), an insulator. *Center*: iron oxide (Fe_3O_4), an insulator at low temperatures, becoming a weak conductor at higher temperatures. *Bottom*: copper (Cu), a metal and very good conductor; in the front left corner of the copper crystal, its detailed atomic arrangement is emphasized. The atom-atom distances in these structures are in the range 0.2 to 0.3 nanometers, where 1 nanometer is a billionth ($1/10^9$) of a meter. The size, shape and color of the atomic balls and bonds are chosen only for easy viewing. (*Source*: Drawn with K.E. Hermann's Balsac software.)

itself with several nearby positive ions; for example, in NaCl, each Na surrounds itself with six Cl ions, and each Cl surrounds itself with six Na ions. This arrangement forms many very strong bonds between the ions.

A characteristic property of ionic materials is that they are mostly **insulators**: this means that electric currents have great difficulty flowing through them. The reason is that the electrons in these substances are in quantum states that are tightly restricted to individual atoms: they can't easily flow to neighboring atoms, so they can't easily conduct electricity through the material. However, a small structural modification in the iron oxide Fe_3O_4 above 125 degrees Kelvin (or –235 degrees Fahrenheit, not shown in Figure 13-7) changes its charge distribution in a way that makes it weakly conducting. We will give a slightly different but equivalent description of insulators in Section 13.7.

A third important type of bonding between atoms is **metallic bonding**. As its name indicates, it is dominant in metals (such as copper, aluminum, nickel, silver, gold, *etc.*). These substances normally also aggregate into larger solid pieces (see the copper crystal at the bottom in Figure 13-7). This increases the number of nearest atoms and allows more overlap of atomic states. For example, in copper, each atom has 12 other atoms as its nearest neighbors. Indeed, **in metals, the electronic states spread out over many atoms**: this enables the **electrical conductivity** that is characteristic of metals.

We will further discuss in Section 13.7 why ionic materials are insulators, while metals are conductors. That will also provide the opportunity to understand semiconductors, which are so important in everyday electronics.

A fourth type of bonding between atoms is **Van der Waals bonding**, named after Dutch physicist **Johannes van der Waals**. This bonding is relatively weak, but sufficient for geckos to hang from ceilings, flat walls and windows, thanks to feet with a large contact area. It is due to deformations of the electron densities of atoms that come close together when the other types of bonding mentioned above are negligible.

A fifth type of bonding between atoms is **hydrogen bonding**. This kind of bonding is present in water ice (frozen water) and, most importantly, in many molecules of living matter, such as DNA. Hydrogen bonds are illustrated in Figure 9-2: **a hydrogen bond consists of one hydrogen atom that connects two other kinds of atoms**; in the case of water ice, a hydrogen atom connects two oxygen atoms in neighboring water molecules. Thus, the hydrogen atom holds two molecules together.

A hydrogen bond can be thought of as a combination of covalent and weaker ionic bonds between the hydrogen atom and its two bonding partners. You can verify that hydrogen bonds are weak by cutting an ice cube with a knife: will the ice cut the knife, or will the knife cut the ice? It is indeed much easier to cut ice than a piece of metal or rock, for example, which have mainly metallic or ionic bonds; covalent bonds are also stronger, as we know from diamond, but also from solid silicon.

Another famous example of hydrogen bonding is the giant **DNA** molecule, a small piece of which is shown in Figure 13-8. DNA consists of two strands that are linked together by hydrogen atoms, shown in the figure as dotted lines. In these hydrogen bonds, hydrogen atoms connect a nitrogen atom (N) to either another nitrogen atom (N-H-N) or an oxygen atom (N-H-O). These hydrogen bonds connect the two possible "base pairs" (adenine-thymine, abbreviated as AT, and guanine-cytosine, GC). The two base pairs act as a simple alphabet, like the digits 0 and 1: the sequence of these base pairs along the long DNA molecule, which is unique to each individual living being or virus, is a message that gives the genetic instructions needed to reproduce that living being or virus.

13.7 Electrical Insulators, Conductors, Semiconductors and Superconductors

We are all familiar with electrical **insulators** such as glass and porcelain, and with electrical **conductors** such as copper wires. We all know that **semiconductors** are central to most of today's electronics and computers. And we have heard of the amazing ability of **superconductors** to allow electricity to flow with no resistance at all. But it is surprising that the cause of **electrical conductivity** was properly understood only after the realization that electrons behave as QM waves: indeed, **electrical conductivity is a quantum property of materials, which can be explained by modern physics**.

This section will attempt to explain electrical conductivity in simple terms. Nevertheless, if you are confused by its combination of unfamiliar

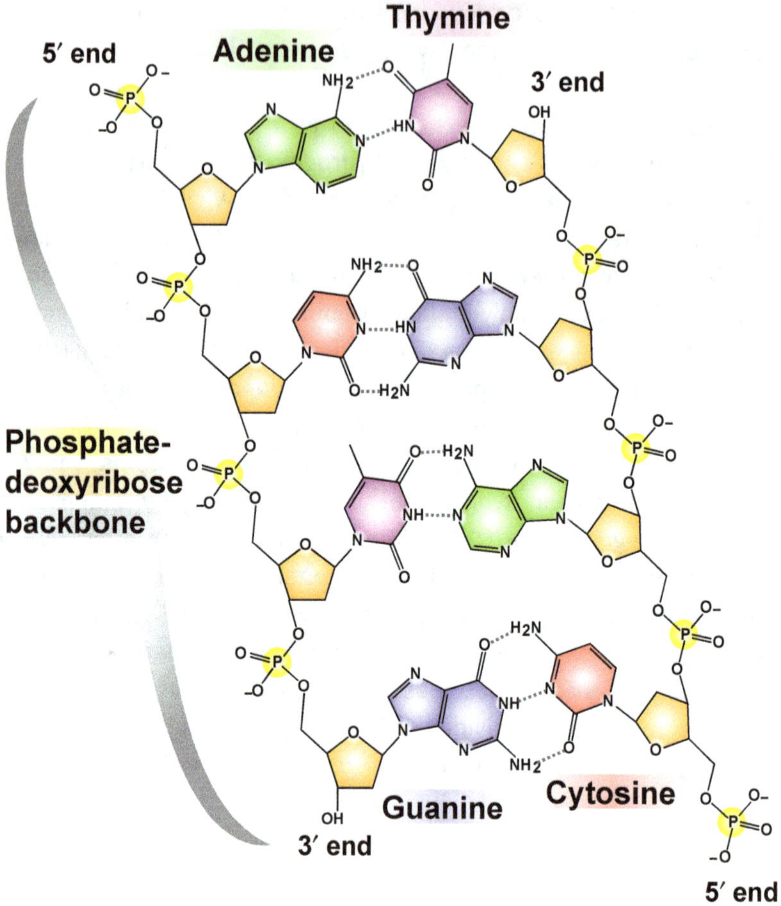

Figure 13-8: A segment of a DNA (deoxyribonucleic acid) molecule. The two strands of DNA are visible at the left and right, respectively; the two strands are linked by hydrogen bonds, shown as dotted lines. In this sketch, the unlabeled "corners" or "kinks" are carbon atoms, while most hydrogen atoms outside the hydrogen bonds are omitted. (*Source*: "DNA chemical structure" by Madeleine Price Ball, in the public domain, https://commons.wikimedia.org/wiki/File:DNA_chemical_structure.svg.)

concepts, you may skip this section in your first reading, or simply read this section to gain a feeling for what quantum physics can do.

Before quantum physics, it was imagined that electrical conductivity (or resistivity, which is a complementary view of the same property) is due to electron particles bumping from atom to atom through a material like copper: each bump slows down the electron, making it harder to pull it along with the help of an electrical potential from a battery or a wall socket. An analogy is a river slowed down by rocks, as it is pulled downstream by gravity; another is wind slowed down by trees, as it is pushed through a forest by atmospheric pressure differences.

However, that "bumping model" of electrical conductivity could not explain a number of observations. One extreme example is that copper conducts electricity one trillion trillion times better than diamond (that is a factor of 10^{24}, or 1 followed by 24 zeroes); diamond is indeed an insulator, but nobody could explain such an enormous difference compared to copper. At the other extreme is superconductivity, found in many metals like niobium, mercury and aluminum (but not copper) at very low temperatures, and other substances at higher pressures or at higher temperatures: why are these materials (literally) infinitely more conducting than copper?

Around 1930, quantum physics stepped in and offered mechanisms of electrical conductivity that explain most of the observations.

The quantum explanation of electrical conductivity is based on the very special behavior of electron waves in crystalline matter, usually simply called **crystals**. By crystals, we mean solids made of well-aligned atoms, such as those shown in Figure 13-7.[19] Water ice, illustrated in Figures 9-1 and 9-2, forms such crystals (as in snowflakes), if we ignore the irregular positions of its hydrogen atoms. Most metals (and many other substances) are composed of well-aligned atoms: usually the crystals are too small to be seen with the naked eye, but to the tiny electrons they appear extremely large. We will use the example of copper because it has a simple crystalline structure and is a very good

[19] Some crystals are transparent, but most are not. Glass pieces in chandeliers are not crystals in the scientific sense of the word crystal: they may have external shapes that look like crystals, but they rarely are made of well-aligned atoms. Glass actually is not crystalline at all but "amorphous", which means "without shape": its atoms, mostly silicon and oxygen, are not well aligned. Crystals, as scientists use the term, do have crystalline external shapes: they have flat faces in directions defined precisely by the atomic alignments. For instance, table salt (sodium chloride, NaCl), shown at the top right in Figure 13-7, can form crystals in the shape of cubes. Many examples are shown at https://en.wikipedia.org/wiki/Crystal.

and commonly used conductor; its structure is illustrated at bottom in Figure 13-7.

The basic structure of a copper crystal is cubic: in Figure 13-7, some cubes are outlined at the bottom left. Each cube contains four copper atoms and is repeated endlessly, forming infinite lines and planes of atoms. This structure is called "face-centered cubic" because atoms are placed at the centers as well as at the corners of the faces of the cubes. It has the highest possible density of atoms, with each atom having 12 near neighbors (if you try to pack balls in a bag, this arrangement takes the least volume): having many neighbors favors metallic bonding, where electrons can roam around many atoms. Other metals that have the same crystal structure include aluminum, nickel, silver, platinum and gold: all are very good conductors of electricity.

In Figure 13-9, we examine the behavior of electron waves in such crystals. We simplify the crystal to be just a 1D line of atoms (drawn as large dots) with equal spacing, continuing infinitely far to the right. We let a wave approach from the left (thick black wave). At the top in Figure 13-9, we give the incoming wave a wavelength that is equal to the atom-atom spacing, so the wave hits each successive atom with the same phase: in the particular snapshot of the figure, the wave has zero

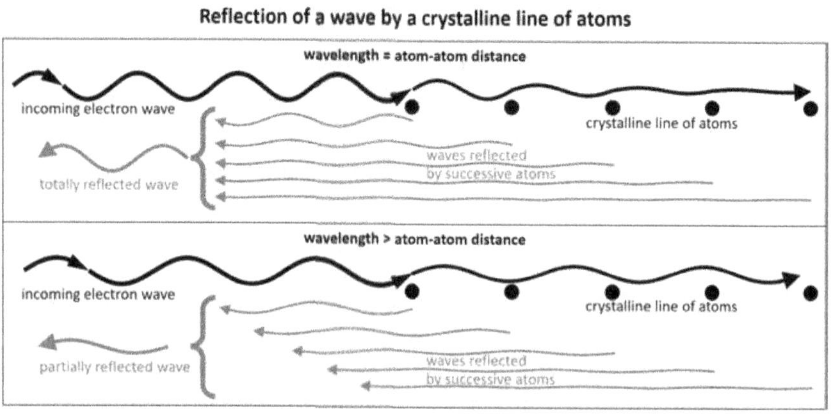

Figure 13-9: Reflection behavior of a wave in a crystalline substance, in 1D with atoms shown as dots. *At the top*: the wavelength equals the atom-to-atom spacing. *At the bottom*: the wavelength is different from the atom-to-atom spacing.

height at each atom. We now look at the reflected waves from each atom, shown as thin gray waves: they all must have the same phase in this situation. Since they are in phase with each other, these reflected waves add up by constructive interference to form one stronger wave going to the left (thick gray wave).

Now, at each reflection, a part of the initial incoming wave is removed, so the incoming wave weakens as it travels to the right: it should therefore die to zero amplitude and result in zero transmission. But if we have zero transmission, we must have total reflection! So, even though the successive reflected waves also become weaker, they must add up to the full amplitude of the incoming wave: we thus should get total reflection when the wavelength equals the atom-atom spacing.

Next, let's increase the wavelength somewhat and see what happens: this is illustrated in the bottom graph of Figure 13-9. We notice that the incoming wave now hits the different atoms with different phases. As a result, the reflected waves (gray) are out of phase with each other. This causes partial destructive interference between them, so that they combine to form only a partial reflection, shown as a thick gray wave with a smaller amplitude than before.

All this seems quite reasonable, until we think about some details. First, we have ignored the fact that all the reflected waves going to the left will in turn be reflected by atoms, forming another set of waves traveling to the right. These in turn can be reflected once more toward the left, and again to the right, and so on without end! We have seen an analogous situation in Figure 10-14 for waves being reflected between double boundaries. It is true that each reflection is weaker, but there are infinitely many such reflections, so what is the net result? That is not easy to tell at this stage!

Second, and more surprising, there is a dilemma in the lower case of Figure 13-9: if we have partial reflection, we must also have partial transmission. This would mean that the incoming wave must continue forever to the right through the line of atoms, with constant amplitude and without ever dying out! In fact, that is how I drew that incoming wave: with a constant amplitude toward the right. How can that be correct? Is it perhaps those infinite numbers of back-and-forth reflections that give this bizarre behavior? The answer is yes!

I know of no way to make these confusing results intuitive and reasonable![20] On the other hand, mathematics gives clear answers and confirmations: this is one among many great examples which show that mathematics can give correct answers to tricky physical problems where our intuition built on our daily experience fails us (nevertheless, mathematical answers must be verified by comparison with nature). Conveniently, the same correct answers are also valid for other kinds of waves, including waves on strings, on water, in air, in liquids and solids, as well as EM waves; but the above-mentioned behavior of electrons in crystals is the most pronounced and important case.

So let's return to the two examples in Figure 13-9: they are drawn correctly, as follows. Inclusion of the infinite numbers of back-and-forth reflections leads, in the upper case, to a total reflection when the wavelength equals the atom-atom spacing; and it leads to a non-zero constant-amplitude transmission in the lower case.

However, there is another surprise (also thanks to mathematics): if we change the wavelength to be only slightly different from the atom-atom spacing, we still get total reflection and zero transmission. This total reflection thus exists in a range or band of wavelengths around the atom-atom spacing (the width of this band increases with the strength of the reflection by individual atoms; and the edge of this band is abrupt, not gradual). Furthermore, the same story repeats itself when we divide the wavelength by two, three or any other whole number: each time we again get a band of possible wavelengths. This means that we get multiple bands of total reflection, separated by bands of partial reflection: this highly counter-intuitive fact is crucial for explaining electrical conductivity!

It is truly amazing that electron waves can travel undiminished over very long distances through the maze of copper atoms pictured in Figure 13-7 (as long as they are well aligned): this is in principle true even all the way from the electrical power plant to your home. By contrast, electron particles would bump constantly and wildly from one atom to the next. Actually, such wild bumping also happens in the quantum

[20] In fact, during most of my professional life, I have worked with such electron waves in this specific situation: their bizarre behavior still is a puzzle to me, but I have to believe quantum physics because it agrees so well with nature!

model, for example when there are misalignments in the copper structure, such as vibrations and defects, so some resistance can take place: at each bump, the electron wave becomes, at least temporarily, a particle, and there is loss of current, reducing the conductivity. **But it is the wave character of electrons that allows conducting electricity over huge distances.**

How are these bands of total reflection connected with the electronic structure of solids? The answer to this question will be the key to explain conductors, insulators and semiconductors.

When we assemble atoms into a crystal, we are combining two distinct quantum wave effects: the internal electronic structure of the atom, on the one hand, and the bands due to total reflection in a crystal, on the other hand. These two effects are quite separate from each other: the atomic levels result from the attraction of the electrons toward the nucleus of an atom, while the bands result from the regular repetition of that atom in the crystal. There exists a wide variety of atoms[21] (there are over 100 types of atoms, from hydrogen to uranium and more), so we get many different electronic structures within atoms; furthermore, we can mix different kinds of atoms, as in NaCl, H_2O and DNA, forming countless different compounds and molecules. And there are 230 different types of crystal structure[22] (a few of which are shown in Figures 9-2 and 13-7), resulting in many distinct band structures. Therefore, there exists an enormous variety of combinations of atom structures and crystal structures (although not all possible combinations actually exist in nature).

Let's take the electronic levels of the atom in Figure 13-5: they are reproduced at left in Figure 13-10. What happens when we line up many such atoms close together, as in a crystal? For simplicity, we will use identical atoms, as in a copper crystal. This is illustrated at right in Figure 13-10. First, the tops of the outer "walls" of the atoms (the curved gray lines) are lowered because the atoms overlap somewhat: this allows some electrons to travel more easily to neighboring atoms by passing over those walls. These electrons will behave like the waves in Figure 13-9. They may be totally reflected, in which case they can't travel far:

---

[21] See the "Periodic table" of elements at https://en.wikipedia.org/wiki/Periodic_table.

[22] See "Crystal structure" at https://en.wikipedia.org/wiki/Crystal_structure.

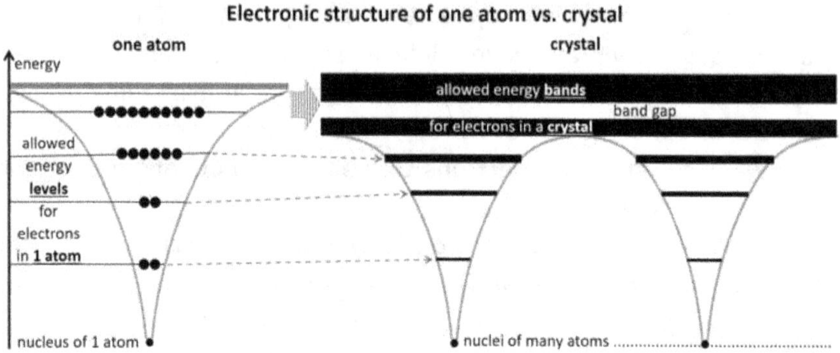

Figure 13-10: Schematic electronic structure of one atom compared to a crystal with many identical atoms (only two such atoms are shown). *At the left* is an atom similar to that at right in Figure 13-5, with dots indicating electrons at different allowed energy levels. *At the right*, multiple identical atoms are lined up close together in a crystal (only two are shown). The allowed energy levels of the single atom are modified by the proximity of the neighboring atoms. The higher-energy levels mix to form many new states that are spread out in allowed energy bands (black). Between those allowed bands are band gaps (white), in which no electron states can exist.

they are then in a band of total reflection, seen as a white strip marked **band gap** in Figure 13-10; no electrons can exist in these band gaps. Or the waves may be transmitted and travel far: these are in a band of partial transmission, marked as allowed **energy bands**. There can be multiple such bands and band gaps; bands can also overlap without a gap between them. These possibilities will become very important in our following discussion.

In Figure 13-11, we show three cases of electronic bands in a crystal: these are more detailed versions of the bands drawn at the top right in Figure 13-10. Now the black bars represent bands that are filled with electrons: they are **filled states**; as with a single atom or molecule, electrons have been filled in the lower-energy states until the whole crystal has become electrically neutral. Gray bars are bands of allowed states without electrons: they are **empty states**.

Now we can discuss the conduction of electrons through crystals (and therefore the flow of electricity through wires). Such an electric current is achieved by an electrical force on the electrons of the material: you do this yourself each time you put an electrical plug in an electrical socket because the electrical potential (voltage) from the

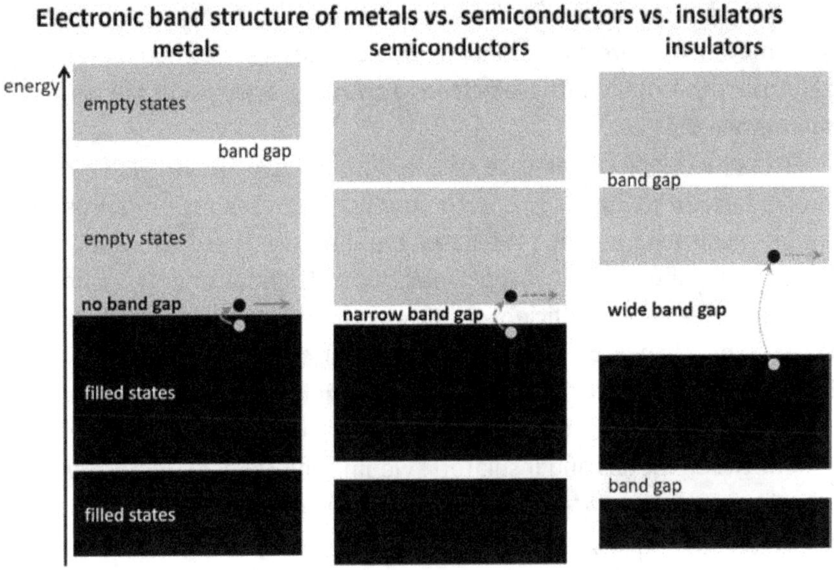

Figure 13-11: Three contrasting sets of electronic bands and band gaps in crystalline solids, showing only the top right part of Figure 13-10. *At the left*: bands forming a metal. *At the center*: bands forming a semiconductor. *At the right*: bands forming an insulator. In the three cases, black bands are filled with electrons, while gray bands have no electrons; white bands are band gaps. The pairs of dots show an electron being lifted from its normal state to an unoccupied state ("empty state" or "empty level"), where it can travel more easily across the crystal. A wide band gap requires giving the electron a much larger energy than a narrow band gap or no band gap: this strongly affects electrical conductivity.

socket pushes electrons along. But there is one problem: an electron cannot easily move to neighboring atoms if their electrons occupy the same energy level (this is due to Pauli's exclusion principle). To move, that electron must gain a bit of energy and go up into a state with a slightly higher energy, as illustrated at the left in Figure 13-11. This is possible if there is no band gap right above that electron. In the higher state, this electron can easily travel far because there are very few electrons at that level. Fortunately, there are always some electrons with that bit of extra energy thanks to the non-zero temperature of the wire, just as there is always some vapor above the surface of water: the vapor can be blown along very far by the wind, much farther than the water molecules inside the water.

We have now sketched the mechanism of electrical conduction in a metal: if there is no band gap right above the highest-energy electrons, thermally-lifted electrons can move easily and far through the metal.

Next, we look at the case of a wide band gap above a filled band, shown at right in Figure 13-11. To conduct electrons now requires that we lift electrons up above the band gap, because there are no allowed states right above the available electrons. Thermal energy at normal temperatures is not sufficient to lift electrons into the empty band, so there is no chance of starting an electrical current. Now we have an **insulator**: **an insulator cannot conduct an electrical current due to a large band gap**.

Between metals and insulators, we find the **semiconductors** (which literally means "half conductors"): their lower conductivity is due to narrower band gaps, as illustrated at the center in Figure 13-11. The narrower gap allows more electrons to thermally jump up to the empty band and from there conduct electricity, but still much less than for metals, so we get a much lower conductivity than for metals. As a result, pure silicon is a semiconductor that conducts electricity a trillion times less well than copper (that is a factor 10^{12} or 1 followed by 12 zeroes).[23]

However, this low conductivity is not in itself very useful: there are easier ways to make poor conductors. Also, it is not the reason why semiconductors are so important in electronics. Their importance comes from the possibility to engineer these materials so as to finely control their conductivity, in particular to rapidly switch their conductivity on or off. This is done by adding foreign atoms to the pure silicon: these

[23] There is another contribution to electrical conductivity: it is complementary to the movement of the electrons, and therefore roughly doubles the conductivity. As is shown in Figure 13-11, an electron that is lifted to fill an empty state leaves behind another empty state: it forms a "hole", as it is called; since this hole is lacking the electron's negative charge, it in effect has a net positive charge. This positively charged hole can equally travel under the force of an electrical potential (but because a hole has the opposite electrical charge to the electron, it travels in the opposite direction). It travels by letting neighboring electrons fill in the hole. A hole moves like a gap in a queue of people: while people move forward to fill the gap, the gap moves backward along the queue.

carefully chosen "impurities" (for example, aluminum and arsenic atoms) are called **dopants** in a process called **doping**.

The function of the dopants is to add electronic states within the narrow band gap, either very close to the filled states below the band gap, or very close to the empty states above the band gap. This creates even narrower band gaps (roughly one tenth of the "narrow" band gap in pure silicon), which make it easier for electrons to jump up to empty levels where they can travel in larger numbers, thereby increasing the possible current.

By using such **extrinsic semiconductors**, as they are called, it is possible to produce a variety of **electronic devices**. These includes the following: **transistors** (further described below); **diodes** (which convert an alternating current into a direct current); **light-emitting diodes** (seen daily in lamps and electronic displays, and used to control television sets and elevator doors); and **lasers** (used at checkout counters and in many other applications).

Let's elaborate the case of the **transistor** a bit further. This device is essentially an electronic switch that can turn a current on or off. An analogy is a light switch operated mechanically with your fingers, which turns on or off an electrical current; another is a water tap, operated with your hand, which turns on or off a stream of water. The transistor is not operated by a finger or hand, but by another electrical current in a semiconductor: thereby one electrical current controls another. This allows computing because such on/off control is a logical operation that can be used to add, subtract, multiply, divide, *etc*. Such logical operations are also used to make many types of automated decisions, for example in elevators, cars, airplanes, cameras, and countless more devices.

Transistors are a major part of the **integrated circuits** or **chips** that modern society depends on more and more. Their impact comes in large part from the fact that transistors (and other electronic devices) can be built on a scale so small that they are capable of immense computing power at a phenomenal speed. Indeed, electrical signals can cross a computer (about 30 centimeters) in about a billionth of a second and go across a chip in a small fraction of that time: this leads to the gigahertz speeds of today's computers, which compute with billions of operations per second ("giga" corresponds to billion).

What are superconductors? In materials called **superconductors**, the resistance to the flow of an electrical current disappears totally at low enough temperatures: the electrical conductivity rises to infinity. This means that an electrical current can flow forever in a superconductor without the input of energy to push the electrons along. Another effect is that a magnetic field cannot exist inside superconductors; as an illustration, if the Earth were a superconductor (which it definitely is not, even if it were cooled to very low temperatures), the Earth's magnetic field would not penetrate below the Earth's surface.

Superconductivity was discovered in 1911 by the Dutch physicist **Heike Kamerlingh Onnes**. In 1957 it was explained by the so-called **BCS theory**, a quantum theory named after the initials of three American physicists: **John Bardeen**, **Leon N. Cooper** and **J. Robert Schrieffer**.

The BCS theory of superconductivity is very different from the model of electrical conductivity that we discussed earlier in this section, although it also uses the concept of electronic band structure and is therefore a quantum effect. It is based on pairs of electrons that move together like a couple of dancers: they stay coupled, even if not close to each other, by interacting through deformations of the material through which they move: an analogy is one person lying on a mattress and causing a dip into which a second person will tend to roll. These pairs are only weakly coupled together, so they can only exist at very low temperatures, but they can travel through the superconducting material with no resistance whatsoever.

The main practical interest in superconductors is the fact that an electrical current can flow through them "for free": once started, such a current will flow forever and infinitely far. If this happened at our normal daily temperatures, we could send electricity from a power station to a user over large distances without losses along the way. Electric power lines in common use today lose roughly 3% of the electric power, while voltage transformers lose a similar amount; another 5% of power is lost in local distribution to customers, adding up to more than 10% loss (the lost energy goes mostly to heat).

We could avoid those energy losses by cooling the wires and related equipment. However, with the present superconductors, cooling is too expensive. The hope, therefore, is to find materials that are superconductors at "room temperature", as scientists like to call our comfortable temperature. The search is ongoing for such **high-temperature superconductors**.

13.8 The Colors of Substances

The substances we see every day are rich in **color**. Amazingly, the origin of color was not understood before the advent of quantum mechanics. In fact, there are multiple mechanisms that create color, most of which rely on quantum effects to create light, including: nuclear reactions (creating sunlight), chemical explosions, electroluminescence (in electronic displays), lasers, lightning, fluorescence and phosphorescence, plasmas (in some displays) and black-body radiation (in old incandescent light bulbs, in electric heaters, in cooking ranges and in ovens).[24]

Most of the colors that we see are in reflected light, rather than in direct light from a source: incoming light (such as white light from the Sun or a lamp) is reflected by most substances, such as paper, furniture, walls, food, drinks, trees, and clouds. In some cases, interference effects create colors, such as Newton rings in oil patches and iridescence in butterfly wings (see Section 12.2).

More frequently, incoming light is partly absorbed by the substance, and only the remaining light is reflected. **Absorption** means that a light photon gives its energy to the substance, after which the photon no longer exists, as we discussed in Section 12.10. Typically, the energy given to the substance by a photon is given to an electron, thereby lifting the electron to a higher energy. In Figure 13-11, we discussed electrons in crystals being lifted to higher levels by thermal energy (see the pairs of dots in that figure). The same can happen with photon energy: a photon's energy can lift an electron to an empty, higher level. However, quantum mechanics says that this can only happen if there is such an empty level at the correct energy, for example in a band above a band gap. Indeed, if the photon energy lifts the electron to a level inside a band gap, there is no possible level for that electron: then the energy transfer from photon to electron is not allowed and it will not happen. The result is then no absorption: the photon is not absorbed and

[24] See also Section 6.3 in Michel A. Van Hove, "*Everyday Physics — Colors, Light and Optical Illusions*", World Scientific Publishing Co., 2022.

therefore continues through the substance. This means that the substance is transparent for photons of that energy, and thus for photons of that color (since a particular energy corresponds to a particular frequency and wavelength and thus color, as seen in Figure 12-4).

We can make the same argument with molecules, such as water. If a photon enters water, it can be absorbed if it can lift an electron from one of the occupied levels shown in Figure 13-6 to a higher empty level, such as level $4a_1$, or $2b_2$, *etc.*, in the obscured part of the graph. If that photon does not have the right energy, it will not be absorbed, and the water will be transparent for photons with that energy. In fact, all visible photons have energies that are insufficient to lift an electron in water molecules. **This is why water is transparent to visible light: the energy of visible photons cannot lift electrons to higher electronic levels, by the rules of quantum mechanics.**

We see that light absorption depends strongly on the wavelength (and therefore frequency and energy) of the light: this means that some colors may be absorbed more than others. For example, plants and leaves in trees are green because they absorb blue and red light much more than green light, as shown in Figure 13-12. It is the molecule of **chlorophyll** that is responsible for this absorption, in the process called **photosynthesis**: chlorophyll absorbs energy from light photons to produce other molecules that are essential to life.[25]

Figure 13-12 illustrates, for each wavelength and color, how much light is absorbed by five different varieties of chlorophyll: an absorbance of 1.0 means that 100% of light with that color is removed from the incoming light and cannot be reflected; an absorbance of 0.0 means that none of that color is absorbed, so it can be fully reflected. We notice strong peaks in the blue and red parts of the color spectrum, meaning that these colors are strongly absorbed by the chlorophyll molecules: they are due to lifting electrons between energy levels of the chlorophyll molecules. The middle range between blue and red is only weakly

[25] In photosynthesis, plants transfer absorbed photon energy to molecules that in turn convert carbon dioxide from the air and water from the rain or ground into a sugar (glucose), from which other molecules are produced that are useful to life. A by-product of photosynthesis is oxygen. We see here the beneficial role of photosynthesis: besides creating useful molecules for life, it removes carbon dioxide (an important cause of global warming) and produces oxygen that we breathe.

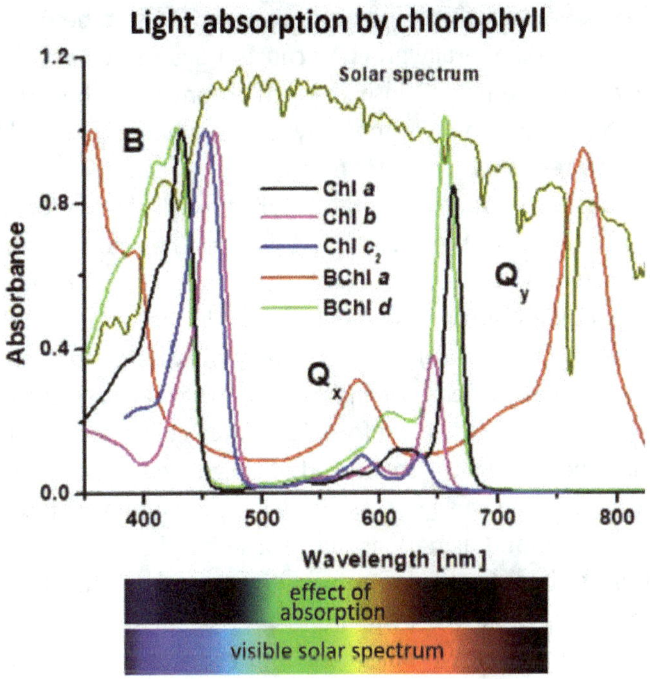

Figure 13-12: Absorption of light by five different chlorophyll molecules as a function of the light's wavelength, shown as lines of five different colors. The orange/brown curve is the intensity of the solar spectrum under typical conditions on the earth's surface, also shown as the simple solar spectrum at the bottom; the spectrum marked "effect of absorption" shows the weakening of the blue and red light due to their absorption. (*Source of top graph*: "Electronic Structure of Chlorophyll Monomers and Oligomers", by Juha Matti Linnanto, in book "Chlorophylls", editors Sadia Ameen, M. Shaheer Akhtar and Hyung-Shik Shin (2022), IntechOpen. Freely available under CC BY-SA 3.0 at https://www.intechopen.com/chapters/81388 or https://10.5772/intechopen.104089.)

absorbed: that is the green range (and to some extent also yellow), so **green is not absorbed and therefore is reflected by other parts of the plant: that gives the green color that we see for plants**. This is shown in the pair of spectra at the bottom in Figure 13-12: they show how the blue and red light are weakened. Our eyes are most sensitive to green light, so we notice the surviving yellow light relatively less. (I have discussed the colors seen in reflection in more detail in another book.[26])

[26] See also Section 4.2 in Michel A. Van Hove, *"Everyday Physics — Colors, Light and Optical Illusions"*, World Scientific Publishing Co., 2022.

Similarly, we can ask why tomatoes are red. In fact, before they are ripe, tomatoes are generally green because they are rich in chlorophyll. When they ripen, tomatoes lose their chlorophyll and become rich in lycopene. In contrast to chlorophyll, lycopene absorbs mostly blue and green light, so red is left to be reflected.

Solid substances such as metals rarely have distinct colors like red, green or blue. Instead, they tend to be grayer. The reason is simple: as we have discussed, if electrons can be lifted to allowed energy levels, the corresponding photons will be absorbed. In many metals, the band structure (see Figure 13-11) consists mainly of wide bands and narrow band gaps: this makes it easy to find electrons that can be lifted to higher energies with visible light. Therefore, light of all colors can be absorbed, so that the reflected light is composed of roughly equal amounts of red, green and blue: this gives gray, as for silver and aluminum (gray can be as bright as white and as dark as black). The band structures of copper and gold make these metals absorb relatively more of the blue light, so they appear more yellow/brown.

The strong reflectivity of metal surfaces (especially silver and aluminum used in mirrors) is due to another factor that we discussed earlier (Section 12.2): their index of refraction is such that much of the light is reflected before it has a chance to be absorbed. This gives the metallic gray its characteristic shiny appearance.

13.9 Can Matter Pass Through Barriers? Radioactivity, Quantum Tunneling

When we walk through a crowd, we have to avoid other people: why? When we cross a mountain, we have to climb over it or go around it, unless there is a tunnel: why? The simple answer is, of course, that substances cannot go through each other.

Waves, however, can go through unexpected places! We already saw an example of this with the transmission of evanescent waves ("dying" waves) through a wall (see Figure 10-17 in Section 10.6). Inside the wall, between its two boundaries, some refracted waves should not be able to travel (this happens if the wave speed is larger inside the wall than outside the wall, and for suitable angles of incidence of the incoming wave outside the wall). These waves should therefore decay and die before they reach the second boundary and emerge behind the wall.

Nevertheless, if the thickness of the wall is small enough, those "dying" waves are not yet "dead" when they reach the second boundary: they can therefore be transmitted behind the wall, where they continue as normal waves (although they are then usually weak), as drawn in Figure 10-17. For this to happen, the wall should have a thickness comparable to the wavelength. Importantly, this mechanism of **evanescent wave coupling** works with any type of wave, including light waves as mentioned in Section 10.6: it should therefore also work with QM waves. That is indeed the case in special circumstances: it is then called **tunneling** or **quantum tunneling** because it resembles tunneling through a mountain.

The existence of quantum tunneling of QM waves means that very light particles can cross sufficiently thin walls or obstacles. Two important examples of such tunneling are **nuclear fusion** and **nuclear fission** (including **radioactivity** or **radioactive decay**), both of which can produce enormous amounts of energy. In nuclear fusion, two **hydrogen** atoms (especially their deuterium and tritium isotopes) need to fuse together despite the obstacle of the strong electric repulsion between their two positive charges: quantum tunneling makes that fusion much more likely. In nuclear fission, particles are emitted from the nucleus of heavy nuclei like uranium and must pass through a barrier: quantum tunneling also greatly enhances this process.

The quantum tunneling of electrons is used extensively in electronic circuits, for example in writing and reading data on our familiar **USB flash drives** (which also have many other names, such as **thumb drives** and **memory sticks**). Usually, two conducting or semiconducting materials are separated by a very thin wall of non-conducting material, typically a metal oxide. Here we again can have an evanescent wave that crosses through the wall and produces a current on the other side.

A spectacular use of quantum tunneling is the **scanning tunneling microscope**, which is one among several modern scanning probe microscopies that are able to image atoms and molecules at least in 2D.[27]

[27] See https://en.wikipedia.org/wiki/Scanning_probe_microscopy.

Figure 13-13 shows an image obtained by this technique which displays hexagonal rings of water molecules frozen on a flat metallic surface; in one "2D ice crystallite", models of water molecules are overlaid to explain how they are arranged.

The principle of scanning tunneling microscopy is to use a very sharp conducting tip (like a needle) to approach the object to be imaged, as illustrated in Figure 13-14. Normally the object is the flat surface of a conducting material which may be covered with a very thin layer of other atoms or molecules. Electrons can tunnel between the tip and the surface when these are very close together, after a voltage is applied between the surface and the tip. The resulting weak electron current

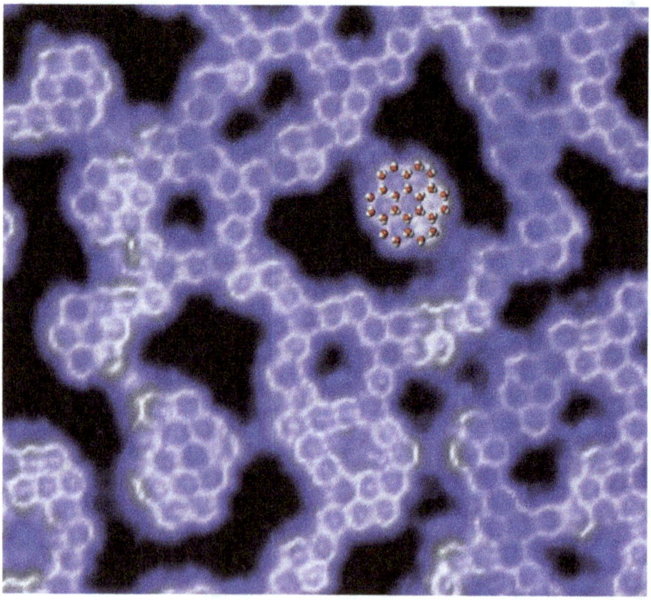

Figure 13-13: Image taken with a scanning tunneling microscope of water molecules partly covering a metal surface (seen as a black background), forming connected "2D ice crystallites". Only one layer of molecules is present on this flat surface of palladium metal at a cold temperature of 50 degrees Kelvin (50 K or –223 °C or –370 °F). In one crystallite, the image is overlaid with models of water molecules, with red oxygen atoms and white hydrogen atoms, showing their hexagonal arrangement. The shortest O-O distances, forming the sides of the hexagons, are about 0.3 nanometer. (*Source*: M. Salmeron, reproduced from Physics Today, February 2010 cover, https://physicstoday.scitation.org/toc/pto/63/2, with the permission of AIP Publishing.)

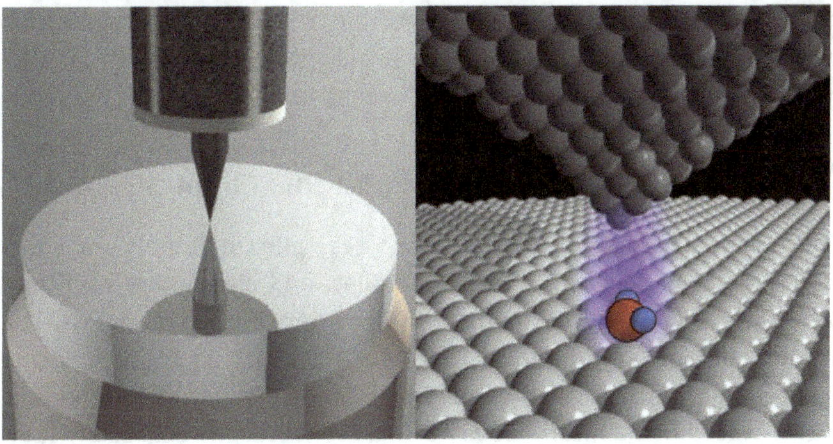

Figure 13-14: Sketch of the geometry of a scanning tunnelling microscope on the human scale (*left*: centimeter scale), and on the atomic scale (*right*: nanometer scale). The balls at right represent metal atoms (gray), hydrogen atoms (blue) and an oxygen atom (red); the blue "haze" represents the tunnelling electrons. (*Source*: Courtesy of M. Salmeron.)

varies with the position of the tip due to the atomic and molecular structures. That current is measured as the tip is scanned across the surface, and displayed to produce the 2D image, which often exhibits atomic-size structures.

The tunneling mechanism in scanning tunneling microscopy can be explained as follows: First, suppose that the surface and the tip are far from each other. Then, electrons that are present in the surface cannot easily escape out of that surface: they see a boundary because they are attracted back into the surface (if a negative electron leaves a surface, a positive charge is left behind that pulls the electron back into the surface). Likewise, electrons in the tip see a boundary that prevents them from leaving the tip.

Now we must remember that these electrons behave like waves: they "spill out" a bit from both the surface and tip, by forming dying evanescent waves like those we saw in Figure 10-4 and at right in Figure 10-6. If we next bring the tip close to the surface, the two boundaries combine to form the wall of Figure 10-17: this wall, if it is thin enough, allows the electrons to cross the wall and conduct charges: they thereby form an electrical current between surface and tip. For this to happen, the wall's thickness must be close to the wavelength of the electron waves, which is roughly a nanometer (a billionth of a meter); this is also close to the size of many molecules, helping to visualize them.

13.10 What have We Learned in this Chapter?

This chapter has described processes and materials on the submicroscopic level of atoms, molecules, and elementary particles, especially electrons. Here, modern quantum physics gives rise to various unfamiliar situations. First, we learn that all those tiny but important particles can also be waves, called quantum mechanical (QM) waves or quantum waves or matter waves. In particular, they behave very much like EM waves in the famous double-slit experiment.

As a consequence, we also find that we can't predict precisely where those particles are or will go: we can only know the probability of finding them somewhere. With a few rules of quantum physics, we can then systematically build up atoms from their nuclei (containing protons and neutrons) and electrons.

Next, we can assemble multiple atoms into molecules as well as crystals. In that process, we find how atoms bond together, using covalent bonds, ionic bonds, metallic bonds, Van der Waals bonds and hydrogen bonds. Our next step is then to understand the electrical properties of insulators, metals, semiconductors and superconductors, by means of the electronic band structures of crystals.

Finally, this chapter explores another intriguing quantum property, tunneling, and describes its importance in radioactivity and its use for imaging molecules.

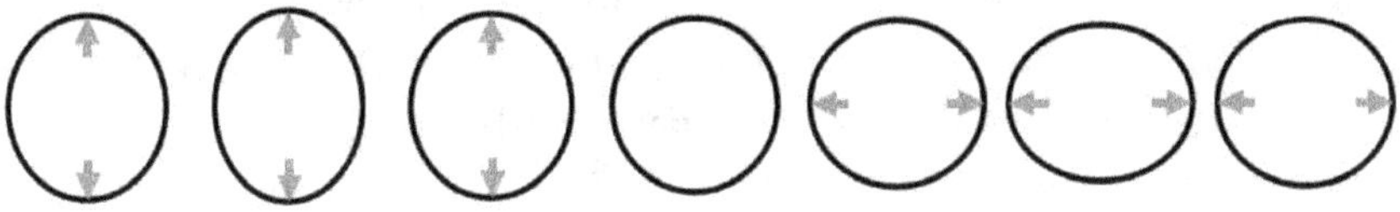

14 Gravitational Waves

Gravitational waves were first detected only recently, in 2015. They had been predicted a century earlier, by Einstein, as part of his theory of general relativity. He was, however, very skeptical that they would ever be seen and initially even thought that they did not exist. This chapter sketches the history and properties of these fascinating waves.

Caution: gravitational waves should not be confused with gravity waves on the surface of water (or other liquids). Although they have the same source (gravity, which is the attraction between masses or objects), they operate on completely different scales of distance, time and speed.

14.1 Origin of Gravitational Waves

The concept of a **gravitational wave** is quite simple: if I move an object repeatedly, another object may feel a varying attraction to it because the distance and direction of the moving object vary repeatedly. This situation is very similar to an EM wave (see Section 12.1): a charge that oscillates (as does an electron in a radio transmitter) causes an EM wave that radiates away and can be felt by another charge as a repeatedly varying force (as does an electron in a radio receiver). **It is now known that gravitational waves and EM waves both travel at the speed of light and can travel through vacuum.**

One gigantic difference, however, between gravity and electromagnetism, and therefore also between gravitational and EM waves, is their strength: **gravity is about 10^{36} times weaker than electromagnetism**; 10^{36} is 1 followed by 36 zeroes,[1] an almost inconceivable number!

Another way to express the relative weakness of gravity and gravitational waves is as follows:[2] The rotation of the **Earth** around the Sun emits gravitational waves that carry away energy. This energy, however, only suffices to power an incandescent light bulb of 200 watts, with which we can only illuminate one large room, and not even heat it up to a comfortable temperature; it is also only a fraction of the power of typical household microwave ovens (600 to 2,000 watts).

If gravity is so incredibly weak compared to electromagnetism, then you may well ask: why do we see the extremely massive Earth rotate around the much more massive Sun, and the very massive Moon rotate around the Earth, all thanks to that weak gravity? Why don't we see vastly stronger EM effects if they are that much stronger than gravity? The only sizable EM effect that we notice is the Earth's weak magnetic field, and we only see it by using a delicately balanced magnet. This seems totally paradoxical!

The explanation is actually quite simple: every electrical charge in the known universe attracts an oppositely charged particle that almost totally cancels its electrical and magnetic effects; for example, atoms contain the same number of protons and electrons, thus becoming electrically neutral; also, the Sun and neon tubes contain almost equal numbers of positive and negative charges flying around at great speed (they form what is called a **plasma** in physics), so they are also practically

---

[1] In other words, 10^{36} is 1 trillion trillion trillion (36 zeroes = 12 + 12 + 12 zeroes).

[2] See https://en.wikipedia.org/wiki/Gravitational_wave.

neutral. Therefore, to each EM effect due to a charge, there will be an almost identical but opposite effect due to an opposite charge nearby: these therefore nearly cancel each other.

Gravity, by contrast, does not have this "neutralizing" feature: there is no negative mass that could cancel out the effect of positive mass. There does exist a small amount of **antimatter** in the universe, but each particle of antimatter has the same mass as a particle of normal **matter**, so its gravitational effect does not cancel out that of normal matter but adds to it. (Antimatter has the opposite electric charge as normal matter, so it almost cancels out the EM effects of the same small amount of normal matter, rather than canceling the gravitational effects.) As a result, in the universe overall, gravity can build up, while electromagnetism largely cancels itself out. But it takes huge amounts of matter (like that in planets and stars) to build up gravity to a noticeable level: our bodies are therefore attracted to the Earth rather than to another human body or to a building.

How much matter is needed so that its gravity can compete with electromagnetism? Remember that gravity is about 10^{36} times weaker than electromagnetism (and that 10^{36} is 1 followed by 36 zeroes). We can use the proton as a convenient measure because it carries a single electric charge. The proton also is a major part of all substances. So, for the gravity of matter to compete with one proton's electric charge, we need matter weighing at least as many as 10^{36} protons.

The matter in a human body weighs about as much as 10^{29} protons, far below 10^{36} protons: gravity is far too weak to attract human bodies to each other; we would need roughly 10^7 or 10 million times more mass! We find such mass in the largest buildings ever constructed by humans, such as the great pyramids or huge dams, weighing roughly as much as 10^{37} protons. The mass of a cubic mountain of 1 kilometer in length, width and height is about 10^{40} protons. The mass of the whole Earth is around 10^{52} protons: we feel the Earth's gravity every time we lift something. The Sun has a mass 100,000 times larger than the Earth, equivalent to about 10^{57} protons.

Now, do we feel attracted gravitationally by huge buildings, dams and mountains? We don't, simply because the entire Earth beneath our feet vastly outweighs all of these. But if we were near such a structure in outer space, far from planets and stars, we would feel such attraction.

The Sun vastly outweighs the Earth, but we don't notice the Sun's gravity because the Sun is so much farther than the Earth under our feet (nevertheless, the Sun causes gravitational tides in the earth's oceans, as does the Moon, as we discussed in Section 11.6).

As a result of the relative weakness of gravity, **it is vastly more difficult to detect gravitational waves than EM waves**. Furthermore, gravitational waves interact only very weakly with matter: indeed, gravitational waves easily travel straight through the Earth despite its huge mass of rocks, soil, magma, *etc*. In fact, detecting a gravitational wave is not like detecting gravity: we easily see the effect of gravity when we fall down or drop objects or watch waves on the sea or the tides in the oceans; this "everyday" gravity, however, only changes very slowly (for example, the tidal attraction of the Moon rotates only once a day around the Earth). By contrast, a gravitational wave can produce a relatively faster oscillation of the strength of the gravitational attraction on an object. But this kind of oscillation can only be created by rapidly shaking a large mass (like a planet or star). How can we shake such a huge object? Only another similarly huge object can do so, as we shall see below with pairs of black holes or neutron stars that orbit very fast around each other.

Before we go further with this story, I recommend that you watch an excellent pair of videos[3] that describe that challenge, as well as the human approach and reaction. The challenging search lasted 60 years: notice the reaction "I couldn't believe it" by the scientist involved in the discovery of the gravitational waves.

14.2 History of Gravitational Waves

A remarkable aspect of gravitational waves is that **Albert Einstein** predicted them as part of his theory of **general relativity** (in 1915), but he very much hesitated to believe their existence throughout the

[3] See two videos: "The Absurdity of Detecting Gravitational Waves", by Veritasium, at https://www.youtube.com/watch?v=iphcyNWFD10, before the observation of the first gravitational wave, and then "How Scientists Reacted to Gravitational Wave Detection", also by Veritasium, https://www.youtube.com/watch?v=ViMnGgn87dg, afterwards.

remaining four decades of his life; afterwards, it took six additional decades for science to discover these waves (in 2015).

Very readable accounts of the history of gravitational waves are available online.[4,5] Here I only give a summary.

Various theories of possible gravitational waves were proposed around 1900. One necessary ingredient was a limited speed, similar to the limited speed of EM waves: the speed of light. An infinite speed causes an infinitely long wavelength, which means no wavy repetition over distance (this is like special tides that would give the same tide height everywhere around the Earth at all times). Newton's classical theory implies an infinite speed, so this theory was not sufficient.

Einstein's theory of general relativity included the limited wave speed that was assumed in his earlier theory of **special relativity** (1905). His more general theory implied the possibility of gravitational waves, traveling at the speed of light. However, Einstein's main challenge was to convince the world of his more general theory: there were many skeptics, including among physicists, so he devoted most of his efforts to defending his general theory. Gravitational waves were then not a main focus of interest, especially as they could not be measured anyway in those days.

At the same time, Einstein himself was a skeptic of his own theory's prediction of the existence of gravitational waves; in fact, in 1906 he wrote: "Thus there are no gravitational waves analogous to light waves."[6] By 1937 (over 30 years later), he had changed his mind: he wrote in a paper that a "Rigorous solution for gravitational cylindrical waves is

---

[4] "The Secret History of Gravitational Waves" by Tony Rothman, *American Scientist*, volume 106, number 2, page 96, March-April 2018, freely available at http://dx.doi.org/10.1511/2018.106.2.96.

[5] "A Brief History of Gravitational Waves" by Jorge L. Cervantes-Cota, Salvador Galindo-Uribarri and George F. Smoot, *Universe*, volume *2,* issue 3, page 22, 2016, freely available at https://doi.org/10.3390/universe2030022.

[6] *"The Collected Papers of Albert Einstein"*, volume 8: "The Berlin Years: Correspondence, 1914–1918 (English Translation Supplement)", (translated by Ann M. Hentschel), page 196, document 194. Available online: https://einsteinpapers.press.princeton.edu/VOL8-trans/224.

provided" and "there are strict solutions".[7] The reason for Einstein's reversal was the complexity of the equations of general relativity: they can only be solved in simple situations, so they are often solved after simplifications that are risky, as Einstein did initially.

An interval of 30 years without significant progress in any field of modern science is highly unusual: it indicates widespread doubts, together with the impossibility of doing experiments to prove the existence of gravitational waves. However, a new experiment can revive a field.

So what can be measured to prove the existence of gravitational waves? **The effect of a gravitational wave is to stretch and compress space itself, together with anything that resides in that space.** Imagine an image drawn on a sheet of rubber: stretching the sheet will also stretch the image. Seeing that happening would thus prove gravitational waves.

In the case of gravitational waves, Figure 14-1 shows what to expect. We use the example of a flexible ring: it can also represent the whole Earth, or a smaller circle drawn on the surface of the Earth; you may also think of a circle drawn on a rubber sheet. A gravitational wave passes through the ring like a finger pushed through a ring. The shape of the ring then oscillates, as illustrated in Figure 14-1, by forming an ellipse (of constant area), elongated one way (such as up-down), and

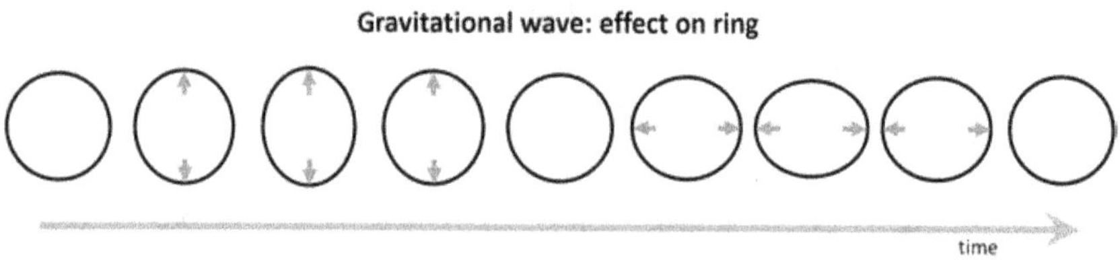

Figure 14-1: Oscillation of a flexible ring in a gravitational wave passing perpendicular to the plane of the figure, showing one period of time from left to right. The oscillation is greatly exaggerated here: in practice, it would be at most about 1 part in 10^{21} (1 followed by 21 zeroes), which is totally invisible to the naked eye or any microscope. In actual experiments, this oscillation is roughly 10^{-18} meter, a thousand times smaller than a small atomic nucleus (such as a single proton).

[7] "On gravitational waves", by A. Einstein and N. Rosen, *Journal of the Franklin Society (Philadelphia)*, volume 223, page 43, 1937.

then the other way (left-right), repeatedly. An analogy would be a horizontal rubber tube that you squeeze horizontally, then vertically, again horizontally, then vertically, and so on. It may be helpful to view this behavior in my Animation 14*1.

ANIMATION 14*1 — See my video WB1 at time 10:11 in its section "**Gravitational waves**" under the title "**Gravitational waves: they travel everywhere, including through Earth**". (See details in the section References and Resources below.)

Gravitational waves are clearly different in their structure from any other waves we have seen so far. Nonetheless, we can imagine a wave of this shape traveling along a rubber tube, empty or filled, so it is not far-fetched. What is very different is the extremely small amplitude of gravitational waves, due to the weakness of gravity: in practice, the stretching would be a thousand times smaller than a small atomic nucleus.

An important question was where to find strong enough sources of gravitational waves. Many physicists during the 20th century, including Einstein for a long time, thought that there was no source in the universe that was powerful enough for detection on Earth. Because of the weakness of gravity, nothing on Earth is large enough and oscillates fast enough to produce detectable gravitational waves. The strongest waves would come from a pair of extremely heavy objects rotating fast around each other: we'll see further below how fast they have to be! As we have seen, even the rotation of the Earth around the Sun produces uselessly weak gravitational waves. So we must look farther into outer space. Good candidates are pairs of neutron stars or black holes[8]

---

[8] Neutron stars are primarily composed of neutrons, compressed together as densely as in atomic nuclei. They therefore are extremely heavy and remarkably small for a star: compared to our Sun, a neutron star may have the same total mass, but a radius of only 10 kilometers (the size of a large city or one 70,000th of the Sun's radius). Two neutron stars may orbit around each other (like a pair of dancers, or the Sun and Earth) at a distance of several Sun diameters, or a few million kilometers, which is only 1% of the Sun-Earth distance. Black holes are stars that have become so dense that light can no longer escape from them. They may have "swallowed" the mass of tens to billions of Sun-sized stars; their size ranges from tens of kilometers to many billions of kilometers (up to about 500 times the Sun-Earth distance).

rotating at close range around each other, because of their amazingly large masses and speeds as they rotate around each other.

Two objects which emit waves and rotate around each other create waves with an expanding spiral shape, as can be seen in the beginning of the video "The Absurdity of Detecting Gravitational Waves" in a footnote of Section 14.1. This spiral shape is frequently shown in the media, but it is not unique to gravitational waves. Indeed, two boats speeding repeatedly around each other also create such an outgoing spiral wave. You can also make such a wave with a food mixer rotating slowly in water or in a viscous liquid. You can in principle create this wave yourself in a basin or bathtub by dragging two objects through the water around each other, but this is easier said than done since our hands can't cross each other: instead, dragging a single object, like your finger, in a circle will give you a similar expanding spiral wave.

In 1957, interest in gravitational waves was revived by an unlikely incentive: funding by the millionaire Roger W. Babson, who thought he owed his fortune to Newton's laws (for instance, he applied the rule "what goes up must come down" to investing on the stock market). After controversial beginnings, Babson funded a conference in 1957 on the subject of "The Role of Gravitation in Physics", which attracted a new generation of physicists and has been repeated many times since, spawning new ideas and approaches.

For the next 25 years, various detectors were built to try to measure the passage of gravitational waves from outer space, but with no success. In 1974, indirect evidence of gravitational waves was obtained due to a pair of neutron stars. Although this work garnered a Nobel Prize, it was not seen as detection of gravitational waves; that breakthrough was still 41 years away. The difficulty was both technical and financial. Detecting the ultraweak waves required ultrasensitive equipment, much of which needed to be invented and built. This led to decades of discussions among scientists and government agencies before sufficient funding could be obtained.

14.3 Detection of Gravitational Waves

Following the idea illustrated in Figure 14-1, scientists aimed to measure changes in the size of objects on the Earth's surface during the passage of gravitational waves coming from outer space. A very precise method of measuring lengths uses laser light: it is now commonly used in measuring distances along roads, on building sites, around plots of land, *etc.*; it is also used to measure the distance between Earth and Moon. The principle is to use wave interference: the wave reflected from a distant mirror interferes with the outgoing laser light, allowing the distance to be measured with great precision. This method is called **interferometry**, meaning measurement by interference.

The principle of interferometry for detecting gravitational waves is sketched at the top of Figure 14-2. A light wave (with a single wavelength) from a laser (at the bottom) is split by a half-transparent mirror into two beams that travel perpendicular to each other. These beams are reflected back by two distant mirrors, and then recombined so they can interfere at a detector (at right). Normally, the distances to the two mirrors are set so that the interference is destructive. Now, if one of the two distances changes by even a fraction of a wavelength compared to the other, the interference is no longer destructive: some laser light will become visible.[9] Note that the laser light waves only serve to measure the stretching effect of the gravitational waves; in particular, the wavelength of the gravitational waves can be almost anything, such as 750 kilometers for a gravitational wave frequency of 100 hertz (cycles per second).

The challenge is to detect changes in distance much smaller than the size of a proton: that requires much invention, time and funding. It helps greatly to use large distances to the mirrors: at the VIRGO detector, shown at the bottom in Figure 14-2, this length is 3 kilometers in both directions; actually, this distance is multiplied by letting the light beam bounce back and forth multiple times. One big challenge is vibrations of all sorts, due to pumps, electric currents, wind, traffic, lightning, earthquakes, sea waves, temperature variations, *etc.* To avoid

---

[9] One intriguing question is the following: if a gravitational wave stretches the distance to a mirror, it also stretches the wavelength of the laser light, so how can the light notice the <u>length</u> difference? It is like measuring the length of a stretched rubber sheet with a stretched ruler: that will still show the same apparent length as before the stretching. The answer is: what is actually measured is the <u>time</u> difference; since the speed of light in vacuum is fixed, the flight time over the changed distance will record that change.

Figure 14-2: *Top*: Principle of the gravitational wave detectors LIGO and VIRGO. *Bottom*: Aerial view of VIRGO near Pisa in Italy, showing its two interfering light beam tunnels pointing to the upper left and upper right. (*Sources*: Top: T. Pyle, Caltech/MIT/LIGO Lab, under CC BY-SA 4.0, https://commons.wikimedia.org/wiki/File:Ligo-interferometer-%28destructive-interference%29.png. Bottom: The Virgo collaboration, in the public domain, https://commons.wikimedia.org/wiki/File:2015_11_06_VirgoAerialView_2.jpg.)

most of these problems, a second detector can be set up thousands of kilometers away (where the same gravitational wave can also be detected), as was done for the LIGO detector (see below): the undesired vibrations will be mostly different there and can then be removed from the measured signals. Furthermore, the light beams must travel through a near-vacuum: therefore, they are made to shine through long pipes with very good vacuum.

In 1988, the US government approved the construction of LIGO, short for Laser Interferometer Gravitational-wave Observatory. It was built at two distant sites, one in the northwest and one in the south of the USA. Construction started in 1994, with the first scientific observations commencing in 2002. By 2015, LIGO had cost a total of about $620 million to build and operate.

In parallel, in 1993 and 1994, respectively, the French and Italian governments agreed to fund the VIRGO project (named after the Virgo constellation of stars, which is a prime target as a source of gravitational waves). The VIRGO laboratory was built in a plain near the historical city of Pisa[10] in Italy, starting in 1997, and scientific observations began in 2007. Also, collaboration with LIGO started to allow beneficial close coordination between their three detectors. By 2017, more than 1,200 scientists and some 100 institutions from around the world participated in the LIGO and VIRGO efforts through the LIGO Scientific Collaboration; most of these scientists are employed in various universities or other scientific institutions.

Finally, in 2015, LIGO's two detectors detected a gravitational wave. Less than one hour after letting LIGO "listen" for gravitational waves, it "heard" one; it had taken over a century to make that measurement possible.

Interestingly, the LIGO scientists only announced their discovery five months later, in a press conference in 2016. During this period of five months, they checked that the detected signal was not due to any fault in the equipment or analysis procedures, or due to local vibrations, earthquakes, electromagnetic fluctuations, *etc.* There is great temptation to announce such "earth-shaking" results as soon as

---

[10] It was in Pisa in the 16th century that Galileo Galilei famously dropped cannonballs from its leaning tower, thereby already investigating gravity.

possible (including for fame, competition, and funder's satisfaction), especially after decades of intense work; however, the price can be high if the announced discovery turns out later to be an error of any kind, as happens occasionally.

What LIGO observed was the merger of two black holes: the two black holes of about 30 solar masses each had been rotating around each other, gradually getting closer together and speeding up as they emitted gravitational waves and energy, until they "melted" into each other to form a single larger black hole. Just before the merger, they were rotating about 100 times per second around each other (try to picture two objects the size of cities circling each other 100 times in a second!). The merger itself took only a few thousandths of a second. It created a gravitational wave that travelled for more than a billion years before reaching Earth, where it hit the two American detectors about 7 milliseconds apart (due to the distance between them). Calculations suggest that the black hole merger itself released energy equivalent to three Suns being totally converted to energy (such as light), as if three Suns blew up completely at once (our Sun contains enough energy to keep shining as it does now for almost 8 billion years!).

Since the initial discovery in 2015, many more observations of gravitational waves have been made by LIGO and VIRGO (but they were interrupted by upgrades and fixes, as well as the Covid-19 pandemic from 2019: new observations have started in 2023.[11]) The observations so far are mainly due to the merger of either two black holes or two neutron stars.

14.4 Future of Gravitational Waves

Gravitational waves have opened a new "window" on the universe. So far, our main window has been electromagnetic radiation: first visible light through optical telescopes; then radio waves through radio telescopes; and, more recently, microwaves, gamma rays, x-rays, ultraviolet and infrared light. So-called cosmic rays have also been used:

[11] See an updated list at https://en.wikipedia.org/wiki/List_of_gravitational_wave_observations. This site includes graphs of observed gravitational waves, entitled "Gravitational Wave Transient Catalog 1".

these are various high-energy particles (especially protons and other atomic nuclei) that reach Earth from the Sun or other stars.

New observatories of gravitational waves are planned in other countries. One European plan, called Laser Interferometer Space Antenna or LISA, is to launch three spacecraft that will form a triangle with three sides of 2.5 million kilometers each (instead of the two sides of 3 kilometers of VIRGO and 4 kilometers of LIGO). These sides will be about 6.5 times longer than the Earth-Moon distance. The triangle of spacecraft, to be launched in 2030, will fly along the Earth's orbit around the Sun, but will trail behind the Earth by 20 degrees (1/18th of the orbit circumference).

14.5 What have We Learned in this Chapter?

This chapter has described the most recently observed type of waves: gravitational waves. Their source is simple: gravitational attraction. But their detection is extremely difficult, due to their weakness. This weakness explains why it has taken a century to detect them for the first time in 2015, since the prediction of their existence. Gravitational waves provide a new "window" on the universe, complementing the main "window" of electromagnetic waves.

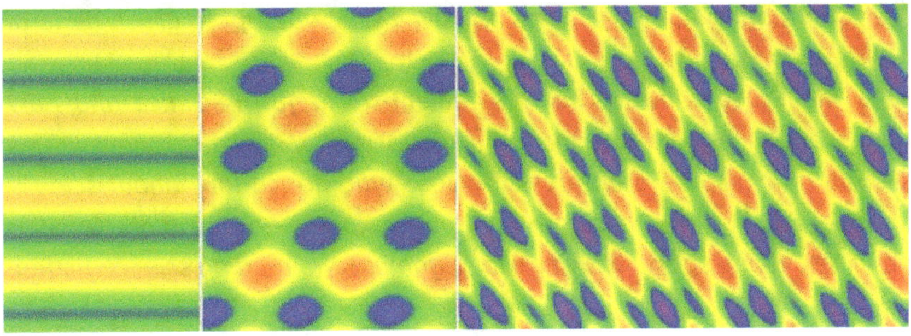

15

Other Waves and Afterthoughts

We have seen a wide variety of waves in this book. But there are still many more, so we will briefly introduce other kinds of waves in this chapter! We will also look at the human side of science, to give a feeling for what it is like to do science as a human being.

The waves which we have discussed so far have very different mechanisms, but they also have very strong similarities, mainly their universal "wavy behavior". In this chapter, we will emphasize such differences and similarities. This will also help us characterize more precisely what is meant by "wavy behavior".

We will consider types of waves which can be called "non-standard" because they stretch the traditional concept of waves. For example, brain waves, traffic waves, human waves and hair waves increasingly deviate from what scientists normally call waves.

Based on the many types of waves that we have covered, we will then summarize general aspects of waves in an attempt to give a general definition of waves.

Finally, we will consider how scientists perform science and how they view such strange concepts as quantum mechanical waves. This section may also be of interest to those who are curious about science as a career. We will therefore compare doing science to more familiar activities: doing sports, playing video games, and reading science fiction.

15.1 String Theory: A Theory of Everything

The so-called **string theory** has been proposed to combine gravity and quantum physics, among other aspects of physics. String theory may become the unified **theory of everything**, or **quantum gravity**, which would bring together all of physics into a single description of all fundamental forces and types of matter. Since this theory is not yet complete, we will not treat it in detail here, but only point out its interesting wave character. Note that string theory is not the same as the description of waves on musical strings, for example, even though string theory borrows important ideas from musical strings.

As its name suggests, string theory assumes that all point-like particles (such as electrons, quarks and photons) act like tiny pieces of string: thus, the internal structure of each of these particles would be a string. Such strings can be thought of as short segments of string that may have two free ends (like a thread of limited length) or may form a closed loop (like a necklace). The size of these strings is exceedingly tiny, around 10^{-35} meter (10^{-35} is 0.00...001 with 34 zeroes after the decimal point).

A string could vibrate in ways that we have discussed for 1D strings, especially in Section 1.3 (for the vibration of bridges), in Chapter 2 (for strings in musical instruments) and in Section 13.5 (as a simple model for electrons in atoms, shown in Figure 13-3). Different particles would be distinguished by the types of vibrations of the string: for instance, whether the string would vibrate with transverse or longitudinal waves, and with how many loops.

15.2 Neural Waves and Brain Waves

Our **nerves** (also called **neurons**) carry signals (information) in the form of waves, generally called **nerve impulses** or **action potentials**. These signals, which are so important in our daily lives, fall into three main categories:

- sensory signals that go from our receptors (detectors for seeing, hearing, smelling, tasting, touching and feeling heat) to our spinal cord and brain
- signals within our brain and spinal cord (for memorizing, thinking and deciding)
- and motor signals from our brain and spinal cord to our muscles (for activating motion).

It is amazing to realize that our brain contains almost 100 billion neurons (individual nerves), each of which connects to 1,000 to 10,000 other neurons, thus forming roughly 10^{15} or a million billion interconnections! The human population on Earth is "only" 8 billion!

How does a nerve signal start and travel? A nerve signal (nerve impulse) starts somewhat like lightning, but on a far smaller and weaker scale: when a sufficient difference of electrical charge happens near a nerve, a small discharge (transfer of charge) activates a wave in this nerve, which carries it along. Such a wave typically has a voltage of a few tens of millivolts, well below the voltage of common batteries (1 to 2 volts) and far below the voltage of wall sockets for household use (110 to 240 volts).

The action of a nerve depends very much on the strength of the stimulation of the nerve. Typically, a weak stimulation may cause one or two spike-shaped impulses in the nerve, each of which may last a millisecond or so before it moves down the nerve. Above a minimum strength, the stimulation causes a repetitive sequence of identical spikes, like a machine gun: the repetition rate, also called frequency or firing rate, increases gradually with the strength of the stimulation. This frequency typically grows from about 50 hertz (repetitions or impulses per second) to over 100 hertz. Frequencies in the range of 50 to

300 hertz are common; they rarely go up to 1,000 hertz. (Some nerves start firing more slowly already for very weak stimulations: there is no minimum stimulation for these nerves.)

A frequency of 50 hertz corresponds to firing every 20 milliseconds, but neural firing is generally not very regular. So, a typical trace of a nerve impulse passing a point along a nerve shows spikes lasting about 1 millisecond and separated by roughly 20 milliseconds.

In humans, nerves can be as long as our body. Single nerves are basically tubes with a diameter of 0.5 to 15 micrometers, too thin to see with the naked eye. Nerves are often bundled together in fibers, some of which are thick enough to be visible.

Despite the importance of nerve signals, there are differing opinions about the mechanism of their motion through nerves. In 1952, **Alan L. Hodgkin** and **Andrew F. Huxley** proposed a very successful model for the motion of nerve signals. This model is electric in nature, as follows.

In the **Hodgkin-Huxley model**, the wall of a single nerve allows ions (positively charged atoms) of sodium and potassium to enter or exit the tube through its wall. In the normal rest state (without nerve impulse), there are more of these ions outside the tube than inside, leaving a positive charge outside and a negative charge inside. A signal arriving at a particular point along the tube reverses this situation: more of these ions go inside, so positive charges are inside and negative charges are outside. This reversal of charge travels along the nerve. After the signal passes, the tube rapidly returns to its rest situation within about 1 millisecond.

It was later found that this Hodgkin-Huxley mechanism allows signals to only travel at about 1 to 2 meters per second. That would be dangerous for us: imagine that you felt pain in your hand with a delay of one second and needed another second to make your hand start to pull back; or imagine playing a ping pong game with 2-second reactions. Fortunately, in reality, nerve signals can travel some 10 times faster. The reason for this faster speed is that many nerves are surrounded by a **sheath** (called **myelin**) that not only protects the nerves but also provides a "short-circuit" in the form of a very thin electrical conductor that lets electricity flow much faster along the nerves, according to recent studies.[1] The Hodgkin-Huxley model can be extended to include this speed-up.[2]

However, there are several observations that do not fit the Hodgkin-Huxley model, which assumes only electrical effects. In particular, there is evidence of changes in shape, internal pressure, optical properties and temperature of the nerve during the passage of an electrical signal. Such effects are not surprising, since they must influence each other to some degree. One could therefore take an opposite view and claim that the elastic shape is the driving mechanism, as in a sound wave through an elastic substance, and that the electrical effects are a consequence of the elastic wave.

Reality more likely lies in between: all these effects contribute. This means that the nerve impulse is an interconnected combination of these different effects: elastic changes in shape, pressure, electrical charges, optical properties and temperature are all intertwined. Of course, including all these aspects at the same time produces a much more complex model.[3]

One intriguing aspect of nerve impulses concerns the basic mechanism of the motion of waves. With all the waves that we have discussed earlier in this book, you could send two waves across each other and find that they don't change each other (except if their amplitudes are very high, as in strong waves on the surface of water, in which case they "break"): waves are simply superposed on top of

[1] This mechanism is called "saltatory conduction". See the freely available article "Saltatory Conduction along Myelinated Axons Involves a Periaxonal Nanocircuit", by Charles C.H. Cohen, Marko A. Popovic, Jan Klooster, Marie-Theres Weil, Wiebke Möbius, Klaus-Armin Nave, and Maarten H.P. Kole, *Cell*, volume 180, page 311, January 23, 2020, https://doi.org/10.1016/j.cell.2019.11.039 .

[2] See two videos by Dr. John Campbell: "Nervous system 1, Motor neuron", at https://www.youtube.com/watch?v=SC2QFEUTQsg; and "Nervous system 2, Sensory neuron", at https://www.youtube.com/watch?v=wIzXajzAPUY.

[3] See, for example, the freely available article "On the complexity of signal propagation in nerve fibres", by Jüri Engelbrecht, Tanel Peets, Kert Tamm, Martin Laasmaa, and Marko Vendelin, *Proceedings of the Estonian Academy of Sciences*, volume 67, number 1, page 28, 2018, https://doi.org/10.3176/proc.2017.4.28.

each other, which is the principle of **superposition** and **interference** of waves. With nerve impulses, however, the electrical mechanism of the Hodgkin-Huxley model does not allow two waves to cross each other (the charge reversal process of one wave prevents the other wave from passing through). On the other hand, the model based on the elastic shape does allow two waves to cross each other (as with sound in air or waves on a string). As pointed out in the article by Engelbrecht and co-workers referenced above, measurements show that nerve impulses moving in opposite directions in fact do cross each other: this supports their argument that a nerve impulse is at least partly an elastic wave similar to sound.

We can ask: ***Does more intense pain cause stronger nerve signals?*** We might think so because we do feel more intense pain more strongly. Actually, each nerve can only carry the same strength of signal each time. Therefore, nature has found a different solution, as mentioned above: **the frequency of the nerve signal tells the strength of the pain (or other sensation).**

How are nerve impulses measured? The mechanism of nerve impulses is studied scientifically on nerves that have been removed from a body, whether human or animal. Sharp electrodes (metallic points) touch different parts of the nerve to measure the local electrical potential and current, thereby evaluating the strength, profile and speed of signals. **Electromyography** and **microneurography** do this on live subjects by inserting very fine electrode needles directly into nerves.

Nerve impulses are also measured medically on living beings in a non-invasive diagnostic procedure called **nerve conduction study**: flat electrodes are then pasted on the skin close to the nerve of interest, so they can measure the electric fields created near the nerve by impulses. This approach is in particular used to study brain waves, since it is much more difficult (and dangerous) to insert needles into a brain than elsewhere in the body.

An example of **brain waves** is shown in Figure 15-1: they are measured at ten different positions on the human scalp, giving the ten curves. This technique is called **electroencephalography** or **EEG** (whose name uses the Greek word for brain: ἐγκέφαλος or enképhalos, meaning "in head").

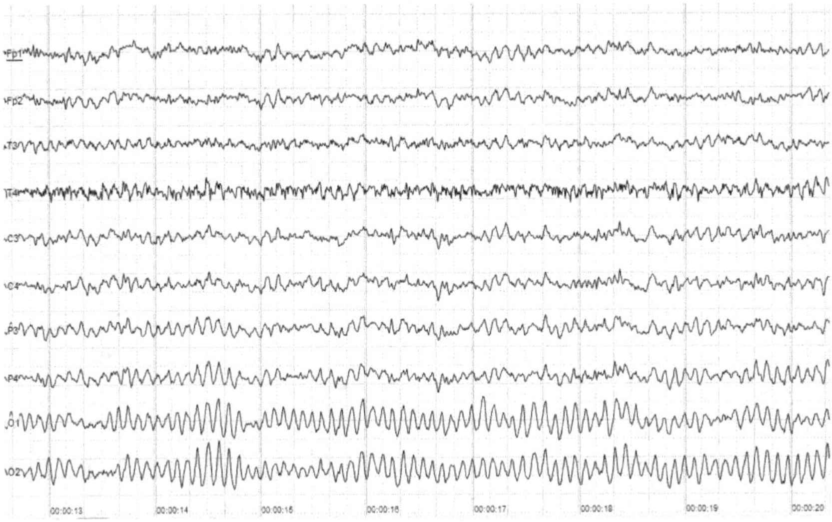

Figure 15-1: Human brain waves recorded as EEG (electroencephalogram) traces at different positions on the scalp; this brain exhibits mainly alpha waves. The horizontal scale gives the time in seconds; the vertical scale shows wave amplitudes. (*Source*: extracted from Human EEG with prominent alpha rhythm.png, by Andrii Cherninskyi, under CC BY-SA 4.0, https://commons.wikimedia.org/wiki/File:Human_EEG_with_prominent_alpha-rhythm.png.)

We notice in Figure 15-1 a few features that differ from the spiked nerve impulses mentioned earlier. The brain waves are not composed of separate, identical spikes of a few milliseconds, but of continuous oscillations that are often irregular. The repetition rate is around 10 per second (~10 hertz) in the lower curves, corresponding to a repetition time of about 100 milliseconds. In the upper curves, we see mostly irregular structures, about 10 milliseconds in duration; they are partly dominated by more regular oscillations of about 10 hertz. We can understand these features as follows.

When one measures brain waves from outside the brain, it is not practical to measure them for a single nerve (neuron). Indeed, the measurement is a sum over the separate activities of many nerves, because of the distance between the brain surface and the electrode outside the skin, which is typically about a centimeter, and because of the similar size of the electrode. This implies that an EEG trace represents the collective electrical activity of many nerves taken together, instead of separate nerve impulses. It's like looking at the long waves on a wheat field due to wind, rather than at the separate oscillations of an individual stalk of wheat.

As a result, the brain waves tend to show the overall degree of brain activity. For example, in Figure 15-1, the dominance of the frequency of 10 hertz indicates that the brain is in a state of relaxed wakefulness, corresponding to relaxing, reflecting, or contemplating. This state was the first to be studied systematically using EEG, and therefore was given the name **alpha waves**, using the first letter of the Greek alphabet. In general, alpha waves lie in the range of 8 to 13 hertz.

Higher frequencies indicate higher brain activity: **beta waves** have frequencies in the range of 13 to 30 hertz; they correspond to active thinking, concentrating, discussing, being alert or anxious.

Lower frequencies correspond to calmer brain activity. **Delta waves** (below 4 hertz) are found in deep dreamless sleep and often in babies, for example: Figure 15-2 shows delta brain waves in an anesthetized mouse. Theta waves (from 4 to 7 hertz) occur during drowsiness, idling, meditation, and repetitive simple actions, and favor fresh ideas.

Typically, every 90 minutes during the sleep of humans, brain waves will speed up from the delta range of deep sleep into the more

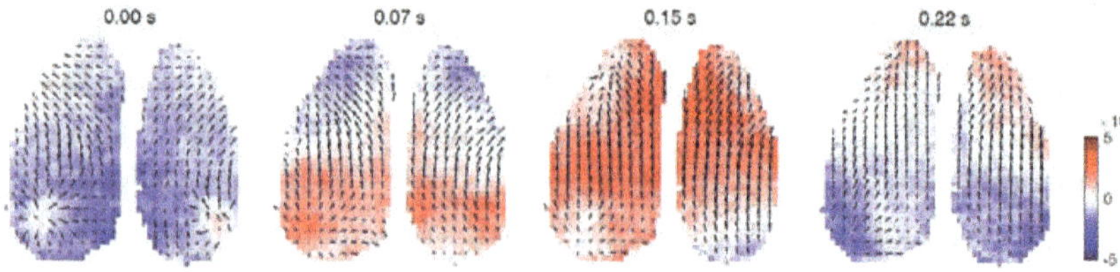

Figure 15-2: Brain waves in a mouse under anesthesia at four different times over a 0.22-second span (front is at the top, back at the bottom, seen from above; the size of each brain hemisphere is about 1 × 0.5 cm). The arrows show the wave direction and speed along the upper surface of the brain; the background colors show the local voltage, using the scale at right between 5 and –5 millivolts. An online animation spanning 3 seconds is available at https://www.jneurosci.org/content/41/16/3665/media-1. (*Source*: "Cortex-Wide Dynamics of Intrinsic Electrical Activities: Propagating Waves and Their Interactions", Yuqi Liang, Chenchen Song, Mianxin Liu, Pulin Gong, Changsong Zhou and Thomas Knöpfel, *Journal of Neuroscience*, volume 41, page 3665, 21 April 2021; freely available under CC BY-SA 4.0 at https://doi.org/10.1523/JNEUROSCI.0623-20.2021.)

active **theta** range: we then dream. Then also, **rapid eye movements (REM)** take place.

After waking up from deep sleep, we may stay in the theta state for many minutes and develop new ideas: this is a creative period for many people!

Low frequencies, however, can also indicate medical problems: in fact, EEG is most frequently used to examine dysfunctions such as brain damage, seizures, epilepsy and tumors.

It is possible to use EEG to control a robot: for example, EEG waves coming from a person observing a scene can direct the motion of a robot.[4]

In 2018, EEG was used to enable two persons to direct some actions of a distant third person, by means of BrainNet, described as "the first multi-person non-invasive direct brain-to-brain interface for collaborative problem solving"[5]. The success rate was about 81%.

In recent years, **artificial intelligence (AI)** and **neural networks** have caught everyone's attention. These are based on mimicking the neural networks of natural brains in computers. As we often find, nature has already tried out and exploited many mechanisms that are remarkably effective: the brain is an outstanding example.

So how does the brain work? We can only sketch a few aspects of the brain here. One essential component of the brain is its networking of different nerves (neurons): each nerve is connected to many others through **synapses**. The main function of a synapse is to transfer a signal from one nerve to another.

An important feature is that multiple signals reaching one nerve from other nerves are combined by the receiving nerve in a way that is not obvious at first sight. Compare this situation with a collection of streams of water joining to form a river: the river will simply receive the totality of all the water from the streams, making the river larger than the individual streams.

---

[4] See "Brain–computer interface", at https://en.wikipedia.org/wiki/Brain-computer_interface.

[5] See freely available article "BrainNet: A Multi-Person Brain-to-Brain Interface for Direct Collaboration Between Brains", by Linxing Jiang, Andrea Stocco, Darby M. Losey, Justin A. Abernethy, Chantel S. Prat, and Rajesh P. N. Rao, *Science Reports*, volume 9, article 6115, 2019, https://doi.org/10.1038/s41598-019-41895-7.

With nerves, this model does not work because different nerves carry impulses of about the same size: remember that stronger signals are sent as faster repetitions of the same impulse, not as larger impulses. What then does a nerve do when it receives multiple impulses at the same time from different nerves? Very simply stated, it will create and send its own impulse only if it has received a sufficiently high number of impulses from the other nerves. Thus, a better analogy than streams flowing into a river is the following: an election where voters can only send one person to a higher office. Another mechanism in the brain is the presence of so-called **inhibitory neurons** which prevent the firing of too many other neurons. One advantage of these methods is that they prevent signals from growing too large, like giant rivers that would flood vast areas.

An interesting aspect of this model of nerve function is that it can make a yes/no decision: a nerve will send an impulse only if it received a sufficient number of other impulses. Yes/no **decisions** are at the basis of **logic** and **thinking**. A river makes no decisions: it blindly accepts all the water that comes from the streams and sends it on.

In the brain, nerves connect to other nerves in a giant network that allows signals to travel through many interconnections between nerves. With billions of nerves involved, this creates a very complex series of multiple decisions that form the thought process of the brain. The details of this thought process are being investigated but are not fully understood; for example, it is not clear how these waves of decisions are related to the alpha pattern seen in EEG.

In any case, **we see that a multitude of individual yes/no decisions in nerves build up to an overall conclusion, the result of the thought process.**

Scientists mimic logical nerve behavior in artificial **neural networks**. These neural networks are not artificial nerves connected together, but computer programs (algorithms) that imitate real neural networks: such a program tells the computer to calculate the results of signals reaching each imaginary nerve, and to make decisions based on those results, in order to reach a final conclusion. This process is called **artificial intelligence**.

For a neural network to reach correct conclusions, it requires that its artificial connections between the imaginary nerves behave correctly. This means telling the computer how to make each individual yes/no decision at the level of each connection between imaginary nerves: in other words, we have to teach the computer which individual yes/no decisions are correct, and which are wrong. Teaching that to a computer is a great challenge which is overcome by letting it learn by itself.

To illustrate this idea of self-learning, consider **face recognition** by computer. Self-learning by the computer means showing it many pictures of faces and telling it which are the face you are seeking, for example your own face. Based on your input of yes/no choices, the computer will adjust each individual decision between imagined nerves to improve its recognition capability. To work well, this requires showing the computer a very large number of images: it will slowly learn to react correctly. Only then will the computer be able to recognize you in a new picture, although probably not with a 100% success rate.

In this manner, artificial intelligence is being developed for many situations: digital assistants, autonomous driving, face recognition, voice recognition, language translation, games, robot vacuum cleaners, online shopping, web searches, chatbots, *etc*. We can say that such artificial intelligence is based on waves of logical decisions in a network that mimics the brain and its nerves.

15.3 Waves in Human Queues, in Road Traffic, and More

We have discussed many types of waves in this book, from waves on strings to brain waves, but there are still many more! These other waves can also have very different mechanisms. It is easy to start a long list of such waves, as follows (I include questions to help understand the mechanisms of these waves):

- **Waves in queues** — After the front person in a queue leaves, the next person makes a step forward to fill the vacancy, while making another vacancy; then the following person steps into that vacancy, leaving a vacancy further back, *etc*. Thereby, a vacancy travels to the back of the queue as a wave pulse (in the direction opposite to each person's motion). This process starts again when the new front person leaves the queue. What happens when a person in the queue does not notice the vacancy reaching him or her? What controls the speed of the wave? What happens if there is no time delay between the reactions of successive people in the queue?

- **Waves in dance groups** — When dancers coordinate their movements, waves are frequently formed, for example when they are lined up and imitate their immediate neighbor with a short delay, thereby making a move repeat and travel down the line. What happens when the delay time is zero? What controls the delay time?
- **Waves in stadiums** — Sometimes, spectators stand up and sit down together in a "Mexican wave" that travels along the stands and around the stadium. This wave has a single up-down pulse, and can be repeated to create a wave train of multiple pulses; what happens when a wave train completes a full tour of the stadium? What happens when some people don't participate in the wave? What controls the wave speed?
- **Waves in falling dominoes** — When one domino falls against the next domino in a chain of dominoes, it starts a wave. However, this process will not start again, unless someone raises the dominoes again. Can you think of a way to make dominoes automatically reposition themselves for a new wave? For example, hanging dominoes could do so, but under what conditions will they allow a wave to travel?
- **Waves in road traffic** — When one car slows down on a busy road, the following cars will slow down as well, forming a wave that usually travels backward. This wave can also travel forward, or remain stationary: under which conditions? The slowing down can bring traffic to a stop in a bottleneck that also travels backward: how does such a bottleneck disappear? This also happens at stoplights.
- **Waves in transportation networks** — Weather frequently causes waves of delays in airline and railway networks. How do these waves spread, and at what speed? How are the disruptions resolved? Why do private vehicles suffer less from such disruptions?
- **Waves in schools of fish** and **flocks of birds** — Some fish and birds move in large, dense groups that exhibit spectacular wave-like shapes. Why? Who or what directs them?

- **Stampedes** and **crushes** — Groups of excited or frightened animals can suddenly run together, forming a wave of collective motion. The word "stampede" is also used for humans rushing together, mostly chaotically and sometimes dangerously (a better term for a human stampede is a "crush"). Members of such groups may not know why they are rushing but simply follow the herd. What starts a stampede? How can a stampede be stopped?
- **Waves in explosions** — When you light the surface of oil, it forms a flame that rushes across the surface as a spreading wave (only the surface of oil can burn, since burning needs oxygen from the air). When you light an explosive like TNT, the explosion penetrates the material as a wave (because it already contains the needed oxygen), igniting more material in less time. What controls the speed of these waves?
- **Waves in epidemics** and **ideas** — When a virus spreads, it creates waves of infections through a population. Viral ideas, images and videos spread rapidly through social media. What can prevent the spreading of viruses? What controls their speed?
- **Waves in fashions** — New clothing styles frequently spread in waves. What causes these waves? How do they spread? At what speed?
- **Waves in politics** — Ideas about governance can spread through a population relatively fast as waves. How? How fast?
- **Waves in ethics** — Philosophical and ethical ideas spread slowly as waves. How? Why slowly?
- And so on... Can you think of other kinds of waves? How about wavy **hair**: how are waves in hair different from other waves? How about the daily **weather** with its waves of sunshine and rain? How about the evolution of the **climate** over centuries?

15.4 General Aspects of Waves

Throughout this book, we have discussed a wide variety of waves. In Section 1.1 we noted, based on a few examples of waves, some common characteristics. After learning about many other types of waves, we can now update and extend these characteristics:

- **Waves are a collective motion of many nearly identical physical elements**, including molecules or atoms in gases, liquids, and solids for waves on strings, sound waves, waves on the surface of water, and neural and brain waves; vacuum in electromagnetic (EM), quantum mechanical (QM) and gravitational waves; humans in queues; vehicles in traffic; and lines of falling dominoes.
- **Waves in general can be created (initiated) by the disturbance of any collection of physical elements that are separately movable or changeable but interconnected.**
- **The motion of physical elements in a wave is communicated to their immediately neighboring elements**, and only indirectly to distant elements. The time needed for this neighbor-to-neighbor communication is the main factor in determining the speed of the wave, which also sets the wavelength. This wavelength can change, as the substance that carries the wave changes.
- **The wave period (repetition time), and therefore also the wave frequency, is set by the time profile of the input disturbance that creates the wave; a wave may also evolve into a resonance with a natural resonance frequency.**
- **The frequency of a wave is constant as it travels**, despite possible changes in the substances that carry it. For example, the frequency of sound and light does not change as they move through different materials.
- **Many waves carry energy over long distances and long times**, such as: waves on strings, sound waves, waves on the surface of water, EM, QM and gravitational waves; in these cases, energy supplied at their start flows with the waves and keeps them going. Waves can gradually lose energy as they move, for example due to friction or scattering, such as sound, waves on the surface of water, EM waves in water or glass and weather systems.
- **Some waves do not carry energy** but must be constantly resupplied with energy, in particular: waves in nerves, queues, and traffic.
- **Waves have a generally smooth motion which is often repeated back and forth as a vibration. Sharp waves are also possible**, such as those waves on water which have sharp crests; such

waves can also break into whitecaps consisting of air bubbles and waterdrops. Sharp waves are also created artificially, for example in the form of electronic square waves, triangular waves and sawtooth waves.

- **Solitary waves and other wave pulses do not repeat their back-and-forth motion.**
- **Waves of similar appearance exist in many supporting substances** (on a string, in air, in water, on a drumhead, in vacuum, in populations, *etc.*).
- **Waves mostly have a sine-like snaking shape** that can travel over long distances with little change of shape. **More complex shapes can be formed by the superposition of multiple sine waves.**
- **Waves do not carry their supporting substance** (string, air, water, vacuum, populations, *etc.*) with them, so **waves move through the substance**, and not with the substance.
- **Many waves can be characterized by wave amplitudes, wavelengths, wave periods, and wave speeds.**
- **Most waves can undergo superposition, interference, refraction, diffraction, and dispersion.**

The features listed above could serve as a definition of the concept of wave, at least from a scientific perspective, although this definition is somewhat vague and open-ended. This vagueness leaves the door open for other kinds of waves, including those that may be discovered or invented in the future.

Also, would you include waves in hair as a kind of wave? If not, what is missing in hair waves compared to other kinds of waves?

15.5 Is Doing Science Like Doing Sports, or Playing Video Games, or Reading Science Fiction?

You may wonder what it is like to do science, and you may ask questions such as: ***What motivates scientists? What character is needed to be a scientist? What are the rewards of doing science? What is the feeling of doing science?***

If you have read Chapter 13 on quantum mechanical waves, you may also have wondered: ***How do scientists deal with such unfamiliar***

concepts? Do you need a special kind of brain to understand such concepts? What kind of training do they require?

There are many ways to address these questions, and the answers will vary from one scientist to another. I have included some opinions of other scientists in the following:

One approach is to compare doing science with activities that you are already familiar with, such as doing sports, playing video games, or reading science fiction. Like science, these activities follow certain rules but also allow creativity for you to go beyond what has been done in the past.

For example, a **sport** has clear rules to ensure fair play, while these rules also allow you to push the human body's abilities as far as nature will allow; maybe you can also use new materials or invent new tactics that give you an advantage, within the rules of the sport. In a **video game**, say a battle simulation, you are limited to certain weapons, but you can develop new tactics and strategies to beat an enemy. In **science fiction**, the writer can imagine a world with different physical laws (such as time travel, teleportation or the ability of humans to fly like birds); these supernatural laws let the story explore new opportunities. The reader may go further than the writer's fiction by imagining other laws or scenarios.

Science, in many ways, is just like those familiar activities: we do science with specific rules and specific goals. These rules of science are primarily **objectivity**, **reproducibility** and the respect of **evidence** from nature in the case of the physical sciences (physics, astronomy, chemistry, materials science, and biology). Science also allows endless creativity within those limits, often pushing the limits of current understanding. Likewise, new sports are frequently created, with new rules or other equipment, but they must stay within the limits of physical reality. Video games and science fiction, while having essentially no limits on contents, must remain within the limits of human capabilities such as reaction speed and comprehension (too rapid action, too much complexity and lack of predictability are discouraging).

Let's compare a specific sport with a specific science: **cross-country motorcycle racing** and **quantum physics**. Through years of practice, motorcyclists learn the behavior of a motorcycle in different terrains, at different speeds, in varying weather, *etc*. Learning means understanding why the motorcycle behaves the way it does, how to steer it reliably and fast through all sorts of situations, and developing strategies for finding the best path that avoids obstacles and other dangers. Learning includes developing an intuition that guides the motorcyclist almost automatically, without needing to rethink each situation or reinvent reactions during a race: this results in rapid and efficient reflexes. The motorcyclist thus becomes intimately familiar with the sport and "lives" automatically within its limits.

Physicists working with quantum mechanics do largely the same thing. They practice for years to familiarize themselves with the various rules imposed by nature on the behavior of electrons and other particles in different circumstances, such as in atoms and molecules. They thereby develop an intuitive feeling for how such particles act or react within the limits set by the laws of nature. Thereby, they can "live" in the quantum world, at least mentally.

Motorcyclists may develop short mental movies of specific situations in a race: they can replay those mental movies repeatedly to develop their reflexes. Physicists can do the same by constructing mental images of electron waves (by analogy with visible waves on water, for example) that interact with other electrons and atomic nuclei; replaying these images in the mind develops an intimate familiarity with the behavior of such waves. This mental imaging becomes very useful when facing new situations, and respects the limits on quantum behavior learned from nature.

Thus, motorcyclists learn to live the various stages of racing in their minds: they can mentally simulate what may happen and prepare themselves before actually racing; they can embed themselves mentally in a virtual race. Physicists similarly learn to mentally embed themselves and live in the world of electrons and other particles: they know how nature will act or react in many situations, so they develop the necessary reflexes to predict more quickly and more accurately what may happen in new circumstances. A scientist may even say: "If I were an electron, what would I do in this new situation?"

We may also compare doing physics with playing a **video game**. Here as with sports, long practice develops reflexes that allow the gamers to react to all manner of circumstances confidently, precisely and rapidly, for example in a virtual battle. Gamers can thereby live in the artificial world of a battle, as if it were real. The gamers' reflexes must necessarily obey the laws and limits of the game: if you have only ten bullets to defend yourself, you will adjust your strategy. Scientists do the same with the tools available to them.

An intriguing question is: ***What is the relation between science and science fiction?*** It is helpful to distinguish between two kinds of science fiction. In one kind of science fiction, the known natural laws are respected: this case focuses on technology which science fiction extends beyond current practical limits: we could call such science fiction "**realistic**", although it was impractical at the time of writing. For instance, in the 19th century, Jules Verne wrote books about travel to the center of the Earth, or deep into the sea, or to the Moon; he also tried to forecast life in the city of Paris as it would be a century later. He extended existing technologies without exceeding known natural laws: his futuristic stories were realistic even if impractical at the time of his writing (for example, he foresaw electric doorbells, silent elevators, pressurized water in homes, giant mechanical calculators, underground railways, telegraphic communication replacing mail, *etc.*).

In the 21st century, we can similarly write about traveling to other stars and galaxies, even though that requires much faster speeds than our engineering can currently deliver. And we can write about cloning ourselves, knowing that cloning humans may become possible sometime in the future: this is also realistic, even though it is impractical now.

The second kind of science fiction goes beyond the known natural laws: it is supernatural because it allows actions that our current knowledge of nature prohibits. We may call this science fiction "fantasy". For example, as far as we know, time travel, teleportation of matter and levitation will never be possible, so they count as "**fantastic**" science fiction.

Existing scientific knowledge and understanding, if they are generally accepted, are clearly not fiction, since they rely on agreement with all observations of nature. But how about scientific **research**? Research can be like science fiction in that it often tries to find and explore new possibilities. Consider the following example: today, we don't have agreed-upon explanations for the collapse of EM waves and QM waves (see Sections 12.10 and 13.2); a new extension beyond existing science may thus be needed.

Similar situations happen quite frequently in physics: today, we don't understand "dark matter" and "dark energy" in the universe; we also don't have an agreed-upon theory that combines gravity and quantum physics (string theory comes closest to that goal). Typically, in such cases, many **theories**, meaning many different speculative explanations, are developed and proposed. However, only one of those theories can fully agree with nature (if two different theories agree, they must be equivalent versions of the same theory): further experiments must be made to find which of all those theories is correct, if any (of course, the correct theory may not yet have been invented).

Before the correct theory for new observations is found, we may ask whether all the proposed theories can be called science fiction. Strictly speaking, the accepted **scientific method** implies that the proposed theories are in fact science fiction. However, scientists avoid the term "science fiction" for their research and use other terms such as **hypotheses** or **models**, in addition to "theories", possibly to emphasize their deep respect for the objective scientific method or to differentiate themselves from the less serious versions of science fiction.

Next, we can ask: do proposed theories belong to the "realistic" kind or of the "fantastic" kind of science fiction? The answer depends on whether the proposed theories are within or beyond existing science, respectively. Both cases are possible. For example, the theory of turbulence in liquids and gases (which describes eddies, tornadoes and other irregular motions) is an application of Newton's laws and thus was "realistic" when it was proposed; similarly, negative refraction (see Section 12.8) is a new application of Maxwell's equations, so it was "realistic" when proposed. But quantum theory and relativity went beyond Newton's laws, Maxwell's equations and thermodynamics: so quantum theory and relativity could be labeled as "fantastic" before they were accepted as being correct; they were accepted only after they explained the observations of the quantized behavior in nature. Recall the century that passed between the suggestion of gravitational waves in Einstein's theory of general relativity and their first observation (see Chapter 14): during that time, gravitational waves could be regarded as fantasy because their existence was not confirmed. In such cases, the difference between realistic and fantastic science fiction may not be clear. There is also the risk that human enthusiasm leads to mistaking one's unproven theory for reality.

Regardless of the labels, scientists often feel close to science fiction because their research has many elements in common with science fiction, such as continually asking "Why? How? What? What if? Where? When?", and frequently imagining novel mechanisms and scenarios to explain seemingly fantastic new observations for which the previous models don't work.

We have so far focused on similarities between science and more familiar human activities. One clear difference is the use of mathematics, especially in the so-called "exact sciences" that include physics. Let's therefore ask: ***What is the role of mathematics in physics?***

Mathematics (math) is used in physics for several purposes. Math started in antiquity with the precise, and therefore fair, counting of objects and money in commerce. It also enabled the comparison of property sizes, such as comparing the areas of two different plots of land with a simple formula giving (area) = (length) × (width). We all know that we can trade length for width, but this formula makes the trade precise and thus fair; similarly, volumes of containers can be compared with **precision**.

Formulas (and more generally, the laws of physics) give us an artificial but realistic mathematical **model** of nature. That model allows us to understand how nature works in every detail. It tells us why a rock that you have thrown lands where it does, or why it goes into space instead.

Formulas also allow **prediction**: this is a very important aspect of math and physics. If you want to buy a piece of land with a particular area and you know its length, you can predict what width will give you the desired area by using the formula for the area. Or suppose you want to know how much time it will take to travel from point A to point B, knowing the distance and your average speed. We have a formula that solves that problem: (average speed) = (distance) / (time). The rules of math allow us to rewrite that equality as (time) = (distance) / (average speed). The power of this new formula is that we can apply it to any new situation that we have not experienced before: that is also prediction.

Knowing a formula also allows the **exploration** of unusual situations. Suppose you throw a rock faster and faster up into the air: it will land farther and farther on the ground, following Newton's formulas of mechanics. But if you throw the rock fast enough, these formulas tell you that the rock will leave Earth. That prediction opens the possibility of space travel. Thus, we can predict the trajectory of spaceships and aim them accurately to reach a particular spot on the Moon, for example.

Engineers must rely extensively on formulas when they design new towers, bridges, vehicles, machines, toys, *etc.*, especially because of the opposing demands of safety and economics. Using math, they can predict the necessary safe design and the corresponding cost.

We have seen four different functions of math: exploration, modeling, prediction and precision in mimicking nature. These functions of math are very effective for achieving further progress in physics.

It is actually remarkable that we can use math as a model or substitute for nature. It appears that nature closely follows mathematical formulas, but it is far from obvious that nature should do so. It's as if nature were a video game based on some mathematical formulas. Indeed, we can simulate natural processes on a computer that mimics the laws of nature in any situation (although not with total precision because our computers are not powerful enough: nature beats us in precision!).

An idea of how precise science can be at present is illustrated by recent experiments[6] on a property of the electron, specifically the electron's g-factor which describes how strongly an electron feels magnetic fields. The measurements of the g-factor agree with theory with a precision of better than 1 part per 10^{12} (or 1 in 1 trillion, or 1 in 1,000,000,000,000). These are among the most precise measurements

---

[6] "Measurement of the bound-electron g-factor difference in coupled ions", by Tim Sailer, Vincent Debierre, Zoltán Harman, Fabian Heiße, Charlotte König, Jonathan Morgner, Bingsheng Tu, Andrey V. Volotka, Christoph H. Keitel, Klaus Blaum and Sven Sturm, freely available in *Nature*, volume 606, page 479, 2022, https://doi.org/10.1038/s41586-022-04807-w. And "Measurement of the Electron Magnetic Moment", X. Fan, T. G. Myers, B. A. D. Sukra, and G. Gabrielse, *Physical Review Letters*, volume 130, number 071801, 2023, https://doi.org/10.1103/PhysRevLett.130.071801.

in science; a more typical precision for important quantities in physics is in the range of 1 in 10^{10} to 1 in 10^5. For comparison:

- 1 in 10^5 is like 1 centimeter (the size of a fingernail) *versus* 1 kilometer, or 1 second *versus* about 1 day.
- 1 in 10^{10} is like a fingernail *versus* 100,000 kilometers, a quarter of the distance from Earth to Moon or 2.5 times the Earth's circumference, or 1 second *versus* 275 years.
- 1 in 10^{12} is like a fingernail *versus* 25 times the distance from Earth to Moon, or *versus* 250 times the Earth's circumference, or 1 second *versus* 27,500 years = 275 centuries = 27.5 millennia.

It is interesting that very often different scientists try to solve the same problem in very different ways. Some scientists are very mathematical in their approach: they are able to manipulate the formulas in a way that exposes new possibilities or reaches higher degrees of accuracy or convenience or insight. There are other scientists who are more qualitative: they gain insight more by feeling how nature behaves. Some scientists excel in simulating nature in computer programs and thereby studying more complex natural behaviors. Experimentalists are adept at manipulating nature to coax it into new behaviors, also with greater accuracies than before; their role is also essential in keeping science objective, based on reality! Great scientists combine several of these capabilities.

Let's now compare the motivations of scientists to explore nature, on the one hand, with the motivations that attract people to play video games, read science fiction, and do sports, on the other hand. This will also help us to further illuminate the mindset of scientists, as well as their careers and impulses.

What are the shared motivations between science and more familiar activities? We can list many shared **motivations**: curiosity, exploration, imagination, discovery, knowledge, understanding, creativity, opportunities, success, excitement, fun, challenge, competition, expertise, rewards, and more...

In this list, I have intentionally omitted power and wealth: very few scientists choose to do science with the goal of becoming powerful or wealthy; moreover, only a few scientists gain power and/or wealth, and that usually happens only late in their careers.

Let's look more closely at the shared motivations. Curiosity is common to all humans, especially youngsters. Think about how babies and children discover the world, expanding their horizons step by step from their mothers to the universe. This exploration extends to imaginary worlds in tales, science fiction, cartoons, comics, video games, movies, as well as dreams. It also includes sports: when we hear of a new sport, we are curious how it works, what advantages it has over other sports, whether it is fun, how challenging it is, whether we are able to outperform other people in that sport, *etc*.

Curiosity, exploration and imagination are central to scientific research: without those motivations, little progress takes place in science. This is like visiting a foreign city and trying to understand how different it is from your own earlier experiences: you can then better understand and even improve yourself and your own city. Scientists continually ask "Why? How? What? What if? Where? When?", just like children. Trying to answer those questions usually leads to new knowledge and new understanding. The same is true in more familiar activities.

One of the most powerful motivations for doing science is the prospect of finding something new. Discovery is often the happy result of curiosity, exploration and imagination. We all love to discover a new trick or capability in sports or in a video game, or a new "fantastic" world in science fiction. This is equally true in scientific discovery, where "eureka moments" count as emotional highs, sometimes after years of investigations at the frontiers of knowledge and understanding.

The quest for understanding is a basic human desire: beyond the pleasure of knowing facts, we frequently ask "How?" and "Why?". Answering the question "How?" leads to an understanding of the workings of nature, which in turn allows us to predict the outcome of new situations, such as forecasting the weather. The question "Why?" leads to deeper understanding of root causes: such understanding is like a ladder with different levels of understanding. In the case of weather, we can ask why there was sunshine yesterday but rain today: we can explain this in terms of high-pressure zones and low-pressure zones. But then we can ask why there are such highs and lows in the atmosphere: that leads to understanding the dynamics of heat and air currents. Next, we can go to the deeper atomic-level understanding of thermodynamics. Then we can think of the structure of atoms, and then the structure of protons, neutrons and electrons. This ladder continues until we no longer have clear answers to the deepest questions, such as "Why does nature behave this way?" and "Where does nature come from, and where is it going?".

Imagination generates creativity. We like to wonder what surprises may lie beyond our horizon, and what imaginary worlds we can dream up; we may also wonder what we can do differently in sports or video games. Creativity is very important in scientific progress. One form of creativity is the new combination of known facts or methods; for example, we can fantasize about humans who fly like birds and live in a gravity-free world, so they can inhabit novel 3D cities in an intergalactic spaceship free of gravity. Another form of creativity is extension beyond our known world, as in establishing new human colonies on other planets.

Creativity is also a central skill for research and development in the industrial and military fields: there the primary aim is to produce effective and efficient devices that achieve specific practical goals, from smartphones to missiles. This can involve a variety of complex ethical and moral issues.

One consequence of creativity is opportunities for new activities and new science. The creation of new opportunities is an important measure of success and progress, in all our endeavors.

Excitement is also a major motivation in many of our activities: we relish the expansion of our horizons and the new opportunities we find. These sentiments are certainly strong motivators in sports, video games, and science fiction, as well as in science.

We all enjoy fun, especially when combined with challenges and success. That is particularly obvious in sports and video games, where difficult accomplishments often give immediate pleasure, especially when there is a high risk of failure. With science fiction, the fun and enjoyment are also present but usually stretched out over the length of the story; the fun and satisfaction are also often more abstract or intellectual in science fiction and in science, compared to sports and video games.

In science, fun is a bit more complex. There certainly is fun in discovering novel behaviors of nature and in showing amazing effects to others (such as exhibiting surprising chemical reactions and explosions, or unexpected mechanical and optical effects); indeed, fun is a strong motivator to attract people to science.

Enjoyment is a clear motivator in science, as it is in our more familiar activities. There is not only great satisfaction in observing a beautiful sunset but also in the realization that a glorious rainbow follows the well-known laws of physics: it's astonishing and admirable that such complex phenomena can be reduced to a few laws of nature.

However, fun and enjoyment are psychological reactions that scientists usually exclude from the scientific results of their work. Indeed, most scientific literature is extremely serious and dull: it is usually made as "non-human" as possible, so as to ensure objectivity[7]. Nevertheless, scientists do try to include fun and enjoyment in the process of doing research, like any other human beings who enjoy camaraderie, jokes, teamwork, travel, *etc*.

Challenges and failure are common features of science as well. While nobody likes failures, they are among the most effective tools of learning, also in science!

Humans often enjoy competition, especially when they gain expertise and can earn rewards, as is very obvious in sports. This trait is also present in science: scientists are strongly motivated by competition, especially when they earn rewards like fame, prizes, titles, promotions, salary raises, and money to do more research. Mostly (but not always), such competition between scientists is collegial and friendly; this often leads to international friendships, thanks to the universality of science and many international conferences, amplified by online communications. As in more familiar activities and professions, competition in science can of course also lead to tensions and pressure. Competition can also drive intense activity in new areas of science, as it does in sports and video games, thereby accelerating progress and innovation.

Competition also plays an important role in verifying scientific results. Being humans, scientists do make mistakes (mostly accidental

---

[7] Objectivity requires that scientific results should not depend on who obtained those results. This ensures that other people will obtain the same results and that the results are universal and agreed. In particular, if performed properly, science does not depend on social factors like language, culture, religion or politics; this also makes possible the close collaboration between scientists across the world. Furthermore, this allows more cross-checking to minimize the risk of errors in scientific results.

but occasionally intentional). They also enjoy winning an argument by proving other scientists wrong and themselves right. This is a natural and effective system for cross-checking results.

There are other rewards for doing science which are less material, but equally satisfying, especially in scientific research. These include the joys of discovery mentioned above, and new knowledge fed by curiosity. It is not only a pleasure to know how something works but also to discover it for yourself (even if it has already been discovered by others), or to discover a new and more elegant explanation. Discussing and arguing about nature with other interested people are also a source of great pleasure. Yet another source of joy is to explain science to others and to reveal how nature does her magic, whether it is in a class, or to family, friends and strangers, or in a serious scientific article, or in a book like this one.

References and Resources

Online videos

- YouTube videos exhibiting central aspects of this book on "*Everyday Physics*" by Michel A. Van Hove:
  - **Waves on Strings:** ***from resonances to music*** **(video WA1) —** https://www.youtube.com/watch?v=eCKigLmmBZc
  - **Waves in Air:** ***sound and musical instruments*** **(video WA2) —** https://www.youtube.com/watch?v=PMR49kj2PeY
  - **Waves and Voices:** ***human and animal*** **(video WA3) —** https://www.youtube.com/watch?v=VwIwJ5Jtiks
  - **Waves on Water:** ***reflection and bending*** **(video WB1) —** https://www.youtube.com/watch?v=03pbq_m_bmU
  - **EM and QM Waves:** ***from electromagnetism to the quantum world*** **(video WB2) —** https://www.youtube.com/watch?v=2jQrphJ8lcg
- YouTube videos featuring Walter Lewin
- YouTube videos featuring Neil deGrasse Tyson

- YouTube video series "Smarter Every Day" by Destin Sandlin: https://www.youtube.com/user/destinws2
- YouTube video series by "Physics Girl" Dianna Cowern: https://www.youtube.com/c/physicsgirl/videos

General literature

- Edward Bryant, *"Tsunami, the Underrated Hazard"*, Springer, Heidelberg, 2014, ISBN-13: 978-3319330969 (paperback), ISBN-13: 978-3319061337 (ebook).
- Norval Fortson, *"Discovering the Nature of Light: the Science and the Story"*, World Scientific Publishing Co., 2022, ISBN-13: 978-9811250293 (paperback), ISBN-13: 978-9811249600 (ebook).
- Robert G. Greenler, *"Rainbows, Halos, and Glories"*, Cambridge University Press, Cambridge, 1999, ISBN-10: 0-5212-3605-3.
- Paul G. Hewitt, *"Conceptual Physics"*, Pearson Addison Wesley, 2015 (12th ed.), ISBN-13: 978-0135205815.
- Steven Holzner, *"Physics I & II for Dummies"*, Wiley, 2011 (2nd ed.), ISBN-10: 1-1192-9359-6 & 0-4705-3806-6.
- Joseph E. Kasper and Steven A. Feller, *"The Complete Book of Holograms: How They Work and How to Make Them"*, Dover, 2001, ISBN-13: 978-0486415802.
- Peter Ladefoged and Keith Johnson, *"A Course in Phonetics"*, Wadsworth Cengage Learning, 2011 (6th ed.), ISBN-13: 978-1428231269.
- Walter Lewin with Warren Goldstein, *"For the Love of Physics: From the End of the Rainbow to the Edge of Time — A Journey Through the Wonders of Physics"*, Free Press, 2012, ISBN-13: 978-1451607130.
- M. Minnaert, *"The Nature of Light and Colour in the Open Air"*, Dover, New York, 1954, ISBN-10: 0-4862-0196-1.
- Richard A. Muller, *"Physics and Technology for Future Presidents: An Introduction to the Essential Physics Every World Leader Needs to Know"*, Princeton University Press, 2010, ISBN-13: 978-0691135045.
- Carl J. Pratt, *"Quantum Physics for Beginners: From Wave Theory to Quantum Computing. Understanding How Everything Works*

by a Simplified Explanation of Quantum Physics and Mechanics Principles", independently published, 2021, ISBN-13: 978-1802356588.

- Adam Rogers, *"Full Spectrum: How the Science of Color Made Us Modern"*, Houghton Mifflin Harcourt, 2021, ISBN-10: 1-3285-1890-6.
- Edward Sapir, *"Language: An Introduction to the Study of Speech"*, Pinnacle Press, 2017, ISBN-13: 978-1374836785.
- Ferdinand de Saussure, *"Course in General Linguistics"*, Open Court, 1986, ISBN-13: 978-0812690231.
- Lauren Tarshis and Scott Dawson, "I Survived the San Francisco Earthquake, 1906", Scholastic Paperbacks, 2012, ISBN-13: 978-0545206990.
- Charles Taylor, *"Exploring Music: The Science and Technology of Tones and Tunes"*, Institute of Physics Publishing, 1992, ISBN-10: 0-7503-0213-5.
- US Army Corps of Engineers, *"Coastal Engineering Manual — Part II"*, Books Express Publishing, 2012, https://books.google.com.hk/books/about/Coastal_Engineering_Manual_Part_II.html?id=FVgKkgEACAAJ&redir_esc=y
- Michel A. Van Hove, *"Everyday Physics — Colors, Light and Optical Illusions"*, World Scientific Publishing Co., 2022, ISBN-13: 978-9811238338 (hardcover), ISBN-13: 978-9811239311 (paperback), ISBN-13: 978-9811238352 (ebook).
- Simon Winchester, *"A Crack in the Edge of the World"*, Harper, 2005, ISBN-13: 978-0060571993.
- Michael M. Woolfson, *"Colour: How We See It and How We Use It"*, World Scientific, 2016, ISBN-13: 978-1786340856.
- J. B. Zirker, *"The Science of Ocean Waves: Ripples, Tsunamis, and Stormy Seas"*, Johns Hopkins University Press, 2013, ISBN-13: 978-1421410784.

Websites

On **physics**:

- Simulations of physics (and other subjects) by PhET Interactive Simulations: https://phet.colorado.edu

- HyperPhysics, exploration of concepts in physics, by C. R. Nave, Georgia State University: http://hyperphysics.phy-astr.gsu.edu/hbase/index.html
- Physclips, platform for learning or teaching physics, by Joe Wolfe, UNSW: https://www.animations.physics.unsw.edu.au/

On **waves and sounds**:

- Music Acoustics, by Joe Wolfe, UNSW: http://newt.phys.unsw.edu.au/music/

Software:

- A.L.M.S. — Audio Literacy for Music Students — freeware, by Prof. Christopher J. Keyes: http://palms.polyu.edu.hk/a-l-m-s-hkbu/
- Praat, feeware: https://www.fon.hum.uva.nl/praat/
- WaveSurfer, freeware: download from sourceforge.net/projects/wavesurfer

Frequently used units and powers of ten

Length:

- 1 nanometer (nm) = 10^{-9} meter (m) = 0.000,000,001 meter (m)
- 1 micrometer (μm) = 10^{-6} meter (m) = 0.000,001 meter (m)
- 1 mil = 0.001 inch (in) = 0.0254 millimeter (mm) = 25.4 micrometers (μm)
- 1 millimeter (mm) = 10^{-3} meter (m) = 0.001 meter (m)
- 1 centimeter (cm) = 10^{-2} meter (m) = 0.01 meter (m) ~ 0.3937 inch (in)
- 1 inch (in) = 2.54 centimeters (cm) = 0.0254 meter (m)
- 1 foot (ft) = 12 inches (in) = 0.3048 meter (m)
- 1 yard = 3 feet (ft) = 0.9144 meter (m)
- 1 meter (m) ~ 1.0936 yards (yd) ~ 3.2808 feet (ft)
- 1 kilometer = 1,000 meters (m) ~ 0.6214 mile* (mi) ~ 0.5340 nautical mile (NM)
- 1 mile* (mi) = 5280 feet (ft) = 1760 yards (yd) = 1.609,344 kilometers (km) ~ 0.8690 nautical mile (NM)
- 1 nautical mile (NM) = 1.852 kilometers (km) ~ 1.1508 miles* (mi)

- 1 astronomical unit (AU) = 149,597,870,700 meters (m) = 149,597,870.7 kilometers (km) ~ 92,955,807.3 miles* (mi) (average distance from Sun to Earth)
- 1 lightyear (ly) = 9,460,730,472,580,800 meters (m) ~ 9.461 trillion kilometers (km) ~ 5.879 trillion miles* (mi) (distance traveled by light in one year)

* the "mile" listed here is today's usual mile, also called statute mile or international mile = 1609.344 meter exactly (it is different from the nautical mile)

Speed:

- 1 meter/second (m/s) = 3.6 kilometers/hour (km/h) ~ 2.2369 miles* per hour (mph)
- 1 kilometer/hour (km/h) ~ 0.2778 meter/second (m/s) ~ 0.6214 mile* per hour (mph)
- 1 mile* per hour (mph) = 1.609,344 kilometers/hour (km/h)
- Speed of sound in air: Mach 1 ~ 343 meters/second (m/s) ~ 1235 kilometers/hour (km/h) ~ 767 miles* per hour (mph)
- Speed of light in vacuum: exactly 299,792,458 meters/second (m/s) ~ 300,000 kilometers/second (km/s) ~ 1,080,000,000 kilometers/hour (km/h) ~ 186,000 miles* per second (mps) ~ 671,000,000 miles* per hour (mph)

Powers of 10 (using the "short scale" names above 10^6, such as 10^9 – billion instead of milliard):

- $10^0 = 1$
- $10^1 = 10$ (deca, da)
- $10^2 = 100$ (hecto, h)
- $10^3 = 1{,}000$ (kilo, k)
- $10^6 = 1{,}000{,}000$ (mega, M): one million
- $10^9 = 1{,}000{,}000{,}000$ (giga, G): one billion
- $10^{12} = 1{,}000{,}000{,}000{,}000$ (tera, T): one trillion
- $10^{15} = 1{,}000{,}000{,}000{,}000{,}000$ (peta, P): one quadrillion
- $10^{18} = 1{,}000{,}000{,}000{,}000{,}000{,}000$ (exa, E): one quintillion
- $10^{21} = 1{,}000{,}000{,}000{,}000{,}000{,}000{,}000$ (zetta, Z): one sextillion
- $10^{-1} = 0.1$ (deci, d)

- 10^{-2} = 0.01 (centi, c)
- 10^{-3} = 0.001 (milli, m)
- 10^{-6} = 0.000,001 (micro, μ)
- 10^{-9} = 0.000,000,001 (nano, n)
- 10^{-12} = 0.000,000,000,001 (pico, p)
- 10^{-15} = 0.000,000,000,000,001 (femto, f)
- 10^{-18} = 0.000,000,000,000,000,001 (atto, a)
- 10^{-21} = 0.000,000,000,000,000,000,001 (zepto, z)

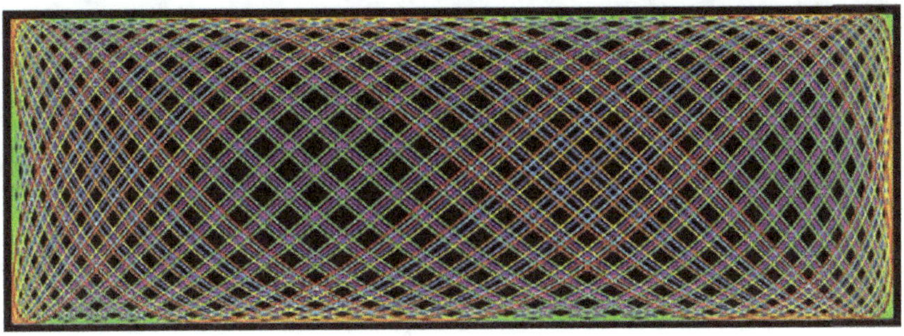

Index

About the Author

Michel A. Van Hove studied physics in Switzerland (ETH) and the UK (Cambridge), and worked in the Netherlands, Germany, USA, as well as Hong Kong, where he retired as Emeritus Chair Professor. His research has focused on the determination of the atomic-scale structure and bonding at solid surfaces and nanostructures. He developed and implemented novel methods of electron scattering theory and computation, such as for nanostructures. He also worked on photoelectron diffraction, scanning tunneling microscopy and atomic-force microscopy, a database of solved surface structures, and molecular machines. He published over 400 articles and 15 books. He taught undergraduate and graduate courses in the USA and Hong Kong. This popularization book in the series "*Everyday Physics*" follows a volume on "*Colors, Light and Optical Illusions*".

www.ingramcontent.com/pod-product-compliance
Lightning Source LLC
LaVergne TN
LVHW010323220926
848484LV00004BB/832
9789819832439